This is an introduction to the theory of affine Lie algebras, to the theory of quantum groups, and to the interrelationships between these two fields that are encountered in conformal field theory.

The description of affine algebras covers the classification problem, the connection with loop algebras, and representation theory including modular properties. The necessary background from the theory of semisimple Lie algebras is also provided. The discussion of quantum groups concentrates on deformed enveloping algebras and their representation theory, but other aspects such as R-matrices and matrix quantum groups are also dealt with.

This book will be of interest to researchers and graduate students in theoretical physics and applied mathematics.

CAMBRIDGE MONOGRAPHS ON MATHEMATICAL PHYSICS

AFFINE LIE ALGEBRAS AND QUANTUM GROUPS

Cambridge Monographs on Mathematical Physics

A. M. Anile *Relativistic Fluids and Magneto-Fluids*
J. Bernstein *Kinetic Theory in the Early Universe*
N. D. Birrell and P. C. W. Davies *Quantum Fields in Curved Space*†
D. M. Brink *Semiclassical Methods in Nucleus–Nucleus Scattering*
J. C. Collins *Renormalization*†
P. D. B. Collins *An Introduction to Regge Theory and High Energy Physics*
M. Creutz *Quarks, Gluons and Lattices*†
F. de Felice and C. J. S. Clarke *Relativity on Curved Manifolds*
B. DeWitt *Supermanifolds, second edition*†
P. G. O. Freund *Introduction to Supersymmetry*†
F. G. Friedlander *The Wave Equation on a Curved Space-Time*
J. A. H. Futterman, F. A. Handler and R. A. Matzner *Scattering from Black Holes*
M. Göckeler and T. Schücker *Differential Geometry, Gauge Theories and Gravity*†
M. B. Green, J. H. Schwarz and E. Witten *Superstring Theory, volume 1: Introduction*†
M. B. Green, J. H. Schwarz and E. Witten *Superstring Theory, volume 2: Loop Amplitudes, Anomalies and Phenomenology*†
S. W. Hawking and G. F. R. Ellis *The Large-Scale Structure of Space-Time*†
F. Iachello and A. Arima *The Interacting Boson Model*
F. Iachello and P. van Isacker *The Interacting Boson–Fermion Model*
C. Itzkyson and J.-M. Drouffe *Statistical Field Theory, volume 1: From Brownian Motion to Renormalization and Lattice Gauge Theory*†
C. Itzkson and J.-M. Drouffe *Statistical Field Theory, volume 2: Strong Coupling, Monte Carlo Methods, Conformal Field Theory, and Random Systems*†
J. I. Kapusta *Finite-Temperature Field Theory*
D. Kramer, H. Stephani, M. A. H. MacCallum and E. Herlt *Exact solutions of Einstein's Field Equations*
N. H. March *Liquid Metals: Concepts and Theory*
L. O'Raifeartaigh *Group Structure of Gauge Theories*†
A. Ozorio de Almeida *Hamiltonian Systems: Chaos and Quantization*†
R. Prenrose and W. Rindler *Spinors and Space-time, volume 1: Two-Spinor Calculus and Relativistic Fields*†
R. Penrose and W. Rindler *Spinors and Space-time, volume 2 : Spinor and Twistor Methods in Space-Time Geometry*
S. Pokorsju *Gauge Field Theories*†
V. N. Popov *Functional Integrals and Collective Excitations*†
R. Rivers *Path Integral Methods in Quantum Field Theory*†
R. G. Roberts *The Structure of the Proton*
W. C. Saslaw *Gravitational Physics of Stellar and Galactic Systems*†
J. M. Stewart *Advanced General Relativity*
R. S. Ward and R. O. Wells Jr *Twistor Geometry and Field Theories*†
J. Fuchs *Affine Lie Algebras and Quantum Groups*

† Issued as a paperback

AFFINE LIE ALGEBRAS AND QUANTUM GROUPS

AN INTRODUCTION, WITH APPLICATIONS IN CONFORMAL FIELD THEORY

JÜRGEN FUCHS

NIKHEF, Amsterdam, The Netherlands

CERN, Geneva, Switzerland

CAMBRIDGE
UNIVERSITY PRESS

CAMBRIDGE UNIVERSITY PRESS
Cambridge, New York, Melbourne, Madrid, Cape Town, Singapore,
São Paulo, Delhi, Dubai, Tokyo, Mexico City

Cambridge University Press
The Edinburgh Building, Cambridge CB2 8RU, UK

Published in the United States of America by
Cambridge University Press, New York

www.cambridge.org
Information on this title: www.cambridge.org/9780521484121

First published 1992
First paperback edition (with corrections) 1995

A catalogue record for this publication is available from the British Library

Library of Congress cataloguing in publication data available

ISBN 978-0-521-41593-4 Hardback
ISBN 978-0-521-48412-1 Paperback

To Gabi

Contents

Preface

It is one of the fundamental aims of science to describe nature in all its tremendous complexity by rules that are both simple and have a general validity. When pursuing this aim, one is inevitably led to the concept of symmetry. In mathematical terms, symmetries are usually described by groups of transformations or possibly, in the case of continuous groups, by the underlying algebras. Discrete (notably finite) groups appear e.g. in solid state physics, in particular in crystallography. Continuous (and hence infinite) groups emerge in quantum physics and field theory; while infinite, they may nevertheless be finite-dimensional, and for many important applications this is indeed the case; an example is provided by the Lorentz group which is the gauge group of general relativity. The standard example in quantum physics arises in the quantum-mechanical description of angular momentum; this leads to the simple Lie algebra $A_1 \cong su_2$, the Lie algebra of the simple Lie group $SU(2)$ which is locally isomorphic to the three-dimensional rotation group. The various gauge groups appearing in the standard model of strong and electroweak interactions and in grand unified theories of particle physics provide a first step of generalization of the structure encountered in the theory of angular momentum; in mathematical terms, the corresponding Lie algebras are direct sums of semisimple and abelian Lie algebras. More recently, a second step of generalization of angular momentum theory has found its way into physics, namely in the form of a class of infinite-dimensional Lie algebras known as affine Kac–Moody algebras. These algebras first emerged in the current algebra approach to the symmetries of elementary particles; more importantly, they proved to play a fundamental role in the quantum theory of relativistic strings (which offers e.g. a possibility to unify the standard model of particle physics with gravity), and in the analysis of critical phenomena in two-dimensional models of statistical mechanics. A qualitatively different second step of generalization arose in the study of integrable quantum systems: this led to the discovery of yet another class of symmetry objects, the so-called quantum groups. Even more recently, it turned out that affine algebras and quantum groups are actually closely related mathematical structures.

The problems to which the theories of affine algebras and of quantum groups have been applied so far are admittedly closer to pure mathematics than to everyday life. In particular, affine Lie algebras (and the more general class of Kac–Moody algebras) enjoy deep links with other areas of mathematics such as combinatorics, finite simple groups, topology, and partial differential equations. Also, in physics so far one has to pay a price for being able to attack these more complicated mathematical structures: the applications are restricted to systems in two space–time dimensions. As an unfortunate consequence, a part of the physical community has been led to believe that such research is more or less non-physical. Of course, the restriction to two dimensions is not a problem if string theory should turn out to be the correct basis for a unified description of particle interactions, because then the two dimensions of "space-time" describe the world sheet swept out by the relativistic string and the physical space-time arises in terms of fields that are defined on this world sheet. In addition, it should not be forgotten that in physics there exist a lot of problems of a principal nature (such as the vanishing of the cosmological constant, the mechanism of quark confinement, or the origin of high temperature superconductivity, to name only a few popular ones) that so far can be described only poorly with "ordinary" physical methods. It is plausible to expect that at least in some cases a complete solution will require a lot more mathematics than what is being applied to these problems so far. In view of the various successful applications of symmetry principles in physics, the required new mathematical structure presumably also involves the description of new symmetry objects. It is therefore quite reasonable to believe that the theories of affine Lie algebras and of quantum groups will make their way into the physical science in one way or another, even if the models in which they show up so far should prove to be of minor relevance to phenomenology. In fact, it has already been possible to apply affine Lie algebras and quantum groups in a three-dimensional setting, namely in the context of the so-called topological field theories.

The present book intends to provide an introduction to the theory of affine Lie algebras and to the theory of quantum groups, as well as illustrate the close relationship between these two kinds of objects. In addition, applications of these mathematical structures to two-dimensional conformal field theories, in particular to Wess–Zumino–Witten theories, are discussed. While there are several other areas of physics where affine Lie algebras play a role, conformal field theory is certainly the most natural field to apply them: associated with any affine Lie algebra is a Virasoro algebra, and the latter is the quantum version of the algebra of two-dimensional conformal transformations. In contrast, in the case of quantum groups the most direct application is in the theory of integrable quantum systems; however, in conformal field theory quantum groups turn out to play an important role as well, and in fact the connection between quantum groups and Lie algebras (which is the reason why these two subjects are treated here in one and the same monograph) was uncovered precisely in the context of conformal field theory.

While the applications cover almost one half of the book, the emphasis is nevertheless on the underlying mathematical structures. These are the subject of chapter 2 (affine Lie algebras) and chapter 4 (quantum groups), and also of chapter 1 where the necessary information about semisimple Lie algebras is provided. The applications in chapter 3 concentrate on the operator product algebra and on the differential equations that arise from the existence of null vectors in affine Lie algebra modules. In chapter 5 the main issues are the fusion rules and their connection with modular transformations, and the duality properties of chiral block functions; these concepts need to be discussed in some detail in order to be able to explain the quantum group structure of conformal field theory, which is done in the sections 5.5 and 5.7.

The principles of two-dimensional conformal field theory are discussed explicitly in section 3.3. However, neither that section nor the remarks in the rest of chapter 3 and chapter 5 are intended to give a complete survey of conformal field theory (this is also not necessary since various excellent reviews of conformal field theory already exist). For example, I do not treat in any detail such important issues as the Coulomb gas representation of conformal field theories, the relations between rational conformal field theory and three-dimensional topological field theory, or conformal field theory on higher genus Riemann surfaces. Also, except for some remarks in section 3.2 the Virasoro algebra is not discussed in its own right, but only in the context of the Sugawara construction.

The theory of semisimple and affine Lie algebras is well developed, and although they are still the subject of active mathematical research, there do not seem to exist many unsolved problems of fundamental importance. As a consequence, the introduction into this subject that is given in chapters 1 and 2 has a good chance of informing the reader about all aspects of semisimple and affine Lie algebras that may be relevant to conformal field theory (and to other areas of physics as well). In contrast, it is almost inevitable that the exposition in chapter 4 is lacking some relevant facts about quantum groups, since the progress in the theory of quantum groups and their application to conformal field theory is still going on rapidly, and surprising features are still being uncovered almost on a day to day basis.

I have decided not to refer to any literature in the main sections of the book; rather, I have added a section containing a survey of relevant references separately at the end of each chapter. It is an enormous task to mention even just the more relevant publications on the broad range of topics covered in these notes. Unjust omissions have therefore occured without doubt; they should be kindly attributed to my ignorance rather than to bad will.

These notes have been written down during my stay as a Wetenschappelijk Medewerker at the Nationaal Instituut voor Kernfysika en Hogeenergiefysica (NIKHEF) in Amsterdam. The final editing of the LaTeX files was mainly performed at CERN (Geneva). I wish to thank the colleagues at NIKHEF and CERN, and especially J.W. van Holten and K.J.F. Gaemers, for the kind hospitality extended to me.

The bulk of the chapters 1 to 3 has been the subject of lectures that I delivered in 1987 and 1988 at Heidelberg University, and in 1989 at NIKHEF. Parts of chapter 4 as well as section 5.5 were covered by lectures at the Bayrischzell winter school in March 1990. A short version of notes on these lectures had been prepared in partial fulfilment of the requirements of a "Habilitation" at Heidelberg University.

Various colleagues have made a contribution to this work in one way or another. Among them I would like to single out Doron Gepner and Peter van Driel who collaborated with me on the issues described in section 3.5, and in section 4.8 and 5.5, respectively, and to whom I owe much insight in related topics, as well as Michael G. Schmidt from whom I learned how to do physics in the first place and who served as a constant source of advice and encouragement. I also learned a lot from discussions or correspondence with F.A. Bais, G.G.A. Bäuerle, P. Bouwknegt, P. Christe, E.A. de Kerf, K. de Vos, U. Ellwanger, R. Flume, A.Ch. Ganchev, F.M. Goodman, A. Klemm, P. Mathieu, T. Nakanishi, Th. Ohrndorf, C. Ramírez, K.-H. Rehren, A.N. Schellekens, C. Schwingert, S. Theisen, T. Tjin, P. van de Petteflet, J.W. van Holten, D. Verstegen, and M. Walton. To M.G. Schmidt and P. Bouwknegt I am also obliged for their careful reading of the manuscript. Finally it must be mentioned that for being able to write down these notes in a reasonable amount of time, the use of the TeX typesetting system [B23] combined with the LaTeX macro package [B24] was essential.

Preface to the paperback edition

Various typographical errors and misstatements have been corrected. I am indebted to C. Schweigert for bringing many of these problems to my attention. I am also grateful to K. Blaubär, L. Gerritzen, M. Kreuzer, P. Schaller, A.N. Schellekens, and M. Windler for detecting some of the errors.

1
Semisimple Lie algebras

This first chapter does not yet discuss any of the issues that are announced in the title of the book. Rather, it prepares the stage for the following chapters, in particular for chapter 2. Many of the concepts that arise in the theory of affine Lie algebras play a very similar role and actually were developed in the theory of semisimple Lie algebras; it is therefore appropriate to introduce these concepts in the latter context. That the present chapter serves a preparatory task, manifests itself in the fact that the presentation is quite condensed. In addition, many issues are weighted rather differently from what one might expect from a review of semisimple Lie algebras.

Also, the motivation for introducing various concepts will sometimes not become clear in the present chapter, but only when we come to apply them later on, and the proofs for most of the assertions are only sketched or are even omitted completely. These features are the price to be paid for treating the theory in an introductory chapter rather than in a book of its own. It is hoped that this is compensated for to some extent by the bibliographical notes given in section 1.9.

To round off the presentation, a few remarks are also included about issues which are not of prime importance to the following chapters. As a consequence, this first chapter should be sufficiently self-contained to serve also as an introduction to the theory of semisimple Lie algebras in its own right.

In section 1.1 the elementary notions such as Lie algebra, subalgebra and simple Lie algebra are explained. Representations and modules are discussed in section 1.2. In section 1.3 we introduce the Killing form and describe the relation between real and complex Lie algebras. Section 1.4 contains the construction of the Cartan–Weyl and Chevalley basis of a semisimple Lie algebra and introduces such concepts as the Weyl vector, the Cartan matrix and Dynkin labels. From the results of the first four sections, it is then possible to prove the classification of simple Lie algebras, which is sketched in section 1.5. The definition and various properties of highest weight modules are the subject of section 1.6. Section 1.7 describes the Weyl group of a simple Lie algebra and its application to the theory of characters, as well as a simple algorithm for decomposing Kronecker tensor products of highest weight modules. Finally, section 1.8. discusses embeddings of reductive Lie algebras in simple Lie algebras, and the associated decompositions of highest weight modules.

1.1 Basic concepts

This section is intended to provide a few very basic notions and definitions. To begin at the beginning, let us introduce the concept of a *Lie algebra.* The concept of a *vector space* over some *field* F is assumed to be known; then we can define what is meant by an *algebra:*

Definition : An **algebra** a is a vector space endowed with a bilinear operation, (1.1.1)

$$\diamond : \mathsf{a} \times \mathsf{a} \rightarrow \mathsf{a}.$$

Equivalently, one may define an algebra as a *ring* $(\mathsf{a}\,,\diamond,+)$ (that is, as a set a endowed with binary operations that obey the usual axioms of addition and multiplication) together with an action of the field F on a (the *scalar multiplication*) which is compatible with the multiplication "$\diamond$" and addition "+". (Note that if in the place of the field F one uses the ring of integers $\mathbb{Z}$, then the scalar multiplication in fact does not add any new structure; in other words, a ring may also be considered to be an algebra over the integers $\mathbb{Z}$.)

To be of any interest, an algebra usually has to carry some additional structure; e.g. the bilinear operation may be required to be associative, thus leading to an *associative algebra.* The additional structure which will be investigated in the sequel is defined as follows: an algebra is a *Lie algebra* if the bilinear operation – then called the *Lie bracket* and commonly denoted by "$[\,,]$" – possesses two special properties:

Definition : A **Lie algebra** $\mathfrak{g}$ is an algebra such that the map $[\,,]:\ \mathfrak{g}\times\mathfrak{g}\rightarrow\mathfrak{g}$ obeys:

(1.1.2a) $[x,x]=0$ for all $x\in\mathfrak{g}$ (antisymmetry)

and

(1.1.2b) $[x,[y,z]]+[y,[z,x]]+[z,[x,y]]=0$ for all $x,y,z\in\mathfrak{g}$ (Jacobi identity).

(1.1.2)

(The property (1.1.2a) is called antisymmetry because as a consequence of the bilinearity one has $0=[x+y,x+y]=[x,x]+[x,y]+[y,x]+[y,y]=[x,y]+[y,x]$ and hence

$$[x,y]=-[y,x]. \tag{1.1.3}$$

Conversely, since (1.1.3) with $y=x$ amounts to $[x,x]+[x,x]=0$, the property (1.1.2a) follows from (1.1.3) provided that the field F over which g is a vector space has a characteristic different from 2 (i.e. that there does not exist a $\xi\in F$ that obeys $\xi+\xi=0$). In the following we always have in mind that $F=\mathbb{C}$ or $F=\mathbb{R}$ so that F has characteristic zero, but most of the statements in sections 1.1–1.3 hold for arbitrary base field F.) Because of the bilinearity, the Lie bracket is already determined uniquely if it is known on a basis of the vector space $\mathfrak{g}$. We denote a basis $\mathcal{B}$ of $\mathfrak{g}$ as

$$\mathcal{B}=\{T^a \mid a=1,\dots,d\}, \tag{1.1.4}$$

where $d=\dim\mathfrak{g}$ is the dimension of the vector space $\mathfrak{g}$, and refer to the elements of a basis as the *generators* of the algebra. Expanding the bracket of two generators with respect to the basis,

$$[T^a,T^b]=f^{ab}{}_c\,T^c, \tag{1.1.5}$$

one can thus define the Lie bracket through the numbers $f^{ab}{}_c\in F$, which are called the *structure constants* of the Lie algebra $\mathfrak{g}$. (Here it is implied that a summation has to be performed over any symbol appearing once as an upper and once as a lower index. We will use this summation convention in all circumstances where we are sure that it does not create confusion, and write out the summations otherwise.) Expressed through the structure constants, the Lie properties (1.1.2) read

$$f^{aa}{}_b=0 \tag{1.1.6}$$

(no sum on a), respectively,

$$f^{ab}{}_c=-f^{ba}{}_c \tag{1.1.7}$$

if F has characteristic different from 2, and

$$f^{ab}{}_c f^{cd}{}_e+f^{da}{}_c f^{cb}{}_e+f^{bd}{}_c f^{ca}{}_e=0. \tag{1.1.8}$$

From any given associative algebra $(\mathsf{a},\diamond)$, one can construct in a canonical way a Lie algebra $\mathfrak{g}(\mathsf{a})=(\mathsf{a},[\,,\,])$, namely by defining the Lie bracket as the *commutator* with respect to the original multiplication, i.e.

$$[x,y]=x\diamond y-y\diamond x. \tag{1.1.9}$$

The standard example for this construction is the Lie algebra of N×N-matrices, for which the associative product $\diamond$ is just the ordinary matrix multiplication. In physics, Lie algebras appear very often as algebras of matrices; as a consequence, in the physics literature the term commutator, or also *commutation rule*, is often used as a synonym for the term Lie bracket.

The next concept is the notion of a subalgebra:

Definition : A Lie **subalgebra** h of the Lie algebra g is a subvectorspace $\mathsf{h} \subseteq \mathsf{g}$ which itself is a Lie algebra. (1.1.10)

The inclusion $i: \mathsf{h} \stackrel{\subseteq}{\to} \mathsf{g}$ of a (Lie) subalgebra h into the Lie algebra g is called an *embedding* of h into g. A subalgebra $\mathsf{h} \subseteq \mathsf{g}$ is called a *proper* subalgebra of g if $\mathsf{h} \neq \mathsf{g}$.

From now on we will save space by using the notation

$$[\mathsf{h}, \mathsf{k}] \equiv \mathrm{span}_F\{[h,k] \mid h \in \mathsf{h}, k \in \mathsf{k}\}, \tag{1.1.11}$$

where in general $\mathrm{span}_F\{x| \, x \in M\}$ stands for the space spanned by the elements of some set M. For example, in this notation the definition (1.1.10) means that a subvectorspace $\mathsf{h} \subseteq \mathsf{g}$ is a subalgebra iff $[\mathsf{h}, \mathsf{h}] \subseteq \mathsf{h}$.

The following subalgebras are called *invariant* subalgebras or *ideals:*

Definition : An **ideal** h of the Lie algebra g is a subalgebra with the property $[\mathsf{g}, \mathsf{h}] \subseteq \mathsf{h}$. (1.1.12)

A *proper* ideal is an ideal that is equal neither to 0 nor to g itself, which are two obvious ideals of g. It is easily checked that if h and k are ideals, then so are $[\mathsf{h},\mathsf{k}]$, $\mathsf{h} \cap \mathsf{k}$ and

$$\mathsf{h} + \mathsf{k} := \{x \in \mathsf{g} \mid x = y + z,\; y \in \mathsf{h},\, z \in \mathsf{k}\}. \tag{1.1.13}$$

It is natural to analyse the structure of a Lie algebra by looking at its ideals. For any Lie algebra g, an example for an ideal is provided by its *derived* algebra g'; this is defined as the set of all linear combinations of brackets of g,

$$\mathsf{g}' := [\mathsf{g}, \mathsf{g}]. \tag{1.1.14}$$

It is obvious that g' is a subalgebra of g, but it is even an ideal because clearly $[\mathsf{g}, [\mathsf{g},\mathsf{g}]] \subseteq [\mathsf{g},\mathsf{g}]$. In fact, on defining the decreasing sequence

$$\mathsf{g}^{\{1\}} := \mathsf{g}', \qquad \mathsf{g}^{\{i\}} := [\mathsf{g}^{\{i-1\}}, \mathsf{g}^{\{i-1\}}] \quad \text{for } i \geq 2, \tag{1.1.15}$$

each term in this *derived series* is an ideal of g, because by induction $\mathsf{k} = [\mathsf{h}, \mathsf{h}]$ with $\mathsf{h} = \mathsf{g}^{\{j\}} \subseteq \mathsf{g}$ implies $[\mathsf{k}, \mathsf{g}] \subseteq \mathsf{k}$. Likewise, the members of the *lower central series*, defined by

$$\mathsf{g}_{\{1\}} := \mathsf{g}', \qquad \mathsf{g}_{\{i\}} := [\mathsf{g}, \mathsf{g}_{\{i-1\}}] \quad \text{for } i \geq 2, \tag{1.1.16}$$

are ideals of g, since again by induction from $\mathsf{k} = [\mathsf{g}, \mathsf{h}]$ with $\mathsf{h} = \mathsf{g}_{\{j\}} \subseteq \mathsf{g}$ one deduces that $[\mathsf{k}, \mathsf{g}] \subseteq \mathsf{k}$. In words, the derived series is obtained by considering first the elements that can be written as Lie brackets, then those that can be written as brackets among brackets, etc., while for the lower central series one considers first brackets, then brackets between elements of $\mathfrak{g}$ and brackets, etc. It is apparent that there may be Lie algebras for which these procedures lead, after some finite number of steps, to $[.,.] = 0$, i.e. for which the series end up with zero. The latter properties turn out to be a convenient criterion for the classification of Lie algebras. Thus one defines:

Definition : A Lie algebra $\mathfrak{g}$ is called *solvable* if the derived series of $\mathfrak{g}$ ends up with $\{0\}$. (1.1.17)

Definition : A Lie algebra $\mathfrak{g}$ is called *nilpotent* if the lower central series of $\mathfrak{g}$ ends up with $\{0\}$. (1.1.18)

Let us list a few facts involving the notions of solvability and nilpotency.

- $\mathfrak{g}$ is solvable iff its derived algebra $\mathfrak{g}'$ is nilpotent.
- $\mathfrak{g}$ is nilpotent iff there exists a natural number n that depends only on $\mathfrak{g}$ such that $\mathrm{ad}_{x_1} \circ \mathrm{ad}_{x_2} \circ \ldots \circ \mathrm{ad}_{x_n}(y) = 0$ for all $x_1, x_2, \ldots, x_n, y \in \mathfrak{g}$.
- Nilpotency implies solvability, as follows from the observation that $\mathfrak{g}^{\{j\}} \subseteq \mathfrak{g}_{\{j\}}$ for all $j \in \mathbb{Z}_{>0}$.
- If $\mathfrak{g}$ is solvable (nilpotent), then so are all subalgebras $\mathfrak{h}$ of $\mathfrak{g}$, since $\mathfrak{h}^{\{j\}} \subseteq \mathfrak{g}^{\{j\}}$ ($\mathfrak{h}_{\{j\}} \subseteq \mathfrak{g}_{\{j\}}$) for all $j \in \mathbb{Z}_{>0}$.
- If $\mathfrak{h}$ and $\mathfrak{k}$ are solvable (nilpotent) ideals of $\mathfrak{g}$, then so is $\mathfrak{h} + \mathfrak{k}$.
- If $\mathfrak{g}$ is nilpotent, then $\mathcal{Z}(\mathfrak{g}) \neq 0$: if, say, $\mathfrak{g}_{\{n-1\}} \neq 0$ and $\mathfrak{g}_{\{n\}} = 0$, then $[\mathfrak{g}, \mathfrak{g}_{\{n-1\}}] = 0$ so that $\mathcal{Z}(\mathfrak{g}) \supseteq \mathfrak{g}_{\{n-1\}}$.

Now let $\mathfrak{h}$ be a maximal solvable ideal (i.e. one enclosed in no larger solvable ideal) of $\mathfrak{g}$. If $\mathfrak{k}$ is any other solvable ideal, then so is $\mathfrak{h} + \mathfrak{k}$, and thus the maximality requirement implies $\mathfrak{h} + \mathfrak{k} = \mathfrak{h}$ and hence $\mathfrak{k} \subseteq \mathfrak{h}$. Thus the maximal solvable ideal of a Lie algebra is unique, so that the following definition makes sense.

Definition : The **radical** $\mathfrak{g}_{\mathrm{rad}}$ of a Lie algebra $\mathfrak{g}$ is the maximal solvable ideal of $\mathfrak{g}$. (1.1.19)

Clearly, $\mathfrak{g}$ is solvable iff it equals its own radical.

Further examples of subalgebras are obtained by introducing the following notions:

Definition : The **center** $\mathcal{Z}(\mathsf{g})$ of a Lie algebra is the set of all elements of g which possess zero bracket with all of g,
$$\mathcal{Z}(\mathsf{g}) := \{x \in \mathsf{g} \mid [x, \mathsf{g}] = 0\}. \tag{1.1.20}$$

Definition : The **centralizer** $\mathcal{C}_{\mathsf{g}}(\mathsf{k})$ of a subset k of the Lie algebra g is the set of all elements of g which possess zero bracket with all of k,
$$\mathcal{C}_{\mathsf{g}}(\mathsf{k}) := \{x \in \mathsf{g} \mid [x, \mathsf{k}] = 0\}. \tag{1.1.21}$$

Definition : The **normalizer** $\mathcal{N}_{\mathsf{g}}(\mathsf{h})$ of a subalgebra h of the Lie algebra g is the set of all elements of g whose brackets with h lie in h,
$$\mathcal{N}_{\mathsf{g}}(\mathsf{h}) := \{x \in \mathsf{g} \mid [x, \mathsf{h}] \subseteq \mathsf{h}\}. \tag{1.1.22}$$

Clearly, $\mathcal{Z}(\mathsf{g})$ is a subalgebra (and in fact an ideal) of g, not just a subset, and
$$\mathcal{C}_{\mathsf{g}}(\mathsf{g}) = \mathcal{Z}(\mathsf{g}). \tag{1.1.23}$$
But $\mathcal{C}_{\mathsf{g}}(\mathsf{k})$ is a subalgebra of g for any subset $\mathsf{k} \subset \mathsf{g}$ as well. Also, for any subalgebra h of g, $\mathcal{N}_{\mathsf{g}}(\mathsf{h})$ is a subalgebra of g. In fact it is the largest subalgebra of g which contains h as an ideal; in particular $\mathcal{N}_{\mathsf{g}}(\mathsf{h}) = \mathsf{g}$ if h is an ideal of g.

One of the fundamental problems in Lie algebra theory is the classification of Lie algebras; this is certainly only possible up to a rather trivial degeneracy known as isomorphism. The notions of homomorphism and isomorphism of Lie algebras are the following obvious extensions of the corresponding notions for vector spaces:

Definition : A **homomorphism** from the Lie algebra g to the Lie algebra h is a linear map $\varphi : \mathsf{g} \to \mathsf{h}$ carrying brackets to brackets,
$$\varphi :\ [x, y] \mapsto \varphi([x, y]) = [\varphi(x), \varphi(y)] \tag{1.1.24}$$
for all $x \in \mathsf{g}$.

Definition : An **isomorphism** from the Lie algebra g to the Lie algebra h is a homomorphism which is one to one and onto. (1.1.25)
An isomorphism of a Lie algebra onto itself is called an **automorphism.**

If there exists an isomorphism (homomorphism) from g to h, then h is said to be *isomorphic* (*homomorphic*) to g, denoted by $\mathsf{h} \cong \mathsf{g}$ ($\mathsf{h} \sim \mathsf{g}$). For any member of the derived and lower central series of g one has under a homomorphism that $(\varphi(\mathsf{g}))^{\{j\}} = \varphi(\mathsf{g}^{\{j\}})$ ($(\varphi(\mathsf{g}))_{\{j\}} = \varphi(\mathsf{g}_{\{j\}})$; thus in particular all homomorphic images of a solvable (nilpotent) Lie algebra are solvable (nilpotent) as well.

Two subalgebras h and k of g are said to be *conjugate* to each other iff there exists an automorphism of g which, when restricted to h, provides an isomorphism of h onto k (and vice versa). The set of all automorphisms of a Lie algebra g will be denoted by $Aut(\mathsf{g})$. Obviously, $Aut(\mathsf{g})$ is a group with the multiplication given by the composition of maps and the unit given by the identity map id. If for $\sigma \in Aut(\mathsf{g})$ there exists an integer n such that $\sigma^n = id$ (and n is the smallest positive integer with that property), then n is called the *order* of σ.

Another important class of maps are the derivations: A *derivation* δ of the Lie algebra g is a linear map which obeys the so-called *Leibniz rule*, i.e.

$$[x,y] \mapsto \delta([x,y]) = [x,\delta(y)] + [\delta(x),y] \quad \text{for all } x,y \in \mathsf{g}. \tag{1.1.26}$$

An example for a derivation is the left multiplication with any element of the algebra: for any $x \in \mathsf{g}$, the map

$$\mathrm{ad}_x : \begin{array}{l} \mathsf{g} \to \mathsf{g} \\ y \mapsto \mathrm{ad}_x(y) := [x,y] \ \text{ for all } y \in \mathsf{g} \end{array} \tag{1.1.27}$$

is a derivation; this is so because for the map (1.1.27) the Leibniz rule is just the Jacobi identity. If for some positive integer N one has $(\mathrm{ad}_x)^N = 0$, then the element $x \in \mathsf{g}$ is called *nilpotent*. Similarly, if for any $y \in \mathsf{g}$ there exists a positive integer N_y (that may depend on y) such that $(\mathrm{ad}_x)^{N_y}(y) = 0$, then $x \in \mathsf{g}$ is called *locally nilpotent*; of course, nilpotency implies local nilpotency, and for a finite-dimensional Lie algebra the two concepts are equivalent. If the map ad_x is diagonalizable, then x is called a semisimple element of g.

We close this section by introducing the Lie algebras which are the most important ones for applications in quantum field theory. These are the *reductive* Lie algebras whose building blocks are the *abelian* and the *simple* Lie algebras.

Definition : An **abelian Lie algebra** is a Lie algebra whose derived algebra vanishes, $[\mathsf{g},\mathsf{g}] = 0$. (1.1.28)

In other words, an abelian Lie algebra equals its own center, $\mathcal{Z}(\mathsf{g}) = \mathsf{g}$, and its derived algebra is zero (thus abelianness can be viewed as a very strong version of solvability or nilpotency).

Definition : A **simple Lie algebra** is a Lie algebra which contains no proper ideals and which is not abelian. (1.1.29)

(The second requirement avoids giving undue prominence to the trivial one-dimensional Lie algebra.) Thus simple Lie algebras are in particular equal to their derived algebras, $\mathsf{g}' = \mathsf{g}$, and their center vanishes, $\mathcal{Z}(\mathsf{g}) = 0$.

Reductive Lie algebras are obtained as *direct sums* of these two types of algebras. The direct sum

$$\mathsf{g} = \mathsf{g}_1 \oplus \mathsf{g}_2 \oplus \ldots \oplus \mathsf{g}_n \tag{1.1.30}$$

of Lie algebras g_i, $i = 1, \ldots, n$ (with brackets $[\,,]_i$) is by definition the direct sum of the vector spaces g_i together with the bracket $[\ ,\ ]$ defined by

$$\begin{aligned} &[x,y] = [x,y]_i \quad \text{for} \quad x,y \in \mathsf{g}_i \ (i = 1, \ldots, n) \ \text{ and} \\ &[\mathsf{g}_i, \mathsf{g}_j] = 0 \qquad\ \ \text{for} \quad i \neq j \ (i,j = 1, \ldots, n). \end{aligned} \tag{1.1.31}$$

In particular, if $\mathsf{g} = \mathsf{h} \oplus \mathsf{k}$, both h and k are ideals of g.

Definition : A **semisimple Lie algebra** is an algebra which is the direct sum of simple Lie algebras. (1.1.32)

Definition : A **reductive Lie algebra** is an algebra which is the direct sum of a semisimple and an abelian Lie algebra. (1.1.33)

A Lie algebra is semisimple iff it does not possess a solvable ideal, i.e. iff its radical is zero; often this property is employed to define semisimplicity, with the property (1.1.32) then following as a theorem. Similarly, a Lie algebra is reductive iff its radical equals its center. In particular, a semisimple Lie algebra is not solvable, but rather its derived algebra is equal to g (and hence any member of its derived series equals g as well; thus, in a sense, semisimplicity is the opposite of solvability).

There is a unique (up to isomorphism) one-dimensional abelian Lie algebra which is denoted by u_1; this algebra has a single generator T with bracket $[T,T] = 0$. Moreover, any d-dimensional abelian Lie algebra is isomorphic to the d-fold direct sum of one-dimensional abelian Lie algebras. As a consequence, the non-trivial part of the classification of reductive Lie algebras up to isomorphism is the classification of simple Lie algebras.

For any proper ideal k of g, one defines the *quotient* g/k of g by k as the restriction of g to the part "not containing" k. That is, if $\mathcal{B}$ is a basis of $\mathfrak{g}$ and $\mathcal{B}' \subset \mathcal{B}$ a basis of k, then as a vector space $\mathfrak{g}/\mathsf{k}$ is generated by the basis $\mathcal{B} \setminus \mathcal{B}'$, and the bracket of $\mathfrak{g}/\mathsf{k}$ is obtained from that of $\mathfrak{g}$ by removing the part $\notin \mathfrak{g}/\mathsf{k}$ from the bracket of $\mathfrak{g}$, or in other words

$$[x+\mathsf{k}, y+\mathsf{k}] = [x,y]+\mathsf{k} \quad \text{for all} \quad x,y \in \mathfrak{g}/\mathsf{k} \tag{1.1.34}$$

(this prescription is unambiguous because for $x' = x+u$, $y' = y+v$ with $u,v \in \mathsf{k}$ one has $[x',y'] = [x,y]+w$ with $w = [u,y]+[x,v]+[u,v] \in \mathsf{k}$). By construction, $\mathfrak{g}/\mathsf{k}$ is a subalgebra of $\mathfrak{g}$. In addition, the following properties are valid for quotients. If h is a solvable ideal of $\mathfrak{g}$ and $\mathfrak{g}/\mathsf{h}$ is solvable, then $\mathfrak{g}$ is solvable. If the quotient $\mathfrak{g}/\mathcal{Z}(\mathfrak{g})$ is nilpotent, then so is $\mathfrak{g}$, because then $\mathfrak{g}_{\{n-1\}} \subseteq \mathcal{Z}(\mathfrak{g})$ for some $n \in \mathbb{Z}_{>0}$, and hence $\mathfrak{g}_{\{n\}} = [\mathfrak{g}, \mathfrak{g}_{\{n-1\}}] \subseteq [\mathfrak{g}, \mathcal{Z}(\mathfrak{g})] = 0$.

For any ideal $\mathsf{k} \subseteq \mathfrak{g}$, the Lie algebra $\mathfrak{g}$ can be written as the *semidirect sum*

$$\mathfrak{g} = (\mathfrak{g}/\mathsf{k}) \oplus\!\!\!\!\!\!\;\; \mathsf{k} \tag{1.1.35}$$

of its ideal and the quotient. More generally, we use the symbol " $\oplus$ ", i.e. write

$$\mathfrak{g} = \mathfrak{g}_1 \oplus \mathfrak{g}_2 \oplus \ldots \oplus \mathfrak{g}_n, \tag{1.1.36}$$

if at least one of the $\mathfrak{g}_i$ is an ideal of $\mathfrak{g}$.

If $\mathfrak{g}$ is not solvable (so that its radical is not all of $\mathfrak{g}$), then the quotient of $\mathfrak{g}$ by its radical is semisimple. Thus any non-solvable Lie algebra $\mathfrak{g}$ can be written as a (semidirect, but in general non-direct) sum

$$\mathfrak{g} = \mathsf{s} \oplus \mathfrak{g}_{\mathrm{rad}} \tag{1.1.37}$$

where s is semisimple and $\mathfrak{g}_{\mathrm{rad}}$ is the radical of $\mathfrak{g}$. This splitting is called a *Levi decomposition* of $\mathfrak{g}$, and s a semisimple quotient of $\mathfrak{g}$. s is in fact a maximal semisimple subalgebra of $\mathfrak{g}$; more specifically, the following theorem of Harish–Chandra holds: if $\mathsf{k} \subseteq \mathfrak{g}$ is any semisimple subalgebra of $\mathfrak{g}$, then there exists an automorphism φ such that $\varphi(\mathsf{k}) \subseteq \mathsf{s}$.

The following two types of subalgebras play an important role in the classification of semisimple Lie algebras.

Definition : A **Cartan subalgebra** $\mathfrak{g}_0$ of a Lie algebra $\mathfrak{g}$ is a maximal nilpotent subalgebra consisting of semisimple elements. (1.1.38)

Definition : A **Borel subalgebra** b of a Lie algebra $\mathfrak{g}$ is a maximal solvable subalgebra.

It can be shown that for any Lie algebra a nonzero Cartan subalgebra exists. (To be precise, this holds only if the base field F is infinite, which is true in the cases of our interest.) More specifically, there exists a Cartan subalgebra $\mathfrak{h}$ of $\mathfrak{g}$ such that the radical of $\mathfrak{g}$ obeys

$$[\mathfrak{g}_{\mathrm{rad}},\mathfrak{g}_{\mathrm{rad}}] \subseteq \mathfrak{h} \subseteq \mathfrak{g}_{\mathrm{rad}}. \tag{1.1.39}$$

Also, any Borel subalgebra is equal to its normalizer, and a nilpotent subalgebra $\mathfrak{h}$ is maximal, i.e. a Cartan subalgebra, if it equals its normalizer as well, $\mathcal{N}_{\mathfrak{g}}(\mathfrak{h}) = \mathfrak{h}$.

While the concepts of Cartan and Borel subalgebras are defined for arbitrary Lie algebras $\mathfrak{g}$, they are most important if $\mathfrak{g}$ is semisimple. Namely, it can be shown that for a semisimple Lie algebra, the Cartan subalgebras are precisely the maximal abelian subalgebras. Also, if $\mathfrak{g}$ is not solvable, then the Borel subalgebras of $\mathfrak{g}$ are in a natural one to one correspondence with those of the semisimple Lie algebra $\mathfrak{s} = \mathfrak{g}/\mathfrak{g}_{\mathrm{rad}}$.

If $\mathfrak{h}$, $\mathfrak{k} \subseteq \mathfrak{g}$ are semisimple and finite-dimensional subalgebras of $\mathfrak{g}$, then the reciprocity relation

$$\mathcal{C}_{\mathfrak{g}}(\mathfrak{h}) = \mathfrak{k} \Leftrightarrow \mathcal{C}_{\mathfrak{g}}(\mathfrak{k}) = \mathfrak{h} \tag{1.1.40}$$

between the centralizers of $\mathfrak{h}$ and $\mathfrak{k}$ holds.

As a final elementary definition, we introduce: A *gradation* of a Lie algebra $\mathfrak{g}$ by an abelian group A is by definition a decomposition

$$\mathfrak{g} = \bigoplus_{i\in\mathsf{A}} \mathfrak{g}_{(i)} \tag{1.1.41}$$

such that

$$[\mathfrak{g}_{(i)},\mathfrak{g}_{(j)}] \subseteq \mathfrak{g}_{(i+j)}. \tag{1.1.42}$$

Let us illustrate some of the concepts introduced above with a few simple examples. First consider the three-dimensional Lie algebra with generators U, V and W and non-trivial brackets

$$[U,V] = W, \qquad [U,W] = [V,W] = 0. \tag{1.1.43}$$

The derived algebra is then one-dimensional and hence abelian; as a consequence this algebra is solvable; but in fact it is nilpotent, since any double bracket clearly vanishes. Next take the two-dimensional Lie algebra with generators U and V and nonvanishing bracket

$$[U,V] = V. \tag{1.1.44}$$

Owing to $[V,V] = 0$, this algebra is again solvable, but it is not nilpotent since multiple application of ad_U to V always yields the nonzero result V. Finally take the three-dimensional Lie algebra spanned by E_+, E_-

and H subject to

$$\begin{aligned} [E_+, E_-] &= H, \\ [H, E_\pm] &= \pm 2\, E_\pm. \end{aligned} \tag{1.1.45}$$

By applying $\mathrm{ad}_{E_\pm}$ twice to an arbitrary element $x = \xi_+ E_+ + \xi_- E_- + \zeta H$ of the algebra, one obtains $-2\xi_\pm E_\pm$. Thus for $\xi_+ \neq 0$ ($\xi_- \neq 0$) any subalgebra containing x also contains E_+ (E_-), while for $\xi_+ = \xi_- = 0$ it contains of course H; as a consequence, any ideal containing x is in fact equal to the algebra itself. Since x was arbitrary, it follows that the Lie algebra defined by (1.1.45) is simple; this algebra is denoted by A_1. In fact, for $F = \mathbb{C}$, A_1 is the unique (up to isomorphism) three-dimensional simple Lie algebra, and there are no lower-dimensional simple Lie algebras (any one-dimensional Lie algebra is isomorphic to the abelian algebra u_1, and any two-dimensional Lie algebra is easily seen to be either abelian or isomorphic to the solvable algebra defined by (1.1.44)).

To conclude this introductory section, we mention that the concepts introduced above for Lie algebras, such as subalgebras, solvability, semisimplicity, etc., have their analogues also for general algebras, in particular for associative algebras; their definitions are obtained from the ones given above by just replacing the Lie bracket by the bilinear operation $\diamond$ of the algebra. For example, introducing structure constants as $T^a \diamond T^b = f^{ab}{}_c T^c$, the requirement that the algebra generated by the T^a is associative reads

$$f^{ab}{}_c f^{cd}{}_e = f^{bd}{}_c f^{ac}{}_e. \tag{1.1.46}$$

1.2 Representations and modules

A very important concept is the one of *representations* of a Lie algebra. To introduce this concept, we first define: The *general linear algebra* $gl(V)$ of a vector space V is the algebra

$$gl(V) := \{\varphi : V \to V \mid \varphi \text{ linear}\} \tag{1.2.1}$$

of endomorphisms of V, with bilinear operation

$$[\,,] : \begin{array}{l} gl(V) \times gl(V) \to gl(V) \\ (\varphi, \psi) \mapsto [\varphi, \psi] := \varphi \circ \psi - \psi \circ \varphi . \end{array} \tag{1.2.2}$$

Here "$\circ$" denotes the composition of maps. Note that $gl(V)$ is a vector space over the same field F as V so that the notation "$-$" makes sense. Also, it is easy to check that with this definition, $gl(V)$ is in fact a Lie algebra over F, as is already foreseen by using the bracket notation. The notation $gl(V)$ is the one usually employed in the physics literature; in mathematics, it is more common to write $End(V)$ for $gl(V)$.

The elements φ of $gl(V)$ can be identified in a natural way with matrices in $gl_{\rm N}$. Namely, for any vector space V over F of dimension $\rm N < \infty$, the space of endomorphisms is as a vector space isomorphic to $gl_{\rm N}$, the set of N×N-matrices (with entries $\in F$); this follows immediately by choosing a basis in V so that the endomorphisms are represented by N×N-matrices. This isomorphism of vector spaces extends to an isomorphism of Lie algebras, with the bracket for $gl_{\rm N}$ given by the matrix commutator.

Now we are ready to define:

Definition : A **representation** R of the Lie algebra $\mathfrak{g}$ is a homomorphism of $\mathfrak{g}$ into the general linear algebra of a vector space V (over the same field F as V), i.e.

$$R:\ \mathfrak{g} \to gl(V) \tag{1.2.3}$$

with

$$[x,y] \mapsto R([x,y]) = R(x)\circ R(y) - R(y)\circ R(x)$$

for all $x,y \in \mathfrak{g}$.

Note that this definition allows for $\dim V = \infty$ even if $\dim \mathfrak{g} < \infty$. An N-dimensional *matrix representation* of a Lie algebra is then of course a homomorphism of $\mathfrak{g}$ into $gl_{\rm N}$, and for any finite-dimensional representation of $\mathfrak{g}$ in $gl(V)$ one obtains an isomorphic matrix representation of dimension $\rm N = \dim V$ by by choosing a basis of V.

Also note that, while for $\mathfrak{g}$ there is no definition of a product other than the Lie bracket, for R one has the product $\circ$ denoting the composition of maps (respectively, for matrix representations, the usual matrix multiplication). With the help of this product one may in fact define arbitrary power series in the elements of R. Now for a given power series in $R(x)$, one may of course formally write down the same power series with $R(x)$ replaced everywhere by x. In this way one arrives at the concept of the *universal enveloping algebra* $\mathsf{U}(\mathfrak{g})$ of a Lie algebra $\mathfrak{g}$ which consists of all finite formal power series in the elements of $\mathfrak{g}$. $\mathsf{U}(\mathfrak{g})$ is an associative algebra (and hence not a Lie algebra), with the product $\diamond$ given by (termwise) formal multiplication; it is generated as a vector space by all monomials in the generators of $\mathfrak{g}$, identifying however all monomials which become equal to each other upon use of the bracket relations (interpreted as commutators) of $\mathfrak{g}$.

Let us consider a few simple examples. As is easily verified, the nilpotent Lie algebra (1.1.43) possesses a three-dimensional matrix representation given by

$$
\begin{aligned}
R(U) &= \begin{pmatrix} 0 & 1 & 0 \\ 0 & 0 & 0 \\ 0 & 0 & 0 \end{pmatrix}, \quad R(V) = \begin{pmatrix} 0 & 0 & 0 \\ 0 & 0 & 1 \\ 0 & 0 & 0 \end{pmatrix}, \\
R(W) &= \begin{pmatrix} 0 & 0 & 1 \\ 0 & 0 & 0 \\ 0 & 0 & 0 \end{pmatrix},
\end{aligned} \tag{1.2.4}
$$

and the solvable Lie algebra (1.1.44) has the two-dimensional matrix representation

$$
R(U) = \begin{pmatrix} 1 & 0 \\ 0 & 0 \end{pmatrix}, \quad R(V) = \begin{pmatrix} 0 & 1 \\ 0 & 0 \end{pmatrix}. \tag{1.2.5}
$$

More generally, for arbitrary $\mathrm{N} \in \mathbb{Z}_{>0}$, the Lie algebra of strictly upper triangular $\mathrm{N} \times \mathrm{N}$-matrices, $\mathfrak{g} = \{M \in gl_{\mathrm{N}} \mid M_{ij} = 0 \text{ for } i < j\}$, is nilpotent, and the Lie algebra of upper triangular matrices, $\mathfrak{g} = \{M \in gl_{\mathrm{N}} \mid M_{ij} = 0 \text{ for } i \le j\}$, is solvable.

Finally, consider the unique complex three-dimensional simple Lie algebra A_1; a two-dimensional matrix representation of the Lie brackets (1.1.45) is provided by

$$
\begin{aligned}
R(E_+) &= \begin{pmatrix} 0 & 1 \\ 0 & 0 \end{pmatrix}, \quad R(E_-) = \begin{pmatrix} 0 & 0 \\ 1 & 0 \end{pmatrix}, \\
R(H) &= \begin{pmatrix} 1 & 0 \\ 0 & -1 \end{pmatrix}.
\end{aligned} \tag{1.2.6}
$$

Thus the (complex) algebra A_1 is isomorphic to the Lie algebra $sl_2(\mathbb{C})$ of complex traceless 2×2-matrices.

For arbitrary Lie algebra $\mathfrak{g}$, an example for a representation is provided by the *adjoint* (or *regular*) representation which is given by

$$
R_{\mathrm{ad}} : \begin{aligned} \mathfrak{g} &\to gl(\mathfrak{g}) \\ x &\mapsto \mathrm{ad}_x, \end{aligned} \tag{1.2.7}
$$

with the map ad_x as defined in (1.1.27). In R_{ad}, the Lie brackets are defined by $[\mathrm{ad}_x, \mathrm{ad}_y] := \mathrm{ad}_x \circ \mathrm{ad}_y - \mathrm{ad}_y \circ \mathrm{ad}_x$. Thus one has $([\mathrm{ad}_x, \mathrm{ad}_y])(z) = [x, [y, z]] - [y, [x, z]]$, so that the antisymmetry and Jacobi identity for the brackets of $\mathfrak{g}$ imply that $([\mathrm{ad}_x, \mathrm{ad}_y])(z) = [[x, y], z] \equiv \mathrm{ad}_{[x,y]}(z)$, which verifies the homomorphism property of R_{ad}. Also recall that ad_x is a derivation; thus the image $R_{\mathrm{ad}}(\mathfrak{g})$ of $\mathfrak{g}$ under the adjoint representation is a subalgebra of $Der(\mathfrak{g})$, the space of all derivations of $\mathfrak{g}$.

Introducing a formal variable t, the polynomial in t defined as

$$\det\left(t - R_{\text{ad}}(x)\right) = \sum_{n=0}^{d} p_n(x)\, t^n \tag{1.2.8}$$

determines certain polynomial functions p_n in $\mathfrak{g}$; the smallest integer n for which $p_n(x) \not\equiv 0$ is called the *rank* of the Lie algebra $\mathfrak{g}$.

Notice that

$$\text{ad}_{T^a}(T^b) = [T^a, T^b] = f^{ab}{}_c\, T^c, \tag{1.2.9}$$

which means that in terms of the generators, the adjoint representation is just given by the structure constants. In other words, the matrices $R_{\text{ad}}(T^a)$ have the entries

$$(R_{\text{ad}}(T^a))_b^{\ c} = f^{ac}{}_b. \tag{1.2.10}$$

In particular, the dimensionality of the adjoint matrix representation is equal to the dimension of the algebra. Thus for example the matrix representations for the algebras (1.2.4) and (1.2.5) given above are (isomorphic to) the adjoint representations of these algebras. In contrast, the adjoint representation of the simple Lie algebra A_1 is given by

$$R_{\text{ad}}(E_+) = \begin{pmatrix} 0 & -2 & 0 \\ 0 & 0 & 1 \\ 0 & 0 & 0 \end{pmatrix}, \quad R_{\text{ad}}(E_-) = \begin{pmatrix} 0 & 0 & 0 \\ -1 & 0 & 0 \\ 0 & 2 & 0 \end{pmatrix}, \tag{1.2.11}$$

$$R_{\text{ad}}(H) = \begin{pmatrix} 2 & 0 & 0 \\ 0 & 0 & 0 \\ 0 & 0 & -2 \end{pmatrix}$$

rather than by (1.2.6).

A concept intimately related to the one of a representation is the following:

Definition : A **module** of the Lie algebra $\mathfrak{g}$ is a vector space V (over the same field F as $\mathfrak{g}$) together with an action of $\mathfrak{g}$ on V, i.e. an operation

$$\bullet : \mathfrak{g} \times V \to V,$$

which possesses the properties (1.2.12)

$$\begin{aligned} (\xi x + \zeta y) \bullet w &= \xi (x \bullet w) + \zeta (y \bullet w), \\ x \bullet (\xi v + \zeta w) &= \xi (x \bullet v) + \zeta (x \bullet w), \\ [x, y] \bullet w &= x \bullet (y \bullet w) - y \bullet (x \bullet w) \end{aligned}$$

for all $\xi, \zeta \in F$, $x, y \in \mathfrak{g}$ and $v, w \in V$.

The *dimension* of a $\mathfrak{g}$-module is the dimension of the underlying vector space.

The connection between representations and modules is obviously as follows. Given a representation $R : \mathfrak{g} \to gl(V)$, the vector space V becomes a module of $\mathfrak{g}$ via the definition

$$x \bullet w := (R(x))(w). \tag{1.2.13}$$

Conversely, given a $\mathfrak{g}$-module V, a representation R of $\mathfrak{g}$ is defined by

$$(R(x))(w) := x \bullet w. \tag{1.2.14}$$

The notation $(R(x))(w)$ used here is adapted to the fact that $R(x)$ is an endomorphism acting on $w \in V$; remembering that $R(x)$ can be identified with a matrix, we write this in the following as $R(x) \cdot w$.

The term *representation space* instead of module is also often used. Owing to a slight abuse of terminology, in the physics literature one usually finds simply "representation" instead of "representation space". We will follow this custom only insofar as we will use the symbol V_R for the module associated to the representation R in just a few circumstances where we think that it is necessary to avoid confusion, while otherwise we employ the letter R for both representations and the associated modules.

The modules which are the object of our main interest are the so-called highest weight modules; they will be discussed in section 1.6. Here we only add some general remarks. A *submodule* is of course a subvectorspace possessing the properties of a module. Clearly, modules which do not possess any (proper) submodules are of particular interest; thus one defines:

Definition : An **irreducible module** of a Lie algebra is a module which contains no proper submodules. (1.2.15)

(The zero-dimensional vector space is of course not regarded as an irreducible module; thus to be precise, one should in fact require that an irreducible module contains exactly two submodules, namely itself and $\{0\}$.) For any irreducible $\mathfrak{g}$-module R, *Schur's lemma* is valid which states that the only endomorphisms of R commuting with $R(x)$ for all $x \in \mathfrak{g}$ are the scalar multiples of the identity map. Any irreducible finite-dimensional module of a solvable Lie algebra (e.g. an abelian algebra) is one-dimensional.

Definition : A **fully reducible module** of a Lie algebra is a module which is the direct sum of irreducible modules. (1.2.16)

(The term "completely reducible" instead of "fully reducible" is also common.) The latter definitions translate to representations as follows. A representation R of $\mathfrak{g}$ is irreducible (fully reducible, not fully reducible) iff the image $R(\mathfrak{g})$ is simple (semisimple, non-semisimple). In particular the adjoint representation of a Lie algebra $\mathfrak{g}$ is irreducible (fully reducible, not fully reducible) iff $\mathfrak{g}$ is simple (semisimple, non-semisimple).

Clearly, a *reducible* (i.e. non-irreducible) module need not be fully reducible. But, from a module which is reducible but not fully reducible, one can build a fully reducible module by the procedure of *quotienting* which is defined in analogy to the quotienting of algebras; i.e., the fully reducible quotient $\tilde{V}$ of V is the module associated to that representation $\tilde{R}$ whose image is the semisimple quotient of the image $R(\mathsf{g})$ (R being the representation having V as its module); if the image $\tilde{R}(\mathsf{g})$ of the quotient representation is in fact simple, then the quotient module is irreducible. Of course, it may well be that the irreducible quotient module obtained this way is the trivial one-dimensional module, and it is also possible that the quotient is $\{0\}$, i.e. not a proper module at all.

The representation theory of abelian Lie algebras is quite trivial: If g is abelian, the representation matrices $R(x)$ for a fully reducible representation are diagonal matrices, and hence any fully reducible module is the direct sum of one-dimensional modules. In particular, on an irreducible module of $\mathsf{g} \cong u_1$, the single generator T acts multiplicatively,

$$R(T) \cdot v = qv\,; \tag{1.2.17}$$

the number q is called the u_1*-charge* of the module. Investigating the representation theory of reductive Lie algebras, one can therefore concentrate on the semisimple part.

Given any two modules, one can define a new module as their tensor product:

Definition : The **Kronecker product** $V_R \times V_S$ of two g-modules V_R and V_S is the vector space $V_R \otimes_F V_S$ (the tensor product over F of the vector spaces underlying the modules) together with the action of g defined as (1.2.18)

$$(R \times S)(x) \cdot (v \otimes w) = (R(x) \cdot v) \otimes w + v \otimes S(x) \cdot w$$

for all $v \in V_R$, $w \in V_S$, $x \in \mathsf{g}$.

An immediate consequence of the definition of the Kronecker tensor product is its associativity: one has

$$(R \times R') \times R'' \cong R \times (R' \times R'') \tag{1.2.19}$$

for all g-modules R, R', R''. For a fully reducible triple Kronecker product, the matrices that implement this isomorphism for a given choice of bases of the modules on the left and right sides are called *Wigner–Racah coefficients* or $6j$*-symbols.*

If g is semisimple, then the Kronecker product of finite-dimensional irreducible modules is fully reducible, and all finite-dimensional modules can be obtained as irreducible components of tensorial (Kronecker) powers of a small set of basic modules (these will be listed at the end of section 1.6; see p. 69). Thus, for semisimple algebras, it is possible to construct the whole representation theory of finite-dimensional modules using only these basic modules and the concept of the Kronecker product. Also, the set of finite-dimensional irreducible modules of a semisimple Lie algebra and their Kronecker product may be viewed as describing the generators and the multiplication, respectively, of an associative algebra over the integers $\mathbb{Z}$. More generally, one can analyse Kronecker products via the so-called Brauer–Weyl theory: one constructs an algebra $\mathcal{C}$ as the centralizer of the action of $\mathfrak{g}$ on the product $R^{\otimes n} \equiv R \times R \times \ldots \times R$, i.e. as the set of matrices that commute with the representation matrices of $\mathfrak{g}$ on $R^{\otimes n}$; the full reducibility of $R^{\otimes n}$ is then equivalent to the semisimplicity of $\mathcal{C}$.

Another possibility for getting a new module from a given one is the following. Given a matrix representation R of $\mathfrak{g}$, define

$$R^+(x) := -(R(x))^t \quad \text{for all } x \in \mathfrak{g}, \tag{1.2.20}$$

where "^{t}" stands for the transposition of matrices. Then one has

$$[R^+(x), R^+(y)] = -[R(x), R(y)]^t = -R([x,y])^t = R^+([x,y]), \tag{1.2.21}$$

and so R^+ is a representation of $\mathfrak{g}$, too. R^+ is called the representation *conjugate* (or *contragredient*) to R. It is a homomorphism of $\mathfrak{g}$ into $gl(V_R^\star)$ where $V_R^\star$ is the vector space dual to V_R, and hence has the same dimension as R. The Kronecker product of any irreducible module R with its conjugate R^+ contains precisely once the *singlet*, i.e. the trivial one-dimensional module, as a submodule. The module conjugate to a u_1-module of charge q possesses the charge $-q$.

For any $\mathfrak{g}$-module R, we may describe the vectors $v \in R$ by their components v_i with respect to some chosen basis. The elements $w \in R^+$ of the conjugate module are then described by the components with respect to the dual basis, i.e. by components w^i with an upper index, and the representation matrices act as follows:

$$\begin{aligned} v_i &\mapsto v'_i = (R(x))_i^{\;j}\, v_j, \\ w^i &\mapsto w'^i = (R^+(x))^i_{\;j}\, w^j = -w^j\, (R(x)^t)_j^{\;i}. \end{aligned} \tag{1.2.22}$$

A tensor t with m upper and n lower indices is called an *invariant tensor* of the representation R iff it obeys

$$\begin{aligned} 0 \;=\; & (R(T^a))_{j_1}^{\;\ell}\, t^{i_1\ldots i_m}_{\ell j_2\ldots j_n} + \ldots + (R(T^a))_{j_n}^{\;\ell}\, t^{i_1\ldots i_m}_{j_1\ldots j_{n-1}\ell} \\ & + (R^+(T^a))^{i_1}_{\;\ell}\, t^{\ell i_2\ldots i_m}_{j_1\ldots j_n} + \ldots + (R^+(T^a))^{i_m}_{\;\ell}\, t^{i_1\ldots i_{m-1}\ell}_{j_1\ldots j_n}. \end{aligned} \tag{1.2.23}$$

For any representation, an invariant tensor is provided by the Kronecker symbol $\delta_i^{\;j}$; for this tensor the requirement (1.2.23) simply reduces to the identity (1.2.20). Sometimes R^+ turns out to be isomorphic to R; this happens iff there exists an invariant tensor e^{ij} (and its inverse e_{ij} such that $e_{ij}e^{jk} = \delta_i{}^k$) which can be used to relate $v \in R$ and $w \in R^+$ via $w^i = e^{ij}v_j$. Such modules are called *selfconjugate*; if the tensor e^{ij} is symmetric, they are also referred to as *orthogonal* modules, while if e^{ij} is antisymmetric, they are called *symplectic* modules. The adjoint module of any semisimple Lie algebra is an orthogonal module; the relevant symmetric two-index tensor is provided by the Cartan–Killing metric κ^{ab}, which will be defined shortly. Also note that, as a result of the existence of an antisymmetric two-index tensor, any symplectic module must be even-dimensional.

The concepts of representations and modules can also be defined analogously for algebras which are not Lie algebras. A representation of the algebra a is a linear map $R : \mathsf{a} \to gl(V)$ from the algebra to the general linear algebra of some vector space (the associated module) V which preserves the bilinear operation of a, $R(x\diamond y) = R(x)R(y)$. It must be noted, however, that unlike in the case of Lie algebras, there is no natural way of defining tensor products of representations, or in other words of identifying an a-module structure on the tensor product (in the vector space sense) of any two given a-modules. One might think that for R_i, $i = 1, 2$, representations of a on V_i, it should be possible to define a tensor product via $x \mapsto R(x) := R_1(x) \otimes R_2(x)$, i.e. $R(x)\cdot(v_1 \otimes v_2) = (R_1(x)\cdot v_1) \otimes (R_2(x)\cdot v_2)$; however, this map is not linear because e.g. $R(\xi x) = R_1(\xi x) \otimes R_2(\xi x) = \xi^2\, R_1(x) \otimes R_2(x) = \xi^2\, R(x)$. Another tentative definition which comes to mind is

$$x \mapsto R(x) := R_1(x) \otimes id_2 + id_1 \otimes R_2(x), \tag{1.2.24}$$

with id_i the identity map on V_i, i.e. $R(x) \cdot (v_1 \otimes v_2) = (R_1(x) \cdot v_1) \otimes v_2 + v_1 \otimes (R_2(x) \cdot v_2)$. But this implies that

$$\begin{aligned} R(x\diamond y) &= R_1(x)R_1(y) \otimes id_2 + id_1 \otimes R_2(x)R_2(y) \\ \neq R(x)R(y) &= R_1(x)R_1(y) \otimes id_2 + R_1(x) \otimes R_2(y) \\ &\quad + R_1(y) \otimes R_2(x) + id_1 \otimes R_2(x)R_2(y), \end{aligned} \tag{1.2.25}$$

so that R is a representation only if the bilinear operation $\diamond$ is antisymmetric (as is in particular the case for Lie algebras).

The restrictions an algebra must obey in order for the definition of a tensor product to make sense can also be put in a more formal setting. This leads to the concept of Hopf algebras that will be introduced in section 4.1.

1.3 The Killing form. Real and complex Lie algebras

For some applications, it is preferable to be able to treat the three indices of the structure constants on an equal footing. This is certainly only possible if one need not distinguish between upper and lower indices, in other words, if there is a metric relating upper and lower indices. Such a metric should be an intrinsic property of the Lie algebra and hence be encoded in the structure constants or, equivalently, be expressible through the adjoint representation. A candidate metric is given by the *Killing form* (or *Cartan–Killing form*) of the Lie algebra:

Definition : The **Killing form** of the Lie algebra $\mathfrak{g}$ is the map

$$\kappa: \begin{array}{l} \mathfrak{g}\times\mathfrak{g}\to F, \\ (x,y)\mapsto\kappa(x,y):=\mathrm{tr}\,(\mathrm{ad}_x\circ\mathrm{ad}_y). \end{array} \tag{1.3.1}$$

Here "$\circ$" denotes the composition of maps, and "tr" the trace of linear maps. It is obvious that the Killing form is symmetric and bilinear. Moreover, it is also associative in the sense that

$$\kappa([x,y],z)=\kappa(x,[y,z]), \tag{1.3.2}$$

(sometimes this property is called *invariance*) and it is preserved by automorphisms of $\mathfrak{g}$,

$$\kappa(\sigma(x),\sigma(y))=\kappa(x,y)\quad\text{for all}\quad\sigma\in Aut(\mathfrak{g}). \tag{1.3.3}$$

In fact these properties define κ up to an overall multiplicative constant. Given a basis $\{T^a\}$, the Killing form is represented by the matrix

$$\kappa^{ab}:=\frac{1}{I_{\mathrm{ad}}}\cdot\mathrm{tr}\,(\mathrm{ad}_{T^a}\circ\mathrm{ad}_{T^b}). \tag{1.3.4}$$

The normalization constant I_{ad} introduced in this formula will be identified in (1.6.42) as the so-called Dynkin index of the adjoint representation. From the identity

$$\mathrm{ad}_{T^a}\circ\mathrm{ad}_{T^b}(T^d)=[T^a,[T^b,T^d]]=f^{ae}{}_c f^{bd}{}_e T^c, \tag{1.3.5}$$

one obtains by summation over $c=d$ an expression for the matrix (1.3.4) in terms of the structure constants,

$$\kappa^{ab}=\frac{1}{I_{\mathrm{ad}}}f^{ae}{}_c f^{bc}{}_e. \tag{1.3.6}$$

This result is of course not surprising: It is just the simplest possibility of forming a tensor with two upper indices out of the structure constants.

The definition (1.3.1) expresses the Killing form through the adjoint representation. Since the adjoint representation is finite-dimensional, one can identify it with a matrix representation; with the help of the representation matrices $R_{\rm ad}(x)$, (1.3.1) can be rewritten as

$$\kappa(x,y) = \mathrm{tr}\,(R_{\rm ad}(x)R_{\rm ad}(y)), \tag{1.3.7}$$

with "tr" now denoting the trace of matrices. Instead of the adjoint, one could in fact use any other finite-dimensional irreducible representation R: Writing

$$\kappa_R(x,y) = \mathrm{tr}\,(R(x)\,R(y)), \tag{1.3.8}$$

one has for simple g

$$\kappa_R(x,y) \propto \kappa(x,y) \tag{1.3.9}$$

where the constant of proportionality depends on the chosen representation (its value I_R will be given in (1.6.43) below).

While the Killing form is bilinear and symmetric, it is generically degenerate (and hence does not provide a proper metric), i.e. there may exist elements $x \neq 0$ of g obeying $\kappa(x,x) = 0$. The investigation of the degeneracy of the Killing form leads to an important result. Namely, the properties of solvability and semisimplicity can be expressed through the degeneracy properties of the Killing form in a simple way (these are the so-called *Cartan criteria*):

- g is solvable iff $\kappa(x,x) = 0$ for all $x \in \mathsf{g}'$.
- g is semisimple iff the Killing form is non-degenerate, i.e. iff the determinant of κ^{ab} does not vanish.

Some related facts about the Killing form are the following. (Recall from section 1.1 that a Lie algebra is called solvable iff its derived algebra ends up with zero, and that the radical of a Lie algebra is its maximal solvable ideal.)

- The radical of g is the orthogonal complement with respect to κ of the derived algebra g'.
- If $\mathsf{g} = \mathsf{s} \oplus \mathsf{h}$ is a Levi decomposition of g into its radical h and its semisimple part s, and if k is the maximal nilpotent subalgebra of h, then κ is non-degenerate on s and on h mod k, and is identically zero on k. In particular, if g is nilpotent, then $\kappa \equiv 0$; conversely, if $\kappa \equiv 0$, then g is solvable.
- If h is any ideal of g, κ the Killing form on g, and κ_{h} the Killing form on h, then $\kappa_{\mathsf{h}} = \kappa\,|_{\mathsf{h}}$.
- The different ideals in a direct sum $\mathsf{g} \oplus \mathsf{h}$ are mutually orthogonal with respect to the Killing form, i.e. $\kappa(x,y) = 0$ for all $x \in \mathsf{g}$, $y \in \mathsf{h}$.

As a consequence, the Killing form of a semisimple Lie algebra is already determined by the Killing forms of its simple ideals.

In the semisimple case the Killing form (and its inverse) can be used as a metric which raises (or lowers) indices. For example, structure constants with only upper indices are defined as

$$f^{abc} := f^{ab}{}_d \, \kappa^{dc} . \tag{1.3.10}$$

It is possible to choose the basis of $\mathfrak{g}$ such that these structure constants with only upper indices are completely antisymmetric; such a basis is often useful in explicit calculations.

Let us also note that for many applications it is very helpful to know the expression of the tensor product $\kappa_{ab} R(T^a) \otimes R(T^b)$ through (independent) invariant tensors $t_{(\ell)}$ of the irreducible module R, i.e. the coefficients c_ℓ in

$$\begin{aligned} (\kappa_{ab}\, R(T^a) \otimes R(T^b))_{i,j}^{\;k,l} &\equiv \kappa_{ab}\, (R(T^a))_i^{\;k}\, (R(T^b))_j^{\;l} \\ &= \sum_\ell c_\ell \, (t_{(\ell)})_{ij}^{\;kl} . \end{aligned} \tag{1.3.11}$$

For the lowest-dimensional modules of the simple Lie algebras, this decomposition will be given in the table (1.6.80) below.

Since for any simple Lie algebra over $\mathbb{R}$, κ is non-degenerate, an appropriate choice of basis brings κ^{ab} to the canonical diagonal form

$$\begin{pmatrix} \mathbf{1}_p & \mathbf{0} \\ \mathbf{0} & -\mathbf{1}_q \end{pmatrix} . \tag{1.3.12}$$

It turns out that the q-dimensional subspace with $\kappa < 0$ is a subalgebra. If $\mathfrak{g}$ is a simple Lie algebra over $\mathbb{C}$, one can associate with it several distinct Lie algebras over $\mathbb{R}$, called *real forms* of $\mathfrak{g}$. There is always a unique (up to isomorphism) real Lie algebra, called the *compact* real form of $\mathfrak{g}$, for which $p = 0$ in (1.3.12), so that in an appropriate basis the Killing form is negative definite,

$$\kappa^{ab} = -\delta^{ab} . \tag{1.3.13}$$

This terminology arises as follows. Recall that a group is by definition a set G together with a multiplication "$\cdot$": $\mathsf{G} \times \mathsf{G} \to \mathsf{G}$ that is associative and possesses a unit (i.e. there exists an element $E \in \mathsf{G}$ such that $E \cdot X = X = X \cdot E$ for all $X \in \mathsf{G}$) and an inverse (i.e. for any $X \in \mathsf{G}$ there exists an element $X^{-1} \in \mathsf{G}$ such that $X^{-1} \cdot X = E = X \cdot X^{-1}$). Now to any simple Lie algebra $\mathfrak{g}$, one can associate a simple *Lie group* G; a Lie group is by definition a smooth manifold which is also a group, with the two structures being compatible; in other words, the group multiplication and inversion are continuous maps. The term compactness then has the following natural meaning: if G is a compact manifold, then the underlying real Lie algebra $\mathfrak{g}$ is called a compact real form.

The connection between $\mathfrak{g}$ and G is provided via an appropriately defined exponential map "exp" which defines a chart of G around its identity element. Thus any element $X \in \mathsf{G}$ in the neighborhood of the identity is written as $X = \exp(x)$ for some $x \in \mathfrak{g}$, or in terms of the generators,

$$X = \exp\left(\sum_a \lambda_a T^a\right) \tag{1.3.14}$$

for a suitable choice of $\lambda_a \in F$, $a = 1, \dots, \dim\mathfrak{g}$, and the λ_a can be used as coordinates on the Lie group manifold in the neighborhood of the identity element. Stated differently, $\mathfrak{g}$ is isomorphic to the space of (left or right) invariant vector fields on G, or in other words, to the tangent space to G at the identity element. Also, as a set, G is contained in the completion (i.e. including also infinite power series) of the universal enveloping algebra $\mathsf{U}(\mathfrak{g})$.

This abstract connection between $\mathfrak{g}$ and G is most easily made explicit if the elements of the Lie algebra $\mathfrak{g}$ are matrices. Then the function exp is the usual exponential power series, $\exp(x) = \sum_{n=0}^{\infty} x^n/n!$, with x^n denoting the n-fold matrix product; also, the unit element E is the unit matrix, and the inverse of $X = \exp(x)$ is $X^{-1} = \exp(-x)$.

Another large class of examples is obtained as follows. The automorphisms $Aut(\mathfrak{g})$ of any finite-dimensional Lie algebra $\mathfrak{g}$ form a Lie group whose Lie algebra is the algebra $Der(\mathfrak{g})$ of derivations of $\mathfrak{g}$; the group $Aut(\mathfrak{g})$ contains a subgroup $Int(\mathfrak{g})$ which is generated by the automorphisms of the form $\exp(\mathrm{ad}_x)$ with $x \in \mathfrak{g}$ nilpotent (so that the exponential power series terminates) whose Lie algebra is the image of $\mathfrak{g}$ under the adjoint representation; $Int(\mathfrak{g})$ is called the *adjoint group* of $\mathfrak{g}$, and its elements are called *inner automorphisms.*

Via the exponential map the finite-dimensional Lie algebras are in one to one correspondence to those finite-dimensional Lie groups which are connected and simply connected (such a group G is called the *universal covering group* associated with $\mathfrak{g}$). Thus the only information about a finite-dimensional Lie group G which is not contained in its Lie algebra concerns properties that depend either on the set $\pi_0(\mathsf{G})$ of different connected components or on the fundamental group $\pi_1(\mathsf{G})$. For example, if $\mathfrak{g}$ is semisimple, all derivations are inner automorphisms so that $Int(\mathfrak{g})$ and $Aut(\mathfrak{g})$ possess the same Lie algebra, and $Aut(\mathfrak{g})$ is the covering group of $Int(\mathfrak{g})$.

Let us now consider the connection between real and complex Lie algebras in more detail. For any complex Lie algebra $\mathfrak{g}$, denote by $\mathfrak{g}_{\mathbb{R}}$ the algebra obtained by restricting the base field of $\mathfrak{g}$ from $\mathbb{C}$ to $\mathbb{R}$; also, for a real Lie algebra h, define its *complexification* $\mathsf{h}_{\mathbb{C}}$ as

$$\mathsf{h}_{\mathbb{C}} := \mathsf{h} \otimes_{\mathbb{R}} \mathbb{C} \equiv \mathsf{h} \oplus \sqrt{-1}\,\mathsf{h}. \tag{1.3.15}$$

Note that $(\mathsf{g}_{\mathbb{R}})_{\mathbb{C}} = \mathsf{g} \oplus \mathsf{g}$. The real forms of a complex Lie algebra g are then $\mathsf{g}_{\mathbb{R}}$ and those subalgebras h of $\mathsf{g}_{\mathbb{R}}$ for which g is the complexification $\mathsf{g} = \mathsf{h}_{\mathbb{C}}$. It is also clear that a real Lie algebra h is semisimple iff its complexification is semisimple (because, relative to a fixed basis of g, the matrix for the Killing form is the same for $\mathsf{g}_{\mathbb{C}}$ as for g) while if h is simple then $\mathsf{h}_{\mathbb{C}}$ is either simple or is the direct sum of two isomorphic simple algebras.

As an example, consider the algebra $sl_2(\mathbb{C})$ of complex traceless 2×2-matrices (which is a Lie algebra, with the Lie bracket given by the matrix commutator). According to (1.2.6) this algebra is isomorphic to A_1, and as generators one can choose

$$J_+ = \begin{pmatrix} 0 & 1 \\ 0 & 0 \end{pmatrix}, \quad J_- = \begin{pmatrix} 0 & 0 \\ 1 & 0 \end{pmatrix}, \quad J_3 = \begin{pmatrix} 1 & 0 \\ 0 & -1 \end{pmatrix}. \tag{1.3.16}$$

But of course one could also use any nonsingular linear combination of these generators as a basis, such as

$$\begin{aligned} J_x &= -i\,(J_+ + J_-) = -i \begin{pmatrix} 0 & 1 \\ 1 & 0 \end{pmatrix}, \\ J_y &= (J_- - J_+) = -i \begin{pmatrix} 0 & -i \\ i & 0 \end{pmatrix}, \\ J_z &= -iJ_3 = -i \begin{pmatrix} 1 & 0 \\ 0 & 1 \end{pmatrix}, \end{aligned} \tag{1.3.17}$$

in which the commutators take the form

$$[J_x, J_y] = 2\,J_z, \quad [J_y, J_z] = 2\,J_x, \quad [J_z, J_x] = 2\,J_y. \tag{1.3.18}$$

The Lie algebras generated by these bases over the real numbers are the two real forms of $sl_2(\mathbb{C})$, namely $sl_2(\mathbb{R})$ and su_2 in the case of the generators (1.3.16) and (1.3.17), respectively. As real Lie algebras, these are not isomorphic because the relation between the two sets of generators involves complex numbers. Also note that, although su_2 is a real Lie algebra, the entries of its elements are generically complex.

1.4 The Cartan–Weyl basis

Many aspects of Lie algebras are best considered after choosing a special type of basis. E.g. one would like to write down the structure constants in a canonical way. For semisimple Lie algebras, the most convenient choice is the so-called *Cartan–Weyl basis.* In contrast, for non-semisimple Lie algebras, e.g. solvable Lie algebras, a canonical form of the structure constants is not known. Thus from now on we consider only semisimple Lie algebras. A Cartan–Weyl basis is defined as follows. First choose a maximal set of linearly independent generators H^i of $\mathfrak{g}$ possessing zero brackets among themselves,

$$[H^i, H^j] = 0 \text{ for } i, j = 1, \dots, r. \tag{1.4.1}$$

Then

$$\mathfrak{g}_0 := \text{span}_F\{H^i \mid i = 1, \dots, r\} \tag{1.4.2}$$

is a maximal abelian subalgebra of $\mathfrak{g}$ and hence a Cartan subalgebra. It can be shown that for $F = \mathbb{C}$ all Cartan subalgebras are conjugate under the adjoint group $Int(\mathfrak{g})$; their common dimension turns out to be equal to the rank r of $\mathfrak{g}$ as defined in section 1.2,

$$r \equiv \text{rank}\,(\mathfrak{g}) = \dim\,(\mathfrak{g}_0). \tag{1.4.3}$$

In contrast, for $F = \mathbb{R}$ it is not in general true that all Cartan subalgebras are conjugate; similarly, when proving some other results cited below one makes use of tools from linear algebra which assume that the base field F of $\mathfrak{g}$ is algebraically closed, which is not the case for $\mathbb{R}$. Thus in the following we assume that $F = \mathbb{C}$.

Because the Cartan subalgebras $\mathfrak{g}_0$ are abelian and $\mathbb{C}$ is algebraically closed, it follows that upon restriction to $\mathfrak{g}_0$, the adjoint representation of $\mathfrak{g}$ splits up as a direct sum of one-dimensional representations, i.e.

$$\mathfrak{g} \rightsquigarrow \bigoplus_\alpha \mathfrak{g}^\alpha, \quad \mathfrak{g}^\alpha = \{x \in \mathfrak{g} \mid \text{ad}_H(x) = \alpha(H) \cdot x \text{ for all } H \in \mathfrak{g}_0\}. \tag{1.4.4}$$

In other words, for a given Cartan subalgebra $\mathfrak{g}_0$, the remaining generators E^α of $\mathfrak{g}$ can be chosen such that they are eigenvectors of $\mathfrak{g}_0$ in the sense that

$$[H^i, E^\alpha] = \alpha^i E^\alpha \quad \text{for } i = 1, \dots, r. \tag{1.4.5}$$

The r-dimensional vector $(\alpha^i)_{i=1,\dots,r}$ of eigenvalues is called a *root* of $\mathfrak{g}$, and the splitting (1.4.4) is called the *root space decomposition* of $\mathfrak{g}$ relative to $\mathfrak{g}_0$. The origin of this terminology is the fact that for a general element H of $\mathfrak{g}_0$, the complex numbers $\alpha(H)$ are the nonzero eigenvalues of ad_H, i.e. the nonzero roots of the characteristic equation

$$\det\,(t - \text{ad}_H) = 0. \tag{1.4.6}$$

The set of all roots of a semisimple Lie algebra will be denoted by Φ. We then have:

Definition : A basis of the semisimple Lie algebra $\mathfrak{g}$ of the form

$$\mathcal{B} = \{H^i \mid i = 1, \ldots, r\} \cup \{E^\alpha \mid \alpha \in \Phi\} \tag{1.4.7}$$

with H^i and E^α obeying (1.4.1) and (1.4.5) is called a **Cartan–Weyl basis** of $\mathfrak{g}$.

The generators E^α and $E^{-\alpha}$ are often called the *step operators* or *ladder operators* associated to the root α. The reason for these terms is the particular form of the action of $R(E^{\pm\alpha})$ on the associated $\mathfrak{g}$-module V (which is given by the formulae (1.6.3) and (1.6.12) below).

An important relation is obtained by considering the restriction κ^{ij} of the Killing form to the Cartan subalgebra. For $\mathrm{ad}_{H^i} : x \mapsto [H^i, x]$ one has

$$\begin{aligned} \mathrm{ad}_{H^i} \circ \mathrm{ad}_{H^j}(H^\ell) &= 0, \\ \mathrm{ad}_{H^i} \circ \mathrm{ad}_{H^j}(E^\alpha) &= \alpha^i \alpha^j E^\alpha, \end{aligned} \tag{1.4.8}$$

and hence

$$\kappa^{ij} \equiv \frac{1}{I_{\mathrm{ad}}} \mathrm{tr}\,(\mathrm{ad}_{H^i} \circ \mathrm{ad}_{H^j}) = \frac{1}{I_{\mathrm{ad}}} \sum_{\alpha\in\Phi} \alpha^i \alpha^j . \tag{1.4.9}$$

Thus for $H \in \mathfrak{g}_0$, i.e. $H = \xi_i H^i$ for some numbers $\xi_i \in \mathbb{C}$, one has $\xi_i \kappa^{ij} \xi_j = \sum_{\alpha\in\Phi} (\xi_i \alpha^i)^2 / I_{\mathrm{ad}}$ which is positive definite if the ξ_i are real (and $H \neq 0$). Hence κ^{ij} is a positive definite matrix in the sense that when the base field is restricted to $\mathbb{R}$, any nonzero element of the Cartan subalgebra has positive length. This means that the Cartan subalgebra is the complexification of a euclidean space. Moreover, it can be seen that the *root space* $\mathrm{span}_F \Phi$ is dual to the Cartan subalgebra (see section 1.6, p. 50). Thus the root space of a semisimple Lie algebra is (the complexification of) a euclidean space, too. The rest of the matrix κ^{ab} in a Cartan–Weyl basis is computed analogously; the result is, for a suitable normalization of the step operators,

$$\kappa^{ab} = \begin{pmatrix} \kappa^{ij} & 0 \\ 0 & \delta^{\alpha,-\beta} \end{pmatrix}. \tag{1.4.10}$$

Accordingly, $\mathfrak{g}_0$ is orthogonal to all the $\mathfrak{g}^\alpha$, and hence from the nondegeneracy of κ^{ab} it follows again that κ^{ij} is nondegenerate.

Choosing any fixed basis for the root space, we call α a *positive root* iff the first component of α in this basis is positive; otherwise we call α a *negative root*; if α is positive (negative), we write $\alpha > 0$ ($\alpha < 0$). For roots α, β we use the notation $\alpha > \beta$ iff $\alpha - \beta$ is a positive root; this defines a partial order on the root system. The sets of positive respectively negative roots are denoted by

$$\Phi_+ := \{\alpha \in \Phi \mid \alpha > 0\}, \qquad \Phi_- := \Phi \setminus \Phi_+. \tag{1.4.11}$$

The step operators E^α associated to the root α are also called *raising* operators, and the $E^{-\alpha}$ *lowering* operators if $\alpha > 0$.

Two fundamental properties of the root system Φ are the following: the roots are not degenerate (i.e. each root vector appears exactly once), and if α is a root, then also $-\alpha$ is a root, but no other multiple of α is a root. As a consequence, one has $\Phi_- = -\Phi_+$, i.e. $\alpha \in \Phi_+ \Leftrightarrow (-\alpha) \in \Phi_-$. This shows that

$$\{E^\alpha \mid \alpha \in \Phi\} = \{E^\alpha \mid \alpha > 0\} \cup \{E^{-\alpha} \mid \alpha > 0\}, \tag{1.4.12}$$

and that the number of elements of Φ_+ is

$$|\Phi_+| = \tfrac{1}{2}\,(d - r) \tag{1.4.13}$$

(in particular $d - r$ is always even). Given a Cartan subalgebra $\mathfrak{g}_0$ of $\mathfrak{g}$, the subalgebras of $\mathfrak{g}$ generated by the step operators for positive and negative roots are denoted by $\mathfrak{g}_+$ and $\mathfrak{g}_-$, respectively,

$$\mathfrak{g}_\pm := \mathrm{span}_F\{E^{\pm\alpha} \mid \alpha \in \Phi_+\}. \tag{1.4.14}$$

These subalgebras are nilpotent. According to the decompositions (1.4.7) and (1.4.12), $\mathfrak{g}$ can be written as the (non-direct) sum

$$\mathfrak{g} = \mathfrak{g}_+ \oplus \mathfrak{g}_0 \oplus \mathfrak{g}_-. \tag{1.4.15}$$

This is called the *Cartan decomposition* of $\mathfrak{g}$. As $\mathfrak{g}_0$ is abelian, the semi-direct sums

$$\mathfrak{b}_\pm := \mathfrak{g}_0 \oplus \mathfrak{g}_\pm \tag{1.4.16}$$

have $\mathfrak{g}_\pm$ as their derived algebras, and since the algebras $\mathfrak{g}_\pm$ are nilpotent, it follows that $\mathfrak{b}_\pm$ are solvable. In fact, they are Borel subalgebras of $\mathfrak{g}$. Namely, let $\mathfrak{h} \subseteq \mathfrak{g}$ be any subalgebra properly including, say, $\mathfrak{b}_+$; then it contains a step operator $E^{-\alpha} \in \mathfrak{g}_-$ for some positive root α; as a consequence, it contains the Lie algebra $\mathrm{span}\{E^\alpha, E^{-\alpha}, [E^\alpha, E^{-\alpha}]\}$; but it is easily checked that this three-dimensional algebra is isomorphic to the simple Lie algebra A_1. Thus $\mathfrak{h}$ is certainly not solvable, and therefore $\mathfrak{b}_\pm$ is maximal among the solvable subalgebras of $\mathfrak{g}$, and hence a Borel subalgebra.

Given the set of positive roots with respect to some chosen basis, one defines: A *simple root* is a positive root which cannot be obtained as a linear combination of other positive roots with positive coefficients. It turns out that, independently of the chosen basis, there are exactly $r = \mathrm{rank}(\mathfrak{g})$ simple roots; they will be denoted by $\alpha^{(i)}$. Hence the set of simple roots is

$$\Phi_s := \{\alpha^{(i)} \mid i = 1, \dots, r\}. \tag{1.4.17}$$

It can be shown that the simple roots provide a basis for the root space, i.e. they are linearly independent and $\mathrm{span}_F\, \Phi_s = \mathrm{span}_F\, \Phi$. Generically this basis is not orthonormal. The non-orthonormality is encoded in the *Cartan matrix*; denoting by "(,)" the scalar product in root space induced by its euclidean metric, this is the following matrix:

Definition : The **Cartan matrix** $A(\mathfrak{g})$ of the semisimple Lie algebra $\mathfrak{g}$ is the $r \times r$-matrix with elements

$$A^{ij}(\mathfrak{g}) := 2\,\frac{(\alpha^{(i)}, \alpha^{(j)})}{(\alpha^{(j)}, \alpha^{(j)})}. \tag{1.4.18}$$

Because the simple roots form a basis of the root space, the Cartan matrix (of a semisimple Lie algebra) is not degenerate. It will turn out (see section 1.5, p. 40) that all elements of the Cartan matrix are integers; therefore the elements A^{ij} are also called the *Cartan integers.* It must also be noted that often the Cartan matrix is defined as the transpose of the matrix (1.4.18).

Combinations of roots and their length such as in A^{ij} appear quite often; thus it makes sense to introduce a new symbol,

$$\alpha^\vee := \frac{2\alpha}{(\alpha, \alpha)}. \tag{1.4.19}$$

For $\alpha \in \Phi$, $\alpha^\vee$ is called the *dual root* or *coroot* of α. With the help of the notation (1.4.19), the Cartan integers can be written as

$$A^{ij} = (\alpha^{(i)}, \alpha^{(j)\vee}). \tag{1.4.20}$$

The space dual to the root space (i.e. the space of 1-forms) is called the *weight space.* It turns out to be useful to choose as a basis of the root space the simple coroots,

$$\mathcal{B} := \{\alpha^{(i)\vee} \mid i = 1, \dots, r\}. \tag{1.4.21}$$

Then, introducing the *fundamental weights* $\Lambda_{(i)}$ of a semisimple Lie algebra $\mathfrak{g}$ as the r 1-forms which obey

$$\Lambda_{(i)}(\alpha^{(j)\vee}) = \delta_i^{\,j}, \tag{1.4.22}$$

the basis of the weight space which is dual to $\mathcal{B}$ is

$$\mathcal{B}^\star := \{\Lambda_{(i)} \mid i = 1, \dots, r\}. \tag{1.4.23}$$

This basis $\mathcal{B}^\star$ is called the *Dynkin basis*; the components of a weight in the Dynkin basis are called *Dynkin labels.* The Dynkin basis is very convenient because, as we will see in section 1.6, the Dynkin labels of the most interesting weights are integers. It is also sometimes convenient to use the notation $\Lambda_{(0)} \equiv 0$ for the zero weight.

The integer span of the roots (coroots) is called the *root lattice* (*coroot lattice*) $L(\mathfrak{g})$ ($L^\vee(\mathfrak{g})$), the integer span of the fundamental weights the *weight lattice* $L_w(\mathfrak{g})$. Obviously the weight lattice is dual to the coroot lattice,

$$L_w(\mathfrak{g}) = (L^\vee(\mathfrak{g}))^\star \equiv \{\lambda \mid \lambda(\alpha^\vee) \in \mathbb{Z} \text{ for all } \alpha \in \Phi\}. \tag{1.4.24}$$

As will be explained in section 1.6, the elements of the weight lattice can be used to label the vectors in the finite-dimensional modules of $\mathfrak{g}$. In particular, the roots are the nonzero weights of the adjoint module. As a consequence, the weight lattice contains the root lattice, $L(\mathfrak{g}) \subset L_w(\mathfrak{g})$. The quotient $L_w(\mathfrak{g})/L(\mathfrak{g})$ is in fact a finite group; its elements are called the *conjugacy classes* of $\mathfrak{g}$-weights.

Any vector space endowed with a (nondegenerate) metric can be identified with its dual via the scalar product induced by the metric. Accordingly we will from now on identify the root and weight spaces of $\mathfrak{g}$ and treat roots and weights on an equal footing. For β an element of the root space, the associated element $\beta^\star$ in the weight space is the 1-form obeying $\beta^\star(\gamma) = (\beta, \gamma)$ for all γ in the root space. In order to describe explicitly the metric tensor in the chosen basis, we observe that the components of a vector (lower indices) respectively a 1-form (upper indices) can be obtained by forming scalar products with the basis elements. Thus we can write the decomposition of a weight into its components with respect to $\mathcal{B}$ and $\mathcal{B}^\star$ as

$$\lambda = \lambda_i\, \alpha^{(i)\vee} = \lambda^j \Lambda_{(j)} \tag{1.4.25}$$

with

$$\begin{aligned} \lambda_i &= (\lambda, \Lambda_{(i)}), \\ \lambda^i &= (\lambda, \alpha^{(i)\vee}) \end{aligned} \tag{1.4.26}$$

(recall that we use the summation convention, i.e. matching upper and lower indices are summed from 1 to r). In particular, as a consequence of (1.4.22) we have

$$\begin{aligned} (\alpha^{(j)\vee})_i &= \delta_i^j = (\Lambda_{(i)})^j, \\ (\alpha^{(j)})_i &= \tfrac{1}{2}(\alpha^{(j)}, \alpha^{(j)})\, \delta_i^j. \end{aligned} \tag{1.4.27}$$

Combining the above equations, it follows that the metrics G which raise and lower indices,

$$\lambda_i = G_{ij}\lambda^j, \quad \lambda^i = G^{ij}\lambda_j, \quad G_{ij}G^{jk} = \delta_i^k \tag{1.4.28}$$

so that the scalar product may be written in various forms as

$$(\lambda,\mu) = \lambda_i\mu^i = G_{ij}\lambda^i\mu^j = G^{ij}\lambda_i\mu_j, \tag{1.4.29}$$

are given by

$$\begin{aligned} G_{ij} &= (\Lambda_{(i)},\Lambda_{(j)}),\\ G^{ij} &= (\alpha^{(i)\vee},\alpha^{(j)\vee}) = \frac{2}{(\alpha^{(i)},\alpha^{(i)})}\,A^{ij}. \end{aligned} \tag{1.4.30}$$

As a consequence, the Dynkin labels of the simple roots are given by the rows of the Cartan matrix,

$$(\alpha^{(i)})^j = A^{ij}. \tag{1.4.31}$$

(As will be seen in the following section, the Cartan integers can be negative; thus the Dynkin basis does not share the property of the basis of simple (co-)roots that the components of a root all have like signs.) The metric (G_{ij}) with lower indices will henceforth be called the *quadratic form matrix* of $\mathfrak{g}$.

We now investigate further the form of the bracket relations in a Cartan–Weyl basis. We already know the brackets (1.4.1) and (1.4.5). Imposing the Jacobi identity, they imply

$$[H^i,[E^\alpha,E^\beta]] = (\alpha+\beta)^i\,[E^\alpha,E^\beta], \tag{1.4.32}$$

so that conformity with (1.4.5) requires

$$[E^\alpha,E^\beta] = e_{\alpha,\beta}E^{\alpha+\beta} \ \text{ if } \ \alpha+\beta\in\Phi \tag{1.4.33}$$

and

$$[E^\alpha,E^{-\alpha}] = \tilde\alpha_i\,H^i \tag{1.4.34}$$

with some numbers $e_{\alpha,\beta}$ and $\tilde\alpha_i$, as well as

$$[E^\alpha,E^\beta] = 0 \ \text{ if } \ \alpha+\beta\neq 0 \ \text{ and } \ \alpha+\beta\notin\Phi. \tag{1.4.35}$$

From the bracket (1.4.34) it follows in particular that $[E^\alpha,[E^\alpha,E^{-\alpha}]] = \tilde\alpha_i\alpha^i\,E^\alpha$; thus the scalar product $(\alpha,\tilde\alpha)$ must depend only on α, but not on any choice of basis, and hence must be proportional to the length squared of α; in other words, $\tilde\alpha^i$ equals α^i up to an overall constant. By normalizing the generators E^α appropriately, the constant of proportionality can be set to any desired value for each root α. A convenient choice is used in the following definition:

Definition : A **Chevalley basis** of a semisimple Lie algebra is a Cartan–Weyl basis with the normalizations chosen such that

$$[E^\alpha, E^{-\alpha}] = H^\alpha \tag{1.4.36}$$

with

$$H^\alpha := (\alpha^\vee, H) \equiv \alpha_i^\vee H^i .$$

With this normalization, the Lie bracket (1.4.5) gives in particular

$$[H^\alpha, E^{\pm\alpha}] = \pm 2\, E^{\pm\alpha} , \tag{1.4.37}$$

while for simple roots it reads

$$[H^i, E^{\alpha^{(j)}}] = (\alpha^{(j)})^i E^{\alpha^{(j)}} = A^{ji} E^{\alpha^{(j)}} . \tag{1.4.38}$$

It is easily verified that for each root α the step operators E^α and $E^{-\alpha}$ together with H^α generate an *sl_2-subalgebra* $\mathsf{h}_\alpha \cong sl_2(F) \cong A_1$ of g. Namely, one has

$$[H, E_\pm] = \pm 2\, E_\pm , \quad [E_+, E_-] = 2\, H \tag{1.4.39}$$

for

$$E_+ \,\hat{=}\, E^\alpha , \quad E_- \,\hat{=}\, E^{-\alpha} , \quad H \,\hat{=}\, H^\alpha . \tag{1.4.40}$$

This implies that many of the properties of a simple Lie algebra g can be analysed by making use of the properties of sl_2. For example, according to the sl_2 representation theory (which will be discussed in detail at the beginning of section 1.6) the weights of any sl_2-module belong to a "string" of the form $-\Lambda, -\Lambda+2, \ldots, \Lambda-2, \Lambda$ which in particular has no "holes". As a consequence, the weights of the h_α-modules have the form of *root strings*, i.e. for any h_α-module R there exists a root β such that $R = \mathrm{span}\{v_\lambda \mid \lambda \in S_{\alpha;\beta}\}$, where

$$S_{\alpha;\beta} = \{\beta + m\alpha \mid m = -n_-, -n_-+1, \ldots, n_+-1, n_+;\ n_\mp \in \mathbb{Z}_{\geq 0}\}, \tag{1.4.41}$$

which neither have any holes, and for any pair of roots this string is unique. In other words, if $\beta + m\alpha$ and $\beta + m'\alpha$ are roots for some integers m, m' satisfying $m \leq m'$, then $-n_- \leq m, m' \leq n_+$, and for any integer ℓ with $m \leq \ell \leq m'$, $\beta + \ell\alpha$ is a root as well. $S_{\alpha;\beta}$ is called the *α-string through β*.

From the definition of the Chevalley generators H^α, it follows immediately that if α, β and $\alpha + \beta$ are roots, then

$$H^{\alpha+\beta} = [(\alpha+\beta, \alpha+\beta)]^{-1} \left((\alpha,\alpha)\, H^\alpha + (\beta,\beta)\, H^\beta \right) . \tag{1.4.42}$$

This result can be combined with the Jacobi identity for the generators E^α, E^β and E^γ with $\gamma = -(\alpha+\beta)$ (and with the obvious relation $H^\gamma \equiv H^{-\alpha-\beta} = -H^{\alpha+\beta}$) to obtain the following relations among the structure constants $e_{\alpha,\beta}$ of the Chevalley basis:

$$\frac{e_{\beta,\gamma}}{e_{\gamma,\alpha}} = \frac{(\alpha,\alpha)}{(\beta,\beta)}, \quad \frac{e_{\gamma,\alpha}}{e_{\alpha,\beta}} = \frac{(\beta,\beta)}{(\alpha+\beta,\alpha+\beta)}. \tag{1.4.43}$$

From these, one can deduce the symmetry properties

$$e_{\alpha,\beta} = e_{\beta,-\alpha-\beta} = e_{-\alpha-\beta,\alpha}. \tag{1.4.44}$$

Similarly, if each of $\alpha+\beta$, $\alpha+\gamma$, $\beta+\gamma$ and $\alpha+\beta+\gamma$ is a root, one obtains the restriction

$$e_{\alpha,\beta}e_{\alpha+\beta,\gamma} + e_{\beta,\gamma}e_{\beta+\gamma,\alpha} + e_{\gamma,\alpha}e_{\alpha+\gamma,\beta} = 0. \tag{1.4.45}$$

There is still some freedom left for the signs of the structure constants $e_{\alpha,\beta}$; this can be fixed by requiring e.g. that

$$e_{-\alpha,-\beta} = -e_{\alpha,\beta}. \tag{1.4.46}$$

Up to such sign choices (and possibly up to a relabelling of the roots) the Chevalley basis of a semisimple Lie algebra is unique.

By evaluating the Jacobi identity for the three generators E^α, $E^{-\alpha}$ and $E^{\beta+j\alpha}$ and employing the symmetry properties (1.4.44), one obtains the recursion relation

$$(e_{\alpha,\beta+(j-1)\alpha})^2 - (e_{\alpha,\beta+j\alpha})^2 = (\beta+j\alpha,\alpha^\vee). \tag{1.4.47}$$

This can be solved starting from the initial condition $e_{\alpha,\beta+n_+\alpha} = 0$. From the requirement that the string must terminate (since the number of roots is finite), one can deduce that $-n_+ \le (\beta,\alpha^\vee) \le n_-$ (this implies in particular that $\beta' := \beta - (\beta,\alpha^\vee)\alpha$ is a root). One then arrives at the explicit formula

$$e_{\alpha,\beta} = \left[n_+(n_-+1)\,\frac{(\alpha+\beta,\alpha+\beta)}{(\beta,\beta)} \right]^{1/2}. \tag{1.4.48}$$

(Note that this shows again that the root space is a euclidean space: for $e_{\alpha,\beta}$ to be real, all inner products (α,α), $\alpha\in\Phi$, must have like signs; also, if (α,α) were zero, then so would be $e_{\alpha,\beta-\alpha}$ for all $\beta-\alpha\in\Phi$, and as a consequence E^α would generate an abelian ideal of $\mathfrak{g}$, in contradiction to the semisimplicity of $\mathfrak{g}$.) In addition, one can show that in fact

$$(\beta,\alpha^\vee) = n_- - n_+. \tag{1.4.49}$$

We have already mentioned that the elements of the Cartan matrix of a semisimple Lie algebra are integers. But one can even show that in fact *all* structure constants in a Chevalley basis are integers. In particular, it turns out that for any root β the integers $n_\pm$ defined by the string $S_{\alpha;\beta}$ obey

$$\frac{n_- + 1}{n_+} = \frac{(\alpha+\beta,\alpha+\beta)}{(\beta,\beta)} \tag{1.4.50}$$

if $n_+ \neq 0$; thus the formula (1.4.48) implies that

$$[E^\alpha, E^\beta] = \pm (n_- + 1)\, E^{\alpha+\beta} \tag{1.4.51}$$

for any pair α, β of roots. If $\mathfrak{g}$ is simply laced, then (as follows immediately from $(\alpha,\alpha) = (\beta,\beta) = (\alpha+\beta,\alpha+\beta)$) α, β and $\alpha+\beta$ are roots iff $(\alpha^\vee,\beta) = -1$. Thus one has $n_+ = n_- + 1$; in addition, from the recursion relation one can deduce that the only possibility to obtain an integer value for $e_{\alpha,\beta+(n_+-1)\alpha}$ is that $n_+ = 1$, in which case the value is ± 1. Thus for simply laced $\mathfrak{g}$, one has, for any pair of roots, $n_+ = 1$, $n_- = 0$, and $|e_{\alpha,\beta}| = 1$; in particular, any root string consists at most of two elements. By inspection, one verifies that the latter statements are true for B_r and G_2 as well, and hence for all simple Lie algebras with at most one short simple root. Including also the remaining simple Lie algebras C_r and F_4, one can show that in full generality one has $(\alpha^\vee,\beta) \geq -3$ for any pair of roots, and from this that $n_+ + n_- \leq 3$.

The notation H^α also suggests that for simple roots, i.e. for $\alpha = \alpha^{(i)}$, the Chevalley generators H^α should be just the Cartan subalgebra generators of a suitable Cartan–Weyl basis. This is indeed the case, as one easily deduces that

$$H^{\alpha^{(i)}} = (\alpha^{(i)\vee})_j H^j = \delta^i_j H^j = H^i. \tag{1.4.52}$$

It is then natural to introduce also a special notation for the step operators associated to simple roots,

$$E^i_\pm := E^{\pm\alpha^{(i)}}. \tag{1.4.53}$$

These generators obey

$$[E^i_+, E^j_-] = \delta^{ij} H^i. \tag{1.4.54}$$

Namely, $[E^i_+, E^j_-] = 0$ if $i \neq j$ because (as a consequence of the defining property of simple roots) the linear combination $\alpha^{(i)} - \alpha^{(j)}$ of simple roots is never a root, while $[E^i_+, E^i_-] = H^{\alpha^{(i)}}$ is the normalized version of the bracket relation (1.4.34).

Let us now illustrate the above analysis with a simple example. The easiest example one can think of is the three-dimensional simple Lie algebra $sl_2 \cong A_1$. But for this algebra the results are in fact almost trivial: there is a single positive root α with Dynkin label equal to two, the Cartan matrix is the number 2, and the quadratic form matrix equals $(\alpha,\alpha)/4$;

the Cartan–Weyl basis is $\{H, E_\pm\}$ as used in (1.1.45), and the Chevalley basis is just identical to the Cartan–Weyl basis. To arrive at a less trivial example, let us consider instead of sl_2 the Lie algebra $sl_3(\mathbb{C})$ of traceless complex 3×3-matrices. The dimension of this algebra is obviously $3 \cdot 3 - 1 = 8$. The following eight traceless matrices are linearly independent and hence form a basis of sl_3:

$$\begin{aligned}
&H^1 = \frac{1}{2}\begin{pmatrix} 1 & 0 & 0 \\ 0 & -1 & 0 \\ 0 & 0 & 0 \end{pmatrix}, \quad H^2 = \frac{1}{2\sqrt{3}}\begin{pmatrix} 1 & 0 & 0 \\ 0 & 1 & 0 \\ 0 & 0 & -2 \end{pmatrix}, \\
&E^\alpha = \begin{pmatrix} 0 & 1 & 0 \\ 0 & 0 & 0 \\ 0 & 0 & 0 \end{pmatrix}, \qquad E^{-\alpha} = \begin{pmatrix} 0 & 0 & 0 \\ 1 & 0 & 0 \\ 0 & 0 & 0 \end{pmatrix}, \\
&E^\beta = \begin{pmatrix} 0 & 0 & 0 \\ 0 & 0 & 1 \\ 0 & 0 & 0 \end{pmatrix}, \qquad E^{-\beta} = \begin{pmatrix} 0 & 0 & 0 \\ 0 & 0 & 0 \\ 0 & 1 & 0 \end{pmatrix}, \\
&E^\theta = \begin{pmatrix} 0 & 0 & 1 \\ 0 & 0 & 0 \\ 0 & 0 & 0 \end{pmatrix}, \qquad E^{-\theta} = \begin{pmatrix} 0 & 0 & 0 \\ 0 & 0 & 0 \\ 1 & 0 & 0 \end{pmatrix}.
\end{aligned} \tag{1.4.55}$$

By computing the commutators of these matrices, one finds that in this basis the following brackets are nonzero:

$$\begin{aligned}
&[\vec{H}, E^{\pm\alpha}] = \pm\,(1,0)\,E^{\pm\alpha}, \\
&[\vec{H}, E^{\pm\beta}] = \pm\,\tfrac{1}{2}\,(-1,\sqrt{3})\,E^{\pm\beta}, \\
&[\vec{H}, E^{\pm\theta}] = \pm\,\tfrac{1}{2}\,(1,\sqrt{3})\,E^{\pm\theta}, \\
&[E^\alpha, E^{-\alpha}] = 2\,H^1, \\
&[E^\beta, E^{-\beta}] = -H^1 + \sqrt{3}\,H^2, \\
&[E^\theta, E^{-\theta}] = H^1 + \sqrt{3}\,H^2, \\
&[E^{\pm\alpha}, E^{\pm\beta}] = \pm\,E^{\pm\theta}, \\
&[E^{\pm\alpha}, E^{\mp\theta}] = \mp\,E^{\mp\beta}, \\
&[E^{\pm\beta}, E^{\mp\theta}] = \pm\,E^{\mp\alpha}.
\end{aligned} \tag{1.4.56}$$

Here we have used $\vec{H}$ as a shorthand notation for the pair (H^1, H^2). The abstract Lie algebra possessing this system of brackets is denoted by A_2.

From this result one also reads off that the chosen basis is actually a Cartan–Weyl basis, with the Cartan decomposition $\mathfrak{g} = \mathfrak{g}_+ \oplus \mathfrak{g}_0 \oplus \mathfrak{g}_-$, namely $\mathfrak{g}_+ = \text{span}\{E^\alpha, E^\beta, E^\theta\}$, $\mathfrak{g}_- = \text{span}\{E^{-\alpha}, E^{-\beta}, E^{-\theta}\}$, and $\mathfrak{g}_0 =$

span$\{H^1, H^2\}$. Thus in particular the rank of sl_3 is two, and in an orthonormal basis of the weight space $\mathbb{R}^2$, the roots are given up to overall normalization by $\pm\alpha$, $\pm\beta$, and $\pm\theta$ with

$$\begin{aligned} \alpha = (1,0), \qquad \beta = \tfrac{1}{2}\,(-1,\sqrt{3}), \\ \theta = \tfrac{1}{2}\,(1,\sqrt{3}) = \alpha + \beta. \end{aligned} \tag{1.4.57}$$

The root system of sl_3 may thus be summarized as in figure (1.4.58).

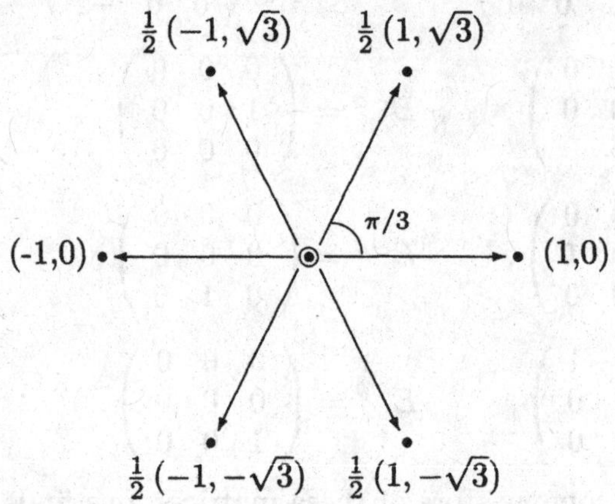

(1.4.58)

Obviously, one may choose α, β and θ as positive roots; the simple roots are then $\alpha^{(1)} = \alpha$ and $\alpha^{(2)} = \beta$. Also, the scalar products of the roots are given by $(\alpha,\alpha) = (\beta,\beta) = (\theta,\theta)$ and $(-\alpha,\beta) = (\alpha,\theta) = (\beta,\theta) = (\alpha,\alpha)/2$. Thus in particular the Cartan matrix and the quadratic form matrix of sl_3 read

$$\begin{aligned} A(sl_3) = \begin{pmatrix} 2 & -1 \\ -1 & 2 \end{pmatrix}, \\ G(sl_3) = \tfrac{1}{6}\,(\alpha,\alpha) \begin{pmatrix} 2 & 1 \\ 1 & 2 \end{pmatrix}. \end{aligned} \tag{1.4.59}$$

Also, the fundamental weights are given up to the normalization factor $(\alpha,\alpha)/2$ by

$$\begin{aligned} \Lambda_{(1)} = (1, \tfrac{1}{\sqrt{3}}), \\ \Lambda_{(2)} = (0, \tfrac{2}{\sqrt{3}}), \end{aligned} \tag{1.4.60}$$

in agreement with the general formula $G_{ij} = (\Lambda_{(i)}, \Lambda_{(j)})$ for the quadratic form matrix. Finally, a Chevalley basis for sl_3 is obtained by introducing $H^\alpha \equiv [E^\alpha, E^{-\alpha}] = H^1$ and analogously $H^\beta = \frac{1}{2}\,(-H^1 + \sqrt{3}H^2)$ and

$H^\theta = \frac{1}{2}(H^1 + \sqrt{3}H^2) = H^\alpha + H^\beta$. The bracket relations of the Chevalley basis can thus be summarized by

$$\begin{aligned}
[H^\alpha, E^{\pm\alpha}] &= \pm 2\, E^{\pm\alpha}, \\
[H^\beta, E^{\pm\beta}] &= \pm 2\, E^{\pm\beta}, \\
[H^\alpha, E^{\pm\beta}] &= \mp E^{\pm\beta}, \\
[H^\beta, E^{\pm\alpha}] &= \mp E^{\pm\alpha}, \\
[E^\alpha, E^{-\alpha}] &= H^\alpha, \\
[E^\beta, E^{-\beta}] &= H^\beta, \\
[E^\theta, E^{-\theta}] &= H^\theta, \\
[E^{\pm\alpha}, E^{\pm\beta}] &= \pm E^{\pm\theta}, \\
[E^{\pm\alpha}, E^{\mp\theta}] &= \mp E^{\mp\beta}, \\
[E^{\pm\beta}, E^{\mp\theta}] &= \pm E^{\mp\alpha}.
\end{aligned} \tag{1.4.61}$$

One of the rare circumstances where one makes use of the basis of simple roots is in the following definition: The *height* $\mathrm{ht}(\alpha)$ of a root (or more generally, of a root lattice vector) α of $\mathfrak{g}$ is the sum of its components in the basis of simple roots,

$$\mathrm{ht}(\alpha) := \sum_{i=1}^{r} b_i \quad \text{for} \quad \alpha = \sum_{i=1}^{r} b_i\, \alpha^{(i)}. \tag{1.4.62}$$

Via the height, one can introduce a natural gradation of $\mathfrak{g}$, the so-called *root space gradation*. This is defined as the $\mathbb{Z}$-gradation

$$\begin{aligned}
\mathfrak{g}_{(0)} &= \mathfrak{g}_0, \\
\mathfrak{g}_{(j)} &= \mathrm{span}\,\{E^\alpha \mid \mathrm{ht}(\alpha) = j\} \qquad \text{for } j \in \mathbb{Z} \setminus \{0\}.
\end{aligned} \tag{1.4.63}$$

Let us for the moment assume that $\mathfrak{g}$ is simple. Then it turns out that there is a unique root θ such that

$$\mathrm{ht}(\theta) > \mathrm{ht}(\alpha) \quad \text{for all} \quad \alpha \in \Phi \setminus \{\theta\}. \tag{1.4.64}$$

The root θ is called the *highest root* of $\mathfrak{g}$. It also possesses the following properties: Firstly,

$$(\theta, \theta) \geq (\alpha, \alpha) \qquad \text{for all} \quad \alpha \in \Phi. \tag{1.4.65}$$

Secondly, $\theta - \alpha$ is a positive root for any $\alpha \in \Phi \setminus \{\theta\}$, i.e. using for α the expansion of (1.4.62) and writing

$$\theta =: \sum_{i=1}^{r} a_i\, \alpha^{(i)}, \tag{1.4.66}$$

one has

$$a_i \geq b_i, \; i = 1, \dots, r \quad \text{for all} \quad \alpha \in \Phi \setminus \{\theta\}. \tag{1.4.67}$$

Owing to the existence of a highest root, the height concept allows for a quick check that the subalgebras $\mathfrak{g}_\pm$ generated by the step operators for the positive and negative roots, respectively, are indeed nilpotent: for, say, $x, y \in \mathfrak{g}_+$ the application of ad_x to y increases the height of the associated root at least by one. Since there exists a maximal height, one therefore ends up with zero after a finite number of applications of ad_x.

As a consequence of the inequality (1.4.67), the expansion coefficients of θ in the basis of simple roots are strictly positive (not just non-negative as for any positive root); the same is then true for the components $a_i^\vee \equiv (\theta^\vee)_i$, i.e.

$$\theta^\vee =: \sum_{i=1}^{r} a_i^\vee \alpha^{(i)\vee}. \tag{1.4.68}$$

The numbers a_i and $a_i^\vee$, $i = 1, \dots, r = \mathrm{rank}(\mathfrak{g})$ are called the *Coxeter labels* and *dual Coxeter labels* of $\mathfrak{g}$, respectively. If $a_i^\vee = 1$, the corresponding fundamental weight $\Lambda_{(i)}$ is called a *minimal* fundamental weight, and if $a_i = 1$, $\Lambda_{(i)}$ is called *cominimal*. The *Coxeter number* and *dual Coxeter number* of $\mathfrak{g}$ are the sums

$$g := 1 + \sum_{i=1}^{r} a_i, \quad g^\vee := 1 + \sum_{i=1}^{r} a_i^\vee. \tag{1.4.69}$$

The dual Coxeter number will appear later on in a variety of circumstances. For now we just mention one example. Defining the *Weyl vector* ρ of a semisimple Lie algebra $\mathfrak{g}$ as half the sum of its positive roots,

$$\rho := \frac{1}{2} \sum_{\alpha \in \Phi_+} \alpha, \tag{1.4.70}$$

we can express the length of this vector through the dimension and the dual Coxeter number of $\mathfrak{g}$ via the so-called *strange formula*

$$(\rho, \rho) = \tfrac{1}{24}\, d g^\vee \, (\theta, \theta). \tag{1.4.71}$$

As will be seen later (p. 76), the Weyl vector also obeys

$$(\rho, {\alpha^{(i)}}^\vee) = 1 \quad \text{for all} \;\; i, \tag{1.4.72}$$

i.e. the decomposition of ρ with respect to the fundamental weights reads

$$\rho = \sum_{i=1}^{r} \Lambda_{(i)}. \tag{1.4.73}$$

As a consequence, the strange formula can be rewritten as

$$\sum_{i,j=1}^{r} G_{ij} = \tfrac{1}{24}\, d g^\vee\, (\theta,\theta)\,, \tag{1.4.74}$$

i.e. as an identity for the quadratic form matrix. It also makes sense to define the *dual Weyl vector* $\rho^\vee$ as half the sum of positive coroots,

$$\rho^\vee := \tfrac{1}{2} \sum_{\alpha\in\Phi_+} \alpha^\vee . \tag{1.4.75}$$

Note that in general this differs from $2\rho/(\rho,\rho)$. For the algebras listed in table (1.5.16), the components of the dual Weyl vector are listed in table (1.4.76).

$\mathfrak{g}$	$\frac{1}{2}(\theta,\theta)(\rho^\vee)^i$	$(\rho^\vee)_i$
A_r	1	$\frac{1}{2}\, i(r-i+1)$
B_r	$\begin{cases} 1 & \text{for } i=1,\ldots,r-1 \\ 2 & \text{for } i=r \end{cases}$	$\begin{cases} \frac{1}{2}\, i(2r-i+1) & \text{for } i=1,\ldots,r-1 \\ \frac{1}{4}\, r(r+1) & \text{for } i=r \end{cases}$
C_r	$\begin{cases} 2 & \text{for } i=1,\ldots,r-1 \\ 1 & \text{for } i=r \end{cases}$	$\frac{1}{2}\, i(2r-i)$
D_r	1	$\begin{cases} \frac{1}{2}\, i(2r-i-1) & \text{for } i=1,\ldots,r-2 \\ \frac{1}{4}\, r(r-1) & \text{for } i=r-1,r \end{cases}$
E_6	1	$(8,15,21,15,8,11)$
E_7	1	$(17,33,48,\frac{75}{2},52,\frac{27}{2},\frac{49}{2})$
E_8	1	$(29,57,84,110,135,91,46,68)$
F_4	$(1,1,2,2)$	$(11,21,15,8)$
G_2	$(1,3)$	$(5,3)$

(1.4.76)

The results of table (1.4.76) for $((\rho)^\vee)^i$ imply that the inner product of $\rho^\vee$ with a simple root is $(\rho^\vee,\alpha^{(i)})=1$ for any $i=1,\ldots,r$; in other words, just like the one-form dual to ρ is $\rho^\star=\sum_i \alpha^{(i)\vee}$, the one-form dual to $\rho^\vee$ is $(\rho^\vee)^\star=\sum_{i=1}^{r}\alpha^{(i)}$. Also note that for a weight λ, the Dynkin labels λ^i are normalization independent so that the components λ_i are normalization dependent, whereas for coweights such as $\rho^\vee$ which are scaled by the root lengths, the Dynkin labels are normalization dependent but the $(\rho^\vee)_i$ are normalization independent.

1.5 Classification of simple Lie algebras

In this section we classify the semisimple Lie algebras over the complex numbers. At the end of the section we add a few remarks on real Lie algebras, but, until further notice, the base field F are the complex numbers. Because semisimple algebras are direct sums of simple ones, we can restrict ourselves to the case of simple Lie algebras. To begin, we observe that for any two complex semisimple Lie algebras $\mathfrak{g}$ and $\tilde{\mathfrak{g}}$ with respective Cartan subalgebras $\mathfrak{g}_0$ and $\tilde{\mathfrak{g}}_0$ and root systems Φ and $\tilde{\Phi}$, any isomorphism $\varphi : \mathfrak{g}_0 \to \tilde{\mathfrak{g}}_0$ which induces a bijection of Φ onto $\tilde{\Phi}$ can be extended to a Lie algebra isomorphism of $\mathfrak{g}$ onto $\tilde{\mathfrak{g}}$. This implies that in order to classify semisimple Lie algebras up to isomorphism, we only have to classify the possible Cartan subalgebras (which is trivial) and root systems. It turns out that in fact the whole information about the root system Φ is already contained in the set Φ_s of simple roots. Namely, it is possible to characterize a semisimple Lie algebra completely by giving the Lie brackets for the generators associated to simple roots in a Chevalley basis, together with the Jacobi identity and a few further relations:

Theorem : The semisimple Lie algebra $\mathfrak{g}(\Phi_s)$ associated to a set

$$\Phi_s = \{\alpha^{(i)} \mid i = 1, \ldots, r\}$$

of r simple roots $\alpha^{(i)}$ is uniquely determined as follows:

(1.5.1a) there are $3r$ generators $\{E^i_\pm, H^i \mid i = 1, \ldots, r\}$ with brackets

$$\begin{aligned} &[H^i, H^j] = 0, \\ &[H^i, E^j_\pm] = \pm A^{ji} E^j_\pm, \\ &[E^i_+, E^j_-] = \delta^{ij} H^i; \end{aligned} \qquad (1.5.1)$$

(1.5.1b) these generators obey the Jacobi identity;

(1.5.1c) $(\mathrm{ad}_{E^i_\pm})^{1-A^{ji}} E^j_\pm = 0$ for $i, j = 1, \ldots, r,\ i \neq j$.

Here $E^i \equiv E^{\alpha^{(i)}}$, A is the $r \times r$ Cartan matrix that according to (1.4.18) is associated to the simple roots Φ_s, and $(\mathrm{ad}_x)^n$ is used as a shorthand notation for $\underbrace{\mathrm{ad}_x \circ \mathrm{ad}_x \circ \ldots \circ \mathrm{ad}_x}_{n \text{ times}}$, so that e.g. $(\mathrm{ad}_x)^2(y) \equiv [x, [x, y]]$.

The relations (1.5.1c) which fix the length of the $\alpha^{(i)}$-string through $\alpha^{(j)}$ are called the *Serre relations*, and the full set of relations (1.5.1) is known as the *Chevalley–Serre relations.* The proof of the theorem (1.5.1) amounts to providing an algorithm for the reconstruction of the whole algebra using only the above relations; this way in particular all ladder operators $E^{\pm\alpha}$ are obtained as multiple commutators of the $E_i^{\pm}$. Such an algorithm (the *Serre construction*) is obtained by induction on the height of the roots: one starts with the simple roots, i.e. the roots of height 1; from the Serre relations one gets the $\alpha^{(i)}$-string through $\alpha^{(j)}$ for $i \neq j$ and hence in particular finds all roots of height 2, and thus all ladder operators which can be written in the form $[E_{\pm}^{i_1}, E_{\pm}^{i_2}]$ for some $i_1, i_2 \in \{1, \ldots, r\}$. Afterwards the Serre relations can be used once again to find the $\alpha^{(i)}$-strings through the roots of height 2, in particular all roots of height 3, and so on.

Let us consider an example. For $\mathfrak{g} = A_1$ there are no Serre relations at all. The simplest non-trivial example is then $\mathfrak{g} = A_2 \cong sl_3$. According to the result (1.4.59) for the Cartan matrix of A_2, the Serre relations for this algebra read

$$\begin{aligned} &[E_{\pm}^1, [E_{\pm}^1, E_{\pm}^2]] = 0, \\ &[E_{\pm}^2, [E_{\pm}^2, E_{\pm}^1]] = 0. \end{aligned} \tag{1.5.2}$$

From this one learns that $\pm(\alpha^{(1)} + \alpha^{(2)})$ are roots of height 2, but that the possible vectors of height three, namely $\pm(2\alpha^{(1)} + \alpha^{(2)})$ and $\pm(\alpha^{(1)} + 2\alpha^{(2)})$, are not roots. In fact, from the Serre relations it follows at once that $\mathrm{ad}_{E_{\pm}^i} \circ \mathrm{ad}_{E_{\pm}^j} (E_{\pm}^k) = 0$ for any triple $i, j, k \in \{1, 2\}$, and as a consequence there are no roots at all of height larger than two. Thus the roots of A_2 are given by the linear combinations $\pm\alpha^{(1)}$, $\pm\alpha^{(2)}$, and $\pm(\alpha^{(1)} + \alpha^{(2)})$ of simple roots, in agreement with the results of section 1.4 (p. 34).

It is important to note that the only parameters occuring in (1.5.1) are the Cartan integers. Thus what we have to do is to classify the Cartan matrices of simple Lie algebras. This then gives the classification of simple Lie algebras, provided that the Cartan matrix of a simple Lie algebra is independent (up to a renumbering of rows and columns) of the choice of the basis of simple roots. That the latter condition is indeed fulfilled can be shown most easily by using the concept of the Weyl group of a simple Lie algebra; see section 1.7, p. 76.

Therefore we now list several properties of Cartan matrices. First, by setting $i = j$ in the definition (1.4.18) of the matrix elements A^{ij}, one obtains

$$\text{(C1)} \quad A^{ii} = 2 \ \text{ for } i = 1, \ldots, r. \tag{1.5.3}$$

Next, by the symmetry of the scalar product in root space,

$$\text{(C2)}\quad A^{ij}=0 \iff A^{ji}=0 \quad \text{for } i,j=1,...,r. \tag{1.5.4}$$

Now according to (1.4.39) for each root α the generators E^{α}, $E^{-\alpha}$ and H^{α} generate an sl_2-subalgebra h_α of $\mathfrak{g}$, namely $[H,E_\pm]=\pm E_\pm$, $[E_+,E_-]=2\,H$ with $E_+\hat{=}E^{\alpha}$, $E_-\hat{=}E^{-\alpha}$, $H\hat{=}H^{\alpha}$. The representation theory of $sl_2(F)$ is rather simple (see the following section); in particular in any finite-dimensional representation the eigenvalues of H are integers. It will also be seen in the next section that the eigenvalues of H^{α} in the adjoint representation of $\mathfrak{g}$ (which is finite-dimensional and hence decomposes into finite-dimensional representations of h_α for any $\alpha\in\Phi$) are just given by $\{(\alpha^\vee,\beta)\mid\beta\in\Phi\}$ (together with r times the eigenvalue zero). It follows that the inner product $(\alpha^\vee,\beta)$ is an integer for any pair of roots $\alpha,\beta\in\Phi$, and hence in particular

$$A^{ij}\in\mathbb{Z} \quad \text{for } i,j=1,...,r. \tag{1.5.5}$$

Moreover, one can show that for any pair α,β of distinct roots one has

$$\begin{aligned}(\alpha,\beta)>0 \;&\Rightarrow\; \alpha-\beta\in\Phi,\\ (\alpha,\beta)<0 \;&\Rightarrow\; \alpha+\beta\in\Phi.\end{aligned} \tag{1.5.6}$$

Since the difference of any two simple roots is never a root, it follows that $(\alpha^{(i)},\alpha^{(j)})\le 0$ for $i\neq j$, and hence the integrality property (1.5.5) can in fact be strengthened to

$$\text{(C3)}\quad A^{ij}\in\mathbb{Z}_{\le 0} \quad \text{for } i,j=1,...,r,\ i\neq j. \tag{1.5.7}$$

Since the simple roots provide a basis for the root space, the matrix of their inner products, and therefore also the Cartan matrix, is non-degenerate,

$$\text{(C4)}\quad \det A\neq 0. \tag{1.5.8}$$

In terms of the root system, the splitting of a semisimple Lie algebra into its simple parts is expressed through the fact that Φ can be written as the sum of "irreducible" subsystems,

$$\Phi=\Phi_{(1)}\oplus\Phi_{(2)}\oplus\ldots\oplus\Phi_{(n)} \tag{1.5.9}$$

such that

$$(\alpha,\beta)=0 \quad \text{for all } \alpha\in\Phi_{(i)},\ \beta\in\Phi_{(j)},\ i\neq j \tag{1.5.10}$$

(and such that none of the $\Phi_{(i)}$ can itself be written as such a sum). The $\Phi_{(i)}$ are of course the root systems of the simple Lie algebras $\mathfrak{g}_{(i)}$ in the decomposition $\mathfrak{g} = \mathfrak{g}_{(1)} \oplus \ldots \oplus \mathfrak{g}_{(n)}$. The restriction from semisimple to simple Lie algebras thus means that the Cartan matrix must be *indecomposable* (or *irreducible*) in the following sense:

(C5) A renumbering of the simple roots which could bring A to the block diagonal form

$$\begin{pmatrix} A_{(1)} & 0 \\ 0 & A_{(2)} \end{pmatrix} \tag{1.5.11}$$

does not exist.

Our task now consists in classifying the matrices obeying (C1) to (C5), implementing in addition the definition (1.4.18) of the Cartan matrix which includes in particular that the root space is euclidean. This is not too difficult. First consider the triangle inequality $(\alpha^{(i)}, \alpha^{(j)})^2 \le (\alpha^{(i)}, \alpha^{(i)})(\alpha^{(j)}, \alpha^{(j)})$. Because of (C1) this implies A^{ij}, $A^{ji} \le 4$ with equality iff $i = j$, i.e.

$$A^{ij} A^{ji} \in \{0, 1, 2, 3\} \quad \text{for } i \neq j. \tag{1.5.12}$$

Taking into account (C2) and (C3), this leaves us only with the possibilities

$$\begin{aligned} &A^{ij} = A^{ji} = 0 && \text{or} \\ &A^{ij} = A^{ji} = -1 && \text{or} \\ &A^{ij} = -1, \quad A^{ji} = -2 && \text{or} \\ &A^{ij} = -1, \quad A^{ji} = -3. && \end{aligned} \tag{1.5.13}$$

The rest of the work consists in finding further restrictions, thereby implementing in particular also the requirement (C4). Note that the fact that the root space is euclidean shows that the property (C4) can be strengthened to

$$\det A > 0 \tag{1.5.14}$$

(see table (1.5.17) below for the explicit values of the determinants of the Cartan matrices for simple Lie algebras). It should be stressed that this does not follow from the properties (C1) to (C5) alone. Conversely, combined with the other properties (C1) to (C3), the inequality (1.5.14) in fact allows to derive the restrictions (1.5.12) on $A^{ij} A^{ji}$ without any reference to a root space with positive definite metric. Thus the classification of simple Lie algebras can be obtained by axiomatically requiring A to possess the following properties:

$$(\mathsf{C})\qquad \begin{aligned} &a)\quad A^{ii}=2,\\ &b)\quad A^{ij}=0\Leftrightarrow A^{ji}=0,\\ &c)\quad A^{ij}\in\mathbb{Z}_{\le 0}\ \text{ for } i\neq j\\ &d)\quad \det A>0,\\ &e)\quad \text{indecomposability.}\end{aligned}\qquad (1.5.15)$$

Note that with this formulation of the constraints on the Cartan matrix, one needs not refer to root spaces at all.

The enumeration of all possible solutions to the set (C) of equations is a somewhat lengthy procedure, but it is purely combinatorial. Here we state only the final result. To do so, it is most convenient to introduce the notion of a *Dynkin diagram*: to each Cartan matrix one associates a diagram consisting of vertices and lines connecting them. Each vertex of the diagram represents a simple root; the vertices for $\alpha^{(i)}$ and $\alpha^{(j)}$ ($i\neq j$) are connected by $\max\{|A^{ij}|,\,|A^{ji}|\}$ lines, i.e. vertices connected by a single, double or triple bond correspond to simple roots spanning an angle of $\frac{2}{3}\pi$, $\frac{3}{4}\pi$ and $\frac{4}{5}\pi$, respectively, while simple roots not connected by a line are mutually orthogonal. Furthermore, an arrowhead ">" is added to the lines from the ith to the jth node if $A^{ij}\neq 0$ and $|A^{ij}|>|A^{ji}|$ (which is equivalent to $(\alpha^{(i)},\alpha^{(i)})>(\alpha^{(j)},\alpha^{(j)})$). Alternatively, one may specify "long" roots by open dots ($\circ$) and "short" roots by filled dots ($\bullet$); this qualification makes sense because the analysis shows that at most two different lengths are allowed for the roots of any given simple Lie algebra. For Cartan matrices which differ only by renumbering the simple roots (and, as a consequence, describe the same Lie algebra), one obtains identical Dynkin diagrams.

The Dynkin diagrams are in fact multi-purpose graphs: the ith node may not only be thought of as representing the ith simple root $\alpha^{(i)}$, but likewise also the ith fundamental weight $\Lambda_{(i)}$, or the Cartan subalgebra generator H^i, or simply the row index of the Cartan matrix (A_{ij}) (yet another possibility is the ith generator $\sigma_{(i)}$ of the Weyl group; see section 1.7).

The result of the classification of simple Lie algebras is displayed in table (1.5.16) in terms of Dynkin diagrams.

name	numbering	Coxeter labels
A_r	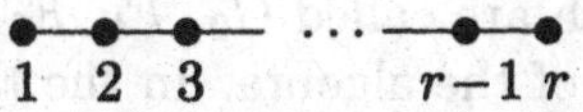	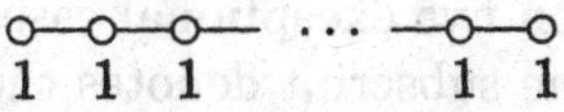
B_r	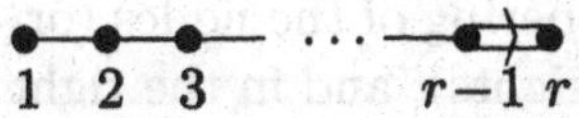	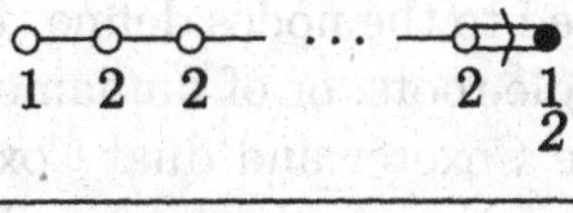
C_r	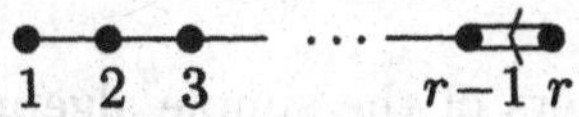	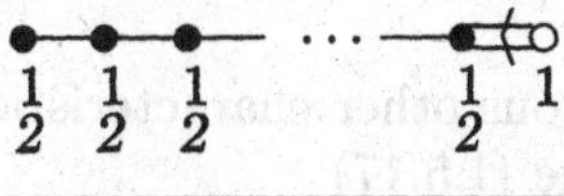
D_r	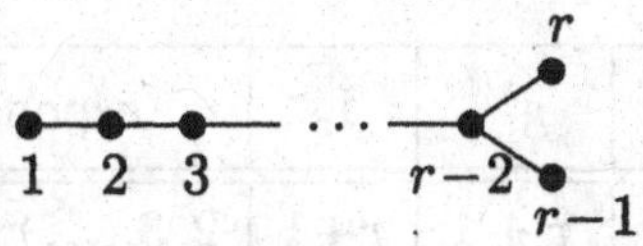	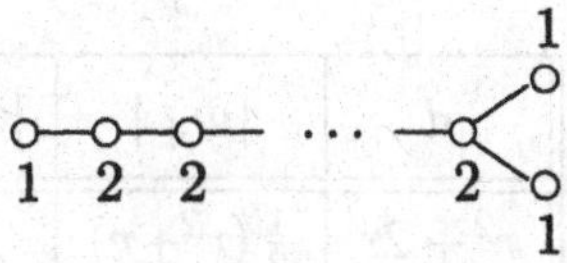
E_6	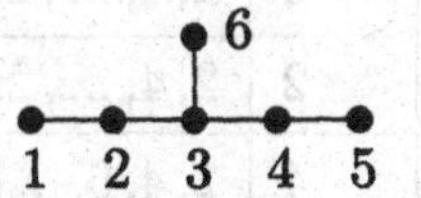	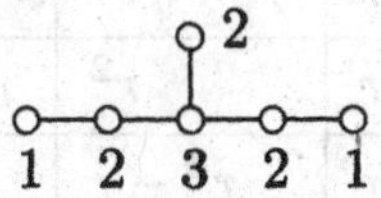
E_7		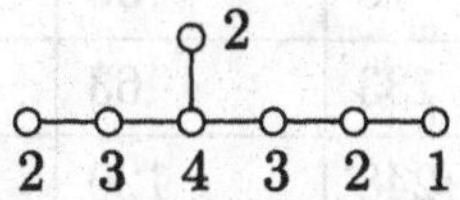
E_8	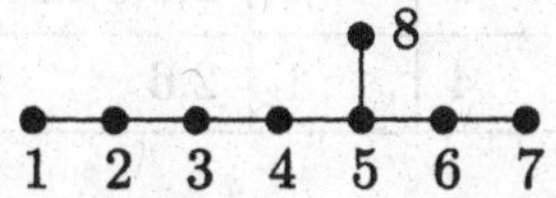	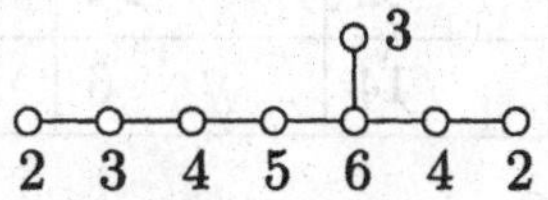
F_4		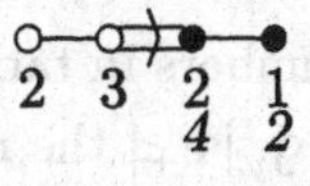
G_2		

(1.5.16)

Thus from table (1.5.16) we see that there are four infinite series, which are denoted by A_r ($r \geq 1$), B_r ($r \geq 3$), C_r ($r \geq 2$) and D_r ($r \geq 4$), and in addition five exceptional cases, which are called G_2, F_4, E_6, E_7 and E_8; here the subscript denotes the rank of the algebra. In the table we have displayed each Dynkin diagram twice. In the left column the numbers adjoined to the nodes define the numbering of the nodes (or, equivalently, of simple roots, or of fundamental weights), and in the right column they are the Coxeter and dual Coxeter labels; the Coxeter labels are given in italics, and are only displayed if they are different from the dual Coxeter labels.

Various other characteristic numbers of the simple algebras are listed in table (1.5.17).

$\mathfrak{g}$	d	$\lvert\Phi_+\rvert$	$\lvert\Phi_+^{(<)}\rvert$	$g^\vee$	$\mathrm{I_c}$	exponents + 1
A_r	r^2+2r	$\frac{1}{2}(r^2+r)$	–	$r+1$	$r+1$	$2,3,\dots,r+1$
B_r	$2r^2+r$	r^2	r	$2r-1$	2	$2,4,\dots,2r$
C_r	$2r^2+r$	r^2	r^2-r	$r+1$	2	$2,4,\dots,2r$
D_r	$2r^2-r$	r^2-r	–	$2r-2$	4	$2,4,6,\dots,2r-2,\ r.$
E_6	78	36	–	12	3	2,5,6,8,9,12
E_7	133	63	–	18	2	2,6,8,10,12,14,18
E_8	248	120	–	30	1	2,8,12,14,18,20,24,30
F_4	52	24	12	9	1	2,6,8,12
G_2	14	6	3	4	1	2,6

(1.5.17)

The various numbers in table (1.5.17) have the following meaning: d is the dimension of $\mathfrak{g}$, $|\Phi_+|$ the number of positive roots, $|\Phi_+^{(<)}|$ the number of short positive roots, and $g^\vee$ the dual Coxeter number. $\mathrm{I_c}$, called the *index of connection*, is the number of conjugacy classes of $\mathfrak{g}$-weights, which by definition is the maximal number of weights of finite-dimensional $\mathfrak{g}$-modules that do not differ by an element of the root lattice, i.e. $\mathrm{I_c} = |L_w/L|$. Finally, the exponents, augmented by one, are the orders of the independent Casimir operators of $\mathfrak{g}$ (compare p. 61). By inspection one finds that the index of connection $|L_w/L|$ of $\mathfrak{g}$ is equal to the determinant of the Cartan matrix, and also equals one plus the number of minimal fundamental weights.

Let us also list the group $\Gamma(\mathfrak{g})$ of automorphisms of the Dynkin diagram of $\mathfrak{g}$; $\Gamma(\mathfrak{g})$ is trivial except for the cases of A_r, D_r and E_6 which are listed in table (1.5.18).

$\mathfrak{g}$	Γ
A_r	$\mathbb{Z}_2$ $(r \geq 2)$
D_r	$\mathcal{S}_3$ $(r=4)$; $\mathbb{Z}_2$ $(r>4)$
E_6	$\mathbb{Z}_2$

(1.5.18)

In table (1.5.18), $\mathbb{Z}_n$ denotes the cyclic group of order n, and $\mathcal{S}_n$ the symmetric group of n objects. The automorphisms of the Dynkin diagram of $\mathfrak{g}$ correspond to those maps $\dot{\omega} : \alpha^{(i)} \mapsto \alpha^{(i')}$ which leave inner products invariant (so that $A^{i'j'} = A^{ij}$). Via

$$\omega : \quad E^i_\pm \mapsto E^{i'}_\pm, \quad H^i \mapsto H^{i'}, \tag{1.5.19}$$

these induce automorphisms of $\mathfrak{g}$. In fact, it is possible to write any automorphism $\sigma \in Aut(\mathfrak{g})$ uniquely as

$$\sigma = \sigma_{\text{int}} \circ \omega \tag{1.5.20}$$

with $\sigma_{\text{int}} \in Int(\mathfrak{g})$ (that is, $\sigma_{\text{int}} = \exp(\text{ad}_x)$ for some nilpotent $x \in \mathfrak{g}$, see section 1.3) and ω of the form (1.5.19). In other words, the symmetry group $\Gamma(\mathfrak{g})$ of the Dynkin diagram of $\mathfrak{g}$, which was listed in the table (1.5.18), satisfies

$$\Gamma(\mathfrak{g}) \cong Aut(\mathfrak{g})/Int(\mathfrak{g}). \tag{1.5.21}$$

It has already been mentioned (p. 29) that in the Dynkin basis the simple roots of a simple Lie algebra $\mathfrak{g}$ are given by the rows of its Cartan matrix. The full root system can also be easily calculated, namely as the nonzero weights of the adjoint module, which can be found by standard methods that will be described in the following section. A convenient way of summarizing the result is by means of a suitable orthonormal basis. Such a basis, which exists because the weight space is a euclidean space isomorphic to $\mathcal{E} = \mathbb{R}^r$, is not too interesting from a purely theoretical point of view, but it is sometimes helpful for practical calculations, e.g. for a quick computation of inner products. For all simple Lie algebras except A_r, E_6, E_7, and G_2 it is possible to choose an orthonormal basis $\{e_i\}$ in $\mathbb{R}^r$ such that the coordinates of any root are all half integers between -2 and 2. In contrast, for the latter algebras the expressions in any orthonormal basis of the weight space involve square roots (for A_2 this is already apparent from the results (1.4.57) for the roots). However, in these cases it is possible to choose an embedding of the weight space in $\mathcal{E} = \mathbb{R}^{r+1}$ (for A_r, E_7, and G_2) or in $\mathcal{E} = \mathbb{R}^{r+2}$ (for E_6), and a basis in $\mathcal{E}$ such that with respect to the latter basis the coordinates of any root are again half integers between -2 and 2. The resulting expressions for the simple roots $\alpha^{(i)} \in \Phi_s$ and arbitrary roots $\alpha \in \Phi$ are listed in table (1.5.22).

$\mathfrak{g}$	$\dim \mathcal{E}$	Φ_s		Φ_+	
A_r	$r+1$	$e_i - e_{i+1}$,	$1 \le i \le r$	$e_i - e_j$,	$1 \le i < j \le r+1$
B_r	r	$e_i - e_{i+1}$, e_r	$1 \le i \le r-1$;	$e_i \pm e_j$, e_i,	$1 \le i < j \le r$; $1 \le i \le r$
C_r	r	$e_i - e_{i+1}$, $2e_r$	$1 \le i \le r-1$;	$e_i \pm e_j$, $2e_i$,	$1 \le i < j \le r$; $1 \le i \le r$
D_r	r	$e_i - e_{i+1}$, $e_{r-1} + e_r$	$1 \le i \le r-1$;	$e_i \pm e_j$,	$1 \le i < j \le r$
E_6	8	$e_{i+1} - e_i$, $\frac{1}{2}(e_8 + e_1 - \sum_{j=2}^{7} e_j)$, $e_1 + e_2$	$1 \le i \le 4$;	$e_i \pm e_j$, $\frac{1}{2}(e_8 - e_7 - e_6 + \sum_{j=1}^{5} (\pm) e_j)$, even number of minus signs	$1 \le j < i \le 5$;
E_7	8	$e_{i+1} - e_i$, $\frac{1}{2}(e_8 + e_1 - \sum_{j=2}^{7} e_j)$, $e_1 + e_2$	$1 \le i \le 5$;	$e_i \pm e_j$, $\frac{1}{2}(e_8 - e_7 + \sum_{j=1}^{6} (\pm) e_j)$, odd number of minus signs; $e_8 - e_7$	$1 \le j < i \le 6$;
E_8	8	$e_{i+1} - e_i$, $\frac{1}{2}(e_8 + e_1 - \sum_{j=2}^{7} e_j)$, $e_1 + e_2$	$1 \le i \le 6$;	$e_i \pm e_j$, $\frac{1}{2}(e_8 + \sum_{j=1}^{7} (\pm) e_j)$, even number of minus signs	$1 \le j < i \le 8$;
F_4	4	$e_2 - e_3$, $e_3 - e_4$, e_4, $\frac{1}{2}(e_1 - e_2 - e_3 - e_4)$		$e_i \pm e_j$, e_i, $\frac{1}{2}(e_1 \pm e_2 \pm e_3 \pm e_4)$	$1 \le i < j \le 4$; $1 \le i \le 4$;
G_2	3	$e_1 - e_2$, $e_2 + e_3 - 2e_1$		$e_1 - e_2$, $e_3 - e_1$, $e_3 - e_2$, $e_2 + e_3 - 2e_1$, $e_1 + e_3 - 2e_2$, $-e_1 - e_2 + 2e_3$	

(1.5.22)

For the cases with $\dim \mathcal{E} > r$ the relevant subspace which is isomorphic to the weight space is identified as the space orthogonal to the vector $\sum_{i=1}^{r+1} e_i$ for A_r and G_2, orthogonal to $e_7 + e_8$ for E_7, and orthogonal to $e_6 - e_7$ and to $e_7 + e_8$ for E_6. Also, the normalization is such that for C_r and G_2 the *short* roots of $\mathfrak{g}$ have length squared equal to two.

Note that the results for the E series nicely display that there exists the embedding $E_6 \subset E_7 \subset E_8$. Also, one can see that there is an embedding $D_8 \subset E_8$, which becomes clear if one makes the replacement $i \mapsto 9 - i$ for one of the algebras, and similarly one finds that $A_7 \subset E_7$.

Let us also present the highest root θ of the algebra in the above orthogonal basis:

$\mathfrak{g}$	A_r	B_r, D_r, F_4	C_r	E_6
θ	$e_1 - e_{r+1}$	$e_1 + e_2$	$2e_1$	$e_8 - e_6 - e_7 + \sum_{j=1}^{5} e_j$

E_7	E_8	G_2
$e_8 - e_7$	$e_7 + e_8$	$2e_3 - e_1 - e_2$

(1.5.23)

The algebras in the infinite series are called the *classical* algebras; they are isomorphic to the matrix algebras $sl_{r+1}(\mathbb{C}), so_{2r+1}(\mathbb{C}), sp_r(\mathbb{C})$, and $so_{2r}(\mathbb{C})$, respectively; here sl_n consists of the traceless $n \times n$-matrices, so_n of the skew-symmetric $n \times n$-matrices, and sp_n of the $2n \times 2n$-matrices M satisfying

$$M\,\epsilon + \epsilon\,M^t = 0, \qquad \epsilon \equiv \begin{pmatrix} \mathbf{0} & \mathbf{1} \\ -\mathbf{1} & \mathbf{0} \end{pmatrix}, \tag{1.5.24}$$

with $\mathbf{1}$ the n-dimensional unit matrix.

For the simple Lie algebras A_r, D_r, E_6, E_7 and E_8, all roots have the same length; these algebras are therefore called *simply laced.* In the other cases there are roots of two different lengths, the length of the short roots being $\sqrt{1/2}$ times the length of the long roots for B_r, C_r, F_4, and $\sqrt{1/3}$ for G_2, respectively. The *dual root system* $\Phi^\vee(\mathfrak{g}) := \{\alpha^\vee \mid \alpha \in \Phi(\mathfrak{g})\}$ of $\mathfrak{g}$ is isomorphic to the root system of a simple Lie algebra which is called the *dual Lie algebra* $\mathfrak{g}^\vee$ of $\mathfrak{g}$, $\Phi^\vee(\mathfrak{g}) \cong \Phi(\mathfrak{g}^\vee)$. Of course, $\mathfrak{g} = \mathfrak{g}^\vee$ if $\mathfrak{g}$ is a simply laced algebra, but this is also true for C_2, G_2 and F_4; in contrast,

$$(B_r)^\vee = C_r \tag{1.5.25}$$

(and vice versa) for $r \geq 3$.

The restrictions on the rank r of the classical algebras are imposed to avoid double-counting; if one had $r \geq 1$ for all four series, then there would be the isomorphisms

$$\begin{aligned} A_1 &\cong B_1 \cong C_1 \cong D_1 \\ B_2 &\cong C_2 \\ D_2 &\cong A_1 \oplus A_1 \\ D_3 &\cong A_3. \end{aligned} \tag{1.5.26}$$

Similarly one could extend the E-series by

$$\begin{aligned} E_5 &\cong D_5, \\ E_4 &\cong A_4, \\ E_3 &\cong A_1 \oplus A_2. \end{aligned} \tag{1.5.27}$$

So far we have assumed that the base field of $\mathfrak{g}$ is $\mathbb{C}$. This was necessary because $\mathbb{R}$ is not algebraically closed. But once one has established the classification of semisimple Lie algebras over $\mathbb{C}$, one gets the classificaton of the semisimple Lie algebras over $\mathbb{R}$ by simply constructing the various real forms corresponding to a given complex $\mathfrak{g}$. This can be done with the help of the involutive automorphisms of $\mathfrak{g}$ (i.e. automorphisms σ of order two). The defining relations of the various real forms of a complex Lie algebra $\mathfrak{g}$ are obtained from those of $\mathfrak{g}$ by imposing the appropriate reality conditions. Different real forms for $\mathfrak{g}$ are related to each other essentially by analytic continuation. For example, the possible real forms of $A_r(\mathbb{C}) = sl_{r+1}(\mathbb{C})$ are the algebras $sl_{r+1}(\mathbb{R})$, $su_{r+1}(\mathbb{C})$ (the traceless skew-hermitian complex $(r+1) \times (r+1)-$matrices), $su_{p,r-p+1}(\mathbb{C})$ for $p = 1, \ldots, [\frac{1}{2}(r+1)]$ (here the square bracket stands for the integer part of a real number), and (for r odd) $su^*_{r+1} \equiv sl_{(r+3)/2}(\mathbb{H})$, with $\mathbb{H}$ denoting the quaternions. The algebra $su_n(\mathbb{C})$ may be characterized by the requirement that its elements leave a positive definite symmetric bilinear form invariant, and for $su_{n,m}(\mathbb{C})$ an analogous statement holds with the bilinear form being non-positive definite, but rather of the signature $\underbrace{++\ldots+}_{n \text{ times}}\underbrace{--\ldots-}_{m \text{ times}}$.

One can show that the compact real forms of the classical Lie algebras are given by $su_{r+1}(\mathbb{C})$, $so_{2r+1}(\mathbb{R})$, $sp_{2r}(\mathbb{H})$ and $so_{2r}(\mathbb{R})$ for the complex algebras A_r, B_r, C_r and D_r, respectively.

1.6 Highest weight modules

For finite-dimensional modules of a semisimple Lie algebra $\mathfrak{g}$, one can always find a basis such that the Cartan subalgebra acts diagonally. This is best understood by first summarizing some facts on the representation theory of $A_1 \cong sl_2$. Consider the generators $E_\pm$ and H of A_1; according to (1.4.39) they obey

$$[H, E_\pm] = \pm 2\, E_\pm, \quad [E_+, E_-] = H. \tag{1.6.1}$$

From this one concludes that it is consistent to assume that H acts diagonally on any module, $H \cdot v = \lambda v$ for some complex number λ. Thus for each module R of A_1 we get a decomposition of R into *weight spaces* $R_{(\lambda)}$ of the form

$$R = \bigoplus_\lambda R_{(\lambda)}, \quad R_{(\lambda)} = \{v \in R \mid R(H) \cdot v = \lambda v,\ \lambda \in F\}. \tag{1.6.2}$$

The numbers λ appearing in (1.6.2) are called the *weights* of the module R, and the dimensionality of $R_{(\lambda)}$ is called the *multiplicity* of the weight λ. It also follows from (1.6.1) that

$$v \in R_{(\lambda)} \Rightarrow R(E_\pm) \cdot v \in R_{(\lambda \pm 2)}. \tag{1.6.3}$$

In particular, for finite-dimensional modules there must exist a weight Λ such that $R_{(\Lambda)} \neq \emptyset$ but $R_{(\Lambda+2)} = \emptyset$; in this case Λ is called a *maximal weight*, and any element $v_\Lambda \in R_{(\Lambda)}$ a maximal weight vector (the term primitive weight vector is also used). For R to be finite-dimensional, repeated application of $R(E_-)$ to v_Λ must yield zero after a finite number of steps; making use of the relations (1.6.1), it follows that this happens iff $\Lambda \in \mathbb{Z}_{\geq 0}$.

Moreover, if R is irreducible, there is exactly one maximal weight, and the different weights of R form an arithmetic progression with difference 2,

$$\lambda = -\Lambda, -\Lambda + 2, \ldots, \Lambda - 2, \Lambda, \tag{1.6.4}$$

with each weight occuring with multiplicity one. In this case Λ is called the *highest weight* and R the *highest weight module* R_Λ. The vector v_Λ is then unique up to scalar multiplication; it is called the *highest weight vector.* The dimension of R_Λ is $\Lambda + 1$. Finally, for each $\Lambda \in \mathbb{Z}_{\geq 0}$, there exists (for $F = \mathbb{R}$ or $F = \mathbb{C}$) exactly one (up to isomorphism) irreducible module of dimension $\Lambda + 1$, namely the highest weight module R_Λ.

Now we can return to the case of arbitrary simple $\mathfrak{g}$. From the remarks above together with the fact that each generator H^i of the Cartan subalgebra $\mathfrak{g}_0$ of $\mathfrak{g}$ corresponds to the generator H of an sl_2-subalgebra of

$\mathfrak{g}$, it follows that $\mathfrak{g}_0$ acts diagonally on any module R of $\mathfrak{g}$, i.e. there is a decomposition into weight spaces analogous to (1.6.2),

$$R = \bigoplus_{\lambda} R_{(\lambda)}, \tag{1.6.5}$$

such that for all $v_\lambda \in R_{(\lambda)}$ and for all $i = 1, ..., r$

$$R(H^i) \cdot v_\lambda = \lambda^i \, v_\lambda. \tag{1.6.6}$$

The r-dimensional vectors $(\lambda^i)_{i=1,...,r}$ are called the *weights* of the module R.

If R is finite-dimensional, then the module with respect to the subalgebra $\mathfrak{h}_\alpha$ generated by $E^\alpha, E^{-\alpha}$ and H^α for any $\alpha \in \Phi$ must be finite-dimensional, too. But then $H^\alpha \equiv (\alpha^\vee, H)$ has integer eigenvalues only; thus

$$(\alpha^\vee, \lambda) \in \mathbb{Z} \ \text{ for all } \ \alpha \in \Phi. \tag{1.6.7}$$

This just means that λ lies in the weight lattice $L_w(\mathfrak{g})$ defined in (1.4.24) (in particular the term "weight" for the vectors (λ^i) is chosen in accordance with the previous use of this term); also, because of the action (1.6.6) of $H^i \in \mathfrak{g}_0$ on weight vectors, the Cartan subalgebra $\mathfrak{g}_0$ can be identified with the weight space, and hence its dual $\mathfrak{g}_0^\star$ with the root space of $\mathfrak{g}$. It is also clear that the integrality property (1.6.7) holds iff that property holds for the simple roots,

$$(\alpha^{(i)\vee}, \lambda) \in \mathbb{Z} \ \text{ for } i = 1, \ldots, r. \tag{1.6.8}$$

Comparison with the definition (1.4.22) of fundamental weights $\Lambda_{(i)}$ then shows that λ is a weight of a finite-dimensional module iff it is an integral linear combination of the fundamental weights,

$$\lambda = \sum_{i=1}^{r} \lambda^i \Lambda_{(i)} \quad \text{with} \quad \lambda^i \in \mathbb{Z} \ \text{ for } i = 1, \ldots, r. \tag{1.6.9}$$

In other words, for any weight of a finite-dimensional module of a simple Lie algebra the Dynkin labels are integers; such weights are called *integral weights*.

Proceeding in analogy with the sl_2 case, it follows that for every finite-dimensional module of $\mathfrak{g}$, there exist maximal weights Λ such that

$$R(E^\alpha) \cdot v_\Lambda = 0 \ \text{ for all } \ \alpha > 0. \tag{1.6.10}$$

If the module is also irreducible, there is exactly one weight with this property; this weight is then called the *highest weight*, and the irreducible finite-dimensional module with highest weight Λ is called the *highest weight module* R_Λ. The Dynkin labels of a highest weight are non-negative integers, $\Lambda^i \in \mathbb{Z}_{\geq 0}$, $i = 1, \dots, r$; weights with this property are called *dominant integral* (more generally weights λ with $\lambda^i \geq 0$ for all i are called *dominant* weights; they are called *strictly dominant* iff $\lambda^i > 0$ for all i). Thus an irreducible finite-dimensional $\mathfrak{g}$-module is characterized by the fact that it has a unique maximal weight and that this weight is dominant integral. Conversely, each dominant integral weight is the highest weight of a finite-dimensional module. *The irreducible finite-dimensional modules of $\mathfrak{g}$ are thus precisely the highest weight modules with dominant integral highest weight.* The elements of R_Λ can all be obtained up to scalar multiplication by applying step operators for negative roots to v_Λ, i.e. any $v \in R_\Lambda$ is of the form $R(x) \cdot v_\Lambda$ for some x in the enveloping algebra of $\mathfrak{g}_-$,

$$v = R(E^{-\beta_1}) R(E^{-\beta_2}) \dots R(E^{-\beta_m}) \cdot v_\Lambda \tag{1.6.11}$$

for appropriate roots $\beta_i \in \Phi_+$ (making use of the bracket relations between the step operators, we may assume without loss of generality that these roots obey $\beta_p > \beta_q$ if $p > q$ and if $\beta_p - \beta_q$ is a root). Conversely, any vector of the form (1.6.11) is in R_Λ provided, of course, that this expression does not vanish identically (the fact that certain naively present vectors v which are defined through the formal expression (1.6.11) actually vanish, implements the irreducibility of the module R_Λ; for more details see section 2.5).

From the formula (1.4.5) for the bracket $[H^i, E^\alpha]$ it follows that $R(H^i) \cdot (R(E^\alpha) \cdot v_\lambda) = R([H^i, E^\alpha] - E^\alpha H^i) \cdot v_\lambda = (\alpha^i + \lambda^i)\, R(E^\alpha) \cdot v_\lambda$, or in other words,

$$R(E^\alpha) \cdot v_\lambda \propto v_{\alpha+\lambda}. \tag{1.6.12}$$

Together with (1.6.11) this implies that the weights of the module R_Λ are all of the form

$$\lambda = \Lambda - \mu, \tag{1.6.13}$$

with μ a sum of positive roots. Let us make this explicit by presenting the weights for a few simple examples. First consider the adjoint module of $\mathfrak{g} = A_2$. As follows immediately from the definition of roots and weights, for arbitrary simple $\mathfrak{g}$ the nonzero weights of the adjoint module are the roots, with the highest weight equal to the highest root. Each of these weights has multiplicity one; in addition, the weight system contains the zero weight with multiplicity $r = \operatorname{rank}\mathfrak{g}$. For the case of A_2 for which the roots are given by (1.4.57), this yields the picture shown in figure (1.6.14). There each weight is symbolized by a box containing the Dynkin labels of the weight, with multiple boxes indicating the multiplicity of the weight. Also, we display which simple roots are to be subtracted when going from one weight in the picture to another.

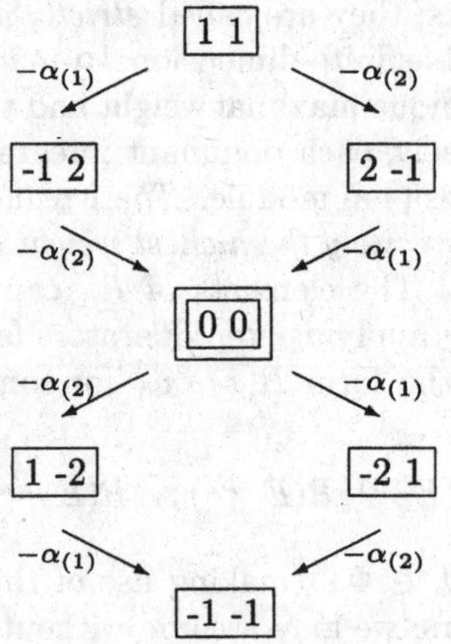

$$(1.6.14)$$

As other examples, we list in figure (1.6.15) the weight diagrams of the two lowest-dimensional non-trivial modules of the Lie algebra $A_2 \cong sl_3$. These are the highest weight modules which have highest weight $\Lambda_{(1)}$ and $\Lambda_{(2)}$, respectively; they are conjugate to each other, and are of dimension three.

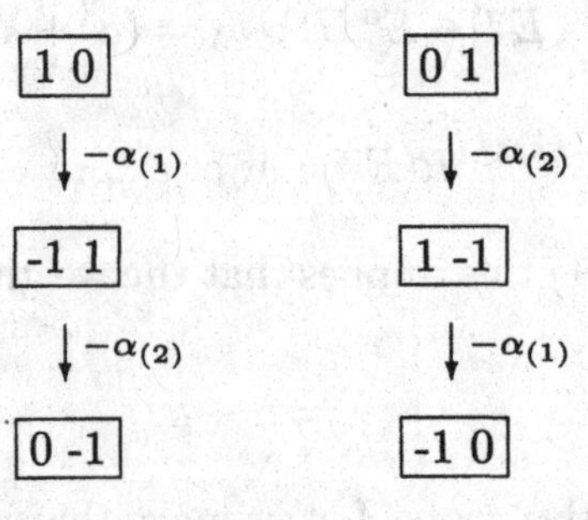

$$
\boxed{1\ 0} \qquad\qquad \boxed{0\ 1}
$$
$$
\downarrow {-\alpha_{(1)}} \qquad\qquad \downarrow {-\alpha_{(2)}}
$$
$$
\boxed{-1\ 1} \qquad\qquad \boxed{1\ -1} \qquad (1.6.15)
$$
$$
\downarrow {-\alpha_{(2)}} \qquad\qquad \downarrow {-\alpha_{(1)}}
$$
$$
\boxed{0\ -1} \qquad\qquad \boxed{-1\ 0}
$$

As a final example, we take one of the lowest-dimensional non-trivial modules of E_6, namely the highest weight module $R_{\Lambda_{(1)}}$. This module is 27-dimensional. Its weight diagram looks as shown in figure (1.6.16). E_6 possesses another 27-dimensional module, which is conjugate to $R_{\Lambda_{(1)}}$; it is the highest weight module with highest weight $\Lambda_{(5)}$.

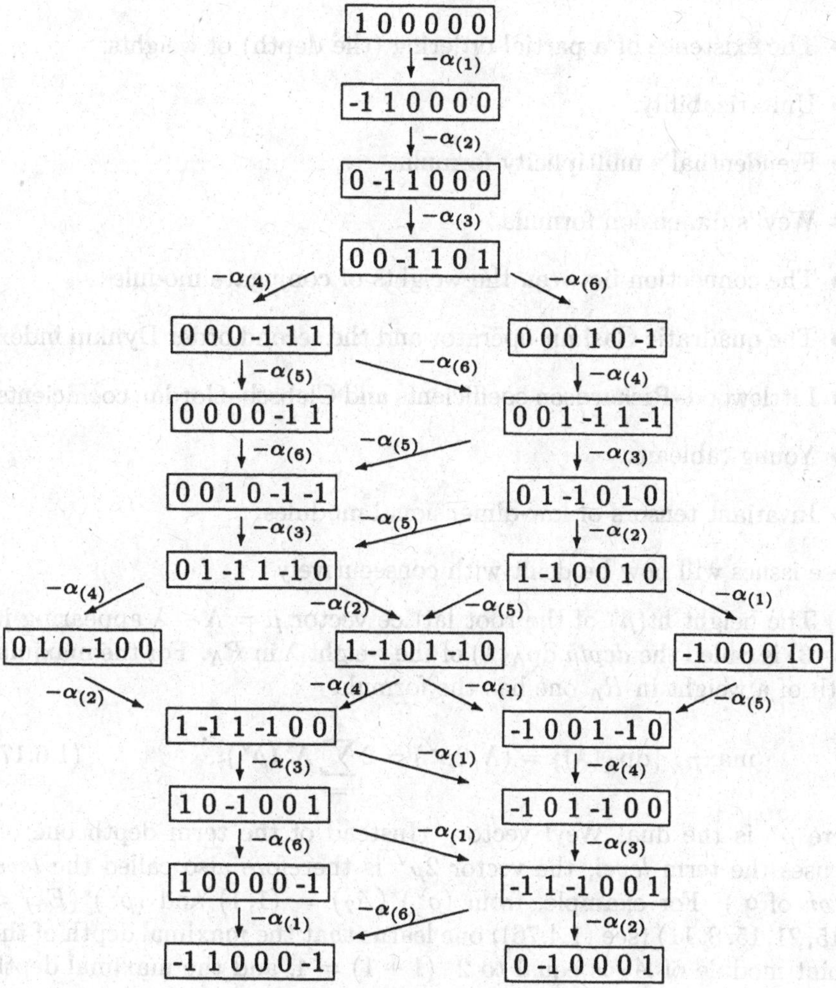
1 0 0 0 0 0
−α(1)
-1 1 0 0 0 0
−α(2)
0 -1 1 0 0 0
−α(3)
0 0 -1 1 0 1
−α(4)
−α(6)
0 0 0 -1 1 1
0 0 0 1 0 -1
−α(5)
−α(6)
−α(4)
0 0 0 0 -1 1
0 0 1 -1 1 -1
−α(6)
−α(5)
−α(3)
0 0 1 0 -1 -1
0 1 -1 0 1 0
−α(3)
−α(5)
−α(2)
0 1 -1 1 -1 0
1 -1 0 0 1 0
−α(4)
−α(2)
−α(5)
−α(1)
0 1 0 -1 0 0
1 -1 0 1 -1 0
-1 0 0 0 1 0
−α(2)
−α(4)
−α(1)
−α(5)
1 -1 1 -1 0 0
-1 0 0 1 -1 0
−α(3)
−α(1)
−α(4)
1 0 -1 0 0 1
-1 0 1 -1 0 0
−α(6)
−α(1)
−α(3)
1 0 0 0 0 -1
-1 1 -1 0 0 1
−α(1)
−α(6)
−α(2)
-1 1 0 0 0 -1
0 -1 0 0 0 1
−α(2)
−α(6)
0 -1 1 0 0 -1
−α(3)
0 0 -1 1 0 0
−α(4)
0 0 0 -1 1 0
−α(5)
0 0 0 0 -1 0

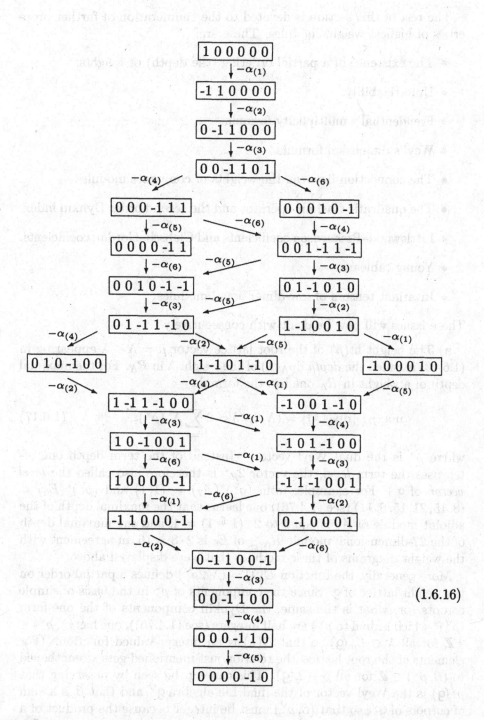

(1.6.16)

The rest of this section is devoted to the enumeration of further properties of highest weight modules. These are:

- The existence of a partial ordering (the depth) of weights.
- Unitarizability.
- Freudenthal's multiplicity formula.
- Weyl's dimension formula.
- The connection between the weights of conjugate modules.
- The quadratic Casimir operator and the second order Dynkin index.
- Littlewood–Richardson coefficients and Clebsch–Gordan coefficients.
- Young tableaux.
- Invariant tensors of low-dimensional modules.

These issues will now be dealt with consecutively.

a) The height $\mathrm{ht}(\mu)$ of the root lattice vector $\mu = \Lambda - \lambda$ appearing in (1.6.13) is called the *depth* $\mathrm{dp}_\Lambda(\lambda)$ of the weight λ in R_Λ. For the maximal depth of a weight in R_Λ one has the formula

$$\max_{R_\Lambda}\{\mathrm{dp}_\Lambda(\lambda)\} = (\Lambda, 2\rho^\vee) = 2\sum_{i=1}^{r} \Lambda^i\,(\rho^\vee)_i, \qquad (1.6.17)$$

where $\rho^\vee$ is the dual Weyl vector. (Instead of the term depth one often uses the term *level*; the vector $2\rho^\vee$ is therefore also called the *level vector* of $\mathfrak{g}$.) For example, from $(\rho^\vee)^\star(A_2) = (1,1)$ and $(\rho^\vee)^\star(E_6) = (8,15,21,15,8,11)$ (see (1.4.76)) one learns that the maximal depth of the adjoint module of A_2 is equal to $2\cdot(1+1) = 4$, and the maximal depth of the 27-dimensional module $R_{\Lambda_{(1)}}$ of E_6 is $2\cdot 8 = 16$, in agreement with the weight diagrams of these modules that were displayed above.

More generally, the function $D(\lambda) = (\lambda, 2\rho^\vee)$ defines a partial order on the weight lattice of $\mathfrak{g}$. Since the components of $\rho^\vee$ in the basis of simple coroots (or, what is the same, the Dynkin components of the one-form $(\rho^\vee)^\star$ which is dual to $\rho^\vee$) are half integers (see (1.4.76)), one has $(\lambda, \rho^\vee) \in \frac{1}{2}\mathbb{Z}$ for all $\lambda \in L_w(\mathfrak{g})$ so that $D(\lambda)$ is an integer-valued function. (For elements of the root lattice, the relation just mentioned gets strengthened to $(\beta, \rho^\vee) \in \mathbb{Z}$ for all $\beta \in L(\mathfrak{g})$. This can e.g. be seen by observing that $\rho^\vee(\mathfrak{g})$ is the Weyl vector of the dual Lie algebra $\mathfrak{g}^\vee$ and that β is a sum of coroots of $\mathfrak{g}^\vee$, so that $(\beta, \rho^\vee)$ must be integer because the product of a weight and of a coroot of $\mathfrak{g}^\vee$ is always an integer. Also, for all simple Lie algebras except A_{4n+2}, A_{4n+3}, B_{2n+1}, C_{2n+1} and E_7 this already follows from the fact that these algebras have an even number of positive roots.)

b) Finite-dimensional highest weight modules of complex and of compact real algebras are *unitarizable.* A module is called unitarizable or, more sloppily, unitary, iff it is possible to define a positive definite inner product

$$(\,|\,):\; R\times R \to \mathbb{C} \tag{1.6.18}$$

with the property that

$$(v \mid R(x)\cdot w) = -(R(\omega_0(x))\cdot v \mid w) \tag{1.6.19}$$

for any $x \in \mathfrak{g}$ and $v, w \in R$. Here ω_0 denotes the *Weyl automorphism*; for a real Lie algebra, this is defined as

$$\omega_0:\; \begin{array}{l} \mathfrak{g} \to \mathfrak{g}, \\ E^i_{\pm} \mapsto -E^i_{\mp},\;\; H^i \mapsto -H^i,\;\; i = 1,\dots,r, \end{array} \tag{1.6.20}$$

while for a complex Lie algebra $\mathfrak{g}$ one defines ω_0 on its compact real form $\mathfrak{g}_{\mathbb{R}}$ as above and extends it to $\mathfrak{g} = \mathfrak{g}_{\mathbb{R}} \oplus i\mathfrak{g}_{\mathbb{R}}$ via

$$\omega_0(x+iy) = \omega_0(x) - i\,\omega_0(y). \tag{1.6.21}$$

In other words, a unitary module of a complex Lie algebra is a module possessing a Hilbert space structure for which one can choose a basis such that the generators are anti-selfadjoint operators in the Hilbert space,

$$R(T^a) = -(R(T^a))^\dagger, \tag{1.6.22}$$

with $(R(T^a))^\dagger$ defined via

$$(v \mid (R(T^a))^\dagger \cdot w) = (R(T^a)\cdot v \mid w). \tag{1.6.23}$$

(In physics, one usually considers a basis of selfadjoint rather than anti-selfadjoint generators which is obtained via $T^a \to i\,T^a$.) In such a basis, the Killing matrix takes the particular form (1.3.13),

$$\kappa^{ab} = -\delta^{ab}. \tag{1.6.24}$$

Note that we often use the notation v loosely for all vectors ξv with $\xi \in F\setminus\{0\}$ (e.g. we speak of *the* highest weight vector v_Λ of R_Λ rather than of the one-dimensional weight space $R_{(\Lambda)}$); the inner product can be used to set the normalization ξ to a fixed value, say such that $(v \mid v) = 1$. In this normalization (and with a convenient choice of phase factors), one has e.g.

$$\left.\begin{array}{l} R_\Lambda(E^{-\alpha})\cdot v_\Lambda = \sqrt{(\Lambda,\alpha)}\, v_{\Lambda-\alpha} \\ R_\Lambda(E^{\alpha})\cdot v_{\Lambda-\alpha} = \sqrt{(\Lambda,\alpha)}\, v_\Lambda \end{array}\right\} \text{ for all } \alpha \in \Phi_+ \tag{1.6.25}$$

as a special example of the formula (1.6.12).

c) The "quantum numbers" λ^i do not characterize v_λ completely. First, of course, one has to specify to which module R the vector v_λ belongs, and also to fix its normalization (e.g. by specifying its length with respect to the inner product (1.6.18)). But even for a given module a weight λ may appear more than once. The analysis shows that for a simple Lie algebra of rank r and dimension d, a complete specification of the weight vectors requires $\frac{1}{2}(d-r) = |\Phi_+|$ different labels, or in other words $\frac{1}{2}(d-3r)$ quantum numbers in addition to the Dynkin labels.

For the multiplicity of a weight λ in the highest weight module R_Λ, one has the *Freudenthal recursion formula*

$$\begin{aligned} \mathrm{mult}_\Lambda(\lambda) = \; & 2\,[(\Lambda+\rho,\Lambda+\rho)-(\lambda+\rho,\lambda+\rho)]^{-1} \\ & \cdot \sum_{\alpha\in\Phi_+} \sum_{\substack{m>0 \\ v_{\lambda+m\alpha}\in R_\Lambda}} (\lambda+m\alpha,\alpha)\cdot \mathrm{mult}_\Lambda(\lambda+m\alpha) \end{aligned} \tag{1.6.26}$$

for $\lambda \neq \Lambda$ together with $\mathrm{mult}_\Lambda(\Lambda) = 1$. Here ρ is the Weyl vector of $\mathfrak{g}$ which was defined in (1.4.70), $\rho = \frac{1}{2}\sum_{\alpha\in\Phi_+}\alpha$. The Freudenthal formula can easily be implemented in a computer program; in this way extensive lists of the weight systems for irreducible highest weight modules of simple Lie algebras have been compiled in the literature.

We also note the following simple algorithm for calculating the weights of a highest weight module, not counting the multiplicities. From the highest weight $\Lambda = \sum_i \Lambda^i \Lambda_{(i)}$, which is the unique weight of depth zero, one gets all weights of depth one by subtracting $\alpha^{(i)}$ if $\Lambda^i > 0$. More generally, from any weight λ of the module one may subtract the simple root $\alpha^{(i)}$ λ^i times if $\lambda^i > 0$. Proceeding inductively, one finds in this way all distinct weights, and finally arrives at some value of the depth for which all weights have non-positive Dynkin labels so that the procedure terminates. The validity of this prescription is easily verified for the A_2- and E_6-examples given above.

d) The dimension of R_Λ can be calculated from the *Weyl dimension formula*

$$d_\Lambda \equiv \dim(R_\Lambda) = \prod_{\alpha\in\Phi_+} \frac{(\Lambda+\rho,\alpha)}{(\rho,\alpha)}. \tag{1.6.27}$$

To apply this formula, it is best to take the roots in the basis of simple coroots, and Λ and $\rho = \sum_i \Lambda_{(i)}$ in the Dynkin basis. E.g. take $g = A_2$ or G_2; then one has

$$d_\Lambda(A_2) = (\Lambda^1+1)\,(\Lambda^2+1)\,(\tfrac{1}{2}\,(\Lambda^1+\Lambda^2)+1) \tag{1.6.28}$$

and

$$d_\Lambda(G_2) = (\Lambda^1+1)\,(\Lambda^2+1)\,(\tfrac{1}{2}\,(\Lambda^1+\Lambda^2)+1) \\ \cdot(\tfrac{1}{3}\,(2\Lambda^1+\Lambda^2)+1)\,(\tfrac{1}{4}\,(3\Lambda^1+\Lambda^2)+1) \\ \cdot(\tfrac{1}{5}\,(3\Lambda^1+2\Lambda^2)+1), \tag{1.6.29}$$

respectively, where Λ^i, $i=1,2$, are the Dynkin labels of Λ (similar expressions will be encountered in the context of quantum groups, see (4.4.15) and (4.4.16)). Another simple example is obtained for $\Lambda=\rho$, for which the Weyl formula (1.6.27) gives

$$d_\rho = 2^{|\Phi_+|} = 2^{(d-r)/2}. \tag{1.6.30}$$

The dimensions of Kronecker products and of direct sums of irreducible highest weight modules are of course

$$d_{\Lambda\times\Lambda'} = d_\Lambda \cdot d_{\Lambda'} \tag{1.6.31}$$

and

$$d_{\Lambda\oplus\Lambda'} = d_\Lambda + d_{\Lambda'}, \tag{1.6.32}$$

respectively (we use here $\Lambda\times\Lambda'$ as a shorthand version of $R_\Lambda\times R_{\Lambda'}$).

e) The weights of the module R^+ which is conjugate to R are obtained from the weights of R by changing their sign. Thus in particular the highest weights of R_Λ and R_{Λ^+} are related through

$$\Lambda^+ = -\lambda_{\min}, \tag{1.6.33}$$

where $\lambda_{\min}$ is the lowest weight of R_Λ, which is the unique weight of R_Λ with maximal depth $\mathrm{dp}_\Lambda(\lambda)$. A module is self-conjugate iff its weight system is invariant under the change of sign; a self-conjugate module with $\mathrm{dp}_\Lambda(\lambda_{\min})\in 2\mathbb{Z}$ is orthogonal, while for $\mathrm{dp}_\Lambda(\lambda_{\min})\in 2\mathbb{Z}+1$ it is symplectic. Inspection shows that all highest weight modules of the simple Lie algebras are self-conjugate except for most of the highest weight modules of A_r, D_{2n+1} and E_6.

f) The *quadratic Casimir operator* $\mathcal{C}_2$ defined as

$$\mathcal{C}_2 := \sum_{a,b} \kappa_{ab}\, T^a T^b \tag{1.6.34}$$

acts as a constant in any irreducible finite-dimensional module R_Λ,

$$R_\Lambda(\mathcal{C}_2)\cdot v = C_\Lambda\, v \quad \text{for all } \ v\in R_\Lambda. \tag{1.6.35}$$

(Note that $\mathcal{C}_2$ is not an element of $\mathfrak{g}$ but rather an element of the universal enveloping algebra $\mathsf{U}(\mathfrak{g})$ of $\mathfrak{g}$; in fact it belongs to the center of $\mathsf{U}(\mathfrak{g})$; see p. 266. The notation $R(\mathcal{C}_2)$ thus stands for $\kappa_{ab}\, R(T^a)R(T^b)$.) In a basis with $\kappa^{ab} = -\delta^{ab}$, this is usually written as

$$\sum_{a=1}^{d} T^a_{(\Lambda)} T^a_{(\Lambda)} = -C_\Lambda \, \mathbf{1}. \qquad (1.6.36)$$

The numbers C_Λ are obtained most easily by applying the quadratic Casimir operator to the highest weight vectors. To do so, first use the result (1.4.10) for the Killing matrix to rewrite the expression (1.6.34) as

$$\mathcal{C}_2 = H^i \kappa_{ij} H^j + \sum_{\alpha \in \Phi} E^\alpha E^{-\alpha}. \qquad (1.6.37)$$

As we will see shortly, in a Chevalley basis κ^{ij} coincides with the inverse quadratic form matrix G^{ij}, and hence in such a basis (1.6.37) is nothing but

$$\mathcal{C}_2 = (H, H) + \sum_{\alpha \in \Phi} E^\alpha E^{-\alpha}. \qquad (1.6.38)$$

Using $(H, H) \cdot v_\Lambda = (\Lambda, \Lambda) \, v_\Lambda$ and (1.6.25), one then finds

$$C_\Lambda = (\Lambda, \Lambda + 2\rho). \qquad (1.6.39)$$

For the adjoint module $R_{\mathrm{ad}} \equiv R_\theta$, the quadratic Casimir eigenvalue can also be obtained by inserting the relation (1.2.10) between the generators $R_\theta(T^a)$ and the structure constants into the definition (1.6.34); this gives

$$C_\theta \, \delta_c^d = \sum_{a,b} f^{ab}{}_c \, f_{ba}{}^d, \qquad (1.6.40)$$

respectively, in a basis with $\kappa^{ab} = -\delta^{ab}$,

$$C_\theta \, \delta^{cd} = \sum_{a,b} f^{abc} f^{abd}. \qquad (1.6.41)$$

Comparing (1.6.40) with the expression (1.3.6) for κ^{ab} (contracted with its inverse κ_{ca}), it follows that the normalization constant I_{ad} introduced in (1.3.4) is nothing but the Casimir eigenvalue $C_{\mathrm{ad}} \equiv C_\theta$,

$$I_\theta = C_\theta. \qquad (1.6.42)$$

Similarly, the constant appearing in the proportionality relation (1.3.9) between the Killing form κ and its analogue $\kappa_\Lambda \equiv \kappa_{R_\Lambda}$ is denoted by I_Λ and called the (second order) *Dynkin index* of the module R_Λ. Thus one has

$$\kappa_\Lambda = (I_\Lambda / I_\theta) \, \kappa, \qquad (1.6.43)$$

or in other words

$$\mathrm{tr}\,(R_\Lambda(T^a) R_\Lambda(T^b)) = I_\Lambda \, \kappa^{ab}. \qquad (1.6.44)$$

One can show that

$$I_\Lambda := \frac{d_\Lambda}{d} C_\Lambda . \tag{1.6.45}$$

As is clear from the formula (1.6.39), the eigenvalues C_Λ are rational numbers; but typically they are not integers. In contrast, it turns out that the second order Dynkin index I_Λ, divided by (θ,θ), is a half integer for any dominant integral weight Λ, and is in fact an integer if the highest weight module is an orthogonal module (half integer values do appear for some symplectic or non-selfconjugate highest weight modules). It is also possible to show that

$$I_\Lambda = \frac{1}{r} \sum_{\lambda:\, v_\lambda \in R_\Lambda} (\lambda,\lambda). \tag{1.6.46}$$

Via this formula, the Dynkin index can also be defined for direct sums of irreducible modules; similarly, for Kronecker products one has

$$I_{\Lambda\times\Lambda'} = I_\Lambda d_{\Lambda'} + I_{\Lambda'} d_\Lambda . \tag{1.6.47}$$

It should also be noted that for applications in physics, the Dynkin index is actually a more natural concept than the quadratic Casimir. E.g. in four-dimensional gauge theories, the Dynkin index I_Λ gives (up to a constant) the contribution of particles carrying the representation R_Λ to the renormalization group β-function; and it also is equal to the number of zero modes of (spin-$\frac{1}{2}$) fermions carrying the representation R_Λ in the background of an instanton.

From (1.6.46) we also learn that the quadratic Casimir eigenvalue in the adjoint representation can be written as

$$C_\theta = \frac{1}{r} \sum_{\alpha\in\Phi} (\alpha,\alpha) = \frac{1}{r}\,(\theta,\theta)(n_L + (\tfrac{S}{L})^2\, n_S) \tag{1.6.48}$$

where n_L and n_S are the numbers of long and short roots, respectively, and S/L the ratio of their lengths (this result can be checked case by case using the numbers given in table (1.5.17)). In particular for simply laced $\mathfrak{g}$ one gets

$$C_\theta = \left(\frac{d}{r} - 1\right)(\theta,\theta). \tag{1.6.49}$$

Now because of the definition (1.4.68) of the dual Coxeter labels and and the property (1.4.72) of the Weyl vector, the dual Coxeter number can be written as

$$g^\vee \equiv 1 + \sum_i a_i^\vee = 1 + (\rho,\theta^\vee) = 1 + \frac{2(\rho,\theta)}{(\theta,\theta)} = \frac{1}{(\theta,\theta)}\,(\theta,\theta+2\rho). \tag{1.6.50}$$

Hence one has the relation

$$C_\theta = (\theta,\theta)\cdot g^\vee \tag{1.6.51}$$

between the quadratic Casimir eigenvalue in the adjoint representation and the dual Coxeter number. (The explicit values for $g^\vee$ in table (1.5.17) then show that the normalization of the long roots usually used in physics is $(\theta,\theta)=1$; in contrast, in mathematics the normalization $(\theta,\theta)=2$ is common.)

As claimed above, in a Chevalley basis the restriction κ^{ij} of the Killing form to the Cartan subalgebra is numerically equal to the metric G^{ij}. To see this, observe that in the Chevalley basis κ^{ij} has to be intrinsically defined by the root system, and hence for symmetry reasons the quantities κ^{ij} and G^{ij} must be proportional; the constant of proportionality is obtained by combining the expression (1.4.9) for κ^{ij} with (1.6.48):

$$\begin{aligned} rC_\theta &= \sum_{\alpha\in\Phi}(\alpha,\alpha) = G_{ij}\sum_{\alpha\in\Phi}\alpha^i\alpha^j = G_{ij}I_\theta\kappa^{ij} \\ &\overset{!}{=} G_{ij}I_\theta\cdot\xi G^{ij} = \xi rC_\theta. \end{aligned} \tag{1.6.52}$$

Thus $\xi=1$, i.e. indeed

$$\kappa^{ij} = G^{ij}. \tag{1.6.53}$$

Note that from this result (and, in fact, already from the expression (1.4.9) of κ^{ij} in terms of the $\mathfrak{g}$-roots) it is obvious that in the Chevalley basis the normalization of the Killing form is just given by the normalization of the inner product in weight space; this may be expressed as

$$\kappa = \kappa_0\cdot(\theta,\theta), \tag{1.6.54}$$

with κ_0 some fixed reference metric.

It is also possible to define elements of $\mathsf{U}(\mathfrak{g})$ which are higher than quadratic in the generators and which share the property of the quadratic Casimir operator to act as a constant in any highest weight module; these are called higher order Casimir operators. They are of the form

$$C_n = d_{a_1a_2\dots a_n}T^{a_1}\dots T^{a_n}, \tag{1.6.55}$$

with $d_{a_1a_2\dots a_n}$ an invariant tensor of the adjoint module. For example, the third order Casimir operator reads

$$C_3 = f_{ad}{}^e f_{be}{}^f f_{cf}{}^d\, T^aT^bT^c. \tag{1.6.56}$$

The number of independent Casimir operators of a simple Lie algebra $\mathfrak{g}$ is equal to the rank of $\mathfrak{g}$. The orders of these Casimir operators have been listed in table (1.5.17); note e.g. that for D_r with r even, there are two independent Casimir operators both having order r. The numbers obtained by subtracting one from these orders of Casimir operators are called the *exponents* of $\mathfrak{g}$.

g) The weight system of a Kronecker product $R_\Lambda \times R_{\Lambda'}$ consists of all weights of the form $\lambda + \lambda'$ with λ and λ' weights of R_Λ and of $R_{\Lambda'}$, respectively. From this it is obvious that this Kronecker product contains $R_{\Lambda+\Lambda'}$ as an irreducible submodule. There exists also a simple rule for finding a second irreducible submodule in the Kronecker product, but in order to write out the whole decomposition

$$R_\Lambda \times R_{\Lambda'} = \bigoplus_i \mathcal{L}_i \, R_{\Lambda_i} \tag{1.6.57}$$

of the Kronecker product into a direct sum of its irreducible submodules, one usually has to look explicitly at the weight space decomposition of the Kronecker product. Another algorithm for decomposing Kronecker products will be described at the end of section 1.7. Except for low-dimensional modules, both algorithms require the use of a computer. In (1.6.57) the summation is meant in the sense that each module R_{Λ_i} appears only once in the sum, i.e. eventual multiplicities larger than one are taken care of by the integers $\mathcal{L}_i$. These integers are called *Littlewood–Richardson coefficients*; for $\mathfrak{g} = A_r$, and also to some extent for the other classical algebras, combinatorial formulae for these numbers can be given, whereas for exceptional $\mathfrak{g}$ no general formula is known (the term Littlewood–Richardson coefficient is sometimes reserved for the case of A_r). If none of the Littlewood–Richardson coefficients of a Kronecker product is larger than unity, then the product is said to be *simply reducible.* The multiplicity of $R_{\Lambda+\Lambda'}$ in $R_\Lambda \times R_{\Lambda'}$ (with R_Λ, $R_{\Lambda'}$ irreducible) is always equal to one. Instead of (1.6.57), often the shorthand notation

$$\Lambda \times \Lambda' = \bigoplus_i \mathcal{L}_i \, \Lambda_i \tag{1.6.58}$$

will be employed.

While in the general case the computation of Kronecker products is tedious, for low-dimensional modules many Kronecker products can be uniquely determined using various sum rules (together with the fact that $R_{\Lambda+\Lambda'} \subset R_\Lambda \times R_{\Lambda'}$), the most important ones being the dimension sum rule

$$d_\Lambda \cdot d_{\Lambda'} = \sum_i \mathcal{L}_i \, d_{\Lambda_i} \tag{1.6.59}$$

and the Dynkin index sum rule

$$I_\Lambda d_{\Lambda'} + I_{\Lambda'} d_\Lambda = \sum_i \mathcal{L}_i \, I_{\Lambda_i} \tag{1.6.60}$$

which follow immediately from the properties (1.6.31) and (1.6.32) of d_Λ

and the corresponding properties (cf. (1.6.47)) of I_Λ. An other general result is that the singlet occurs in the decomposition of $\Lambda\times\Lambda'$ iff $\Lambda'=\Lambda^+$, and in this case its multiplicity is equal to one.

Two natural choices of bases of the Kronecker product are, in a self-explanatory notation, provided by $\{v_{\lambda;\lambda'}\equiv v_{\Lambda,\lambda;\Lambda',\lambda'}\mid v_\lambda\in R_\Lambda,\ v_{\lambda'}\in R_{\Lambda'}\}$ and by $\{v_{\Lambda_i,\lambda_i}\mid v_{\lambda_i}\in R_{\Lambda_i},\ \mathcal{L}_i\neq 0\}$. The coefficients $C^{\lambda,\lambda';\lambda_i}_{\Lambda,\Lambda';\Lambda_i}$ which are obtained when the former basis is expanded in the latter,

$$v_{\lambda;\lambda'}=\sum_{\Lambda_i,\lambda_i} C^{\lambda,\lambda';\lambda_i}_{\Lambda,\Lambda';\Lambda_i}\, v_{\Lambda_i,\lambda_i}, \tag{1.6.61}$$

are called the *Clebsch–Gordan coefficients* of the Kronecker product (here for notational simplicity we use only the weights in order to characterize the weight vectors, i.e. we suppress all multiplicity indices which could possibly be needed). Similarly, one defines Clebsch–Gordan coefficients $C^{\lambda_1,\lambda_2\ldots\lambda_n;\lambda}_{\Lambda_1,\Lambda_2\ldots\Lambda_n;\Lambda}$ describing how R_Λ is obtained in a multiple Kronecker product $R_{\Lambda_1}\times\ldots\times R_{\Lambda_n}$.

Explicit general formulae for the Clebsch–Gordan coefficients are known only for $\mathfrak{g}=A_1$. A property which holds for arbitrary simple $\mathfrak{g}$ is

$$C^{\lambda,\lambda';\lambda''}_{\Lambda,\Lambda';\Lambda''}\propto\delta_{\lambda+\lambda',\lambda''}. \tag{1.6.62}$$

Also, as the Clebsch–Gordan coefficients effect a change of basis, they satisfy the corresponding orthogonality relations. Furthermore, since the weight vector corresponding to the largest weight $\Lambda+\Lambda'$ in the tensor product is unique, one has

$$C^{\Lambda,\Lambda';\Lambda+\Lambda'}_{\Lambda,\Lambda';\Lambda+\Lambda'}=1. \tag{1.6.63}$$

If the Kronecker product is simply reducible, this generalizes to

$$C^{\lambda,\lambda';\Lambda''}_{\Lambda,\Lambda';\Lambda''}\neq 0\Rightarrow C^{\lambda,\lambda';\Lambda''}_{\Lambda,\Lambda';\Lambda''}=1, \tag{1.6.64}$$

and the coefficients involving non-highest weights are unique up to phases; in contrast, for non-simply reducible Kronecker products the Clebsch–Gordan coefficients are defined only up to more general unitary transformations.

h) For $\mathfrak{g}=sl_n\equiv A_r$, a convenient description of irreducible highest weight modules is in terms of *Young tableaux.* These sl_n-modules are completely determined by specifying their symmetry properties as submodules of the Kronecker product of a suitable number of defining (n-dimensional) modules. In other words, denoting the weight vectors of the defining module by v_i, $i=1,\ldots,n$, the weight vectors of any other irreducible highest weight module are given by $v_{i_1i_2\ldots i_N}$ where $i_1,i_2,\ldots,i_N\in\{1,2,\ldots,n\}$ are restricted to a particular behavior under permutations of the indices, i.e.

under the symmetric group $\mathcal{S}_N$. The Young tableau Y_Λ of the module R_Λ provides a graphic description of these symmetry properties. Each index i_p is represented by a quadratic box, and the boxes representing indices which are to be symmetrized respectively antisymmetrized, are glued together in horizontal rows and vertical columns, respectively. Thus the Young tableau for the defining module is a single box, while the Kronecker product of two defining modules contains two irreducible submodules, the *symmetric tensor* (with $v_{ij} = v_{ji} \propto v_i \otimes v_j + v_j \otimes v_i$) and the *antisymmetric tensor* (with $v_{ij} = -v_{ji} \propto v_i \otimes v_j - v_j \otimes v_i$) with Young tableaux consisting of two boxes in a single row and column, respectively. The highest weights of these modules are $2\Lambda_{(1)}$ and $\Lambda_{(2)}$, respectively:

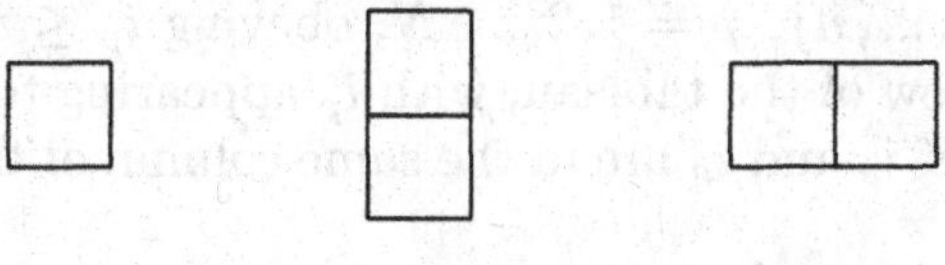

(1.6.65)

$\Lambda_{(1)}$ $\Lambda_{(2)}$ $2\Lambda_{(1)}$

More generally, the fundamental modules $R_{\Lambda_{(j)}}$, $j = 1, \dots, n-1$, are the j-fold antisymmetric tensor modules so that their Young tableaux consist of a single column of j boxes. As a consequence, for an arbitrary irreducible highest weight module of highest weight $\Lambda = \sum_{i=1}^{n-1} \Lambda^i \Lambda_{(i)}$, the Young tableau has $\sum_i \Lambda^i$ columns, with the first Λ^{n-1} columns consisting of $n-1$ boxes, the next Λ^{n-2} columns consisting of $n-2$ boxes, etc. For example, the Young tableaux Y_θ of the adjoint module and Y_ρ of the irreducible module having the Weyl vector ρ as highest weight are as shown in figure (1.6.66).

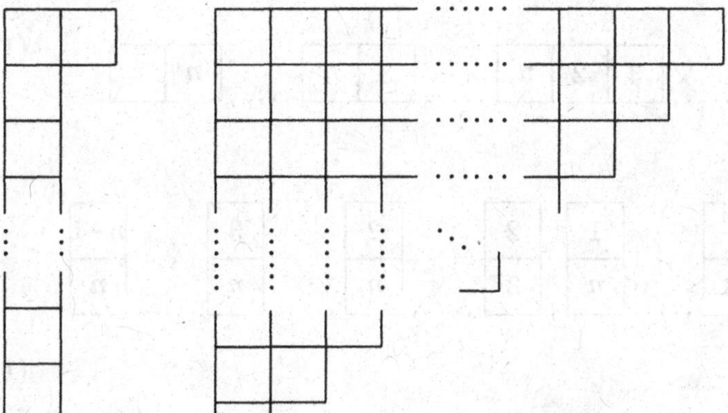

(1.6.66)

θ ρ

Note that the conjugacy class of $R_{\Lambda_{(j)}}$ is j (more precisely, it is given by ζ^j with ζ the generator of $L_w/L \cong \mathbb{Z}_n$); thus the conjugacy class of any module R_Λ is equal to the number of boxes of Y_Λ modulo n.

The dimensionality of an sl_n-module is equal to the number of independent n-tuples $(i_1, i_2, ..., i_n)$ corresponding to the N-index weight vectors $v_{i_1 i_2 \dots i_N}$, $i_1, i_2, \dots, i_N \in \{1, 2, \dots, n\}$. The symmetrizations and antisymmetrizations which connect different tuples of indices effectively impose the requirements that for $1 \le p < q \le N$ one has $i_p \le i_q$ if the indices i_p and i_q are symmetrized, respectively $i_p < i_q$ if i_p and i_q are antisymmetrized. As a consequence, the dimensionality of an sl_n-module is equal to the number of distinct possibilities to fill the boxes of its Young tableau with integers $i_p \in \{1, 2, \dots, n\}$, $p = 1, 2, \dots, N$, obeying $i_p \le i_q$ if i_p and i_q are put in the same row of the tableau, with i_p appearing to the left of i_q, and obeying $i_p < i_q$ if i_p and i_q are in the same column of the tableau, with i_p above i_q.

Let us make this explicit with a few simple examples. First consider the defining module whose Young tableau consists of a single box. Obviously the allowed fillings of the box are just

$$\boxed{1}\,,\ \boxed{2}\,,\ \boxed{3}\,, \dots,\ \boxed{n} \tag{1.6.67}$$

Thus the defining module of sl_n is n-dimensional. Next take the symmetric and antisymmetric tensor modules. The allowed fillings of the boxes are then as given in figures (1.6.68) and (1.6.69), respectively.

$$\begin{array}{|c|c|}\hline 1&1\\\hline\end{array}\,,\ \begin{array}{|c|c|}\hline 1&2\\\hline\end{array}\,,\dots,\ \begin{array}{|c|c|}\hline 1&n\\\hline\end{array}\,,\ \begin{array}{|c|c|}\hline 2&2\\\hline\end{array}\,,$$

$$\begin{array}{|c|}\hline 2\\\hline\end{array}\,,\dots,\ \begin{array}{|c|c|c|}\hline 3&2&n\\\hline\end{array}\,,\dots,\ \begin{array}{|c|c|}\hline 3&n\\\hline\end{array}\,,\dots,\ \begin{array}{|c|c|}\hline n&n\\\hline\end{array} \tag{1.6.68}$$

$$\begin{array}{|c|}\hline 1\\\hline 2\\\hline\end{array}\,,\ \begin{array}{|c|}\hline 1\\\hline 3\\\hline\end{array}\,,\dots,\ \begin{array}{|c|}\hline 1\\\hline n\\\hline\end{array}\,,\ \begin{array}{|c|}\hline 2\\\hline 3\\\hline\end{array}\,,\dots,\ \begin{array}{|c|}\hline 2\\\hline n\\\hline\end{array}\,,\dots,\ \begin{array}{|c|}\hline 3\\\hline n\\\hline\end{array}\,,\dots,\ \begin{array}{|c|}\hline n{-}1\\\hline n\\\hline\end{array} \tag{1.6.69}$$

From the figures one reads off that the dimensionalities of these modules are $n(n+1)/2$ and $n(n-1)/2$, respectively. Finally we display the possible fillings for the adjoint module $R_\theta = R_{\Lambda_{(1)}+\Lambda_{(2)}}$ of $\mathfrak{g} = sl_3$, which show that this module is eight-dimensional (figure (1.6.71)).

$$\begin{array}{|c|c|}\hline 1 & 1 \\ \hline 2 \\ \cline{1-1}\end{array}\,,\quad \begin{array}{|c|c|}\hline 1 & 2 \\ \hline 2 \\ \cline{1-1}\end{array}\,,\quad \begin{array}{|c|c|}\hline 1 & 3 \\ \hline 2 \\ \cline{1-1}\end{array}\,,\quad \begin{array}{|c|c|}\hline 1 & 1 \\ \hline 3 \\ \cline{1-1}\end{array}\,,$$
$$\begin{array}{|c|c|}\hline 1 & 2 \\ \hline 3 \\ \cline{1-1}\end{array}\,,\quad \begin{array}{|c|c|}\hline 1 & 3 \\ \hline 3 \\ \cline{1-1}\end{array}\,,\quad \begin{array}{|c|c|}\hline 2 & 2 \\ \hline 3 \\ \cline{1-1}\end{array}\,,\quad \begin{array}{|c|c|}\hline 2 & 3 \\ \hline 3 \\ \cline{1-1}\end{array} \tag{1.6.70}$$

For more complicated Young tableaux the enumeration of all possible fillings becomes rather lengthy. However, it is possible to shortcut the calculation and express the dimensionality of an irreducible highest weight module R_Λ of sl_n rather directly in terms of the associated Young tableau Y_Λ. Namely, one can show that the following so-called *hook formula* is valid:

$$d_\Lambda = \prod_{(i,j)\in Y_\Lambda} \frac{n-i+j}{h_{ij}}. \tag{1.6.71}$$

Here the pairs (i,j) label the boxes of Y_Λ, with i numbering the rows from top to bottom, and j numbering the columns from left to right. The numbers h_{ij}, the so-called hook lengths, are the numbers of boxes of the hooks with (i,j) as upper left hand corner, i.e. of the boxes (i',j') such that $i'=i,\ j'\geq j$ or $i'\geq i,\ j'=j$.

In terms of the lengths a_j of the jth column and b_i of the ith row, the hook lengths read

$$h_{ij} = a_j + b_i - i - j + 1. \tag{1.6.72}$$

Since the number of columns of length i of Y_Λ is Λ^i, the lengths of the rows are simply

$$b_i = \sum_{j=i}^{n-1} \Lambda^j, \tag{1.6.73}$$

whereas the length of the columns may be expressed through the Dynkin labels Λ^i as

$$a_i = \max\{j \mid \sum_{k=j}^{n-1} \Lambda^k \geq i\}. \tag{1.6.74}$$

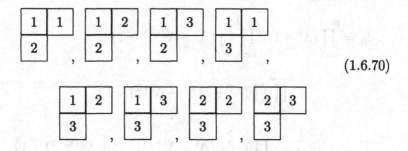

Combining the latter formulae, one sees that the numerator and denominator of the hook formula contain a huge number of equal factors; cancelling (most of) them between numerator and denominator, one arrives at the formula

$$\begin{aligned} d_\Lambda = \prod_{i=1}^{n-1} (\Lambda^i + 1) \cdot \prod_{i=1}^{n-2} \left(\tfrac{1}{2} (\Lambda^i + \Lambda^{i+1}) + 1 \right) \\ \cdot \prod_{i=1}^{n-3} \left(\tfrac{1}{3} (\Lambda^i + \Lambda^{i+1} + \Lambda^{i+2}) + 1 \right) \ldots \\ \ldots \cdot \prod_{i=1}^{2} \left(\tfrac{1}{n-2} (\Lambda^i + \Lambda^{i+1} + \ldots + \Lambda^{i+n-3}) + 1 \right) \\ \cdot \left(\tfrac{1}{n-1} (\Lambda^1 + \Lambda^2 + \ldots + \Lambda^{n-1}) + 1 \right). \end{aligned} \tag{1.6.75}$$

This result for the dimension of R_Λ is in turn nothing but the Weyl dimension formula (1.6.27) for the case of sl_n.

Young tableaux are an important tool in many other explicit calculations as well. A major application is the following simple recipe for the calculation of Kronecker products. The Young tableaux of the irreducible submodules in the Kronecker product $R_\Lambda \times R_{\Lambda'}$ are obtained by gluing the boxes of (say) $Y_{\Lambda'}$ to Y_Λ in all possible ways which are compatible with the symmetry properties (and such that the top left corner of Y_Λ remains the top left corner of the new tableau). If a column of length n is produced in this procedure, it can be discarded because it corresponds to the n-fold antisymmetric tensor which is just the trivial one-dimensional module (in terms of weight vectors, one has $v_{i_1 i_2 \ldots i_n} = \epsilon_{i_1 i_2 \ldots i_n} v$ with $\epsilon_{i_1 i_2 \ldots i_n}$ the Levi-Civita symbol). Let us consider as an example the Kronecker product of two adjoint modules of $A_2 = sl_3$. We mark the three boxes of the Young tableau for one of the modules by the letters a, b and c such that a and b correspond to symmetrized indices, and a and c to antisymmetrized ones. These three boxes now have to be added to the other Young tableau such that a is not in the same column as b (because symmetrized indices must not be antisymmetrized) nor in the same row as c (because antisymmetrized indices must not be symmetrized). Also, a must come "before" b and c (in the natural ordering of boxes, i.e. proceeding row after row from top to bottom, and from left to right within each row); in particular, if a and b are in the same row, then a must be to the left of b, and if a and c are in the same column, then a has to be above c. This recipe leaves six possibilities of gluing the boxes together, leading to the decomposition shown in figure (1.6.76). In that figure the filled dot in the last line stands for the trivial one-dimensional module (the singlet). Also, when proceeding to the second equality the labels a, b

and c have been suppressed, and the irrelevant columns of length three have been discarded.

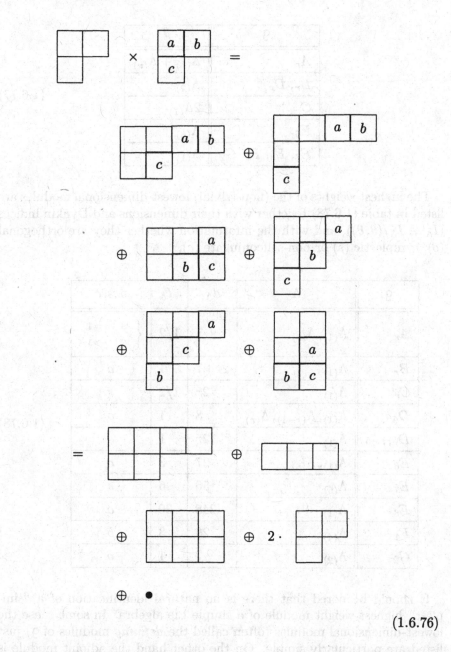

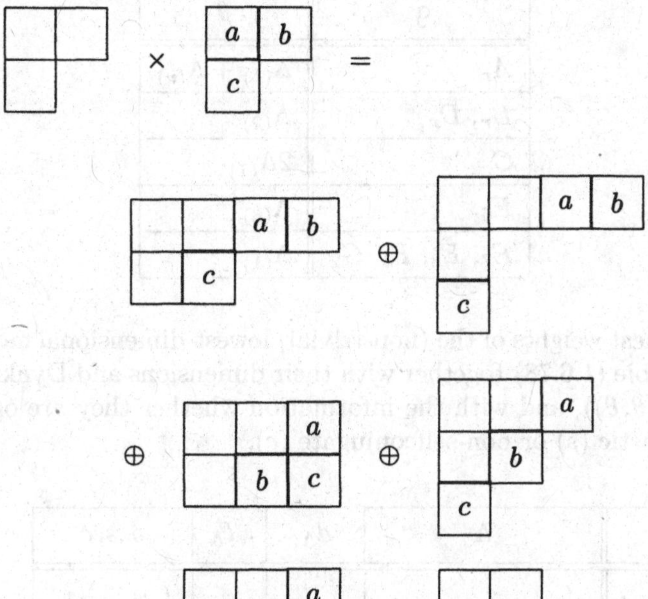

(1.6.76)

j) Finally we list the highest weights of some frequently occurring modules of the simple complex Lie algebras. The highest weight of the adjoint module (i.e. the highest root of the Lie algebra $\mathfrak{g}$) is as displayed in table (1.6.77).

$\mathfrak{g}$	θ
A_r	$\Lambda_{(1)}+\Lambda_{(r)}$
B_r, D_r	$\Lambda_{(2)}$
C_r	$2\Lambda_{(1)}$
E_6	$\Lambda_{(6)}$
E_7, E_8, F_4, G_2	$\Lambda_{(1)}$

(1.6.77)

The highest weights of the (non-trivial) lowest-dimensional modules are listed in table (1.6.78) together with their dimensions and Dynkin indices ($\tilde{I}_\Lambda \equiv I_\Lambda/(\theta,\theta)$), and with the information whether they are orthogonal (o), symplectic (s) or non-selfconjugate (c).

$\mathfrak{g}$	Λ	d_Λ	$\tilde{I}_\Lambda$	o, s, c
A_r	$\Lambda_{(1)}, \Lambda_{(r)}$	$r+1$	$1/2$	$\begin{cases} s & r{=}1 \\ c & r{>}1 \end{cases}$
B_r	$\Lambda_{(1)}$	$2r{+}1$	1	o
C_r	$\Lambda_{(1)}$	$2r$	$1/2$	s
D_4	$\Lambda_{(1)}, \Lambda_{(r-1)}, \Lambda_{(r)}$	8	1	o
$D_r, {\scriptstyle r\geq 5}$	$\Lambda_{(1)}$	$2r$	1	o
E_6	$\Lambda_{(1)}, \Lambda_{(5)}$	27	3	c
E_7	$\Lambda_{(6)}$	56	6	s
E_8	$\Lambda_{(1)}$	248	30	o
F_4	$\Lambda_{(4)}$	26	3	o
G_2	$\Lambda_{(2)}$	7	1	o

(1.6.78)

It should be noted that there is no natural identification of a "simplest" highest weight module of a simple Lie algebra. In some sense the lowest-dimensional modules (often called the *defining* modules of $\mathfrak{g}$) just listed are particularly simple. On the other hand the adjoint module is of course most directly related to the algebra. Finally one might regard those modules as the simplest ones from which all others can be obtained by building Kronecker products. These are sometimes referred to as the *basic* modules of $\mathfrak{g}$; they are the lowest-dimensional (non-trivial) modules in the case of the algebras A_r and C_r and of all exceptional algebras, while for B_r it is the module of highest weight $\Lambda_{(r)}$ (the so-called *spinor* module, which has dimension 2^r); finally, in the D_r case two irreducible modules, with highest weights $\Lambda_{(r)}$ and $\Lambda_{(r-1)}$, respectively, are needed (the *spinor* and *conjugate spinor* modules, each of dimension 2^{r-1}).

We also list the invariant tensors for the modules (1.6.78). For any simple $\mathfrak{g}$, one has the Levi-Civita tensor $\delta_i^{\ j}$ and the completely antisymmetric tensor $\epsilon^{i_1 i_2 \ldots i_{d_\Lambda}}$; in addition, there are the *primitive* invariants (i.e. invariants through which any other invariant can be obtained via contraction with $\delta_i^{\ j}$) shown in table (1.6.79).

$\mathfrak{g}$	invariants
A_r	–
$B_r,\ D_r$	δ_{ij}
C_r	f_{ij}
E_6	d_{ijk}
E_7	$f_{ij},\ d_{ijkl}$
E_8	$\delta_{ab},\ f_{abc},\ ?$
F_4	$\delta_{ij},\ d_{ijk}$
G_2	$\delta_{ij},\ f_{ijk}$

(1.6.79)

In table (1.6.79), $f_{i_1 \ldots i_n}$ stands for completely antisymmetric, and $d_{i_1 \ldots i_n}$ for completely symmetric and traceless tensors. The question mark in the case of E_8 means that there are further primitive invariants which are not yet known explicitly (it is only known that they must be higher than quartic); this lack of knowledge is related to the fact that in the E_8 case the lowest-dimensional module is the adjoint (which we have indicated in the table by using indices $a, b, \ldots$ rather than $i, j, \ldots$). The decomposition (1.3.11) of $t_{ik}^{jl} \equiv \kappa_{ab} R_\Lambda(T^a)_i^{\ j} R_\Lambda(T^b)_k^{\ l}$ (Λ the highest weight of the defining module) in terms of invariant tensors is as shown in table (1.6.80). There we have also listed the quadratic Casimir eigenvalues of the respective modules because these fix the overall normalization (the normalization used here is the "physics" normalization $(\theta, \theta) = 1$). Also, the primitive invariant tensors are normalized such that $t_{ij_1 \ldots j_n} t^{kj_1 \ldots j_n} = \delta_i^{\ k}$

where t stands generically for d or f.

$\mathfrak{g}$	expansion of $t_{ik}^{\ jl}$	C_Λ
A_r	$\frac{1}{2}\left(\frac{1}{r+1}\,\delta_i^{\ j}\delta_k^{\ l}-\delta_i^{\ l}\delta_k^{\ j}\right)$	$\frac{r(r+2)}{2(r+1)}$
B_r	$\frac{1}{2}\left(\delta_{ik}\delta_{jl}-\delta_{il}\delta_{jk}\right)$	r
C_r	$-\frac{1}{4}\left(\delta_i^{\ l}\delta_k^{\ j}+f_{ik}f^{jl}\right)$	$\frac{r}{2}+\frac{1}{4}$
D_r	$\frac{1}{2}\left(\delta_{ik}\delta_{jl}-\delta_{il}\delta_{jk}\right)$	$r-\frac{1}{2}$
E_6	$-\frac{1}{6}\,\delta_i^{\ j}\delta_k^{\ l}-\frac{1}{2}\,\delta_i^{\ l}\delta_k^{\ j}+5\,d_{ikm}d^{jlm}$	$\frac{26}{3}$
E_7	$-\frac{1}{4}\left(\delta_i^{\ l}\delta_k^{\ j}+f_{ik}f^{jl}+d_{ikmn}f^{mj}f^{nl}\right)$	$\frac{57}{4}$
F_4	$\frac{1}{3}\left(\delta_{ik}\delta_{jl}-\delta_{il}\delta_{jk}\right)+\frac{7}{3}\left(d_{ikm}d_{jl}^{\ \ m}-d_{ilm}d_{jk}^{\ \ m}\right)$	6
G_2	$\frac{1}{2}\left(\delta_{ik}\delta_{jl}-\delta_{il}\delta_{jk}\right)-f_{ijm}f_{kl}^{\ \ m}$	2

(1.6.80)

1.7 The Weyl group. Characters

As already mentioned several times, for any $\alpha\in\Phi$ the weights of R_Λ decompose into highest weight modules with respect to the sl_2-subalgebra $\mathfrak{h}_\alpha$ generated by $E^\alpha, E^{-\alpha}$ and H^α. The weights of the adjoint module are just the roots (in addition to the r-fold weight zero). Thus in the adjoint representation, the step operators act as

$$R_\theta(E^\alpha)\cdot v_\beta \propto v_{\alpha+\beta}. \tag{1.7.1}$$

In section 1.4 we have seen that the weights of the $\mathfrak{h}_\alpha$-modules have the form of root strings $S_{\alpha;\beta}$, i.e. for any $\mathfrak{h}_\alpha$-module R there exists a root β such that $R=\{v_\lambda\mid\lambda\in S_{\alpha;\beta}\}$ where

$$S_{\alpha;\beta}=\{\beta+m\alpha\mid m=-n_-,-n_-+1,...,n_+-1,n_+;\ n_\mp\in\mathbb{Z}_{\geq 0}\}. \tag{1.7.2}$$

The root $-\beta$ is contained in the string $S_{\alpha;\beta}$ iff $\beta=\pm\alpha$, as can be seen as follows. Assuming that $-\beta=\beta+m\alpha$, the integer m is obtained by multiplying the formula $-\beta=\frac{1}{2}m\alpha$ by $\alpha^\vee$:

$$-(\beta,\alpha^\vee)=\tfrac{1}{2}\,m\,(\alpha,\alpha^\vee)=m. \tag{1.7.3}$$

This means that

$$-\beta=\beta+(-(\beta,\alpha^\vee)\alpha)=\beta-(\alpha,\beta)\,\alpha^\vee. \tag{1.7.4}$$

But the right hand side of (1.7.4) means geometrically just a reflection of β at the hyperplane (through zero) perpendicular to α; since upon this reflection only vectors parallel to α go into their negative, and since among these vectors only $\pm\alpha$ are roots, the assertion follows.

It turns out that the reflection encountered in this argument is of interest for arbitrary pairs α, β of roots. In particular any such reflection $\sigma_\alpha, \alpha \in \Phi$ takes roots into roots,

$$\sigma_\alpha : \begin{array}{c} \Phi \to \Phi \\ \sigma_\alpha(\beta) = \beta - (\beta, \alpha^\vee)\, \alpha. \end{array} \tag{1.7.5}$$

This means that $(\beta, \alpha^\vee)$ (which is an integer because β is a weight so that the integrality rule (1.6.7) applies) lies between $-n_+$ and n_-. Now as seen above,

$$(\beta, \alpha^\vee) v_\beta \equiv \alpha_i^\vee \beta^i v_\beta = \alpha_i^\vee R_\theta(H^i) \cdot v_\beta = R_\theta(H^\alpha) \cdot v_\beta, \tag{1.7.6}$$

so that the sl_2-representation theory (from which we know that for any weight λ also $-\lambda$ is a weight, and that any two weights differ by an even integer) implies that β and $\sigma_\alpha(\beta)$ are weights of the same h_α-module. Thus $\sigma_\alpha(\beta)$ belongs to the α-string through β and hence is indeed a root.

The operations σ_α can be interpreted geometrically as reflections in the root space which is an r-dimensional euclidean space; thus in particular any reflection possesses an inverse (namely itself), and the composition of reflections leads again to reflections as well as rotations. As a consequence, the maps σ_α, $\alpha \in \Phi(\mathfrak{g})$, generate a discrete group; this group is called the *Weyl group* of $\mathfrak{g}$ and denoted by $W(\mathfrak{g})$. From the geometric interpretation of the reflections, one can also see that the Weyl group is a finite group, and that its elements are automorphisms of Φ, i.e. W is contained in the group $Aut(\Phi)$ of automorphisms of the root system (which in turn is a subgroup of the symmetric group of $d = \dim \mathfrak{g}$ objects). More precisely, W is a normal (i.e. invariant) subgroup of $Aut(\Phi)$ and hence $Aut(\Phi)$ has the structure of a semidirect product

$$Aut(\Phi) = W \ltimes \Gamma; \tag{1.7.7}$$

it turns out that the subgroup Γ appearing here is isomorphic to the group of automorphisms of the Dynkin diagram of $\mathfrak{g}$. Further properties of W are:

- For any basis Φ_s of simple roots and any $\sigma \in W$, the image $\sigma(\Phi_s)$ is again a basis of simple roots. W acts transitively and freely, i.e. any basis of simple roots can be obtained from a given one by applying a suitable element σ of W, and this σ is unique; also, $\sigma(\Phi_s) = \Phi_s$ iff $\sigma = id$.

- More generally, W permutes transitively and freely the *Weyl chambers*. The Weyl chambers are those open simplicial cones which are obtained by deleting all hyperplanes perpendicular to roots from the root space; they are all congruent and are in one to one correspondence to the bases of simple roots. In particular there is one chamber whose points have only positive Dynkin labels; this is called the *fundamental* (or the *dominant*) Weyl chamber.

- For any root α, $W(\alpha)$ spans the whole root space. For any basis Φ_s of simple roots, one generates the whole root system as its image under the Weyl group,

$$W(\Phi_s) = \Phi. \tag{1.7.8}$$

- For any basis of simple roots $\alpha^{(i)}$, W is generated by the *fundamental reflections*

$$\sigma_{(i)} \equiv \sigma_{\alpha^{(i)}}, \tag{1.7.9}$$

for $i = 1, ..., r$. Thus any $\sigma \in W$ can be written as a "word" in the fundamental reflections. Using the relations among the fundamental reflections, a given Weyl reflection may be described by different words; the minimal possible number of "letters" in the word for σ is called the *length* $\ell(\sigma)$ of the Weyl group element. A decomposition of σ consisting of such a minimal number of letters, i.e. $\sigma = \sigma_{(i_1)} \circ \sigma_{(i_2)} \circ \ldots \circ \sigma_{(i_{\ell(\sigma)})}$, is called a *reduced* one.

From the geometric interpretation of the fundamental reflections it is clear that for any element of $\sigma \in W(\mathfrak{g})$ that does not contain a rotation, and hence in particular for any reflection σ_α, $\alpha \in \Phi(\mathfrak{g})$, the length $\ell(\sigma)$ must be odd.

- The length ℓ obeys $\ell(\sigma) = \ell(\sigma^{-1})$, as well as

$$\ell(\sigma) + \ell(\sigma') \geq \ell(\sigma' \circ \sigma) \geq |\,\ell(\sigma^{-1}) - \ell(\sigma')\,| \tag{1.7.10}$$

for any two Weyl group elements σ, σ'. Also, the length is bounded from above, and the "longest" element is unique, i.e. there exists a $\sigma_{\max} \in W$ such that

$$\ell(\sigma_{\max}) > \ell(\sigma) \quad \text{for all} \quad \sigma \in W \setminus \{\sigma_{\max}\}. \tag{1.7.11}$$

- Another distinguished element of W is the *Coxeter element* σ_c defined as

$$\sigma_c := \sigma_{(1)} \circ \sigma_{(2)} \circ \ldots \circ \sigma_{(r)}. \tag{1.7.12}$$

The order of σ_c, i.e. the lowest integer n such that $\underbrace{\sigma_c \circ \sigma_c \circ \ldots \circ \sigma_c}_{n \text{ times}} = id$, is equal to the Coxeter number of $\mathfrak{g}$.

- The notion of length can be used to define a partial ordering on the Weyl group, the so-called *Bruhat ordering.* This is done by writing "$\sigma_1 \leftarrow \sigma_2$" iff both $\sigma_1 = \sigma_\alpha \circ \sigma_2$ for some positive root α and $\ell(\sigma_1) = \ell(\sigma_2) + 1$; the Bruhat ordering "$\prec$" is then given by saying that $\sigma \prec \sigma'$ iff there exist Weyl group elements $\sigma_1, \sigma_2, ..., \sigma_n$ such that $\sigma \leftarrow \sigma_1 \leftarrow \sigma_2 \ldots \leftarrow \sigma_n \leftarrow \sigma'$.

- The reflection with respect to a simple root $\alpha^{(i)}$ sends this root into its negative and permutes the rest of the positive roots,

$$\sigma_{(i)}(\Phi_+ \backslash \{\alpha^{(i)}\}) = \Phi_+ \setminus \{\alpha^{(i)}\}; \tag{1.7.13}$$

as a consequence, this reflection subtracts $\alpha^{(i)}$ from the Weyl vector,

$$\sigma_{(i)}(\rho) = \rho - \alpha^{(i)}. \tag{1.7.14}$$

On an arbitrary weight λ, the fundamental reflections act as

$$\left(\sigma_{(i)}(\lambda)\right)^j = \lambda^j - A^{ij}\,\lambda^i. \tag{1.7.15}$$

- W not only permutes the roots, i.e. the weights of the adjoint module, but also the weights of any other highest weight module, i.e. for any $\alpha \in \Phi$,

$$\sigma_\alpha : \begin{array}{c} \Phi_\Lambda \to \Phi_\Lambda \\ \lambda \mapsto \sigma_\alpha(\lambda) := \lambda - (\lambda, \alpha^\vee)\,\alpha \end{array} \tag{1.7.16}$$

is an automorphism of the weight system Φ_Λ of R_Λ. This is just a consequence of the fact that σ_α takes any basis of simple roots to another such basis, and hence reflects the arbitrariness of the choice of basis of simple roots. In particular, weights of a module which transform into each other under a Weyl reflection have the same multiplicity.

The longest element $\sigma_{\max} \in W$ takes any highest weight into minus the highest weight of the conjugate module,

$$\sigma_{\max}(\Lambda) = -\Lambda^+. \tag{1.7.17}$$

- Scalar products are invariant under W:

$$(\sigma(\lambda), \sigma(\mu)) = (\lambda, \mu) \tag{1.7.18}$$

for any $\sigma \in W$.

- A modified action of $\sigma \in W$ can be defined via

$$\sigma \star \lambda := \sigma(\lambda + \rho) - \rho, \tag{1.7.19}$$

with ρ the Weyl vector of $\mathfrak{g}$. Because of

$$\begin{aligned}\sigma_1 \star (\sigma_2 \star \lambda) &= \sigma_1 \star (\sigma_2(\lambda+\rho)-\rho) \\ &= \sigma_1(\sigma_2(\lambda+\rho))-\rho = (\sigma_1 \circ \sigma_2) \star \lambda,\end{aligned} \tag{1.7.20}$$

the operations "$\star$" form a group which is isomorphic to W. As an example, consider the modified action for the reflection with respect to the highest root. One has

$$\sigma_\theta(\lambda) = \lambda - \left(\sum_{i=1}^{r} a_i^\vee \lambda^i \right) \theta, \tag{1.7.21}$$

and therefore (recalling that $\rho^i = 1$ for all $i = 1, ..., r$)

$$\sigma_\theta \star \lambda = \sigma_\theta(\lambda) + (1 - g^\vee)\,\theta. \tag{1.7.22}$$

- The points of the weight space that stay fixed under a Weyl reflection are of course the points in the hyperplanes perpendicular to the relevant root. This includes in particular the weights on the boundary of the fundamental Weyl chamber, or in other words those weights λ for which at least one Dynkin label vanishes: one has $\sigma_{(i)}(\lambda) = \lambda$ if $\lambda^i = 0$. Similarly, for the modified action of the Weyl group one obtains

$$\sigma_{(i)} \star \lambda = \lambda \quad \text{if} \quad \lambda^i = -1. \tag{1.7.23}$$

- Finally we list in table (1.7.24) the number $|W|$ of elements of the Weyl groups W of the simple Lie algebras, and also the structure of W except for the E-series where the structure is rather complicated.

$\mathfrak{g}$	$\lvert W\rvert$	W
A_r	$(r+1)!$	S_{r+1}
B_r, C_r	$2^r.r!$	$S_r \ltimes (\mathbb{Z}_2)^r$
D_r	$2^{r-1}.r!$	$S_r \ltimes (\mathbb{Z}_2)^{r-1}$
E_6	$2^7.3^4.5$	
E_7	$2^{10}.3^4.5.7$	
E_8	$2^{14}.3^5.5^2.7$	
F_4	$2^7.3^2$	$S_3 \ltimes S_4 \ltimes (\mathbb{Z}_2)^3$
G_2	$2^2.3$	$\mathcal{D}_6$

(1.7.24)

In (1.7.24), S_n denotes the symmetric group of n objects, and $\mathcal{D}_6$ the dihedral group. For the classical algebras, the structure of the Weyl group becomes particularly evident in the orthonormal basis that was used in the description (1.6.78) of the root system: for A_r, the simple reflection $\sigma_{(i)}$ acts as $\xi_i \leftrightarrow \xi_{i+1}$ on the coordinates of a weight in this basis, and as a consequence any Weyl group element acts as a permutation of these coordinates, so that indeed $W(A_r) \cong \mathcal{S}_{r+1}$; for the other classical algebras, the Weyl group induces again permutations of the coordinates in the orthonormal basis, but in addition also sign changes $\xi_i \mapsto -\xi_i$ (in the case of D_r, the number of sign changes is restricted to be even).

As examples, let us describe the Weyl groups of A_1 and A_2. For A_1, there is a single generator σ acting on weights $\lambda \in \mathbb{Z}$ as multiplication by -1,

$$\sigma(\lambda) = -\lambda. \tag{1.7.25}$$

The Weyl group is therefore isomorphic to $\mathcal{S}_2 = \mathbb{Z}_2$. Also, because of $\rho = 1$ the modified action of σ reads

$$\sigma \star \lambda \equiv \sigma(\lambda+1) - 1 = -\lambda - 2. \tag{1.7.26}$$

For A_2, the Weyl group is already considerably more complicated. It is generated by the simple reflections which act on weights $\lambda = (\lambda^1, \lambda^2)$ as

$$\begin{aligned} \sigma_{(1)}(\lambda^1,\lambda^2) &= (-\lambda^1, \lambda^1+\lambda^2), \\ \sigma_{(2)}(\lambda^1,\lambda^2) &= (\lambda^1+\lambda^2, -\lambda^2). \end{aligned} \tag{1.7.27}$$

It follows that

$$W(A_2) = \{1, \sigma_{(1)}, \sigma_{(2)}, \sigma_{(1)}\circ\sigma_{(2)}, \sigma_{(2)}\circ\sigma_{(1)}, \sigma_{(1)}\circ\sigma_{(2)}\circ\sigma_{(1)}\}, \tag{1.7.28}$$

and that

$$\sigma_\theta = \sigma_{(1)}\circ\sigma_{(2)}\circ\sigma_{(1)} = \sigma_{(2)}\circ\sigma_{(1)}\circ\sigma_{(2)}. \tag{1.7.29}$$

The latter reflection acts as

$$\sigma_\theta(\lambda^1,\lambda^2) = (-\lambda^2, -\lambda^1). \tag{1.7.30}$$

From $\rho = (1,1)$ it follows that $\sigma \star (\lambda^1,\lambda^2) = \sigma(\lambda^1+1, \lambda^2+1) - (1,1)$. In particular,

$$\begin{aligned} \sigma_{(1)} \star (\lambda^1,\lambda^2) &= (-\lambda^1-2, \lambda^1+\lambda^2+1), \\ \sigma_{(2)} \star (\lambda^1,\lambda^2) &= (\lambda^1+\lambda^2+1, -\lambda^2-2), \\ \sigma_\theta \star (\lambda^1,\lambda^2) &= (-\lambda^2-2, -\lambda^1-2). \end{aligned} \tag{1.7.31}$$

The action (1.7.14) of the simple reflections on the Weyl vector and the invariance (1.7.18) of scalar products are the key to the derivation of the identity (1.4.72) for the Weyl vector. Namely, observe that $\sigma_{(i)}(\alpha^{(j)\vee}) =$

$\alpha^{(j)\vee} - A^{ij}\alpha^{(i)\vee}$. Considering (1.7.18) for the special case $\mu = \rho,\ \lambda = \alpha^{(j)\vee}$ and inserting (1.7.14), one thus gets

$$\left(1 - (\alpha^{(i)\vee}, \rho)\right) A^{ij} = 0. \tag{1.7.32}$$

This must be true for arbitrary values of i and j (of course, taking $i = j$ simplifies the argument), and hence (1.4.72) follows as promised.

As another application, we can now complete the proof of the classification theorem of simple Lie algebras. Namely, we can show that the (irreducible) Cartan matrices are in one to one correspondence with the simple Lie algebras (up to simultaneous permutations of rows and columns). Indeed, as W acts transitively on the bases and respects scalar products, the definition (1.4.18) of the Cartan matrix shows that it is independent of the choice of basis up to a renumbering of the simple roots.

An even more important application of the Weyl group is in the calculation of the *characters* of a simple Lie algebra $\mathfrak{g}$. For any highest weight module R_Λ of $\mathfrak{g}$, the character $\chi_\Lambda \equiv \chi_{R_\Lambda}$ is by definition the map

$$\chi_\Lambda : \begin{array}{l} \mathfrak{g}_0 \to F \\ h \mapsto \chi_\Lambda(h) := \operatorname{tr}\exp(R_\Lambda(h)). \end{array} \tag{1.7.33}$$

Here $\mathfrak{g}_0$ is a Cartan subalgebra of $\mathfrak{g}$ (recall that $\mathfrak{g}_0$ can be identified with the weight space, i.e. instead of $h \in \mathfrak{g}_0$ we could also consider weights μ as arguments of χ_Λ); also, tr denotes the trace of matrices, and exp is defined by the exponential power series.

Since $R_\Lambda(0)$ is a $d_\Lambda \times d_\Lambda$-matrix with all entries equal to zero, the character evaluated on the zero weight is just the dimension of R_Λ,

$$\chi_\Lambda(0) = d_\Lambda. \tag{1.7.34}$$

According to (1.6.6) the generator h acts as $R_\Lambda(h) \cdot v_\lambda = (\lambda, h)\, v_\lambda$ for all $v_\lambda \in R_\Lambda$. Thus (1.7.33) can be rewritten as

$$\chi_\Lambda(h) = \sum_\lambda \operatorname{mult}_\Lambda(\lambda) \exp((\lambda, h)) \tag{1.7.35}$$

where the sum goes over all weights of R_Λ. By an analogous formula, the character can then also be defined for direct sums of highest weight modules. Similarly, the character of a quotient module is obtained by subtracting the character of the submodule which is divided out from the character of the original module. Finally, since the weights of conjugate modules are the negative of each other, one has

$$\chi_{\Lambda^+}(\mu) = \chi_\Lambda(-\mu); \tag{1.7.36}$$

in particular, characters of selfconjugate modules are invariant under $\mu \mapsto -\mu$.

Using the behavior of the weights λ under Weyl reflections, it is then possible to derive the *Weyl character formula*

$$\chi_\Lambda(h) = \frac{\sum_{\sigma\in W}(\mathrm{sign}\,\sigma)\,\exp[(\sigma(\Lambda+\rho),h)]}{\sum_{\sigma\in W}(\mathrm{sign}\,\sigma)\,\exp[(\sigma(\rho),h)]}. \tag{1.7.37}$$

Here ρ is the Weyl vector, the summation goes over the Weyl group, and

$$\mathrm{sign}\,\sigma = (-1)^{\ell(\sigma)} \tag{1.7.38}$$

where $\ell(\sigma)$ is the length of σ, i.e. $\mathrm{sign}\,\sigma = 1, -1$ if σ can be written as the composition of an even and odd number, respectively, of fundamental reflections. The map sign defined this way is the unique homomorphism of W onto $\{\pm 1\} \cong \mathbb{Z}_2$.

The denominator of (1.7.37) can also be written in the form

$$\sum_{\sigma\in W}(\mathrm{sign}\,\sigma)\,\exp[(\sigma(\rho),h)] = \prod_{\alpha>0}\left[\exp(\tfrac{1}{2}(\alpha,h)) - \exp(-\tfrac{1}{2}(\alpha,h))\right]$$

$$\equiv \exp((\rho,h))\prod_{\alpha>0}[1-\exp(-(\alpha,h))]\,. \tag{1.7.39}$$

This is known as the *denominator identity*. The character formula and denominator identity are in fact the source of the Weyl dimension formula: using the character formula for $h = \xi\rho$ with $\xi \in \mathbb{C}$ and the invariance of the inner product under the Weyl group, one has

$$\chi_\Lambda(\xi\rho) = \frac{\sum_{\sigma\in W}(\mathrm{sign}\,\sigma)\,\exp[\xi(\Lambda+\rho,\sigma(\rho))]}{\sum_{\sigma\in W}(\mathrm{sign}\,\sigma)\,\exp[\xi(\rho,\sigma(\rho))]}. \tag{1.7.40}$$

With the help of the denominator identity, it then follows that

$$\chi_\Lambda(\xi\rho) = \prod_{\alpha>0}\frac{\exp[\frac{1}{2}\,\xi(\Lambda+\rho,\alpha)] - \exp[-\frac{1}{2}\,\xi(\Lambda+\rho,\alpha)]}{\exp[\frac{1}{2}\xi(\rho,\alpha)] - \exp[-\frac{1}{2}\xi(\rho,\alpha)]}; \tag{1.7.41}$$

in the limit $\xi \to 0$, this leads to the Weyl dimension formula (1.6.27).

As an example, consider the simplest case $\mathfrak{g} = A_1$ so that $\mathfrak{g}_0$ is one-dimensional, i.e. $h = \xi\,J_3$ for some $\xi \in F$; inserting the weight space decomposition (1.6.4) into (1.7.35), the characters become

$$\chi_\Lambda(h) = \sum_{n=0}^{\Lambda} q^{\frac{1}{2}\Lambda-n} = \lfloor\Lambda+1\rfloor_q. \tag{1.7.42}$$

Here q is defined as $q = \exp(\xi)$, and for any complex number μ we introduced the number

$$\lfloor\mu\rfloor \equiv \lfloor\mu\rfloor_q := \frac{q^{\mu/2}-q^{-\mu/2}}{q^{1/2}-q^{-1/2}}, \tag{1.7.43}$$

which because of $\lfloor\mu\rfloor_1 = \mu$ is called a *deformation* of μ, or shortly the *q-number* associated to μ. The Weyl group in this example is $W(A_1) = \{id, \sigma\}$ with σ acting as $\lambda \mapsto -\lambda$, i.e. $W \cong \{\pm 1\}$. Also, $(\Lambda, H) = \Lambda$ and $(\rho, H) = 1$, i.e. in terms of the Weyl group, the first and second term in the numerator or denominator of (1.7.43) are due to the identity map and to the reflection σ, respectively.

When the formula (1.7.35) for the characters is applied to direct sums of irreducible modules, one obtains

$$\chi_{\oplus_i R_{\Lambda_i}} = \sum_i \chi_{\Lambda_i}. \tag{1.7.44}$$

Similarly, the characters of Kronecker products of irreducible modules obey

$$\chi_{R_\Lambda \times R_{\Lambda'}} = \chi_\Lambda \cdot \chi_{\Lambda'}. \tag{1.7.45}$$

From the last two formulae one can learn a lot about the decomposition of a Kronecker product into a direct sum of irreducible modules. Using the Weyl character formula, it is in fact possible to obtain a rather explicit prescription for the decomposition.

Namely, for $\Lambda \times \Lambda' = \bigoplus_i \mathcal{L}_i \Lambda_i$ one has the character sum rule

$$\chi_\Lambda \cdot \chi_{\Lambda'} = \sum_i \mathcal{L}_i \, \chi_{\Lambda_i}. \tag{1.7.46}$$

Expressing χ_Λ and the χ_{Λ_i} through the Weyl formula, this implies

$$\begin{aligned} &\sum_{\sigma\in W} (\operatorname{sign}\sigma) \exp(\sigma(\Lambda+\rho))\, \chi_{\Lambda'} \\ &\qquad = \sum_{\sigma\in W} (\operatorname{sign}\sigma) \sum_i \mathcal{L}_i \exp(\sigma(\Lambda_i+\rho)). \end{aligned} \tag{1.7.47}$$

Combining this with $\chi_{\Lambda'} = \sum \mathrm{mult}_{\Lambda'}(\lambda') \exp(\lambda')$, a comparison of both sides of the equation would immediately lead to the conclusion that

$$\mathcal{L}_{\Lambda\Lambda'}^{\;\Lambda_i} = \mathrm{mult}_{\Lambda'}(\Lambda_i - \Lambda), \tag{1.7.48}$$

provided that it were allowed to forget about the sum over the Weyl group (here we use the notation $\mathcal{L}_{\Lambda\Lambda'}^{\;\Lambda_i}$ instead of $\mathcal{L}_i$ in order to express explicitly how these coefficients depend on all three highest weights involved). Now since the Weyl group permutes the Weyl chambers, it is clear that the equality (1.7.47) must still hold when restricted to any fixed Weyl chamber. As a consequence, if all the weights

$$\lambda'' := \Lambda + \rho + \lambda' \tag{1.7.49}$$

for λ' a weight of $R_{\Lambda'}$ are contained in the fundamental Weyl chamber, then one can indeed remove the summation over the Weyl group, because then on both sides of the equation only the term with the identity element of the Weyl group can give contributions lying in the fundamental chamber. In this case, (1.7.48) is indeed the correct result for the Littlewood–Richardson coefficients. In contrast, if some weight λ'' lies in a chamber other than the fundamental one, then on the left hand side of the equation there is a further contribution lying in the fundamental chamber, because owing to the Weyl invariance of the weight system of $R_{\Lambda'}$, there exists a Weyl group element σ such that $\sigma(\lambda'')$ lies in the fundamental chamber. Owing to $\mathrm{mult}_{\Lambda'}(\lambda') = \mathrm{mult}_{\Lambda'}(\sigma(\lambda'))$, this contribution is simply given by

$$(\mathrm{sign}\,\sigma)\,\mathrm{mult}_{\Lambda'}(\lambda')\,\exp(\sigma(\Lambda+\rho+\lambda')). \tag{1.7.50}$$

This results in a contribution to $\mathcal{L}_{\Lambda\Lambda'}^{\;\Lambda_i}$ with Λ_i such that $\sigma(\Lambda+\rho+\lambda') = \Lambda_i+\rho$, i.e.

$$\lambda' = \sigma \star \Lambda_i - \Lambda. \tag{1.7.51}$$

As a consequence, the multiplicities (1.7.48) get modified as follows:

$$\mathcal{L}_{\Lambda\Lambda'}^{\;\Lambda_i} = \sum_{\sigma\in W}{}' (\mathrm{sign}\,\sigma)\,\mathrm{mult}_{\Lambda'}(\sigma \star \Lambda_i - \Lambda). \tag{1.7.52}$$

Here the sum $\sum'$ is only over those Weyl group elements for which $\sigma \star \Lambda_i - \Lambda$ is a weight of $R_{\Lambda'}$ (alternatively, the sum goes over the whole Weyl group if one uses the convention that $\mathrm{mult}_\Lambda(\mu) = 0$ if μ is not a weight of R_Λ).

This formula is already very explicit. It becomes even more transparent if one observes that it translates into the following recipe, sometimes called the *Racah–Speiser algorithm*: Given Λ and the weight system $\{\lambda'\}$ of $R_{\Lambda'}$, consider the weights $\{\lambda''\} = \{\lambda' + \Lambda + \rho\}$. Any such weight lying on the boundary of the fundamental Weyl chamber can be ignored because upon a fundamental reflection which has $\mathrm{sign}\,\sigma = -1$ it gets mapped to itself so that the sum in (1.7.52) contains two terms which are equal up to sign. For any weight λ'' in a Weyl chamber other than the fundamental one, take the (unique) Weyl reflection σ which maps it to the fundamental chamber and add $(\mathrm{sign}\,\sigma)\cdot\mathrm{mult}(\lambda'')$ to $\mathrm{mult}(\sigma(\lambda''))$. After this has been done for all λ'', the resulting multiplicities of strictly dominant weights μ are the Littlewood-Richardson coefficients $\mathcal{L}_{\Lambda\Lambda'}^{\;\mu-\rho}$.

This algorithm is simple and efficient. The method is best illustrated by working out some non-trivial examples. Let us also mention that in section 4.5 an extension of the algorithm will be considered which will be relevant to Kronecker products of certain quantum group modules.

As an illustration, we give the following examples. First, consider the product of two adjoint modules of $\mathfrak{g} = A_2$, which has already been analysed, with the help of Young tableaux, in (1.6.76). The weights of this module have been displayed in (1.6.14). Adding to them the highest weight $\theta = (1,1)$, one obtains the following weights: (2,2), (0,3), (3,0), twice (1,1), (0,0), $(2,-1)$, and $(-1,2)$. The first six of them lie in the closure of the fundamental Weyl chamber and hence are left untouched. In contrast, the last two weights lie outside the fundamental Weyl chamber; they are mapped onto themselves by the modified action of $\sigma_{(1)}$ and $\sigma_{(2)}$, respectively; hence they have to be discarded. The tensor product decomposition then reads

$$R_\theta \times R_\theta = R_{2\theta} \oplus R_{3\Lambda_{(1)}} \oplus R_{3\Lambda_{(2)}} + 2 \cdot R_\theta + R_0, \qquad (1.7.53)$$

or, in terms of dimensionalities,

$$8 \times 8 = 27 + 10 + 10^+ + 2 \cdot 8 + 1. \qquad (1.7.54)$$

As a less trivial example, consider, still for $\mathfrak{g} = A_2$, the Kronecker product between the adjoint module and the six-dimensional symmetric tensor module $R_{2\Lambda_{(1)}}$. Adding the highest weight (2,0) to the weights of the adjoint module, one obtains five modules that lie in the closure of the fundamental Weyl chamber, namely (3,1), (1,2), (0,1), and twice (2,0), two weights outside the fundamental chamber which are mapped onto themselves by the modified action of $\sigma_{(2)}$, namely $(4,-1)$ and $(1,-1)$, and in addition the weight $(3,-2)$ which obeys $\sigma_{(1)} \star (3,-2) = (2,0)$. Thus the weights $(4,-1)$ and $(1,-1)$ have to be discarded, while the weight $(3,-2)$ cancels one of the two weights (2,0). The resulting tensor product decomposition is thus in agreement with (1.6.76),

$$R_\theta \times R_{2\Lambda_{(1)}} = R_{3\Lambda_{(1)}+\Lambda_{(2)}} \oplus R_{\Lambda_{(1)}+2\Lambda_{(2)}} \oplus R_{2\Lambda_{(1)}} \oplus R_{\Lambda_{(2)}}, \qquad (1.7.55)$$

or, in terms of dimensionalities,

$$8 \times 6 = 24 + 15 + 6^+ + 3. \qquad (1.7.56)$$

Next consider the tensor product of the module $R_{\Lambda_{(1)}}$ of E_6 with its conjugate module $R_{\Lambda_{(5)}}$. According to the weight diagram (1.6.16) there are now 27 weights to be considered. However, the calculation is again very simple: adding the highest weight (0,0,0,0,1,0) to the weights given in (1.6.16) one obtains only three weights that lie in the closure of the fundamental Weyl chamber, namely (1,0,0,0,1,0), (0,0,0,0,0,1), and (0,0,0,0,0,0), while for any other weight at least one Dynkin label λ^i equals -1 so that it is mapped onto itself by the modified action of $\sigma_{(i)}$. Thus one is left with the three weights in the closure of the fundamental chamber, so that

$$R_{\Lambda_{(1)}} \times R_{\Lambda_{(5)}} = R_{\Lambda_{(1)}+\Lambda_{(5)}} \oplus R_{\Lambda_{(6)}} \oplus R_0. \qquad (1.7.57)$$

In terms of dimensionalities, this reads

$$27 \times 27^+ = 650 + 78 + 1. \qquad (1.7.58)$$

Finally take the tensor product of two adjoint modules $R_\theta = R_{\Lambda_{(1)}}$ of E_8. Of the 248 weights of R_θ, there are eleven weights λ such that $\lambda+\theta$ lies in the closure of the fundamental Weyl chamber, namely $\lambda+\theta = (2,0,0,0,0,0,0,0)$, (0,1,0,0,0,0,0,0), (0,0,0,0,0,0,1,0), the zero weight, and eight times the weight θ. For 230 weights, at least one Dynkin label of $\lambda+\theta$ equals -1 so that they cancel against themselves. Finally, there are seven other weights, namely $\lambda+\theta = \theta - \alpha^{(j)}$ for $j = 2,3,\ldots,8$; for these one has $\sigma_{(j)} \star (\theta - \alpha^{(j)}) = \theta$, and hence they cancel all but one of the weights θ. Thus one arrives at the decomposition

$$R_\theta \times R_\theta = R_{2\Lambda_{(1)}} \oplus R_{\Lambda_{(2)}} \oplus R_{\Lambda_{(7)}} \oplus R_\theta \oplus R_0. \tag{1.7.59}$$

Again for convenience we also list the corresponding dimensionalities:

$$248 \times 248 = 27\,000 + 30\,380 + 3\,875 + 248 + 1. \tag{1.7.60}$$

1.8 Branching rules

A problem which in applications appears quite often is to express modules of a Lie algebra $\mathfrak{g}$ in terms of a Lie subalgebra $\mathfrak{h} \subset \mathfrak{g}$. When dealing with this problem, it is sufficient to consider only *maximal subalgebras* $\mathfrak{h}$, i.e. subalgebras such that there does not exist any subalgebra $\mathfrak{k}$ obeying $\mathfrak{h} \subset \mathfrak{k} \subset \mathfrak{g}$. Non-maximal subalgebras can then be dealt with by performing the computations in a series of steps, first considering maximal subalgebras, then their maximal subalgebras, and so on. We will be interested in the case where the Lie algebra $\mathfrak{g}$ is simple and its subalgebra $\mathfrak{h}$ semisimple or possibly reductive (of course, $\mathfrak{g}$ contains many more non-reductive subalgebras, such as its Borel subalgebras).

A subalgebra $\mathfrak{h}$ of a simple Lie algebra $\mathfrak{g}$ is called a *regular* subalgebra iff the roots of $\mathfrak{h}$ are contained in the root system of $\mathfrak{g}$; otherwise $\mathfrak{h}$ is called a *special* subalgebra. More generally, subalgebras which are contained in some regular subalgebra are called *R-subalgebras* while all others are called *S-subalgebras*. Thus in order to classify all semisimple subalgebras of simple Lie algebras, one has to look for the maximal regular subalgebras and for the maximal S-subalgebras.

The root system and simple root system of the subalgebra $\mathfrak{h} \subset \mathfrak{g}$ will be denoted by $\tilde{\Phi}$ and $\tilde{\Phi}_s$, respectively. According to one of the main properties of simple roots, $\tilde{\alpha}, \tilde{\beta} \in \tilde{\Phi}_s$ implies $\tilde{\alpha} - \tilde{\beta} \notin \tilde{\Phi}$. If $\mathfrak{h}$ is a regular subalgebra, this also means that $\tilde{\alpha} - \tilde{\beta} \notin \Phi$ (the root system of $\mathfrak{g}$), because otherwise $E^{\tilde{\alpha}-\tilde{\beta}} \propto [E^{\tilde{\alpha}}, E^{-\tilde{\beta}}]$ would lie in $\mathfrak{h}$, in contradiction to $\tilde{\alpha} - \tilde{\beta} \notin \tilde{\Phi}$. As a consequence, the regular subalgebras of $\mathfrak{g}$ are in one to one correspondence to the subsets $\tilde{\Phi}_s \subset \Phi$, which obey

$$\tilde{\alpha}, \tilde{\beta} \in \tilde{\Phi}_s \subset \Phi \implies \tilde{\alpha} - \tilde{\beta} \notin \Phi. \tag{1.8.1}$$

Given such a subset, the regular subalgebra is spanned by the generators $\{E^{\tilde{\alpha}}, E^{-\tilde{\alpha}}, H^{\tilde{\alpha}} \mid \tilde{\alpha} \in \tilde{\Phi}_s\}$.

There exists a simple algorithm to find all sets $\tilde{\Phi}$ with the required property. Namely, all maximal regular semisimple subalgebras can be obtained by choosing

$$\tilde{\Phi}_s \subset \Phi_s \cup \{-\theta\}. \tag{1.8.2}$$

More precisely, there are no other maximal regular semisimple subalgebras besides the ones with simple root systems $\tilde{\Phi}_s = \Phi_s \cup \{-\theta\} \setminus \{\alpha^{(i)}\}$, $i = 1, ..., r$, and conversely, with very few exceptions each such choice does yield such a subalgebra (the exceptions are encountered when removing the following simple roots from $\Phi_s \cup \{-\theta\}$: the root $\alpha^{(3)}$ of F_4 or of E_7, or the roots $\alpha^{(2)}$, $\alpha^{(3)}$ or $\alpha^{(5)}$ of E_8). However, in the case of $\mathfrak{g} = A_r$, the subalgebra obtained this way is just A_r itself; as a consequence, for A_r the relevant subalgebras are precisely the ones with simple roots $\tilde{\Phi}_s = \Phi_s \cup \{-\theta\} \setminus \{\alpha^{(i)}, \alpha^{(j)}\}$, $i, j = 1, ..., r$, $i \neq j$. These subalgebras are maximal among the semisimple subalgebras, but not among all subalgebras of A_r.

That this prescription always yields a regular subalgebra is easily understood: adding minus the highest root to Φ_s preserves the property that the difference of any two elements is not a root, and by removing one of the simple roots one then recovers a linearly independent set of roots. There are also some non-semisimple maximal regular subalgebras: they are obtained by removing a single generator $E^{\alpha^{(i)}}$ from a Chevalley–Serre basis of $\mathfrak{g}$, and obviously contain the semisimple subalgebra that is obtained by removing $E^{-\alpha^{(i)}}$ and $H^{\alpha^{(i)}}$ as well.

The rule just given is best illustrated in terms of Dynkin diagrams. By adding a node corresponding to $-\theta$ to the Dynkin diagram, one gets the so-called *extended Dynkin diagram* of $\mathfrak{g}$ (this coincides with the Dynkin diagram of the untwisted affine Lie algebra $\mathfrak{g}^{(1)}$ associated to $\mathfrak{g}$, and hence is displayed in table (2.1.12) below). For $\mathfrak{g} \neq A_r$, the Dynkin diagram of the subalgebra $\mathfrak{h}$ is then obtained from the extended Dynkin diagram of $\mathfrak{g}$ by removing the node corresponding to the simple root $\alpha^{(i)}$. As an example, take $\mathfrak{g} = E_6$; then for $i = 2, 4$, or 6, one obtains the embedding $A_1 \oplus A_5 \subset E_6$, while the choice $i = 3$ yields $A_2 \oplus A_2 \oplus A_2 \subset E_6$ (and $i = 1$ or 5 the trivial embedding $E_6 \subseteq E_6$). For $g = A_r$ the removal of a single node from the extended Dynkin diagram just gives back the original Dynkin diagram; this explains why one has to discard two simple roots from $\Phi_s \cup \{-\theta\}$ in order to arrive at a (proper) maximal regular semisimple subalgebra; inspection of the Dynkin diagram immediately shows that these subalgebras are either $\mathfrak{h} = A_{r-1}$ or $\mathfrak{h} = A_s \oplus A_{r-s-1}$

for $s = 1, 2, ..., [\frac{r-1}{2}]$. It is also easily seen that the latter subalgebras are maximal only among the semisimple subalgebras; namely, by removing the generator $E^{\alpha^{(i)}}$ from a Chevalley basis rather than the root $\alpha^{(i)}$ from $\Phi \cup \{-\theta\}$, one gets the non-semisimple subalgebra $\mathsf{h} \oplus u_1 \subset A_r$ with h as given before.

To find also the maximal S-subalgebras, a more complicated procedure must be followed. For embeddings into the classical algebras, one can employ the realization of the corresponding compact Lie groups as groups of matrices to deduce that to any n-dimensional $\mathfrak{g}$-module R one can associate an embedding of $\mathfrak{g}$ into so_n, sp_n or su_n if R is orthogonal, symplectic or complex, respectively. These embeddings are special (if they are proper), and up to a few known exceptions they are maximal. As an example, take the defining module of E_6; this is 27-dimensional and complex (see table (1.6.78)), and accordingly there is a maximal special embedding $E_6 \subset A_{26}$. As an example for an orthogonal module, take the adjoint module of any simple $\mathfrak{g}$; this gives rise to the maximal special embedding

$$\mathfrak{g} \subset so_{\dim \mathfrak{g}}. \tag{1.8.3}$$

The method also works for modules which are not irreducible, but then it does not yield maximal embeddings. For example, the defining module $R = R_{\Lambda_{(1)}}$ of $su_n \cong A_{n-1}$ is complex and hence corresponds to the trivial embedding $su_n \subseteq su_n$, but the reducible module $R \oplus R^+ = R_{\Lambda_{(1)}} \oplus R_{\Lambda_{(n-1)}}$ is orthogonal, yielding the embedding $su_n \subset so_{2n}$; this is however not maximal, but rather contained in the maximal regular non-semisimple embedding

$$su_n \oplus u_1 \subset so_{2n}. \tag{1.8.4}$$

All S-subalgebras obtained this way are simple. There are also a few non-simple S-subalgebras of the classical algebras, and again they can be understood in terms of their realization by matrices; one finds the following non-simple embeddings:

$$\begin{gathered} su_m \oplus su_n \subset su_{mn}, \\ so_m \oplus so_n \subset so_{mn}, \\ sp_m \oplus sp_n \subset so_{4mn}, \\ so_m \oplus sp_n \subset sp_{mn}, \\ B_r \oplus B_s \subset D_{r+s+1}. \end{gathered} \tag{1.8.5}$$

The last embedding is a special case of the more general series

$$so_m \oplus so_n \subset so_{m+n}, \tag{1.8.6}$$

which except for odd m and n describes a regular embedding.

The method just described provides all S-subalgebras of the classical Lie algebras. The S-subalgebras of the exceptional simple Lie algebras require a case-by-case analysis which yields the following result:

$\mathfrak{g}$	maximal S-subalgebras of $\mathfrak{g}$
E_6	$A_1^{[9]},\, C_4^{[1]},\, F_4^{[1]},\, G_2^{[3]},\, A_2^{[2]}\oplus G_2^{[1]}$
E_7	$A_1^{[231]},\, A_1^{[399]},\, A_2^{[21]},\, A_1^{[15]}\oplus A_1^{[24]},\, A_1^{[7]}\oplus G_2^{[2]},$ $A_2^{[3]}\oplus F_4^{[1]},\, C_3^{[1]}\oplus G_2^{[1]}$ (1.8.7)
E_8	$A_1^{[520]},\, A_1^{[760]},\, A_1^{[1240]},\, B_2^{[12]},\, A_1^{[16]}\oplus A_2^{[6]},\, F_4^{[1]}\oplus G_2^{[1]}$
F_4	$A_1^{[156]},\, A_1^{[8]}\oplus G_2^{[1]}$
G_2	$A_1^{[28]}$

In this table, the labels in square brackets denote the so-called *Dynkin index* $I_{\mathfrak{h}\subset\mathfrak{g}}$ of the embedding $\mathfrak{h}\subset\mathfrak{g}$. Up to a normalization factor $c = I_{\rm ad}(\mathfrak{g})/I_{\rm ad}(\mathfrak{h})$, this quantity is defined as the ratio between the bilinear form on $\mathfrak{h}$ that is obtained via the embedding from the Killing form on $\mathfrak{g}$ and the Killing form of $\mathfrak{h}$ (this way $I_{\mathfrak{h}\subset\mathfrak{g}}$ is well defined because all invariant bilinear forms on $\mathfrak{h}$ are proportional). Thus, denoting the embedding map as $i\colon \mathfrak{h}\to\mathfrak{g}$, one has

$$c\, I_{\mathfrak{h}\subset\mathfrak{g}} = \frac{\kappa_{\mathfrak{g}}(i(x),i(y))}{\kappa_{\mathfrak{h}}(x,y)}\,, \tag{1.8.8}$$

with x,y any two elements of $\mathfrak{h}$. Because of the relation (1.6.54) that fixes the normalization of the Killing form in terms of the length of the highest root, the Dynkin index can also be written as

$$I_{\mathfrak{h}\subset\mathfrak{g}} = \frac{(\theta(\mathfrak{g}),\theta(\mathfrak{g}))}{(\theta(\mathfrak{h}),\theta(\mathfrak{h}))}\,. \tag{1.8.9}$$

For regular subalgebras, the highest root $\theta(\mathfrak{h})$ is a root of $\mathfrak{g}$, and hence according to the latter formula the Dynkin index $I_{\mathfrak{h}\subset\mathfrak{g}}$ is equal to one for $\mathfrak{g}$ simply laced, either 1 or 2 for $\mathfrak{g}$ non-simply laced other than G_2, and 1 or 2 or 3 for $\mathfrak{g}=G_2$. More generally, from the geometry of the root system of $\mathfrak{g}$ it follows that the Dynkin index is an integer for any embedding $\mathfrak{h}\subset\mathfrak{g}$.

There exists still another possibility to characterize the Dynkin index of an embedding. Namely, if $R:\mathfrak{g}\to V$ is a representation of $\mathfrak{g}$ on a vector space V, then $R\circ i:\mathfrak{h}\to V$ is a representation of $\mathfrak{h}$, and according to (1.6.43) one has

$$\begin{aligned} &\mathrm{tr}\,((R\circ i)(x)\,(R\circ i)(y)) = (I_{R\circ i}/I_{\mathrm{ad}}(\mathsf{h}))\,\kappa_{\mathsf{h}}(x,y) \\ &\equiv \mathrm{tr}\,(R(i(x))\,R(i(y))) = (I_R/I_{\mathrm{ad}}(\mathsf{g}))\,\kappa_{\mathsf{g}}(i(x),i(y)). \end{aligned} \tag{1.8.10}$$

Thus for any representation R of g, the Dynkin index of an embedding $\mathsf{h}\subset\mathsf{g}$ can be expressed as the ratio of the second order Dynkin indices of R and $R\circ i$,

$$I_{\mathsf{h}\subset\mathsf{g}} = \frac{I_{R\circ i}}{I_R}. \tag{1.8.11}$$

The Dynkin index can be used to distinguish between various inequivalent embeddings of a subalgebra h in g; for example the table above shows that E_8 contains three different S-subalgebras which are all isomorphic to A_1. If also non-maximal embeddings are considered, then there are also some cases where several inequivalent embeddings of h in g possess the same Dynkin index. For example, there are two inequivalent embeddings of A_1 into E_8 possessing Dynkin index 30; for the first embedding, $A_1^{[30]_1} \subset E_8$, the minimal regular subalgebra of E_8 containing the relevant A_1-subalgebra is either D_5 or $A_1\oplus A_1\oplus D_4$, while for the second embedding $A_1^{[30]_2} \subset E_8$, it is either $A_3\oplus A_4$ or $D_3\oplus D_5$.

Having established the classification of subalgebras, we are now in a position to investigate the *branching rules*

$$R(\mathsf{g}) \rightsquigarrow \bigoplus_j R_j(\mathsf{h}). \tag{1.8.12}$$

which describe how an irreducible highest weight module of g becomes a module of $\mathsf{h}\subset\mathsf{g}$. For any branching rule, the obvious character sum rule

$$\chi_R(\mathsf{g}) = \sum_j \chi_{R_j}(\mathsf{h}), \tag{1.8.13}$$

(with the argument of the g-character restricted to the weight space of h) is valid, and from this one gets a dimension sum rule by evaluation on the zero weight; in addition, as a consequence of (1.8.11) there is a sum rule for the second order Dynkin index:

$$I_R(\mathsf{g})\cdot I_{\mathsf{h}\subset\mathsf{g}} = \sum_j I_{R_j}(\mathsf{h}). \tag{1.8.14}$$

Now recall that the weight spaces of g and h can be identified with their Cartan subalgebras g_0 and h_0; because of $i(\mathsf{h}_0)\subseteq\mathsf{g}_0$, the embedding $i:\mathsf{h}\overset{\subseteq}{\to}\mathsf{g}$ thus induces a map between the weight spaces of g and of h; in geometrical terms, this map is just a projection. In particular, for $\mathrm{rank}(\mathsf{h}) = \mathrm{rank}(\mathsf{g})$, this projection is necessarily the identity: from this it follows that the roots (i.e. the weights of the adjoint module) of h are among the roots of g, or in other words, that the embedding is a regular one. Thus for a regular embedding the branching rule of the adjoint module reads

$$R_{\mathrm{ad}}(\mathsf{g}) \rightsquigarrow R_{\mathrm{ad}}(\mathsf{h}) \oplus \big(\bigoplus_{j\in J} R_j \big) , \tag{1.8.15}$$

with the set J non-empty. More generally, it can be shown that for any regular embedding $i:\mathsf{h}\subset\mathsf{g}$ and any representation R of g, the representation $R\circ i$ of h is reducible. A certain inverse of this is true for $\mathsf{g}=A_r$, B_r or C_r : if h is an S-subalgebra of one of these algebras, then it possesses an irreducible module of dimension r, $2r+1$, and $2r$, respectively.

To discuss concrete examples, one has to identify the explicit form of the projection between the weight spaces. For regular embeddings, this is rather simple. Namely, the Dynkin labels of the h-weights λ are just the Dynkin labels $\lambda^i=(\lambda,\alpha^{(i)\vee})$ with respect to those simple g-roots which survive the inclusion (1.8.2) – i.e. just the corresponding Dynkin labels of the g-weights – together with the number $(\lambda,-\theta^\vee)$ (which may be called the Dynkin label with respect to $-\theta$), which is easily calculated by expressing θ in terms of the simple g-roots. For special embeddings the situation is more complicated, but one can still derive a so-called projection matrix which directly maps a g-weight to the associated h-weight. Given the explicit form of the various h-weights, it is then straightforward to group these weights into the weight systems of irreducible highest weight modules. However, there is no need to go into details here because exhaustive tables of branching rules (obtained by computer) are available, and so we will just present as an example the branching rules of the adjoint of E_8 for the three S-embeddings $A_1\subset E_8$. Denoting the modules by their dimensionalities, these branching rules read

$$\begin{aligned}
A_1^{[520]}\subset E_8:\quad & 248 \rightsquigarrow 3+7+11+15+17+19+2\cdot 23\\
&\qquad\qquad +27+29+35+39,\\
A_1^{[760]}\subset E_8:\quad & 248 \rightsquigarrow 3+11+15+19+23+27+29+35\\
&\qquad\qquad +39+47,\\
A_1^{[1240]}\subset E_8:\quad & 248 \rightsquigarrow 3+15+23+27+35+39+47+59.
\end{aligned} \tag{1.8.16}$$

1.9 Literature

Most of the contents of chapter 1 can be found in any standard textbook on Lie algebras. A very concise introduction is the book by Humphreys [B17] which was taken as the basic reference for the sections 1.1 to 1.6, while for the sections 1.7 and 1.8 the exposition is close to the one of Cahn [B8]. For the connection between Lie groups and Lie algebras that is only very briefly mentioned in section 1.3, and for details about the classification of real Lie algebras, a good reference is Gilmore's book [B12]. As examples of other textbooks which cover much of the material presented in chapter 1, we mention [B6, B13, B15, B18, B34, B35, B36], but there are of course many more textbooks on semisimple Lie algebras. In [B9], many of the topics of our interest are described in the context of universal enveloping algebras. Finally, a good introduction for physicists, with applications to the standard model and to grand unified theories, is the review by Slansky [R27].

An important topic of chapter 1 that is not explained in detail in the standard textbooks and reviews is the Racah–Speiser algorithm; this algorithm actually dates back to Weyl [B34] and is sometimes referred to as the method of characters [B36], but to the best of our knowledge it has first been described in full detail in [530, 477, 531]; see also [R10]. Many textbooks also do not present explicit formulæ involving invariant tensors, such as (1.6.80); good references for such formulæ are [B33, B15, 294, 142]. For more details about Young tableaux, see [B33, B13, B15, B25, B31, R4], and for the definition and calculation of Clebsch–Gordan coefficients and other recoupling coefficients, see [B4, R10].

Let us also list a few important original papers on Lie algebras (vast lists of further references can be found in [B12, B35]). The concept of Lie groups was introduced by Lie [386, 387]. The classification of complex simple Lie algebras is due to Killing [337] and Cartan [105]. The classification of real simple Lie algebras was also begun by Cartan [105, 107], and completed by Lardy [369] (for reviews, see [521, 38]). The fundamental weights have been introduced by Cartan [106]. Root diagrams were first used in [553], and the concept of simple roots is due to Dynkin [179, 178]. The most important issues in the representation theory of simple Lie groups and Lie algebras were first treated by Cartan [108, 109] and Weyl [575, 576, B33, B34], leading for example to the concept of the Cartan–Weyl basis. For later developments, see e.g. [111, 66, 295, 282, R22, 179, 59, 255]. The Chevalley basis was introduced in [121]. For the Brauer–Weyl theory of tensor product decompositions see [577, 88, 86, 87, 335]. The reflections σ_α contained in the Weyl group were already known to Killing [337]; the fact that they generate a group was recognized by Weyl [576]. In more generality, groups generated by reflections were studied by Coxeter [135, 137].

Dynkin diagrams were first used by Coxeter [136]; they were independently found by Dynkin in [177] where also the extended Dynkin diagrams are introduced. (Our convention for the numbering of the nodes of the Dynkin diagrams is the one of Dynkin; in the literature other conventions are also used, e.g. in [B6] and [B17].) The classification of semisimple subalgebras of simple Lie algebras is due to Dynkin [177, 178] (the exceptions to the rule for maximal regular semisimple subalgebras were discovered in [264]). The second order Dynkin index and the Dynkin index of an embedding were also defined there; higher order Dynkin indices are analyzed in [462, 463, 418]. The quadratic Casimir operator was introduced in [110]; higher order Casimir operators are discussed in [476, 533, 548, 276, 58, 464, 256]. The classification of irreducible highest weight modules of semisimple Lie algebras was essentially completed by Dynkin [178], with some further results concerning the congruence and conjugacy properties obtained in [406, 419, 72, 371].

The use of Young tableaux dates back to Young [591, 592] and Frobenius [222]. The combinatorial rule for the Littlewood–Richardson coefficients of A_r was given in [388] (for a review, see [B27, B25]); some results for other algebras can be found in [339, 340, 353, 354, 352, 344]. The calculation of Clebsch–Gordan coefficients and of the more general coupling coefficients known e.g. as $6j$- and $9j$-symbols (and which are mainly due to Wigner and Racah [B10, 578]) is described in [61, 59, 325, 150, 2, 97, 89, 182, 98].

The Freudenthal formula for weight multiplicities is derived in [B11, p. 243]; a different multiplicity formula is due to Kostant [358]. Efficient algorithms for the calculation of weight multiplicities and branching rules are described in [451, 430, 431, 432, 564]. The strange formula (1.4.71) is obtained in [B11].

Finally we mention some important tables. For the embeddings of semisimple Lie algebras, the best reference is still [177, 178]. Dimensions, indices and branching rules of highest weight modules are listed in [B28]. Weight multiplicities of highest weight modules can be found in [B7]. Various tables relevant to the application of semisimple Lie algebras to the standard model and grand unified theories are given in [R27], which contains e.g. a list of the maximal subalgebras of the classical Lie algebras of rank up to eight.

2
Affine Lie algebras

In this chapter we come to the first object of our real interest, the theory of affine Lie algebras. In section 2.1, these Lie algebras are classified in terms of their Cartan matrices, or what is the same, in terms of their Dynkin diagrams, and are set into the context of the more general class of Kac–Moody algebras. Just as in the case of semisimple Lie algebras, there are several infinite series as well as a few isolated cases. In addition, there is a distinction between untwisted and twisted affine algebras according to specific properties of the Coxeter labels that are attached to the Dynkin diagrams. A particular realization of affine algebras is presented in section 2.2, namely as central extensions of loop algebras; to this end the concept of the central extension is introduced also for generic Lie algebras. In this realization, the difference between untwisted and twisted algebras arises as a consequence of the analytic properties of the maps which define the associated centerless loop algebras. Section 2.3 describes the root system of affine algebras and analyses the Killing form; also, the possibility of using other gradations than the one employed by us (which is the one most relevant to conformal field theory) is briefly mentioned.

In section 2.4 we discuss the properties of highest weight modules of affine algebras; this involves in particular the important concept of integrability of a module. The irreducibility of highest weight modules is the subject of section 2.5; this leads to the concept of null vectors, which already plays a role for semisimple Lie algebras but was not mentioned explicitly in chapter 1. Having identified the general structure of null vectors, we then also display part of the weight system of the simplest affine irreducible highest weight modules. Section 2.6 describes the Weyl group of an affine algebra and its application to the theory of characters, leading to the Weyl–Kac character formula and to various combinatorial identities. In section 2.7 we discuss the behavior of affine characters under the modular group. This behavior can be employed to analyse the branching rules for embeddings of affine algebras, as is done in section 2.8, where we also list a few conformal embeddings and branching rules explicitly.

2.1 Classification of Kac–Moody algebras

So far we have discussed in some detail only semisimple Lie algebras (or possibly their direct sums with abelian algebras). These algebras are in particular finite-dimensional. In physics, Lie algebras arise in the description of symmetries. Thus, since many interesting systems possess infinitely many independent symmetries, infinite-dimensional Lie algebras are just as important in physics as finite-dimensional ones. There are essentially three classes of infinite-dimensional Lie algebras which have been investigated in some detail, namely

- the Lie algebras of vector fields which generate the diffeomorphisms of some manifold $\mathcal{M}$;
- the Lie algebras of operators in some Hilbert or Banach space;
- the Lie algebras of smooth mappings of some manifold $\mathcal{M}$ to a finite-dimensional Lie algebra.

The *Kac–Moody* Lie algebras which are the subject of our interest here are examples of the latter class, namely with $\mathcal{M}$ being the circle S^1, allowing, however, also for a so-called *central extension* of the algebra. The Kac–Moody algebras have attracted much attention in both mathematics and physics. Mathematically, they are of interest in their own right because at least a large subclass of them (the *affine* Lie algebras) can be classified completely and possesses an interesting representation theory, with both the classification and the representation theory being very similar to those of simple Lie algebras. Moreover, there are numerous connections to other branches of mathematics such as number theory, topology, singularity theory or the theory of finite simple groups. In physics, there are a variety of applications of Kac–Moody algebras, the most important ones being in two-dimensional conformal field theory and in the theory of completely integrable systems.

Recall from section 1.5 that a finite-dimensional simple Lie algebra is completely characterized by $3r$ generators $\{E^i_\pm,\, H^i \mid i = 1, \ldots, r\}$ obeying the Jacobi identity and the relations

$$\begin{aligned} &[H^i, H^j] = 0, \\ &[H^i, E^j_\pm] = \pm A^{ji} E^j_\pm, \\ &[E^i_+, E^j_-] = \delta^{ij} H^j, \\ &(\mathrm{ad}_{E^i_\pm})^{1-A^{ji}} E^j_\pm = 0 \text{ for } i \neq j. \end{aligned} \tag{2.1.1}$$

More precisely, the simple Lie algebras are obtained by requiring that the $r \times r$-matrix A appearing in (2.1.1) is an (irreducible) Cartan matrix, i.e. is indecomposable in the sense of (1.5.11) and obeys

$$\begin{aligned} &A^{ii} = 2, \\ &A^{ij} \leq 0 \text{ for } i \neq j, \\ &A^{ij} = 0 \Leftrightarrow A^{ji} = 0, \\ &A^{ij} \in \mathbb{Z}, \end{aligned} \tag{2.1.2}$$

and

$$\det A > 0. \tag{2.1.3}$$

In particular the rank of A is equal to r. The Kac–Moody Lie algebras are obtained by weakening the conditions on the matrix A. It should be noted that if one weakens the axioms of some given mathematical theory, one can be almost sure that much of the power and beauty of the theory gets destroyed. In the present case, however, the new theory is in fact even more interesting than the original one. What one has to do is to relax the condition (2.1.3) on the determinant of A. When one removes this condition completely, one gets the general class of Kac–Moody algebras. The most important subclass of Kac–Moody algebras is obtained if (2.1.3) is replaced by

$$\det A_{\{i\}} > 0 \text{ for all } i = 0, \ldots, r, \tag{2.1.4}$$

where $A_{\{i\}}$ are the matrices obtained from A by deleting the ith row and column (the determinants $\det A_{\{i\}}$ are called the principal minors of A). From here on we change our labelling convention for the Chevalley generators $E^i_\pm, H^i$: we take $i = 0, 1, \ldots, r$.

In particular for general Kac–Moody algebras the rank of A can be arbitrary, while the validity of (2.1.4) implies that the rank is at least r. A matrix obeying the inequalities (2.1.4) is called degenerate positive semidefinite, while a matrix obeying (2.1.2) is called a *generalized Cartan matrix.* A generalized (irreducible) Cartan matrix which is degenerate positive semidefinite is called an (irreducible) *affine Cartan matrix.* The

Lie algebras defined by generators and relations as in (2.1.1), with A an affine Cartan matrix, are called *affine Lie algebras.*

In the following we essentially consider only affine Lie algebras. These contain the simple Lie algebras which are obtained for $\operatorname{rank} A = r+1$, because (2.1.3) together with (2.1.2) already implies the validity of (2.1.4), i.e. simple Cartan matrices obey (2.1.4) automatically. The rank of the non-simple affine Cartan matrices is

$$\operatorname{rank} A = r. \tag{2.1.5}$$

Instead of requiring (2.1.4), one can characterize an affine Cartan matrix also by the requirement that there exists a non-singular diagonal matrix D such that DA is symmetric and positive semidefinite (if DA is in fact positive definite, then A is the Cartan matrix of a simple Lie algebra). This is so because together with (2.1.2) this requirement already implies that DA can have at most one zero eigenvalue.

Once the classification of simple Lie algebras is known, the classification of affine Lie algebras is straightforward. We first enumerate the cases $r=0$ and $r=1$. For $r=0$ there is only a single possibility, the simple Lie algebra $\mathfrak{g} = A_1$ with Cartan matrix

$$A_1: \qquad A = (\,2\,). \tag{2.1.6}$$

For $r=1$, one finds by inspection the three rank two simple Lie algebras, with Cartan matrices,

$$\begin{aligned} A_2: &\quad \begin{pmatrix} 2 & -1 \\ -1 & 2 \end{pmatrix}, \\ C_2: &\quad \begin{pmatrix} 2 & -1 \\ -2 & 2 \end{pmatrix}, \\ G_2: &\quad \begin{pmatrix} 2 & -3 \\ -1 & 2 \end{pmatrix}, \end{aligned} \tag{2.1.7}$$

and in addition two algebras with rank one Cartan matrices,

$$\begin{aligned} A_1^{(1)}: &\quad \begin{pmatrix} 2 & -2 \\ -2 & 2 \end{pmatrix}, \\ A_1^{(2)}: &\quad \begin{pmatrix} 2 & -4 \\ -1 & 2 \end{pmatrix}. \end{aligned} \tag{2.1.8}$$

For $r > 1$, we can use as a starting point the observation that by deleting the ith row and ith column ($i \in \{0, 1, \ldots, r\}$ arbitrary) from an affine Cartan matrix, one must produce the Cartan matrix of a simple Lie algebra. This means in particular that any 2×2-matrix obtained from A by deleting $r-1$ rows and the corresponding columns must be a rank two

Cartan matrix, i.e. be one of the matrices in (2.1.7). Therefore for $i \neq j$, one has either

$$A^{ij} = A^{ji} = 0, \tag{2.1.9}$$

or else

$$\begin{aligned} \min\{|A^{ij}|, |A^{ji}|\} &= 1, \\ \max\{|A^{ij}|, |A^{ji}|\} &\leq 3 \end{aligned} \tag{2.1.10}$$

(in particular, the submatrices must obey $A^{ij}A^{ji} \leq 3$; as a consequence, the algebras $A_1^{(1)}$ and $A_1^{(2)}$ which have $A^{12}A^{21} = 4$ are rather special; this is related to the fact that the underlying "horizontal" (cf. p. 97) subalgebra A_1 is the only simple Lie algebra with the property that the highest root θ is a simple root). Implementing the constraints listed above, it becomes a matter of straightforward combinatorics to enumerate the affine Cartan matrices for any given value of r. As an illustration, let us take $r = 2$ and consider the case where a rank two 2×2 submatrix is equal to the Cartan matrix of G_2, such that, up to a renumbering of rows and columns,

$$A = \begin{pmatrix} 2 & p & q \\ r & 2 & -3 \\ s & -1 & 2 \end{pmatrix}. \tag{2.1.11}$$

The determinant of this matrix is $2 - p(2r + 3s) - q(r + 2s)$. It is readily seen that there are only two allowed combinations of the integers p, q, r, s for which this number is zero, namely either $p = r = -1$, $q = s = 0$, or else $p = r = 0$, $q = s = -1$. The Lie algebras corresponding to these two sets of integers are denoted by $G_2^{(1)}$ and $G_2^{(3)}$, respectively.

The analysis of the general case proceeds analogously. The results are listed in table (2.1.12), which provides the names of the algebras and their Dynkin diagrams. The numbers which are adjoined to the nodes are the Coxeter and dual Coxeter labels; their precise definition will be presented below.

From the table we read off that the situation is not very different from the one encountered in the classification of simple Lie algebras: there are a few infinite series, supplemented by some exceptional cases. More precisely, we have seven infinite series of (non-simple) affine Lie algebras, which are labelled $A_r^{(1)}$ $(r \geq 2)$, $B_r^{(1)}$ $(r \geq 3)$, $C_r^{(1)}$ $(r \geq 2)$, $D_r^{(1)}$ $(r \geq 4)$, $B_r^{(2)}$ $(r \geq 3)$, $C_r^{(2)}$ $(r \geq 2)$ and $\tilde{B}_r^{(2)}$ $(r \geq 2)$, and in addition nine exceptional affine algebras, which are denoted by $A_1^{(1)}$, $E_6^{(1)}$, $E_7^{(1)}$, $E_8^{(1)}$, $F_4^{(1)}$, $G_2^{(1)}$, $A_1^{(2)}$, $F_4^{(2)}$ and $G_2^{(3)}$.

name	Dynkin diagram
$A_1^{(1)}$	
$A_r^{(1)}$	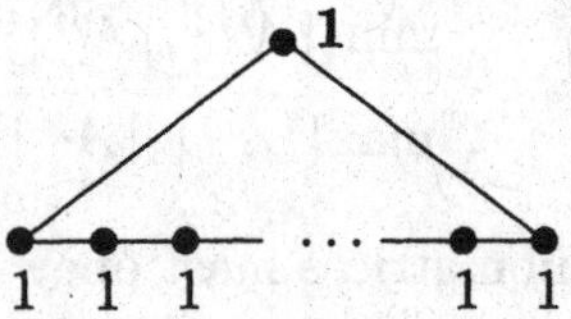
$B_r^{(1)}$	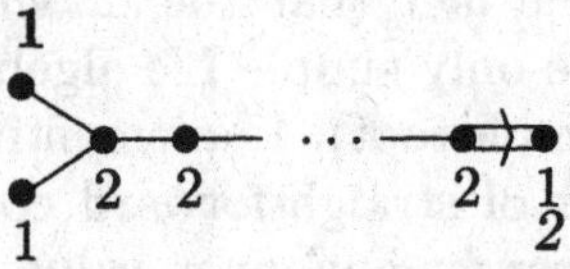
$C_r^{(1)}$	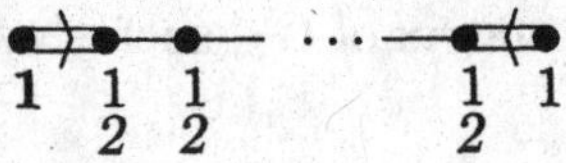
$D_r^{(1)}$	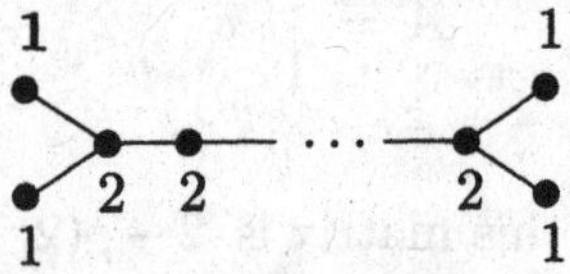
$E_6^{(1)}$	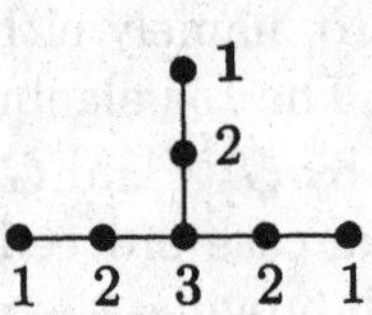
$E_7^{(1)}$	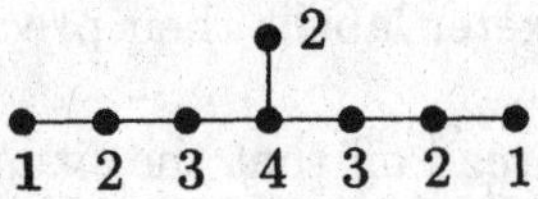
$E_8^{(1)}$	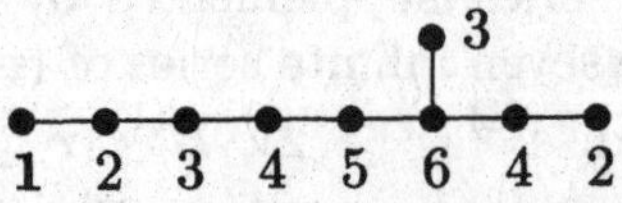

(2.1.12)

name	Dynkin diagram
$F_4^{(1)}$	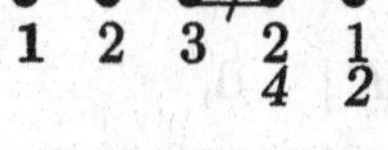
$G_2^{(1)}$	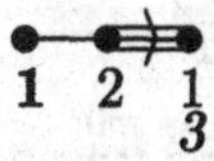
$A_1^{(2)}$	
$B_r^{(2)}$	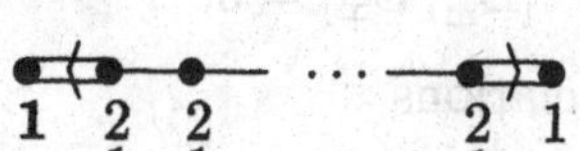
$\tilde{B}_r^{(2)}$	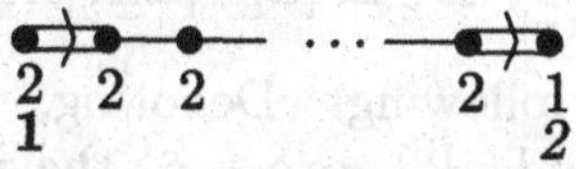
$C_r^{(2)}$	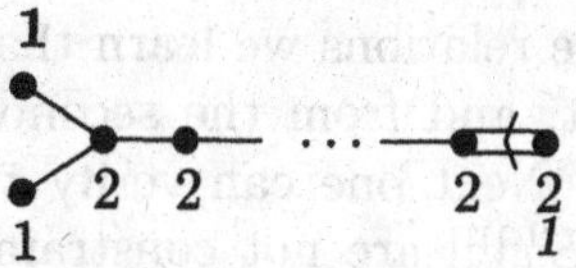
$F_4^{(2)}$	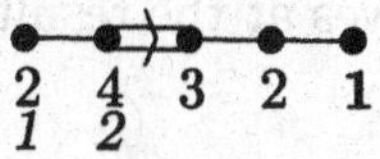
$G_2^{(3)}$	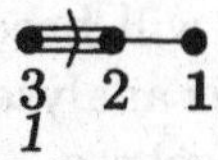

(2.1.12)

(continued)

As an illustration, let us write down the bracket relations (2.1.1) of the Chevalley–Serre basis explicitly for the case of the simplest affine Lie algebra $A_1^{(1)}$. One has

$$\begin{aligned}
&[H^0, H^1] = 0,\\
&[E^0_+, E^0_-] = H^0,\\
&[E^1_+, E^1_-] = H^1,\\
&[H^0, E^0_\pm] = \pm 2\, E^0_\pm,\\
&[H^1, E^1_\pm] = \pm 2\, E^1_\pm,\\
&[H^0, E^1_\pm] = \mp 2\, E^1_\pm,\\
&[H^1, E^0_\pm] = \mp 2\, E^0_\pm,\\
&[E^0_\pm, E^1_\mp] = 0,
\end{aligned} \tag{2.1.13}$$

together with the Serre relations

$$\begin{aligned}
0 = (\mathrm{ad}_{E^1_\pm})^3 E^0_\pm \equiv [E^1_\pm, [E^1_\pm, [E^1_\pm, E^0_\pm]]],\\
0 = (\mathrm{ad}_{E^0_\pm})^3 E^1_\pm \equiv [E^0_\pm, [E^0_\pm, [E^0_\pm, E^1_\pm]]].
\end{aligned} \tag{2.1.14}$$

From this we learn the following. Denoting, with foresight, the simple roots by $\alpha^{(1)} = \bar{\alpha}$ and $\alpha^{(0)} = \delta - \bar{\alpha}$, the vanishing of the bracket $[E^0_\pm, E^1_\mp] \equiv [E^{\pm(\delta-\bar{\alpha})}, E^{\mp\bar{\alpha}}]$ tells us that $\delta - 2\bar{\alpha}$ is not a root. Similarly, from the first of the Serre relations we learn that δ and $\delta + \bar{\alpha}$ are roots, but $\delta + 2\bar{\alpha}$ is not a root, and from the second that δ and $2\delta - \bar{\alpha}$ are roots, but not $3\delta - 2\bar{\alpha}$. Next one can verify that $[E^1_\pm, [E^0_\pm, [E^0_\pm, E^1_\pm]]]$ and $[E^1_\pm, [E^1_\pm, [E^0_\pm, [E^0_\pm, E^1_\pm]]]]$ are not constrained to vanish so that 2δ and $2\delta + \bar{\alpha}$ are roots, but, using again the Serre relations, not $2\delta + 2\bar{\alpha}$. Iterating this procedure, one arrives at the result that the root system of $A_1^{(1)}$ is

$$\Phi = \{\pm\bar{\alpha} + n\delta \mid n \in \mathbb{Z}\} \cup \{n\delta \mid n \in \mathbb{Z} \setminus \{0\}\}. \tag{2.1.15}$$

The meaning of the notations $\bar{\alpha}$ and δ will become clear in section 2.3 where the root system of affine algebras is analysed in the context of their realization as centrally extended loop algebras.

The numbers attached to the vertices in (2.1.12) are the Coxeter labels a_i and dual Coxeter labels $a_i^\vee$ (if a_i differs from $a_i^\vee$, a_i is listed in italics below $a_i^\vee$). For a (non-simple) affine Lie algebra, these are defined by requiring that $(a_i)_{i=0,\dots,r}$ and $(a_i^\vee)_{i=0,\dots,r}$ are left respectively right eigenvectors with eigenvalue zero of the affine Cartan matrix, i.e.

$$\sum_{j=0}^{r} a_j A^{ji} = 0 = \sum_{j=0}^{r} A^{ij} a_j^\vee, \tag{2.1.16}$$

together with the normalization condition

$$\min\{a_i \mid i=0,\ldots,r\} = 1 = \min\{a_i^\vee \mid i=0,\ldots,r\}. \tag{2.1.17}$$

Note that the dual Coxeter labels obey

$$a_i^\vee = \tfrac{1}{2} \sum_{j\neq i} |A^{ij}|\, a_j^\vee. \tag{2.1.18}$$

The sums of the Coxeter and dual Coxeter labels are known as the Coxeter number g and dual Coxeter number $g^\vee$ of $\mathfrak{g}$:

$$g = \sum_{i=0}^{r} a_i, \quad g^\vee = \sum_{i=0}^{r} a_i^\vee. \tag{2.1.19}$$

Inspecting table (2.1.12), one can check that indeed by removing any node from the Dynkin diagram of an affine Lie algebra one gets the Dynkin diagram of some simple Lie algebra. The notations for the affine algebras have been chosen such that for any non-simple affine algebra $X_r^{(\ell)}$, with X denoting one of the letters $A, B, \ldots, G$, there is one node in the Dynkin diagram such that its removal leads to the Dynkin diagram of the simple Lie algebra X_r, with X denoting the same letter as before. The numbering of the nodes of the affine Dynkin diagram is the one inherited from the Dynkin diagram of X_r, with the additional node (the removal of which reduces the Dynkin diagram of $X_r^{(\ell)}$ to that of X_r) identified as the *zeroth node.* We have distinguished the zeroth node in an affine Dynkin diagram by writing its Coxeter label in boldface; table (2.1.12) shows that this Coxeter label is in fact equal to one in all cases,

$$a_0 = 1. \tag{2.1.20}$$

Inspecting also the other Coxeter labels of the Dynkin diagrams, we see that we can distinguish between two classes of non-simple affine algebras $X_r^{(\ell)}$, namely the ones with $\ell = 1$, which are called *untwisted* affine algebras, and those with $\ell > 1$, called *twisted* affine algebras. In the untwisted case, removal of the zeroth node of the Dynkin diagram produces the Dynkin diagram of X_r and at the same time also the correct (dual) Coxeter labels (defined in terms of the highest root of X_r as in (1.4.66) and (1.4.68)) for its nodes; thus the Dynkin diagram of $X_r^{(1)}$ is identical to the extended Dynkin diagram of X_r which was used in section 1.8 to classify regular subalgebras of X_r. In contrast, in the twisted case the Coxeter numbers are not produced correctly (nor would they if any other node with $a_i = 1$ were removed from the affine Dynkin diagram). For any (untwisted or twisted) affine algebra $\mathfrak{g} = X_r^{(\ell)}$, we will denote the simple Lie algebra X_r by $\mathfrak{g}_{\rm hor}$, and call it the *horizontal subalgebra* of $\mathfrak{g}$.

One also observes that for the twisted affine algebras the only possibility of obtaining the Dynkin diagram of $\mathfrak{g}_{\rm hor}$ from that of $\mathfrak{g}$ by deleting a node is to choose a node which has $a_i = 1$ and in addition, if there are several nodes with this property, require that also $a_i^\vee = 1$, which is obeyed only by the zeroth node. In contrast, in the untwisted case the Dynkin diagram of $\mathfrak{g}_{\rm hor}$ can be obtained from that of $\mathfrak{g}$ not only by removing the zeroth node, but also by removing instead any other node having its Coxeter label equal to one. Moreover, all nodes of the $X_r^{(1)}$ Dynkin diagram with $a_i = 1$ are related to each other by automorphisms of the Dynkin diagram. These diagram automorphisms look as follows: there are no nontrivial automorphisms in the cases of $E_8^{(1)}, F_4^{(1)}$ and $G_2^{(1)}$, while in the other cases the automorphisms form a group $\Gamma(\mathfrak{g})$ according to table (2.1.21).

$\mathfrak{g}$	Γ
$A_r^{(1)}$	$\mathcal{D}_{r+1}$
$B_r^{(1)}$, $C_r^{(1)}$, $E_7^{(1)}$	$\mathbb{Z}_2$
$E_6^{(1)}$	$\mathcal{S}_3$
$D_4^{(1)}$	$\mathcal{S}_4$
$D_r^{(1)}$, $r \geq 5$	$\mathcal{D}_4$

(2.1.21)

In table (2.1.21), $\mathcal{S}_n$ denotes the symmetric group of n objects, and $\mathcal{D}_n$ the *dihedral group* of n objects, which is by definition the symmetry group of a regular polygon with n edges (thus e.g. $\mathcal{D}_4$ is the symmetry group of the square). Note that these automorphism groups always contain those of the Dynkin diagrams of the horizontal subalgebra (compare the table (1.5.18)), $\Gamma(\mathfrak{g}_{\rm hor}) \subseteq \Gamma(\mathfrak{g})$, and are in fact larger than $\Gamma(\mathfrak{g}_{\rm hor})$ except for $E_8^{(1)}$, $F_4^{(1)}$ and $G_2^{(1)}$. The gain in symmetry is related in a precise manner to the center $\mathcal{Z}$ of the universal covering group G which has $\mathfrak{g}_{\rm hor}$ as its Lie algebra. Namely, $\mathcal{Z}$ is an invariant subgroup of the automorphism group $\Gamma(\mathfrak{g})$ such that

$$\Gamma(\mathfrak{g}_{\rm hor}) = \Gamma(\mathfrak{g})/\mathcal{Z}. \qquad (2.1.22)$$

Inspection shows that $\mathcal{Z}$ is always a cyclic group, i.e. $\mathcal{Z} \cong \mathbb{Z}_n$ for some integer n, except for $D_r^{(1)}$ with r odd, in which case $\mathcal{Z} \cong \mathbb{Z}_2 \times \mathbb{Z}_2$. In more detail, the situation looks as follows. The universal covering group associated to A_r is isomorphic to $SU(r+1)$ (the group of complex $(r+1)\times(r+1)$-matrices of determinant 1) which has $\mathbb{Z}_{r+1}$ as its center (namely the multiples $c\mathbf{1}$ of the unit matrix with c an $(r+1)$th root of unity); thus the relevant quotient is $\mathcal{D}_{r+1}/\mathbb{Z}_{r+1} \cong \mathbb{Z}_2$ (for $r \geq 2$, and $\cong 1$ for $r = 1$

since $\mathcal{D}_2 \cong \mathbb{Z}_2$), which is indeed the automorphism group of the Dynkin diagram of A_r. Similarly, the covering group of E_6 has center $\mathbb{Z}_3$, and $\mathcal{S}_3/\mathbb{Z}_3 \cong Z_2$ is indeed the symmetry of the E_6 Dynkin diagram. Next consider $\mathfrak{g}_{\rm hor} = D_r$; its covering group $SO(2r)$ has center $\mathbb{Z}_4$ for r even, respectively $\mathbb{Z}_2 \times \mathbb{Z}_2$ for r odd; thus $\mathcal{S}_4/\mathbb{Z}_4 \cong \mathcal{S}_3$ is the symmetry of the Dynkin diagram of D_4, whereas for $r \geq 4$ the Dynkin diagram symmetry is identified as $\mathcal{R}_4/\mathcal{Z} = \mathbb{Z}_2$. Finally, for the remaining algebras in the list one has $\mathcal{Z} \cong \mathbb{Z}_2$ so that there is no symmetry left for the Dynkin diagrams of B_r, C_r, and E_7.

Let us also note that distinct affine Dynkin diagrams never give simultaneously the same rank and the same Coxeter number. This implies that affine Lie algebras possessing distinct Dynkin diagrams are non-isomorphic.

For non-affine Kac–Moody algebras, one is still far from having a complete classification. Partial results are known for the *hyperbolic* Kac–Moody algebras. These correspond to those generalized Cartan matrices A for which there exists a non-singular diagonal matrix D such that DA is symmetric and of signature $(r, 1)$, i.e. possesses precisely one negative eigenvalue in addition to r positive eigenvalues. In the following figure, we display the Dynkin diagrams for some low-rank hyperbolic Kac–Moody algebras.

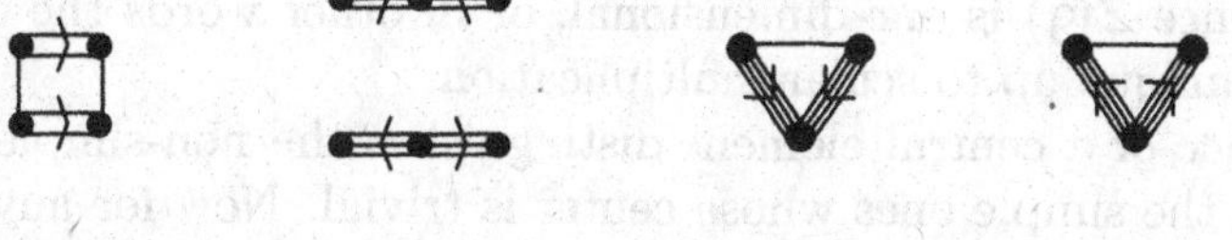

(2.1.23)

It can be shown that there are no hyperbolic algebras of rank larger than 10, and that the number of hyperbolic algebras with rank larger than 2 is finite. The hyperbolic algebras can also be characterized by requiring that their Dynkin diagram is not of affine type, but becomes the Dynkin diagram of an affine algebra, or a direct sum of affine algebras, upon deletion of any of its nodes.

As a final example, consider the following Dynkin diagram:

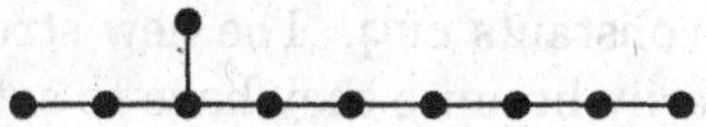

(2.1.24)

This is not an affine Dynkin diagram, but becomes the Dynkin diagram of $E_8^{(1)}$ upon deletion of the rightmost node. Thus the Kac–Moody algebra associated to this Dynkin diagram is hyperbolic. As a comparison of the Dynkin diagram to those of the E series shows, it is natural to denote this algebra by E_{10}; likewise, one could write E_9 for $E_8^{(1)}$, and define E_r

with $r > 10$ in the obvious way by joining consecutively further nodes to the rightmost one of the Dynkin diagram. Moreover, by induction one derives that the determinant of the Cartan matrices for the E series is given by the formula $\det A(E_r) = 9 - r$; thus E_{10} is the only hyperbolic algebra in the E series, and E_9 the only (non-simple) affine one.

2.2 Loop algebras and central extensions

In contrast to simple Lie algebras, non-simple affine Lie algebras $\mathfrak{g}$ possess a nontrivial center. Namely, by construction, for any constant ζ the element

$$K := \zeta \sum_{i=0}^{r} a_i^{\vee} H^i \tag{2.2.1}$$

is a *central element*, i.e. $K \in \mathcal{Z}(\mathfrak{g})$. This is so because the bracket relations (2.1.1) give not only $[K, H^i] = 0$, but also

$$[K, E_{\pm}^i] = \pm \sum_{j=0}^{r} a_j^{\vee} A^{ij} E_{\pm}^i = 0, \tag{2.2.2}$$

owing to the defining property (2.1.16) of the dual Coxeter labels. Also, affine Cartan matrices have by definition only one zero eigenvalue, and as a consequence $\mathcal{Z}(\mathfrak{g})$ is one-dimensional, or in other words the central element K is unique up to scalar multiplication.

The existence of a central element distinguishes the non-simple affine algebras from the simple ones whose center is trivial. Now for any given Lie algebra $\mathfrak{g}$, one can of course construct an (ℓ-dimensional) *central extension* $\hat{\mathfrak{g}}$ by simply adding ℓ central generators K^i to the generators T^a of $\mathfrak{g}$ and imposing

$$[T^a, K^i] = 0 \quad \text{for } i = 1, \ldots, \ell, \ a = 1, \ldots, d. \tag{2.2.3}$$

In addition, the brackets of the original generators then generalize to

$$[T^a, T^b] = f^{ab}{}_c T^c + f^{ab}{}_i K^i, \tag{2.2.4}$$

with $f^{ab}{}_c$ the structure constants of $\mathfrak{g}$. The new structure constants $f^{ab}{}_i$ cannot be chosen arbitrarily because they have to satisfy the Jacobi identity. The dimension of the space of solutions of the resulting equations for $f^{ab}{}_i$ is then the number of independent allowed central extensions of $\mathfrak{g}$.

The choice $f^{ab}{}_i \equiv 0$ is of course always a solution, but this is a rather trivial one because $\hat{\mathfrak{g}}$ is then just the direct sum of $\mathfrak{g}$ and an abelian algebra, $\hat{\mathfrak{g}} \cong \mathfrak{g} \oplus (u_1)^{\ell}$. However, $f^{ab}{}_i \equiv 0$ is in general not the only solution of this type. Namely, $\hat{\mathfrak{g}}$ is also of this form if there exists any choice of basis

$$\tilde{T}^a = T^a + u_i^a K^i \tag{2.2.5}$$

in which the structure constants

$$\tilde{f}^{ab}_{\ \ i} = f^{ab}_{\ \ i} - \sum_c f^{ab}_{\ \ c} u_i^c \tag{2.2.6}$$

are zero. This may be rephrased more abstractly as follows. Writing

$$[x, y]_{\text{new}} = [x, y]_{\text{old}} + \Omega(x, y)\, K \tag{2.2.7}$$

(and $[K, x] = 0$), the Lie properties imply that Ω is bilinear and antisymmetric, and obeys

$$0 = \Omega(x, [y, z]) + \Omega(y, [z, x]) + \Omega(z, [x, y]). \tag{2.2.8}$$

These are the defining properties of a 2-*cocycle*. The central extension is trivial iff this cocycle is in fact a coboundary, i.e. iff $\Omega(x, y)$ is a linear function of $[x, y]_{\text{old}}$.

Now consider the case where $\mathfrak{g}$ is a complex simple Lie algebra. Choose a basis where the structure constants f^{abc} are completely antisymmetric and $\sum_a f^{abc} f^{abd} \propto \delta^{cd}$. Then the Jacobi identity requires

$$f^{ab}_{\ \ i} = \sum_{c,d,e} f^{abc} f^{cde} f^{de}_{\ \ i}, \tag{2.2.9}$$

and therefore the choice

$$u_i^a = \sum_{b,c} f^{abc} f^{bc}_{\ \ i} \tag{2.2.10}$$

leads to

$$\tilde{f}^{ab}_{\ \ i} = \sum_{c,d,e} f^{abc} f^{cde} f^{de}_{\ \ i} - \sum_c f^{abc} u_i^c = 0. \tag{2.2.11}$$

Thus simple (and, more generally, semisimple) Lie algebras possess no nontrivial central extensions.

The close connection between the Dynkin diagrams of simple algebras and of non-simple affine algebras suggests that the non-simple affine algebras may be obtainable as some generalization of simple algebras which allows for a nontrivial central extension. This is indeed the case. For any simple Lie algebra $\bar{\mathfrak{g}}$, consider the space of analytic maps from the circle S^1 to $\bar{\mathfrak{g}}$. If $\{T^a \mid a = 1, \dots, d\}$ is a basis of $\bar{\mathfrak{g}}$ and S^1 is considered as the unit circle in the complex plane with coordinate z, then a basis of this vector space is

$$\mathcal{B} = \{T_n^a \mid a = 1, \dots, d;\ n \in \mathbb{Z}\}, \tag{2.2.12}$$

where

$$T_n^a := T^a \otimes z^n \tag{2.2.13}$$

with "$\otimes$" a formal multiplication. Moreover, this space inherits a natural bracket operation from $\mathfrak{g}$, namely

$$[T_m^a, T_n^b] \equiv [T^a \otimes z^m, T^b \otimes z^n] := [T^a, T^b] \otimes (z^m \cdot z^n), \tag{2.2.14}$$

i.e.

$$[T_m^a, T_n^b] = f^{ab}{}_c\, T^c \otimes z^{m+n} = f^{ab}{}_c\, T^c_{m+n}, \tag{2.2.15}$$

where $f^{ab}{}_c$ are the structure constants of $\mathfrak{g}$. It is readily checked that with this bracket operation the space of analytic maps from S^1 to $\mathfrak{g}$ becomes a Lie algebra itself. This infinite-dimensional Lie algebra is called the *loop algebra* $\mathfrak{g}_{\rm loop}$ over $\mathfrak{g}$.

Since under the Lie bracket (2.2.15) the index n of T_n^a is additive (and hence provides a $\mathbb{Z}$-gradation of the algebra), the subset of $\mathfrak{g}_{\rm loop}$ generated by the generators T_0^a is a subalgebra, the so-called zero mode subalgebra $\mathfrak{g}_{[0]}$, of $\mathfrak{g}_{\rm loop}$.

As in the case of simple Lie algebras, one can now look for central extensions of $\mathfrak{g}_{\rm loop}$. The most general ansatz for the new Lie bracket is

$$[T_m^a, T_n^b] = f^{ab}{}_c\, T^c_{m+n} + (f^{ab}{}_i)_{mn}\, K^i. \tag{2.2.16}$$

Obviously the zero mode subalgebra is isomorphic to the original simple Lie algebra $\mathfrak{g}$, and hence by a suitable redefinition of generators one can put $(f^{ab}{}_i)_{00}$ to zero; in fact this is also possible for all $(f^{ab}{}_i)_{m0}$, $m \in \mathbb{Z}$. It then follows that for any fixed n, the set $\{T_n^a \mid a = 1, \dots, d\}$ transforms in the adjoint representation of $\mathfrak{g}_{[0]}$, and also that with respect to the indices a, b the structure constants $(f^{ab}{}_i)_{mn}$ form an invariant tensor of the adjoint representation of $\mathfrak{g}_{[0]}$. But there is a unique (up to normalization) such tensor, namely the Killing form $\bar\kappa$ of $\mathfrak{g} \cong \mathfrak{g}_{[0]}$. Hence there can be only one-dimensional non-trivial central extensions of $\mathfrak{g}_{\rm loop}$: there is no index i, and $(f^{ab}{}_i)_{mn}$ reduces to

$$(f^{ab})_{mn} = \bar\kappa^{ab} f_{mn}. \tag{2.2.17}$$

Because of the symmetry of the Killing form, f_{mn} must be antisymmteric in m, n; also we already know that $f_{m0} = 0$ for all $m \in \mathbb{Z}$. It turns out that there is exactly one possibility for f_{mn} compatible with these requirements and with the Jacobi identity, namely

$$f_{mn} = m\, \delta_{m+n,0}. \tag{2.2.18}$$

In summary, there is a unique non-trivial central extension $\hat{\mathfrak{g}}$ of $\mathfrak{g}_{\rm loop}$, characterized by the brackets

$$[T^a_m, T^b_n] = f^{ab}{}_c T^c_{m+n} + m\,\delta_{m+n,0}\bar\kappa^{ab} K,$$
$$[K, T^a_n] = 0 \tag{2.2.19}$$

among the generators T^a_n and K. If we write a general element of $\hat{\mathfrak{g}}$ as $x \otimes \mathcal{P}(z) + \zeta K$ with $x \in \mathfrak{g}$ and $\mathcal{P}$ a Laurent polynomial, the Lie bracket of two arbitrary elements of $\hat{\mathfrak{g}}$ reads

$$[x \otimes \mathcal{P}(z) + \zeta K, y \otimes \mathcal{Q}(z) + \eta K]$$
$$= [x, y] \otimes (\mathcal{P}(z)\mathcal{Q}(z)) + \bar\kappa(x, y) K \otimes \mathrm{Res}\,(\mathcal{Q}(z)\partial\mathcal{P}(z)); \tag{2.2.20}$$

here $\partial \equiv \frac{\mathrm{d}}{\mathrm{d}z}$, and Res denotes the residue (i.e. the coefficient of z^{-1}) of a Laurent polynomial. In many applications such as in chapter 3 one is interested in the case where $\mathfrak{g}$ is a compact real Lie algebra. Choosing a basis such that $\bar\kappa^{ab} = -\delta^{ab}$, we can then write the first of the bracket relations (2.2.19) in the form

$$[T^a_m, T^b_n] = f^{abc} T^c_{m+n} - m\delta^{ab}\delta_{m+n,0} K. \tag{2.2.21}$$

The above construction of (complex) $\hat{\mathfrak{g}}$ can be summarized as

$$\hat{\mathfrak{g}} := \mathfrak{g}_{\mathrm{loop}} \oplus \mathbb{C}K \tag{2.2.22}$$

(semidirect sum) with

$$\mathfrak{g}_{\mathrm{loop}} := \mathbb{C}[z, z^{-1}] \otimes_{\mathbb{C}} \mathfrak{g}. \tag{2.2.23}$$

By constructing the root system (see section 2.3) one observes that the untwisted affine algebra $\mathfrak{g}^{(1)}$ is obtained from $\hat{\mathfrak{g}}$ by adding one further generator D, i.e.

$$\mathfrak{g}^{(1)} = \hat{\mathfrak{g}} \oplus \mathbb{C}D = \mathbb{C}[z, z^{-1}] \otimes_{\mathbb{C}} \mathfrak{g} \oplus \mathbb{C}K \oplus \mathbb{C}D, \tag{2.2.24}$$

where the new Lie brackets are defined such that the antisymmetry and Jacobi identity are still obeyed, namely

$$[D, T^a_m] = -[T^a_m, D] = m\,T^a_m,$$
$$[D, K] = 0 \tag{2.2.25}$$

in addition to the relations (2.2.19). To be precise, this is a particular choice among several possible ones, namely the one corresponding to a fixed *homogeneous gradation* (for more details, see p. 109 below). The center of the algebra obtained this way is still one-dimensional; in other words, the generator K introduced here by centrally extending $\mathfrak{g}_{\mathrm{loop}}$ can be identified with K as defined in (2.2.1) if the normalization constant ζ used there is chosen appropriately; the correct choice turns out to be

$\zeta = \frac{1}{2}(\bar\theta,\bar\theta)$. The above realization of the untwisted affine algebras shows in particular that, just as the loop algebras, they are infinite-dimensional (from the abstract definition in terms of generalized Cartan matrices in the previous section, this is not at all obvious; rather, it can only be seen by explicitly constructing the root system).

From the brackets (2.2.15) and (2.2.19) of $\bar{\mathfrak{g}}_{\rm loop}$ and $\mathfrak{g}$ it follows at once that the zero mode subalgebra of $\bar{\mathfrak{g}}_{\rm loop}$ is also the zero mode subalgebra of $\mathfrak{g}$. Moreover it can be characterized as being generated by those generators T^a_m which have vanishing bracket with D, and it is isomorphic to the original finite-dimensional Lie algebra $\bar{\mathfrak{g}}$ and to the horizontal subalgebra $\mathfrak{g}_{\rm hor}$ that was defined in the previous section,

$$\bar{\mathfrak{g}}_{[0]} \cong \mathfrak{g}_{\rm hor} \cong \bar{\mathfrak{g}}. \tag{2.2.26}$$

From now on, we will therefore use barred quantities when referring to the horizontal subalgebra of a (non-simple) untwisted affine algebra. Note that D does not appear on the right hand side of any bracket, and that D is in fact the only generator with this property; thus $\hat{\mathfrak{g}}$ is the derived algebra of $\mathfrak{g}^{(1)}$,

$$\hat{\mathfrak{g}} = [\mathfrak{g}^{(1)},\mathfrak{g}^{(1)}]. \tag{2.2.27}$$

The generator D is therefore, somewhat sloppily, referred to as the *derivation* of $\mathfrak{g}^{(1)}$. On the subalgebra $\bar{\mathfrak{g}}_{\rm loop}$, D acts $z\frac{\rm d}{{\rm d}z}$:

$$[D, x\otimes \mathcal{P}(z)] = x\otimes z\partial\mathcal{P}(z). \tag{2.2.28}$$

Instead, by building the derived algebra, one can obtain $\hat{\mathfrak{g}}$ from $\mathfrak{g}^{(1)}$ also by the following construction which is based on the presentation (2.1.1) of $\mathfrak{g}^{(1)}$ in terms of generators and relations. Any untwisted affine algebra $\mathfrak{g}^{(1)}$ possesses a natural $\mathbb{Z}^{r+1}$-gradation defined by

$$\begin{aligned} &{\rm degree}(E^i_\pm) = (\underbrace{0,0,\ldots,0}_{i-1\ \text{times}},\pm1,\underbrace{0,0,\ldots,0}_{r-i\ \text{times}}),\\ &{\rm degree}(H^i) = (\underbrace{0,0,\ldots,0}_{r\ \text{times}}); \end{aligned} \tag{2.2.29}$$

then $\hat{\mathfrak{g}}$ is the quotient of $\mathfrak{g}^{(1)}$ by its largest $\mathbb{Z}^{r+1}$-graded ideal which intersects ${\rm span}\{H^i \mid i=0,\ldots,r\}$ only in $\{0\}$.

By the above construction, we have only obtained the untwisted affine algebras. To get also a realization of the twisted ones, it is necessary to generalize the construction. What one has to give up is the analyticity of the maps from S^1 to $\bar{\mathfrak{g}}$, i.e. one must no longer require that

$$\mathcal{P}({\rm e}^{2\pi i}z) = \mathcal{P}(z). \tag{2.2.30}$$

Rather, we impose the "twisted" boundary conditions

$$x\otimes\mathcal{P}({\rm e}^{2\pi i}z) = \omega(x)\otimes\mathcal{P}(z) \tag{2.2.31}$$

for all $x \in \bar{\mathfrak{g}}$, where ω is an automorphism of $\bar{\mathfrak{g}}$ of finite order. This induces a $\mathbb{Z}_\ell$-gradation of $\mathfrak{g}$, where ℓ is the order of ω. Namely, $\bar{\mathfrak{g}}$ splits into eigenspaces of ω as

$$\bar{\mathfrak{g}} = \bigoplus_{i=0}^{\ell-1} \bar{\mathfrak{g}}_{[i]} \tag{2.2.32}$$

with

$$\bar{\mathfrak{g}}_{[j]} := \{x \in \bar{\mathfrak{g}} \mid \omega(x) = \mathrm{e}^{2\pi i \cdot j/\ell} x\}, \tag{2.2.33}$$

and

$$[\bar{\mathfrak{g}}_{[i]}, \bar{\mathfrak{g}}_{[j]}] \subseteq \bar{\mathfrak{g}}_{[i+j \bmod \ell]}. \tag{2.2.34}$$

Also, $\bar{\mathfrak{g}}_{[0]}$ is a subalgebra of $\bar{\mathfrak{g}}$, and the $\bar{\mathfrak{g}}_{[j]}, j = 1, \dots, \ell - 1$, transform as some highest-weight representation of $\bar{\mathfrak{g}}_{[0]}$. Now the definition of $\bar{\mathfrak{g}}_{[j]}$ shows that the maps $\mathcal{P}$ in (2.2.31) must behave as

$$\mathcal{P}(\mathrm{e}^{2\pi i} z) = \mathrm{e}^{2\pi i \cdot j/\ell} \mathcal{P}(z) \tag{2.2.35}$$

if they are to be multiplied with $x \in \bar{\mathfrak{g}}_{[j]}$, and hence a basis for the vector space is

$$\{T^a_{m+j/\ell}\} \equiv \{T^a \otimes z^{m+j/\ell} \mid T^a \in \bar{\mathfrak{g}}_{[j]},\ j = 0, 1, \dots, \ell-1;\ m \in \mathbb{Z}\}. \tag{2.2.36}$$

As in the untwisted case, the vector space generated by this basis is naturally endowed with a Lie algebra structure, namely via

$$[T^a_{m+j/\ell}, T^b_{n+j'/\ell}] = f^{ab}{}_c T^c_{m+n+(j+j')/\ell}, \tag{2.2.37}$$

and again there exists a unique non-trivial central extension generated by (2.2.36) together with a central generator K, with brackets

$$\begin{aligned} &[T^a_{m+j/\ell}, T^b_{n+j'/\ell}] \\ &\quad = f^{ab}{}_c T^c_{m+n+(j+j')/\ell} + \left(m + \frac{j}{\ell}\right) \bar{\kappa}^{ab} \delta_{m+n+(j+j')/\ell,0} K \end{aligned} \tag{2.2.38}$$

and $[T^a_{m+j/\ell}, K] = 0$; this Lie algebra is denoted by $\hat{\mathfrak{g}}^\omega$. Again one can adjoin a derivation to this algebra; this way one gets the twisted affine algebra $\mathfrak{g}^\omega$. Note that the algebra $\bar{\mathfrak{g}}_{[0]}$ can then be characterized as that subalgebra of $\mathfrak{g}$ which is generated by $\{T^a_m \mid m = 0\}$ so that the notation $\bar{\mathfrak{g}}_{[0]}$ introduced in (2.2.32) is in accordance with its use in the case of untwisted algebras.

If ω is an inner automorphism, this construction in fact does not lead to anything new: an inner automorphism of $\bar{\mathfrak{g}}$ of order ℓ can always be written as

$$\omega = \exp\left(2\pi i \sum_{j=1}^{r} (\zeta_\omega)_j \, \mathrm{ad}_{H^j}\right) \tag{2.2.39}$$

with the *shift vector* ζ_ω fulfilling

$$\ell\,(\zeta_\omega, \bar{\alpha}) \in \mathbb{Z} \quad \text{for all} \quad \bar{\alpha} \in \bar{\Phi}. \tag{2.2.40}$$

In particular, such automorphisms leave the Cartan subalgebra of $\mathfrak{g}$ invariant so that the corresponding generators appearing in (2.2.38) are "integer moded", i.e. are of the form $\{H^i_n\}_{n\in\mathbb{Z}}$. One may then verify that the following generators satisfy the bracket relations (2.2.19), (2.2.25) of an untwisted affine algebra:

$$\begin{aligned} \tilde{E}^\alpha_n &= E^\alpha_{n+(\zeta_\omega,\alpha)}, \\ \tilde{H}^i_n &= H^i_n + \zeta^i_\omega \, \delta_{n,0} \, K, \\ \tilde{K} &= K, \\ \tilde{D} &= D - (\zeta_\omega)_i H^i_0. \end{aligned} \tag{2.2.41}$$

In contrast, if the automorphism ω is outer, then $\mathfrak{g}^\omega$ is a new algebra. Recalling that any outer automorphism of a simple Lie algebra $\mathfrak{g}$ can be decomposed into an inner automorphism and an automorphism of the Dynkin diagram of $\mathfrak{g}$, it follows that the only nontrivial ones correspond to Dynkin diagram automorphisms and hence are of order $\ell = 2$ in the cases $\mathfrak{g} = A_r$ $(r \geq 2)$, D_r $(r \geq 5)$ and E_6, and of order $\ell = 3$ in the case $\mathfrak{g} = D_4$. The twisted affine algebras found in the previous section are exactly the Lie algebras obtained in these cases from $\hat{\mathfrak{g}}^\omega$ by adding the derivation D. To be precise, there exists a specific homogeneous gradation for which the relation goes as in table (2.2.42).

$\hat{\mathfrak{g}}^\omega \oplus \mathbb{C}D$	$\mathfrak{g}$
$A_1^{(2)}$	A_2
$F_4^{(2)}$	E_6
$B_r^{(2)}$	D_{r+1}
$\tilde{B}_r^{(2)}$	A_{2r}
$C_r^{(2)}$	A_{2r-1}
$G_2^{(3)}$	D_4

(2.2.42)

The twisted algebras $A_1^{(2)}$, $F_4^{(2)}$, $B_r^{(2)}$, $\tilde{B}_r^{(2)}$, $C_r^{(2)}$, $G_2^{(3)}$ are often (in fact, in the larger part of the literature) denoted instead by $A_2^{(2)}$, $E_6^{(2)}$, $D_{r+1}^{(2)}$, $A_{2r}^{(2)}$, $A_{2r-1}^{(2)}$ and $D_4^{(3)}$, respectively, in order to emphasize the simple algebra $\bar{\mathfrak{g}}$ they stem from in this particular realization of the algebra $\mathfrak{g}$ as a central extension of a loop algebra. Note that while in the untwisted case the horizontal subalgebra $\mathfrak{g}_{\rm hor}$ and the algebra $\bar{\mathfrak{g}}$ which is the basis of the loop algebra are isomorphic, this is no longer true for the twisted affine algebras. Rather, the horizontal subalgebra is again identical to the zero mode subalgebra of $\mathfrak{g}$, i.e. to the subalgebra spanned by the generators T^a_m with $m = 0$, and hence

$$\mathfrak{g}_{\rm hor} \cong \bar{\mathfrak{g}}_{[0]} \subset \bar{\mathfrak{g}}. \tag{2.2.43}$$

In the above construction of the affine algebras the finite-dimensional algebra $\bar{\mathfrak{g}}$ has to be simple because the affine Cartan matrix is required to be indecomposable. If the requirement of indecomposability is removed, then of course the same construction works with $\bar{\mathfrak{g}}$ now required to be semisimple. In fact one can also extend the construction easily to the case of reductive $\bar{\mathfrak{g}}$. One then has to introduce an independent central extension for each simple subalgebra and also for each u_1-subalgebra. However, the central extensions of the u_1 algebras can all be identified with each other owing to the fact that the irreducible modules of u_1 are one-dimensional. Writing $\bar{\mathfrak{g}} = \bar{\mathfrak{s}} \oplus \bar{\mathfrak{h}}$ with $\bar{\mathfrak{s}}$ semisimple and $\bar{\mathfrak{h}}$ abelian, the central extension of $\mathfrak{g}_{\rm loop}$ is therefore of the form of a direct sum

$$\hat{\mathfrak{g}} = \hat{\mathfrak{s}} \oplus \hat{\mathfrak{h}}, \tag{2.2.44}$$

with the Lie brackets of $\hat{\mathfrak{s}}$ given by (2.2.19) (respectively, in the twisted case, by (2.2.38)), and those of $\hat{\mathfrak{h}}$ by

$$[t^a_m, t^b_n] = d^{ab}\, m\, \delta_{m+n,0} K. \tag{2.2.45}$$

Here $m, n \in \mathbb{Z}$, $a, b = 1, \ldots, \dim \bar{\mathfrak{h}}$, and d^{ab} a constant symmetric matrix; a Lie algebra with the brackets (2.2.45) is called a *Heisenberg algebra* because it is analogous to the Heisenberg commutation relations in quantum mechanics. Any affine algebra in fact contains a Heisenberg algebra as a subalgebra, namely the one generated by $\{H^i_m\}$ and K with $[H^i_m, K] = 0$ and

$$[H^i_m, H^j_n] = \bar{\kappa}(H^i, H^j)\, m\, \delta_{m+n,0} K. \tag{2.2.46}$$

If there is no index a, the algebra (2.2.45) is also called the $\hat{u}_1$ Kac–Moody algebra. Note that by suitable redefinition of the generators, the matrix d^{ab} in (2.2.45) can be taken to be δ^{ab}; in other words, any Heisenberg algebra is isomorphic to a direct sum of $\hat{u}_1$ algebras (with identified centers).

2.3 The root system

We now construct the analogue of the Cartan–Weyl basis for the affine Lie algebras. For the moment we consider only the untwisted case. The affine Cartan–Weyl basis is most easily found by using the realization of $\mathfrak{g}$ as a centrally extended loop algebra with derivation, as presented in the previous section.

We first have to determine a maximal abelian subalgebra. This certainly contains the Cartan subalgebra of $\mathfrak{g}_{\rm hor}$, generated by H^i_0, $i = 1, \ldots, r$, and the central generator K. Because of

$$\begin{aligned}&[H^i_0, H^j_n] = [K, H^j_n] = 0,\\ &[H^i_0, D] = [K, D] = 0,\end{aligned} \tag{2.3.1}$$

it is however clear that it contains at least one further generator. Choosing D as this generator, the relation

$$[D, H^j_n] \neq 0 \ \text{ for } \ n \neq 0 \tag{2.3.2}$$

shows that it is in fact the only one. Thus the Cartan subalgebra of $\mathfrak{g}$ is

$$\mathfrak{g}_0 = \text{span}\{K, D, H^i_0 \mid i = 1, \ldots, r\}, \tag{2.3.3}$$

or, what is the same,

$$\mathfrak{g}_0 = \text{span}\{D, H^i_0 \mid i = 0, 1, \ldots, r\}. \tag{2.3.4}$$

(If the presentation (2.1.1) of $\mathfrak{g}$ via generators and relations is used, it is considerably more involved to show that the rank of a maximal abelian subalgebra is $r + 2$; namely, one has to perform the analogue of the Serre construction described on p. 39.) The roots with respect to $\mathfrak{g}_0$ are easily found from the observation that for any root $\bar{\alpha}$ of $\bar{\mathfrak{g}}$ and any $n \in \mathbb{Z}$,

$$[H^i_0, E^{\bar{\alpha}}_n] = \bar{\alpha}^i E^{\bar{\alpha}}_n, \quad [K, E^{\bar{\alpha}}_n] = 0, \quad [D, E^{\bar{\alpha}}_n] = n E^{\bar{\alpha}}_n, \tag{2.3.5}$$

and

$$[H^i_0, H^j_n] = [K, H^j_n] = 0, \ \ [D, H^j_n] = n H^j_n. \tag{2.3.6}$$

The roots with respect to (H, K, D) are thus

$$\alpha = (\bar{\alpha}, 0, n), \quad \bar{\alpha} \in \Phi(\bar{\mathfrak{g}}),\ n \in \mathbb{Z} \tag{2.3.7}$$

and

$$\alpha = (0, 0, n), \quad n \in \mathbb{Z} \setminus \{0\}, \tag{2.3.8}$$

corresponding to the generators $E_n^{\bar\alpha}$ and $H_n^j, n \neq 0$, respectively. The roots (2.3.7) are non-degenerate (i.e. have multiplicity one) while the roots (2.3.8) are r-fold degenerate (i.e. have multiplicity r) because they do not depend on the label j of H_n^j.

Note that for the derived algebra $\mathsf{g}' \equiv [\mathsf{g},\mathsf{g}] = \hat{\mathsf{g}}$, all roots are infinitely degenerate because in $\hat{\mathsf{g}}$ there is no Cartan subalgebra generator distinguishing between different labels n. On the other hand, when one defines the affine algebra g via generators and relations as in (2.1.1), then it is rather clear that the simple roots are not degenerate. Thus the inclusion of the derivation D is indeed just what is needed to obtain g from the centrally extended loop algebra $\hat{\mathsf{g}}$. However, there remains in fact a large freedom in the choice of the brackets $[D, H^i]$. This choice is said to define a *gradation* of the algebra g, because any such choice corresponds to the decomposition of an underlying simple Lie algebra $\bar{\mathsf{g}}$ into eigenspaces with respect to some automorphism ω. The choice presented in (2.2.25) above corresponds to the *homogeneous gradation* which is defined by requiring

$$[D, x] = 0 \quad \text{for all} \quad x \in \mathsf{g}_{\text{hor}} \subset \mathsf{g}. \tag{2.3.9}$$

In the untwisted case, this fixes the gradation uniquely; in contrast, for some of the twisted algebras, there is still some further freedom left: the algebras $C_r^{(2)}$, $F_4^{(2)}$, and $G_2^{(3)}$ admit two inequivalent homogeneous gradations, and the algebras $B_r^{(2)}$ admit $[\frac{r}{2}]+1$ inequivalent ones. While the gradation used above corresponds to identifying the zero mode subalgebra $\mathsf{g}_{[0]}$ of g as the horizontal subalgebra g_{hor} (remember that g_{hor} is obtained by deleting a node with $a_i = a_i^\vee = 1$ from the Dynkin diagram of g), for the other homogeneous gradations the zero mode subalgebra is obtained by deleting a node obeying $a_i = 1$ but $a_i^\vee \neq 1$ from the Dynkin diagram. Inspection of the various Dynkin diagrams thus shows that in the case of $C_r^{(2)}, F_4^{(2)}, G_2^{(3)}$ and $B_r^{(2)}$, the zero mode subalgebras in these other homogeneous gradations are, respectively, D_r instead of C_r, C_4 instead of F_4, A_2 instead of G_2, and $B_n \oplus B_{r-n}$ for $n = 1, 2, \ldots, [\frac{r}{2}]$ instead of B_r.

A gradation can actually be associated to each conjugacy class of the Weyl group of $\bar{\mathsf{g}}$. Throughout this book we always work in that homogeneous gradation for which the zero mode subalgebra is obtained by deleting a node with $a_i = a_i^\vee = 1$ from the Dynkin diagram; this corresponds to the class of the identity element of $W(\bar{\mathsf{g}})$. An other gradation that is frequently used in the literature is the one that corresponds to the class of the Coxeter element $\bar\sigma_c \in W(\bar{\mathsf{g}})$; this is called the *principal* gradation.

While the infinite degeneracy of the roots of the derived algebra can be removed by including the derivation D, there is no way to get rid of the remaining degeneracy of the roots (2.3.8).

We denote the set of roots of $\mathfrak{g}$ by Φ. In the untwisted case, we identify the root system of the horizontal subalgebra of $\mathfrak{g}$ with the root system $\bar{\Phi}$ of $\bar{\mathfrak{g}}$ via

$$(\bar{\alpha},0,0) \equiv \bar{\alpha}. \qquad (2.3.10)$$

As in the semisimple case, the next task is to define positive roots and identify a set of simple roots. Given such an identification for $\bar{\mathfrak{g}}$, the positive roots can be defined by

$$\Phi_+ := \{\alpha = (\bar{\alpha},0,n) \in \Phi \mid n > 0 \quad \text{or} \quad (n=0,\, \bar{\alpha} \in \bar{\Phi}^+)\} \qquad (2.3.11)$$

and the negative ones by $\Phi_- = \Phi \setminus \Phi_+$. We will again use the notation $\alpha > 0$ ($\alpha < 0$) for positive (negative) roots. Denoting the subalgebras of $\mathfrak{g}$ generated by the generators corresponding to positive and to negative roots by $\mathfrak{g}_+$ and $\mathfrak{g}_-$, respectively, i.e. $\mathfrak{g}_+ = \text{span}\{E^\alpha \,|\, \alpha > 0\}$, $\mathfrak{g}_- = \text{span}\{E^{-\alpha} \,|\, \alpha > 0\}$, we have again the Cartan decomposition

$$\mathfrak{g} = \mathfrak{g}_+ \oplus \mathfrak{g}_0 \oplus \mathfrak{g}_-, \qquad (2.3.12)$$

and $\mathfrak{g}_\pm$ are nilpotent subalgebras while $\mathfrak{b}_\pm := \mathfrak{g}_0 \oplus \mathfrak{g}_\pm$ are Borel subalgebras of $\mathfrak{g}$.

With the choice (2.3.11), the simple roots are

$$\alpha^{(i)} = (\bar{\alpha}^{(i)},0,0) = \bar{\alpha}^{(i)} \quad \text{for} \quad i = 1,\ldots,r \qquad (2.3.13)$$

and

$$\alpha^{(0)} = (-\bar{\theta},0,1) = \delta - \bar{\theta}. \qquad (2.3.14)$$

Here $\bar{\theta}$ is the highest root of $\bar{\mathfrak{g}}$, and we have set

$$\delta = (0,0,1) \qquad (2.3.15)$$

(i.e. the degenerate roots are $\alpha = n\cdot\delta$, $n \neq 0$). The generators corresponding to the simple roots are denoted by E_+^i, $i = 0,1,\ldots,r$. This notation is in agreement with the presentation (2.1.1) of $\mathfrak{g}$, i.e. the step operators E_+^i are precisely the generators appearing there. Their explicit form can be read off (2.3.7):

$$\begin{aligned} E_+^i &= E_0^{\bar{\alpha}^{(i)}} \quad \text{for} \quad i = 1,\ldots,r, \\ E_+^0 &= E_1^{-\bar{\theta}}. \end{aligned} \qquad (2.3.16)$$

Also, for the special case of $\mathfrak{g} = A_1^{(1)}$, there are two simple roots $\alpha^{(1)} = \bar{\theta} = \bar{\alpha}$ and $\alpha^{(0)} = \delta - \bar{\theta} = \delta - \bar{\alpha}$, with $\bar{\alpha}$ the single positive root of the horizontal subalgebra A_1. Thus the root system Φ is indeed in agreement with the result (2.1.15) which was obtained by studying the Chevalley–Serre basis of $A_1^{(1)}$.

To understand better the appearance of $\bar{\theta}$ in $\alpha^{(0)}$, recall that any root of $\bar{\mathfrak{g}}$ can be obtained by adding simple roots to $-\bar{\theta}$ (see p. 35) so that by adding simple roots $\alpha^{(i)}$, $i = 1, ..., r$, to $\alpha^{(0)}$ one can obtain all roots of the form $(\bar{\alpha}, 0, 1)$. Thus $\alpha^{(0)}$ is just what is needed to have $(\bar{\alpha}, 0, 1) \in \Phi_+$ for any $\bar{\alpha} \in \bar{\Phi}$ (as prescribed by the choice (2.3.11) of Φ_+), and analogously for all other positive roots. In other words, writing

$$\alpha = \sum_{i=0}^{r} b_i \, \alpha^i , \tag{2.3.17}$$

one has

$$\begin{aligned} \alpha > 0 \quad &\Leftrightarrow \quad b_i \geq 0 , \\ \alpha < 0 \quad &\Leftrightarrow \quad b_i \leq 0 . \end{aligned} \tag{2.3.18}$$

Next one should look for an analogue of the Killing form. The definition (1.3.1) used in the simple case can not be taken over to the affine case because ad_x are now infinite-dimensional maps so that the trace over $\mathrm{ad}_x \circ \mathrm{ad}_y$ is not well defined. Instead, recalling that the Killing form could be characterized completely (up to normalization) by the properties of symmetry, bilinearity and associativity, we try to fix it also in the affine case by imposing these requirements. To this end we consider special combinations of the generators appearing in the associativity constraint

$$\kappa([w, x], y) = \kappa(w, [x, y]) \tag{2.3.19}$$

and simplify the resulting formulae with the help of the symmetry and bilinearity of κ. Choosing $w = T^a_m$, $x = D$, $y = T^b_n$, one gets $(m + n)\, \kappa(T^a_m, T^b_n) = 0$, while the choice $w = T^a_m$, $x = T^b_0$, $y = T^c_{-m}$ gives $f^{ab}{}_d \, \kappa(T^d_m, T^c_{-m}) = f^{bc}{}_d \, \kappa(T^a_m, T^d_{-m})$, implying that

$$\kappa(T^a_m, T^b_n) \propto \delta_{m+n,0} \, \bar{\kappa}^{ab} . \tag{2.3.20}$$

The choice $w = T^a_m$, $x = T^b_1$, $y = T^c_{-m-1}$ gives $f^{ab}{}_d \, \kappa(T^d_{m+1}, T^c_{-m-1}) = f^{bc}{}_d \, \kappa(T^a_m, T^d_{-m})$, (2.3.20) is m-independent; therefore one can choose the normalization of κ such that

$$\kappa(T^a_m, T^b_n) = \delta_{m+n,0} \, \bar{\kappa}^{ab} . \tag{2.3.21}$$

Next, consideration of $w = T^a_m$, $x = T^b_n$, $y = K$ and $w = T^a_m$, $x = T^b_{-m}$, $y = K$ shows that

$$\kappa(T^a_m, K) = 0 = \kappa(K, K) . \tag{2.3.22}$$

Finally, the choice $w = T^a_m$, $x = T^b_n$, $y = D$ yields

$$f^{ab}{}_c \, \kappa(T^c_{m+n}, D) + m \, \bar{\kappa}^{ab} \delta_{m+n,0} \kappa(K, D) = -n \, \kappa(T^a_m, T^b_n) , \tag{2.3.23}$$

so that the symmetry of κ shows that

$$\kappa(T^a_m, D) = 0 , \quad \kappa(K, D) = 1 . \tag{2.3.24}$$

Since D does not appear on the right hand side of the bracket relations it is not possible to obtain $\kappa(D,D)$ in this way. However, suppose that $\kappa(D,D)=\eta$ is nonvanishing; then one could consider the generator

$$D' = D - \tfrac{1}{2}\,\eta K \tag{2.3.25}$$

instead of D. D' has the same Lie brackets as D, and using the bilinearity of κ one has $\kappa(D',D')=0$. Therefore without loss of generality we can assume that

$$\kappa(D,D)=0. \tag{2.3.26}$$

Summarizing, we get for $\{T^a_m, K, D\}$ the Killing form

$$\kappa = \begin{pmatrix} \bar\kappa\,\delta_{m+n,0} & 0 & 0 \\ 0 & 0 & 1 \\ 0 & 1 & 0 \end{pmatrix}. \tag{2.3.27}$$

As in the semisimple case, the restriction of κ to the Cartan subalgebra provides a metric for the root space and for its dual, the weight space, namely

$$G^{ij} = \begin{pmatrix} \bar G^{ij} & 0 & 0 \\ 0 & 0 & 1 \\ 0 & 1 & 0 \end{pmatrix}. \tag{2.3.28}$$

Unlike $\bar G^{ij}$, this metric is of lorentzian nature, as it has the real eigenvectors $(0^r,1,\pm 1)$ with eigenvalues ± 1. (To make this explicit, one could diagonalize the 2×2-matrix in the lower right hand corner by using $D+K$ and $D-K$ instead of K and D as generators; however, this would obscure the structure of the bracket relations because then all generators would appear on their right hand sides.)

According to (2.3.28), the scalar product of two elements $\lambda = (\bar\lambda, k, n)$ and $\lambda' = (\bar\lambda', k', n')$ is

$$(\lambda,\lambda') = (\bar\lambda,\bar\lambda') + kn' + k'n. \tag{2.3.29}$$

In particular for roots the scalar product equals the $\bar{\mathfrak g}$-product of their $\bar{\mathfrak g}$-components,

$$(\alpha,\alpha') \equiv ((\bar\alpha,0,n),(\bar\alpha',0,n')) = (\bar\alpha,\bar\alpha'). \tag{2.3.30}$$

Thus for the non-degenerate roots,

$$(\alpha,\alpha) > 0 \quad \text{for } \bar\alpha \in \bar\Phi, \tag{2.3.31}$$

while for the degenerate roots,

$$(\alpha,\alpha) = 0 \quad \text{for } \alpha = n\cdot\delta,\ n \neq 0. \tag{2.3.32}$$

The non-degenerate roots are therefore called *real roots*, whereas the degenerate ones are called *imaginary roots* (this latter term is a bit misleading; the term *lightlike* roots which is also sometimes used is certainly more appropriate).

The simple roots are non-degenerate and hence real; therefore the definition of simple coroots,

$$\alpha^{(i)\vee} := \frac{2}{(\alpha^{(i)}, \alpha^{(i)})} \alpha^{(i)}, \quad i = 0, \ldots, r, \tag{2.3.33}$$

makes sense. Then one can also define the matrix

$$A^{ij} := (\alpha^{(i)}, \alpha^{(j)\vee}), \quad i, j = 0, \ldots, r. \tag{2.3.34}$$

This is in fact just the affine Cartan matrix as it was introduced in section 2.1. As a consequence, we finally can prove that the centrally extended loop algebras with derivation are indeed isomorphic to the untwisted affine algebras defined via their generalized Cartan matrix. To see that the numbers A^{ij} defined by (2.3.34) make up the correct affine Cartan matrix, one just has to check that the matrix corresponds to the correct Dynkin diagram. Let us do this for two examples, namely $\bar{\mathfrak{g}} = A_r$ and $\bar{\mathfrak{g}} = E_6$. In the former case, one has

$$\bar{\theta} = \bar{\Lambda}_{(1)} + \bar{\Lambda}_{(r)} \tag{2.3.35}$$

and hence

$$(\alpha^{(0)}, \alpha^{(i)}) = -(\bar{\theta}, \bar{\alpha}^{(i)}) = -\delta_{i,1} - \delta_{i,r}; \tag{2.3.36}$$

therefore the additional node in the Dynkin diagram must be connected by a single line with the first and last nodes so that we get indeed the Dynkin diagram of $A_r^{(1)}$ (both for $r = 1$ and for $r > 1$) from that of A_r. In the case of E_6, one has $\bar{\theta} = \bar{\Lambda}_{(6)}$ and hence $(\alpha^{(0)}, \alpha^{(i)}) = -\delta_{i,6}$, again leading to the correct Dynkin diagram.

The root system of the twisted affine Lie algebras can be obtained by considerations similar to those above. The relevant outer automorphisms ω correspond to automorphisms $\dot{\omega}$ of the Dynkin diagram of $\bar{\mathfrak{g}}$, and hence act as $\omega(\bar{\alpha}^{(i)}) = \bar{\alpha}^{(\dot{\omega}(i))}$, and analogously on the Chevalley generators as $\omega(H^i) = H^{\dot{\omega}(i)}$, $\omega(E_\pm^i) = E_\pm^{\dot{\omega}(i)}$ for $i = 1, \ldots, r$. From this one can deduce that the ladder operators corresponding to an arbitrary root obey $\omega(E^{\bar{\alpha}}) = \varepsilon_{\bar{\alpha}} E^{\omega(\bar{\alpha})}$ with $\varepsilon_{\bar{\alpha}} \in \{\pm 1\}$ and $\varepsilon_{-\bar{\alpha}} = \varepsilon_{\bar{\alpha}}$. Also one has $\omega(\Phi_+) = \Phi_+$ so that in particular $\omega(\bar{\alpha}) \neq -\bar{\alpha}$ for any root $\bar{\alpha}$.

Now first consider the algebras of type $\mathfrak{g} = X^{(2)}$. For the Cartan subalgebra, one has

$$\begin{aligned} &(\bar\alpha^{(i)}, H) \in \mathfrak{g}_{[0]} && \text{if } \dot\omega(i) = i, \\ &\left.\begin{aligned} &\tfrac12\,(\bar\alpha^{(i)} + \bar\alpha^{(\dot\omega(i))}\,, H) \in \mathfrak{g}_{[0]} \\ &\tfrac12\,(\bar\alpha^{(i)} - \bar\alpha^{(\dot\omega(i))}\,, H) \in \mathfrak{g}_{[1]} \end{aligned}\right\} && \text{otherwise.} \end{aligned} \tag{2.3.37}$$

(Recall that $\mathfrak{g}_{[0]}$ and $\mathfrak{g}_{[1]}$ are the eigenspaces of ω to the eigenvalues $+1$ and -1, respectively.) For the step operators, notice that $\omega(E^{\bar\alpha} \pm E^{\omega(\bar\alpha)}) = \pm\varepsilon_{\bar\alpha}(E^{\bar\alpha} \pm E^{\omega(\bar\alpha)})$, and hence one has

$$\begin{aligned} &E^{\bar\alpha} \in \mathfrak{g}_{[0]} && \text{if } \omega(\bar\alpha) = \bar\alpha \text{ and } \varepsilon_{\bar\alpha} = 1, \\ &E^{\bar\alpha} \in \mathfrak{g}_{[1]} && \text{if } \omega(\bar\alpha) = \bar\alpha \text{ and } \varepsilon_{\bar\alpha} = -1, \\ &\left.\begin{aligned} &(E^{\bar\alpha} + \varepsilon_{\bar\alpha} E^{\omega(\bar\alpha)}) \in \mathfrak{g}_{[0]} \\ &(E^{\bar\alpha} - \varepsilon_{\bar\alpha} E^{\omega(\bar\alpha)}) \in \mathfrak{g}_{[1]} \end{aligned}\right\} && \text{if } \omega(\bar\alpha) \neq \bar\alpha. \end{aligned} \tag{2.3.38}$$

In other words, the roots of $\mathfrak{g}_{[0]}$ are

$$\begin{aligned} \Phi(\mathfrak{g}_{[0]}) = \{&\bar\alpha \mid \omega(\bar\alpha) = \bar\alpha,\ \varepsilon_{\bar\alpha} = 1\} \\ &\cup \{\tfrac12(\bar\alpha + \omega(\bar\alpha)) \mid \omega(\bar\alpha) \neq \bar\alpha,\ \varepsilon_{\bar\alpha} = 1\}, \end{aligned} \tag{2.3.39}$$

and the weights of $\mathfrak{g}_{[1]}$ are

$$\begin{aligned} \{&\bar\alpha \mid \omega(\bar\alpha) = \bar\alpha,\ \varepsilon_{\bar\alpha} = -1\} \\ &\cup \{\tfrac12(\bar\alpha + \omega(\bar\alpha)) \mid \omega(\bar\alpha) \neq \bar\alpha,\ \varepsilon_{\bar\alpha} = -1\}. \end{aligned} \tag{2.3.40}$$

In particular, a basis of simple roots for $\mathfrak{g}_{[0]}$ is given by the vectors $\bar\alpha^{(i)}_{[0]} = \frac12(\bar\alpha^{(i)} + \bar\alpha^{(\dot\omega(i))})$ (thus for each pair of indices i_1, i_2 with $i_1 \neq i_2 = \dot\omega(i_1)$ the rank gets reduced by one). The simple roots of $\mathfrak{g}$ can then be chosen to be the simple roots $\bar\alpha^{(i)}_{[0]}$ of $\mathfrak{g}_{[0]}$ together with $\frac12\delta - \bar\Lambda_1$ where $\delta = (0,0,1)$ is as defined in (2.3.15) and $\bar\Lambda_1$ is the highest weight of the $\mathfrak{g}_{[0]}$-module furnished by $\mathfrak{g}_{[1]}$, i.e. the highest vector of the form (2.3.40). The real roots are then either of the form

$$\alpha = \bar\alpha + n\,\delta \tag{2.3.41}$$

with $\bar\alpha \in \bar\Phi(\mathfrak{g}_{[0]})$ and $n \in \mathbb{Z}$, or of the form

$$\alpha = \bar\lambda_1 + (n + \tfrac12)\,\delta \tag{2.3.42}$$

with $\bar\lambda_1$ a weight of $R_{\bar\Lambda_1}$ and $n \in \mathbb{Z}$. The imaginary roots are

$$\alpha = \tfrac12\,n\,\delta \tag{2.3.43}$$

with $n\in\mathbb{Z}\setminus\{0\}$. The real roots are not degenerate, while the multiplicity of the imaginary roots is

$$\text{mult}\,(\tfrac{1}{2}\,n\delta)=\begin{cases}\text{rank}(\mathfrak{g}_{[0]}) & \text{for}\quad n\in 2\mathbb{Z}\\ r-\text{rank}(\mathfrak{g}_{[0]}) & \text{for}\quad n\in 2\mathbb{Z}+1.\end{cases}\tag{2.3.44}$$

In the case of the triply twisted algebra $G_2^{(3)}$, the algebra $\mathfrak{g}$ on which ω acts is D_4. By an analysis analogous to the one performed for the doubly twisted algebras, one finds that the roots of $\mathfrak{g}_{[0]}$ are

$$\begin{aligned}\Phi(\mathfrak{g}_{[0]})=\{\bar\alpha\mid\omega(\bar\alpha)=\bar\alpha\}&\\ \cup\,\{\tfrac{1}{3}\,(\bar\alpha+\omega(\bar\alpha)+\omega^2(\bar\alpha))\mid\omega(\bar\alpha)\neq\bar\alpha\}&.\end{aligned}\tag{2.3.45}$$

From $(\bar\alpha,\omega(\bar\alpha))=0$ for $\bar\alpha\neq\omega(\bar\alpha)$ it follows that the roots in the two subsets appearing here have relative length squares 1 : 3, and hence $\mathfrak{g}_{[0]}$ is to be identified with G_2. A basis of simple roots for G_2 is given by $\{\bar\alpha^{(4)}\,,\tfrac{1}{3}\,(\bar\alpha^{(1)}+\bar\alpha^{(2)}+\bar\alpha^{(3)})\}$, and the simple root that one must add to obtain a basis of $G_2^{(3)}$ is $\tfrac{1}{3}\delta-\bar\Lambda$ with $\bar\Lambda$ the highest vector of the form $\tfrac{1}{3}\,(\bar\alpha+\omega(\bar\alpha)+\omega^2(\bar\alpha))$ with $\omega(\bar\alpha)\neq\bar\alpha$; this turns out to be $\bar\Lambda=\tfrac{1}{3}\,(\bar\alpha^{(1)}+\bar\alpha^{(2)}+\bar\alpha^{(3)})=\bar\Lambda_{(2)}$. Thus the roots of $G_2^{(3)}$ are of one of the three forms

$$\alpha=\bar\alpha+n\,\delta\quad\text{or}\quad\alpha=\bar\lambda+(n\pm\tfrac{1}{3})\,\delta,\tag{2.3.46}$$

where $\bar\alpha$ are the roots of G_2 and $\bar\lambda$ the weights of the lowest-dimensional module $R_{\bar\Lambda_{(2)}}$ of G_2, and where $n\in\mathbb{Z}$. Also, the real roots are non-degenerate, while the imaginary roots are of the form

$$\alpha=\tfrac{1}{3}\,n\,\delta\tag{2.3.47}$$

with $n\in\mathbb{Z}\setminus\{0\}$, and their multiplicity is one if $n/3\in\mathbb{Z}$, and two otherwise.

Inspection shows that in almost all cases the nonzero weights of the modules furnished by $\mathfrak{g}_{[1]}$ (and, for $G_2^{(3)}$, by $\mathfrak{g}_{[2]}$ as well), are the short roots of $\mathfrak{g}_{[0]}$. The only exceptions are $A_1^{(2)}$ and $\tilde B_r^{(2)}$ $(r\geq 2)$, for which the nonzero weights of $\mathfrak{g}_{[1]}$ are the roots of $\mathfrak{g}_{[0]}$ together with the short roots multiplied by two.

As already mentioned, for some of the twisted algebras there exist more than one independent homogeneous gradation, and hence different choices of the zero mode subalgebra $\mathfrak{g}_{[0]}$ are possible. The above results correspond to the gradation in which $\mathfrak{g}_{[0]}$ coincides with the horizontal subalgebra $\mathfrak{g}_{\text{hor}}$. As an example for a different gradation, consider $G_2^{(3)}$; then one can have A_2 instead of G_2 as zero mode subalgebra. The real roots are in this case

$$\alpha = \bar\alpha + n\delta \quad\text{or}\quad \alpha = \bar\lambda + (n+\tfrac{1}{3})\,\delta \\ \text{or}\quad \alpha = \bar\lambda^{+} + (n-\tfrac{1}{3})\,\delta, \tag{2.3.48}$$

where $\bar\alpha$ are now the roots of A_2, $\bar\lambda$ the weights of the 10-dimensional A_2-module $R_{3\bar\Lambda_{(1)}}$ and $\bar\lambda^{+}$ the weights of its conjugate module $R_{3\bar\Lambda_{(2)}}$. The real roots are not degenerate, and the imaginary roots are again of the form $\frac{1}{3}n\delta$ with n a nonzero integer, and with multiplicity equal to one if n is divisible by three, and equal to two otherwise.

We summarize some of the above results in table (2.3.49)

$\bar{\mathfrak g}$	$\mathfrak g$	$\bar{\mathfrak g}_{[0]}$	$\bar{\mathfrak g}_{[1]}$	$\bar{\mathfrak g}_{[2]}$	$\dim\bar{\mathfrak g}_{[1]}$
A_2	$A_1^{(2)}$	A_1	$4\bar\Lambda_{(1)}$	–	5
E_6	$F_4^{(2)}$	F_4	$\bar\Lambda_{(4)}$	–	26
		C_4	$\bar\Lambda_{(4)}$	–	42
D_{r+1}	$B_r^{(2)}$	B_r	$\bar\Lambda_{(1)}$	–	$2r+1$
		$B_n\oplus B_{r-n}$	$(\bar\Lambda_{(1)},\bar\Lambda_{(1)})$	–	$(2n+1)(2r-2n+1)$
A_{2r}	$\tilde B_r^{(2)}$	B_r	$2\bar\Lambda_{(1)}$	–	$r(2r+3)$
A_{2r-1}	$C_r^{(2)}$	C_r	$\bar\Lambda_{(2)}$	–	$(2r+1)(r-1)$
		D_r	$2\bar\Lambda_{(1)}$	–	$(2r-1)(r+1)$
D_4	$G_2^{(3)}$	G_2	$\bar\Lambda_{(2)}$	$\bar\Lambda_{(2)}$	7
		A_2	$3\bar\Lambda_{(1)}$	$3\bar\Lambda_{(2)}$	10

(2.3.49)

2.4 Highest weight modules

Among the modules of an affine algebra, the highest weight modules R_Λ are again the most interesting ones. Analogously to the case of simple Lie algebras, they are defined by two types of requirements: first, there is a highest weight vector $v_\Lambda \in R_\Lambda$ which is annihilated by all generators corresponding to positive roots, i.e.

$$R_\Lambda(E_n^{\bar\alpha})\cdot v_\Lambda = 0 \quad\text{for all}\quad \bar\alpha > 0, \\ R_\Lambda(T_n^{a})\cdot v_\Lambda = 0 \quad\text{for all}\quad n > 0; \tag{2.4.1}$$

second, any vector in the module can be obtained by applying an appropriate number of generators corresponding to negative roots to v_Λ, i.e.

$$v = R_\Lambda(T^{a_1}_{-n_1}) R_\Lambda(T^{a_2}_{-n_2}) \ldots R_\Lambda(T^{a_m}_{-n_m}) \cdot v_\Lambda. \tag{2.4.2}$$

Here each $T^{a_p}_{-n_p}$ is either $E^{-\bar\alpha_p}_{-n_p}$ with $\bar\alpha_p > 0$ and $n_p \geq 0$, or $H^{i_p}_{-n_p}$ or $E^{\bar\alpha_p}_{-n_p}$ with $n_p > 0$ so that we may rewrite this as

$$v = \prod_j E^{-\bar\alpha_j}_{-n_j} \prod_p H^{i_p}_{-n_p} \prod_q E_0^{-\bar\alpha_q} \cdot v_\Lambda, \tag{2.4.3}$$

with n_j and n_p positive integers, and $\bar\alpha_j$ and $\bar\alpha_q$ positive $\bar{\mathfrak{g}}$-roots.

The highest weight vector (and hence any vector in the highest weight module) is an eigenvector of the Cartan subalgebra of $\mathfrak{g}$. The eigenvalues are given by the highest weight Λ, i.e.

$$\Lambda = (\bar\Lambda, k, n) \tag{2.4.4}$$

with

$$\begin{aligned} R_\Lambda(H^i_0) \cdot v_\Lambda &= \bar\Lambda^i \, v_\Lambda, \\ R_\Lambda(K) \cdot v_\Lambda &= k \, v_\Lambda, \\ R_\Lambda(D) \cdot v_\Lambda &= n_0 \, v_\Lambda. \end{aligned} \tag{2.4.5}$$

The highest weight module may thus also be defined in a compact way as the module obtained by acting on v_Λ with the quotient of the universal enveloping algebra $\mathsf{U}(\mathfrak{g})$ by the ideal that is generated by the generators of $\mathfrak{g}_+$ together with $H^i_0 - \bar\Lambda_{(i)}$, $i = 1, \ldots, r$, $K - k$ and $D - n_0$.

Because the Cartan subalgebra generators H^i_0, $i = 1, \ldots, r$, just generate the Cartan subalgebra of $\mathfrak{g}_{\rm hor} \cong \bar{\mathfrak{g}}$, their eigenvalues are the components of a highest weight of $\bar{\mathfrak{g}}$. The eigenvalue with respect to the derivation is in fact rather irrelevant because by a redefinition of $R_\Lambda(D)$ it can be brought to zero (namely by $R_\Lambda(D) \mapsto R_\Lambda(D) - n_0 R_\Lambda(K)/k$); therefore from now on we assume $n_0 = 0$. Thus a highest weight module of $\mathfrak{g}$ is characterized by a highest weight $\bar\Lambda$ of $\bar{\mathfrak{g}}$ together with an eigenvalue k of the central generator K. The normalized eigenvalue

$$k^\vee := k \cdot \frac{2}{(\bar\theta, \bar\theta)} \tag{2.4.6}$$

is called the *level* of R_Λ.

As follows immediately from $[K, T^a_n] = 0$, k is in fact also the K-eigenvalue of all other vectors in R_Λ,

$$R_\Lambda(K) \cdot v_\lambda = k \, v_\lambda \quad \text{for all } v_\lambda \in R_\Lambda. \tag{2.4.7}$$

Similarly, from $[D, T^a_n] = nT^a_n$ one gets

$$R_\Lambda(D) \cdot v_\lambda = -(\textstyle\sum_p n_p)\, v_\lambda. \tag{2.4.8}$$

Here n_p are the numbers appearing in the relation (2.4.2) between v_λ and the highest weight vector v_Λ; their sum $\sum_p n_p$ is called the *grade* of the weight λ.

Owing to the existence of the imaginary roots, all nontrivial modules of $\mathfrak{g}$ are infinite-dimensional: already the submodules with respect to the sl_2-subalgebra of $\mathfrak{g}$ corresponding to an imaginary root are infinite-dimensional. In general the same statement will then also hold for real roots. In special cases, however, the submodules with respect to the sl_2-subalgebras of all real roots are finite-dimensional (loosely speaking, such modules are "less infinite-dimensional" than a generic $\mathfrak{g}$-module). To investigate when this happens, we first recall that the sl_2-subalgebra of $\mathfrak{g}$ corresponding to the root $\alpha = (\bar\alpha, 0, n)$ is generated by $E^\alpha = E^{\bar\alpha}_n$, $E^{-\alpha} = E^{-\bar\alpha}_{-n}$ and

$$H^\alpha \equiv \frac{2}{(\alpha,\alpha)}\,[E^\alpha, E^{-\alpha}]. \tag{2.4.9}$$

The Chevalley basis version of the brackets (2.2.19) reads

$$\begin{aligned}
&[H^i_m, H^j_n] = m\,\bar G^{ij}\,\delta_{m+n,0}\,K,\\
&[H^i_m, E^{\bar\alpha}_n] = \bar\alpha^i E^{\bar\alpha}_{m+n},\\
&[E^{\bar\alpha}_m, E^{\bar\beta}_n] = e_{\bar\alpha,\bar\beta} E^{\bar\alpha+\bar\beta}_{m+n} \qquad \text{for } \bar\alpha+\bar\beta \text{ a } \bar{\mathfrak{g}}\text{-root},\\
&[E^{\bar\alpha}_n, E^{-\bar\alpha}_{-n}] = \textstyle\sum_{i=1}^r \bar\alpha_i H^i_0 + nK,
\end{aligned} \tag{2.4.10}$$

where it has been used that in the Chevalley basis of $\bar{\mathfrak{g}}$ one has $\bar\kappa^{ij} = \bar G^{ij}$ (cf. (1.6.53)). Thus the generator (2.4.9) can be expressed as

$$H^\alpha = \frac{2}{(\alpha,\alpha)}\,[E^{\bar\alpha}_n, E^{-\bar\alpha}_{-n}] = \frac{2}{(\bar\alpha,\bar\alpha)}\left(\sum_{i=1}^r \bar\alpha_i H^i_0 + nK\right). \tag{2.4.11}$$

Thus for any weight $\lambda = (\bar\lambda, k, m)$ of a highest weight module R_Λ,

$$R_\Lambda(H^\alpha)\cdot v_\lambda = \frac{2}{(\bar\alpha,\bar\alpha)}\left(\sum_{i=1}^r \bar\alpha_i \bar\lambda^i + nk\right) v_\lambda = \frac{2}{(\bar\alpha,\bar\alpha)}\,((\bar\alpha,\bar\lambda)+nk)\, v_\lambda, \tag{2.4.12}$$

i.e.

$$R_\Lambda(H^\alpha)\cdot v_\lambda = (\lambda, \alpha^\vee)\, v_\lambda. \tag{2.4.13}$$

Now from the representation theory of sl_2 we know that the finite-dimensional modules are precisely those for which the eigenvalue with respect to the Cartan subalgebra generator H is an integer for any weight vector, and for which in addition this integer is non-negative in the case of a highest weight vector. Hence what we are looking for are highest weight modules R_Λ such that for any real root α and any weight λ of R_Λ, one has

$$(\lambda, \alpha^\vee) \in \mathbb{Z}, \tag{2.4.14}$$

and also

$$(\Lambda, \alpha^\vee) \in \mathbb{Z}_{\geq 0}. \tag{2.4.15}$$

Since there are infinitely many real roots, these are an infinite number of conditions. However, from the definition of the simple roots it is clear that only finitely many of the conditions are independent, namely those where α is a simple root. Also, because of the relation (2.4.2) between v_λ and v_Λ, the integrality property (2.4.14) is a consequence of (2.4.15). Therefore the set of independent conditions is that the highest weight should be dominant integral, i.e.

$$\Lambda^i \equiv (\Lambda, \alpha^{(i)\vee}) \in \mathbb{Z}_{\geq 0} \quad \text{for all} \quad i = 0, 1, \ldots, r. \tag{2.4.16}$$

The modules obeying the constraint (2.4.16) are called *integrable* highest weight modules. The reason for this name is the following. For an infinite-dimensional representation, it is generically not possible to go from a representation of the Lie algebra g to a representation of a corresponding Lie group G via the exponential map

$$R(x) \mapsto \mathrm{e}^{R(x)}, \tag{2.4.17}$$

simply because the formal power series $\mathrm{e}^{R(x)}$ cannot be given a well defined meaning. The construction does work, however, if this formal power series terminates for all $x \in \mathsf{g}$ for which the weight vectors are not eigenstates of $R(x)$, i.e. for all elements which are not in the Cartan subalgebra of g (the group is then *defined* to be the one generated by $\mathrm{e}^{tR(E^i_\pm)}$, $i = 0, \ldots, r$, $t \in F$). This happens if and only if the $R(E^i_\pm)$ are locally nilpotent which is equivalent to the integrability requirement (2.4.16).

Let us now consider the constraint (2.4.16) in more detail. Taking $i = 1, \ldots, r$, it follows immediately that integrability of the highest weight module R_Λ with $\Lambda = (\bar\Lambda, k, 0)$ requires $\bar\Lambda$ to be a dominant integral weight of $\bar{\mathsf{g}}$, i.e. $\bar\Lambda^j \geq 0$ for all $j = 1, \ldots, r$. Before taking $i = 0$, first consider more generally (2.4.14) for the root $\alpha = (-\bar\alpha, 0, 1)$ with $\bar\alpha > 0$; we have

$$\begin{aligned} R_\Lambda(H^\alpha) \cdot v_\lambda &\equiv \tfrac{2}{(\bar\alpha,\bar\alpha)}\, R_\Lambda\left(-\sum_{i=1}^{r} \bar\alpha_i H^i_0 + K\right) \cdot v_\lambda \\ &= (-(\bar\lambda, \bar\alpha^\vee) + \tfrac{2}{(\bar\alpha,\bar\alpha)}\, k)\, v_\lambda, \end{aligned} \tag{2.4.18}$$

so that $\frac{2}{(\bar{\alpha},\bar{\alpha})}k$ must be an integer for any root $\bar{\alpha}$ of $\bar{\mathfrak{g}}$. Specializing to $\bar{\alpha}=\bar{\theta}$, i.e. taking $\alpha=\alpha^{(0)}$, we learn that the level (2.4.6) must be an integer,

$$k^\vee \in \mathbb{Z}, \tag{2.4.19}$$

in any integrable highest weight module. Finally, also specializing to the highest weight, $\lambda=\Lambda$, we see that this integer is bounded from below,

$$k^\vee \geq (\bar{\Lambda},\bar{\theta}^\vee), \tag{2.4.20}$$

or, what is the same,

$$k \geq (\bar{\Lambda},\bar{\theta}). \tag{2.4.21}$$

In particular, as $\bar{\Lambda}$ is dominant and hence $(\bar{\Lambda},\bar{\theta}) \geq 0$, the level of an integrable module must be a non-negative integer,

$$k \geq 0 \tag{2.4.22}$$

(the special value $k=0$ is possible only for $\bar{\Lambda}=0$; in this case the highest weight is zero, $\Lambda=(0,0,0)$, i.e. R_Λ is the trivial one-dimensional module. Thus one has in fact $k>0$ for any nontrivial integrable highest weight module).

We have thus seen that (2.4.19) and (2.4.21) are necessary conditions for integrability. It turns out that taken together they are also sufficient, provided that (as will be discussed in more detail in the following section) one considers the irreducible quotient modules. This provides a simple prescription for enumerating all integrable highest weight modules. Namely, writing $\bar{\Lambda}=\sum_{i=1}^r \bar{\Lambda}^i\bar{\Lambda}_{(i)}$ and recalling the defining equation for the dual Coxeter labels, i.e. $\bar{\theta}^\vee=\sum_{i=1}^r a_i^\vee \alpha^{(i)\vee}$, the inequality (2.4.20) reads

$$k^\vee \geq \sum_{i=1}^r a_i^\vee \bar{\Lambda}^i. \tag{2.4.23}$$

Since the dual Coxeter labels $a_i^\vee$ are positive and the Dynkin labels $\bar{\Lambda}^i$ non-negative, this means that for any fixed value of the level only a finite number of highest weight modules is integrable. The simplest example for this is given by $k=0$ where, as already mentioned, only the trivial module is integrable. More interesting is the situation at $k^\vee=1$; then in addition to the weight $\bar{\Lambda}=0$, the allowed highest weights of $\bar{\mathfrak{g}}$ are precisely the minimal fundamental weights, i.e. the fundamental weights $\bar{\Lambda}_{(i)}$ for which the dual Coxeter labels $a_i^\vee$ are equal to one. These may be read off the table (1.5.16), but for convenience we also list them explicitly in table (2.4.24).

$\bar{\mathfrak{g}}$	$\bar{\Lambda}_{(i)}$ minimal
A_r	$i = 1, 2, ..., r$
B_r	$i = 1,\ r$
C_r	$i = 1, 2, ..., r$
D_r	$i = 1,\ r-1,\ r$
E_6	$i = 1,\ 5$
E_7	$i = 6$
E_8	–
F_4	$i = 4$
G_2	$i = 2$

(2.4.24)

In other words, the minimal fundamental weights are all fundamental weights of A_r and C_r, the fundamental weights of the spinor and vector modules of B_r and D_r, and the fundamental weights of the lowest-dimensional modules of E_6, E_7, F_4 and G_2, while none of the fundamental weights of E_8 is allowed. In the case of A_r and C_r where all dual Coxeter labels are equal to one, the allowed highest weights of $\bar{\mathfrak{g}}$ for arbitrary values of the level are obtained by simply requiring that $\bar{\Lambda}$ is a dominant integral weight and that the sum of its Dynkin labels $\bar{\Lambda}^i$ is smaller than $k^\vee$,

$$\bar{\mathfrak{g}} = A_r,\, C_r: \qquad \sum_{i=1}^{r} \bar{\Lambda}^i \le k^\vee, \tag{2.4.25}$$

while for the other cases one gets formulae which are slightly more complicated,

$$\begin{aligned}
\bar{\mathfrak{g}} = B_r:&\quad \bar{\Lambda}^1 + \bar{\Lambda}^r + 2\textstyle\sum_{i=2}^{r-1} \bar{\Lambda}^i \le k^\vee,\\
D_r:&\quad \bar{\Lambda}^1 + \bar{\Lambda}^{r-1} + \bar{\Lambda}^r + \textstyle\sum_{i=2}^{r-2} \bar{\Lambda}^i \le k^\vee,\\
E_6:&\quad \bar{\Lambda}^1 + \bar{\Lambda}^5 + 2\,(\bar{\Lambda}^2 + \bar{\Lambda}^4 + \bar{\Lambda}^6) + 3\,\bar{\Lambda}^3 \le k^\vee,\\
E_7:&\quad \bar{\Lambda}^6 + 2\,(\bar{\Lambda}^1 + \bar{\Lambda}^5 + \bar{\Lambda}^7) + 3\,(\bar{\Lambda}^2 + \bar{\Lambda}^4) + 4\,\bar{\Lambda}^3 \le k^\vee,\\
E_8:&\quad 2\,(\bar{\Lambda}^1 + \bar{\Lambda}^7) + 3\,(\bar{\Lambda}^2 + \bar{\Lambda}^8) + 4\,(\bar{\Lambda}^3 + \bar{\Lambda}^6) + 5\,\bar{\Lambda}^4 + 6\,\bar{\Lambda}^5 \le k^\vee,\\
F_4:&\quad \bar{\Lambda}^4 + 2\,(\bar{\Lambda}^1 + \bar{\Lambda}^3) + 3\,\bar{\Lambda}^2 \le k^\vee,\\
G_2:&\quad \bar{\Lambda}^2 + 2\,\bar{\Lambda}^1 \le k^\vee.
\end{aligned} \tag{2.4.26}$$

As in the case of semisimple algebras, one defines the *fundamental weights* $\Lambda_{(i)}$, $i = 0, 1, \ldots, r$ by the property that they are dual to the simple coroots,

$$(\Lambda_{(i)}, \alpha^{(j)\vee}) = \delta_i^j \tag{2.4.27}$$

for all $j = 0, \ldots, r$. The explicit form of $\Lambda_{(0)}$ follows immediately from the formulae (2.3.13), (2.3.14) for the simple roots,

$$\Lambda_{(0)} = (0, \tfrac{1}{2}(\bar\theta, \bar\theta), 0), \tag{2.4.28}$$

while for $i = 1, \ldots, r$ one easily sees that

$$\Lambda_{(i)} = (\bar\Lambda_{(i)}, m_i, 0) \tag{2.4.29}$$

with appropriate numbers m_i. These numbers are obtained by examining the scalar product of (2.4.29) with $\alpha^{(0)}$, $0 = (\Lambda_{(i)}, \alpha^{(0)}) = (\bar\Lambda_{(i)}, -\bar\theta) + m_i$ for $i \neq 0$, from which one concludes

$$m_i = (\bar\Lambda_{(i)}, \bar\theta) = \bar\theta_i = \tfrac{1}{2}\,(\bar\theta, \bar\theta)\, a_i^\vee . \tag{2.4.30}$$

Thus

$$\Lambda_{(i)} = (\bar\Lambda_{(i)}, \tfrac{1}{2}(\bar\theta, \bar\theta) a_i^\vee, 0) \quad \text{for} \quad i = 1, \ldots, r. \tag{2.4.31}$$

(Because of $a_0^\vee = 1$ this formula also holds for $i = 0$ if we use the convention that $\bar\Lambda_{(0)} \equiv 0$.) For any weight of the form

$$\lambda = \sum_{i=0}^{r} \lambda^i \Lambda_{(i)}, \tag{2.4.32}$$

we therefore have

$$\lambda = (\bar\lambda, \tfrac{1}{2}\,(\bar\theta, \bar\theta) \textstyle\sum_{i=0}^{r} a_i^\vee \lambda^i, 0). \tag{2.4.33}$$

Comparing this with the general form $\lambda = (\bar\lambda, k, 0)$ of the weight, one learns that $k^\vee = \sum_{i=0}^{r} a_i^\vee \lambda^i$. But since k is already determined by the highest weight alone, this is identical to

$$k^\vee = \sum_{i=0}^{r} a_i^\vee \Lambda^i, \tag{2.4.34}$$

or in other words

$$k^\vee = \Lambda^0 + (\bar\theta^\vee, \bar\Lambda). \tag{2.4.35}$$

Note that this is compatible with the definition (2.2.1) of K (and in fact fixes the constant appearing there to $\zeta = \frac{1}{2}(\bar\theta, \bar\theta)$); this also shows that the definition $\Lambda = \sum_{i=0}^{r} \Lambda^i \Lambda_{(i)}$ of Λ in terms of Dynkin labels is consistent with its definition via the eigenvalue equations (2.4.5). In terms of the

Dynkin labels Λ^i, the requirement (2.4.20) of integrability of R_Λ reads $\Lambda^0 \geq 0$, so the integrability constraints (2.4.19) and (2.4.20) can be combined to

$$\Lambda^i \in \mathbb{Z}_{\geq 0}, \quad i = 0, 1, \ldots, r, \tag{2.4.36}$$

which is just a rephrasing of (2.4.16). Also note that if one defines

$$\theta := (\bar\theta, 0, 1), \tag{2.4.37}$$

one can write

$$k = (\Lambda, \theta) - (\bar\Lambda, \bar\theta). \tag{2.4.38}$$

Of course the weight (2.4.37) has not the meaning of something like a highest root of $\mathfrak{g}$; the discussion of the root system of $\mathfrak{g}$ in section 2.3 shows clearly that an affine algebra does not possess a highest (or lowest) root (in particular the adjoint module is no highest (and no lowest) weight module).

One may also try to extend the definition of the quadratic Casimir operator to affine algebras, but the following consideration shows that this is quite pointless. First, one cannot take over the definition (1.6.34) to the affine case because now there is an infinite number of roots so that the sum over the root system as in (1.6.37) is no longer well defined. A way out of this dilemma is to define the Casimir eigenvalues by the eigenvalue equation (1.6.39), with the affine analogue of the Weyl vector defined through the requirement that its properties are analogous to those of the Weyl vector $\bar\rho$ of simple Lie algebras, which means (compare (1.4.73))

$$\begin{aligned} \rho := \textstyle\sum_{i=0}^r \Lambda_{(i)} &= (\bar\rho, \tfrac{1}{2}(\bar\theta,\bar\theta)(\textstyle\sum_{i=1}^r a_i^\vee + 1), 0) \\ &= (\bar\rho, \tfrac{1}{2}\, C_{\bar\theta}, 0). \end{aligned} \tag{2.4.39}$$

Here $C_{\bar\theta}$ is the Casimir eigenvalue in the adjoint module of the horizontal subalgebra. One then has

$$\begin{aligned} (\Lambda, \Lambda + 2\rho) &\equiv ((\bar\Lambda, k, n), (\bar\Lambda + 2\bar\rho, k + C_{\bar\theta}, n)) \\ &= C_{\bar\Lambda} + n(2k + C_{\bar\theta}). \end{aligned} \tag{2.4.40}$$

Thus the conventional choice $n = 0$ just reproduces the Casimir eigenvalue with respect to the horizontal subalgebra (and hence in particular is independent of the level of the module R_Λ).

Let us also discuss briefly the highest weight modules of the twisted affine algebras. The vectors in the module that possess grade zero now no longer furnish a module of the horizontal subalgebra $\mathfrak{g}_{\rm hor}$, but rather of the zero mode subalgebra $\mathfrak{g}_{[0]} \neq \mathfrak{g}_{\rm hor}$. Thus they are labelled by a highest weight $\bar\Lambda$ of $\mathfrak{g}_{[0]}$ together with a value of the level $k^\vee$. For integrability, one needs again

$$(\Lambda, \alpha^\vee) \in \mathbb{Z}_{\geq 0} \quad \text{for all} \quad \alpha \in \Phi_r(\mathfrak{g}). \tag{2.4.41}$$

Fundamental weights are again defined by requiring that $(\Lambda_{(i)}, \alpha^{(j)\vee}) = \delta_i^j$, where now the simple roots are

$$\alpha^{(i)} = (\bar\alpha^{(i)}, 0, 0), \quad i = 1, \ldots, r = \operatorname{rank} \mathfrak{g}_{[0]}, \tag{2.4.42}$$

with $\bar\alpha^{(i)}$ the simple roots of $\mathfrak{g}_{[0]}$, and

$$\alpha^{(0)} = (-\bar\Lambda_1, 0, \tfrac{1}{\ell}) \tag{2.4.43}$$

with ℓ the order of the relevant automorphism ω and $\bar\Lambda_1$ the highest weight of the $\mathfrak{g}_{[0]}$-module furnished by $\mathfrak{g}_{[1]}$. Combining these equations, one finds

$$\begin{aligned} \Lambda_{(i)} &= \left(\bar\Lambda_{(i)}, \frac{\ell}{2} (\bar\Lambda_1, \bar\Lambda_1) \frac{a_i^\vee}{a_0^\vee}, 0 \right), \quad i = 1, \ldots, r, \\ \Lambda_{(0)} &= \left(0, \frac{\ell}{2} (\bar\Lambda_1, \bar\Lambda_1), 0 \right). \end{aligned} \tag{2.4.44}$$

It turns out that the dual Coxeter labels $a_i^\vee$ of $\mathfrak{g}$ that appear in these expressions coincide up to an overall factor $2a_0^\vee/(\bar\theta, \bar\theta)$ with the expansion coefficients of $\bar\Lambda_1$ in the basis of simple coroots of $\mathfrak{g}_{[0]}$. Here $\bar\theta$ is the highest root of the algebra $\bar{\mathfrak{g}}$ on which the automorphism ω is defined; also recall that $a_0^\vee = 1$ except for $A_1^{(2)}$ and $\tilde B_r^{(2)}$ for which $a_0^\vee = 2$.

The general solution to the constraint (2.4.41) is then again that the highest weight $\Lambda = (\bar\Lambda, k, 0)$ must be dominant integral, i.e. that it is a linear combination of the fundamental weights $\Lambda_{(i)}$, $i = 0, 1, \ldots, r = \operatorname{rank} \mathfrak{g}_{[0]}$, with non-negative integer coefficients Λ^i. Now for the level of such a highest weight module one finds the formula

$$k^\vee = k_0 \sum_{i=0}^{r} a_i^\vee \Lambda^i, \tag{2.4.45}$$

where

$$k_0 = \frac{\ell}{a_0^\vee} \cdot \frac{(\bar\Lambda_1, \bar\Lambda_1)}{(\bar\theta, \bar\theta)}, \tag{2.4.46}$$

with $\bar\theta$ the highest root of $\bar{\mathfrak{g}}$. Inspection shows that $k_0 = 4$ for $A_1^{(2)}$, $k_0 = 2$ for $\tilde B_r^{(2)}$, and $k_0 = 1$ else. Thus for an integrable module the level can be any non-negative integer for all twisted affine algebras except for $\tilde B_r^{(2)}$ and

$A_1^{(2)}$, in which cases it must in addition be even respectively a multiple of four.

Let us stress that the latter results are valid in the particular homogeneous gradation that we decided to choose, i.e. the gradation for which the zero mode subalgebra of $\mathsf{g} = X_r^{(\ell)}$ is isomorphic to the simple Lie algebra X_r. If a different gradation is chosen, then generically one obtains for the same g-module R_Λ a different highest weight with respect to the zero mode subalgebra. (For example, the four level one modules of $D_r^{(1)}$ which in the homogeneous gradation carry the singlet, vector, and spinor modules of D_r, respectively, become isomorphic in the principal gradation.) In contrast, the level of a module is an intrinsically defined quantity and hence independent of the choice of gradation.

2.5 Null vectors

We now discuss in detail the irreducibility of highest weight modules which is implicit in the discussions of the previous section (and also already in section 1.6). Recall that a highest weight module can be constructed from some highest weight vector v_Λ by applying lowering operators as in (2.4.2) (or (1.6.11)). If the set of vectors obtained this way, which may be described as $\mathsf{U}(\mathsf{g}_-) \cdot v_\Lambda$, is formally taken to be independent, then the module is called a *Verma module*. Thus the vector space underlying the Verma module is obtained by construction rather than being given a priori. Verma modules are in general not irreducible nor even fully reducible. To get an irreducible module R_Λ from some given Verma module V_Λ, one must divide out any relations which exist among this set of vectors.

The generic form of such a relation is

$$R(x) \cdot v_\Lambda = 0 \tag{2.5.1}$$

where x is an element of the enveloping algebra $\mathsf{U}(\mathsf{g}_-)$ of the nilpotent subalgebra g_- of g which is generated by the lowering operators $E^{-\alpha}$, $\alpha > 0$. In terms of the Verma module, the left hand side of (2.5.1) is just a vector v_μ in the module,

$$v_\mu = R(x) \cdot v_\Lambda \in V_\Lambda. \tag{2.5.2}$$

Thus the irreducible quotient of V_Λ is obtained by setting an appropriate set of elements of V_Λ to zero. Such vectors, which are actually zero although formally they are not, are called *null vectors*; they are vectors in the sense of weight vectors attached to an element of the weight space only in terms of the formally defined Verma module; in terms of the irreducible quotient module they are the zero vector (but *not* attached to the zero weight). Any linear combination of null vectors is of course also a null vector. Also, if v_μ is a null vector, then so is $R(x) \cdot v_\mu$ for any x in the enveloping algebra $\mathsf{U}(\mathsf{g}_-)$ of g_-. In many cases the latter is also true for some $x \in \mathsf{U}(\mathsf{g}_+)$; on the other hand, one may identify a maximal linearly independent set of null vectors none of which leads to new null vectors via the action of $R(x)$, $x \in \mathsf{U}(\mathsf{g}_+)$; these vectors are called *primitive* (or *singular*) null vectors. Acting on a primitive null vector with $R(E^\alpha)$ for any raising operator $E^\alpha \in \mathsf{g}_+$ must give zero already in the Verma module,

$$R(E^\alpha) \cdot v_\mu = 0 \quad \text{for all} \quad \alpha > 0, \tag{2.5.3}$$

because otherwise $R(E^\alpha) \cdot v_\mu$ would be a null vector, in contradiction to the primitivity of v_μ. Therefore a primitive null vector v_μ fulfills the defining requirement of a highest weight vector, and hence V_Λ contains the Verma module V_μ as a submodule. Conversely, any highest weight vector in the Verma module V_Λ other than v_Λ is a primitive null vector either of V_Λ or of one of its Verma submodules. It can be shown that any Verma module contains a unique maximal proper Verma submodule $V_{\mu,\mathrm{max}}$ (here "proper" is meant in the sense that the submodule must not be equal to the module itself, but that the trivial submodule $\{0\}$ is still allowed). Thus for any Verma module there is a unique irreducible highest weight module R_Λ which is obtained by quotienting V_Λ by $V_{\mu,\mathrm{max}}$. Also note that a reducible Verma module is never fully reducible. This is so because no proper Verma submodule V' of V_Λ can contain the highest weight vector v_Λ (otherwise V' would have to be equal to V_Λ by definition of the Verma module).

In the case of the highest weight modules discussed in the sections 1.6 and 2.4, the explicit form of the null vectors is found rather easily. Consider first the case of a simple Lie algebra $\bar{\mathsf{g}}$ and a dominant integral $\bar{\mathsf{g}}$-weight $\bar\Lambda$. Then the relations which have to be quotiented out are just those which require that R_Λ splits into finite-dimensional modules with respect to the sl_2-subalgebras of $\bar{\mathsf{g}}$ corresponding to any root $\bar\alpha$. Of these, only the relations involving the highest weight and the simple $\bar{\mathsf{g}}$-roots are independent; thus the primitive null vectors are

$$v_{\mu,i} := R(E^{-\bar\alpha^{(i)}})^{\bar\Lambda^i+1} \cdot v_{\bar\Lambda}, \quad i = 1, \ldots, r \tag{2.5.4}$$

($r \equiv \mathrm{rank}(\bar{\mathsf{g}})$). In section 1.6 we were mainly interested in finite-dimensional $\bar{\mathsf{g}}$-modules, which are by far the most important modules for applications in both mathematics and physics. Working only with finite-dimensional modules, one never has to worry about any Verma module or null vector; this is the reason why we did not have to mention these concepts when we discussed the highest weight modules of simple Lie algebras in section 1.6.

In the case of affine Lie algebras, there are no longer any nontrivial finite-dimensional modules. When constructing infinite-dimensional highest weight modules, it is then more natural to form first the Verma module V_Λ and afterwards go to its irreducible quotient R_Λ. As we saw in the previous section, the irreducible quotient is integrable iff it decomposes into finite-dimensional modules with respect to the sl_2-subalgebras corresponding to any of the simple $\mathfrak{g}$-roots. The null vector equations corresponding to the simple roots $\alpha^{(i)}$ with $i = 1, \dots, r$ are simply implemented by considering only irreducible modules rather than Verma modules with respect to the horizontal subalgebra $\mathfrak{g}_{\mathrm{hor}}$. As a consequence, there is only one new primitive null vector given by

$$v_{\mu,0} := (R(E^{-\alpha^{(0)}}))^{\Lambda^0+1} \cdot v_\Lambda, \tag{2.5.5}$$

or, more explicitly, expressing $\alpha^{(0)}$ and Λ^0 through the formulae (2.3.14) and (2.4.35),

$$v_{\mu,0} = (R(E^{\bar{\theta}}_{-1}))^{k^\vee - (\bar{\theta}^\vee, \bar{\Lambda})+1} \cdot v_\Lambda. \tag{2.5.6}$$

The existence of null vectors is closely tied to the unitarizability of a module. Namely, one has for any two elements $v_\lambda, v_\nu \in V_\Lambda$ that $v_\lambda = R(x) \cdot v_\Lambda$, $v_\nu = R(y) \cdot v_\Lambda$ for some elements x, y in the enveloping algebra of $\mathfrak{g}_-$; hence, if one tries to define an inner product as in (1.6.18), one gets

$$(v_\lambda \mid v_\mu) = (v_\Lambda \mid (R(x))^\dagger R(y) \cdot v_\Lambda). \tag{2.5.7}$$

Now for x in the enveloping algebra of $\mathfrak{g}_-$, one has $(R(x))^\dagger = R(x')$ for some x' in the enveloping algebra of $\mathfrak{g}_+$. The explicit form of x' follows with the help of linearity and of $(R(xy))^\dagger = (R(y))^\dagger (R(x))^\dagger$ from the corresponding relation for the generators which reads

$$(R(E^i_+))^\dagger = R(E^i_-), \qquad (R(H^i))^\dagger = R(H^i); \tag{2.5.8}$$

alternatively, if one chooses, rather than taking the Cartan–Weyl basis, the generators of the horizontal subalgebra to be anti-hermitian, then

$$(R(T^a_n))^\dagger = -R(T^a_{-n}). \tag{2.5.9}$$

Because according to (2.5.3) a primitive null vector is annihilated by any operator in the enveloping algebra of $\mathfrak{g}_+$, it then follows that the inner product of a primitive null vector v_μ with any other vector v_λ vanishes,

$$(v_\lambda \mid v_\mu) = (v_\Lambda \mid R(x') \cdot v_\mu) = 0. \tag{2.5.10}$$

The result (2.5.10) is the reason why v_μ is called a null vector. This relation implies in particular that on the Verma module V_Λ containing v_μ the inner product is necessarily degenerate. A necessary condition for the existence of a non-degenerate inner product in a highest weight module is therefore that the null vectors are projected out, i.e. that one considers the irreducible quotient module R_Λ. On the other hand, starting from $(v_\Lambda \,|\, v_\Lambda) \neq 0$ one can calculate all inner products with the help of (2.5.7) (commuting $(R(x))^\dagger$ through $R(y)$ so that it acts on v_Λ which it annihilates); this way it turns out that for Λ dominant integral, irreducibility is also sufficient for the existence of a positive definite inner product. For example, the length squared of $R(E^i_-) \cdot v_\Lambda$ is

$$\begin{aligned}(R(E^i_-) \cdot v_\Lambda \mid R(E^i_-) \cdot v_\Lambda) &= (v_\Lambda \mid R(E^i_+) R(E^i_-) \cdot v_\Lambda) \\ &= (v_\Lambda \mid [R(E^i_+), R(E^i_-)] \cdot v_\Lambda) = (v_\Lambda \mid R(H^i) \cdot v_\Lambda) \\ &= \Lambda^i \, (v_\Lambda \mid v_\Lambda).\end{aligned} \tag{2.5.11}$$

The same sort of calculation also shows immediately that no positive definite inner product can be defined for highest weights which are not dominant integral. Thus the unitarizable highest weight modules of an affine Lie algebra are precisely the irreducible integrable ones.

Having identified the null vectors, we are now in a position to determine explicitly the weight system of the integrable irreducible highest weight modules of untwisted affine algebras. One just has to act with the lowering operators for simple roots on the highest weight vector, then on the weight vectors obtained this way, and so on, and to remove any primitive null vector encountered during this procedure.

Let us explicate this for the lowest grades in two simple cases. We take $\mathfrak{g} = A_1^{(1)}$ so that there are just two simple roots, with Dynkin labels

$$\alpha^{(0)} = (2, -2, -1), \qquad \alpha^{(1)} = (-2, 2, 0) \tag{2.5.12}$$

(here we write the first two Dynkin labels as the eigenvalues of H^0 and H^1 rather than the eigenvalues of H^1 and K). The first example is the level one module with highest weight $\Lambda = \Lambda_{(0)} = (1, 0, 0)$; the highest weight vector is then annihilated by $R_\Lambda(E^{-\alpha^{(1)}})$, while acting with $R_\Lambda(E^{-\alpha^{(0)}})$ produces the weight vector for the weight $(-1, 2, -1)$ which is annihilated by $R_\Lambda(E^{-\alpha^{(0)}})$ but not by $R_\Lambda(E^{-\alpha^{(1)}})$, etc. One then arrives at the weight system in figure (2.5.13) for grades 0 to 6 (here the weights are enclosed in boxes, and the index on a box indicates its multiplicity (it is omitted if it is equal to one), and the right- and downwards arrows denote the action of $-\alpha^{(1)}$ and $-\alpha^{(0)}$, respectively).

$$\boxed{1\ 0\ 0}$$

$$-\alpha^{(0)} \swarrow$$

$$\boxed{-1\ 2\ -1} \xrightarrow{-\alpha^{(1)}} \boxed{1\ 0\ -1} \xrightarrow{-\alpha^{(1)}} \boxed{3\ -2\ -1}$$

$$-\alpha^{(0)} \swarrow \qquad -\alpha^{(0)} \swarrow$$

$$\boxed{-1\ 2\ -2} \rightarrow \boxed{1\ 0\ -2}_2 \rightarrow \boxed{3\ -2\ -2}$$

$$\boxed{-1\ 2\ -3}_2 \rightarrow \boxed{1\ 0\ -3}_3 \rightarrow \boxed{3\ -2\ -3}_2$$

$$\boxed{-3\ 4\ -4} \rightarrow \boxed{-1\ 2\ -4}_3 \rightarrow \boxed{1\ 0\ -4}_5 \rightarrow \boxed{3\ -2\ -4}_3 \rightarrow \boxed{5\ -4\ -4}$$

$$\boxed{-3\ 4\ -5} \rightarrow \boxed{-1\ 2\ -5}_5 \rightarrow \boxed{1\ 0\ -5}_7 \rightarrow \boxed{3\ -2\ -5}_5 \rightarrow \boxed{5\ -4\ -5}$$

$$\boxed{-3\ 4\ -6}_2 \rightarrow \boxed{-1\ 2\ -6}_7 \rightarrow \boxed{1\ 0\ -6}_{11} \rightarrow \boxed{3\ -2\ -6}_7 \rightarrow \boxed{5\ -4\ -6}_2$$

$$\cdots \qquad \cdots \qquad \cdots \qquad \cdots \qquad \cdots$$

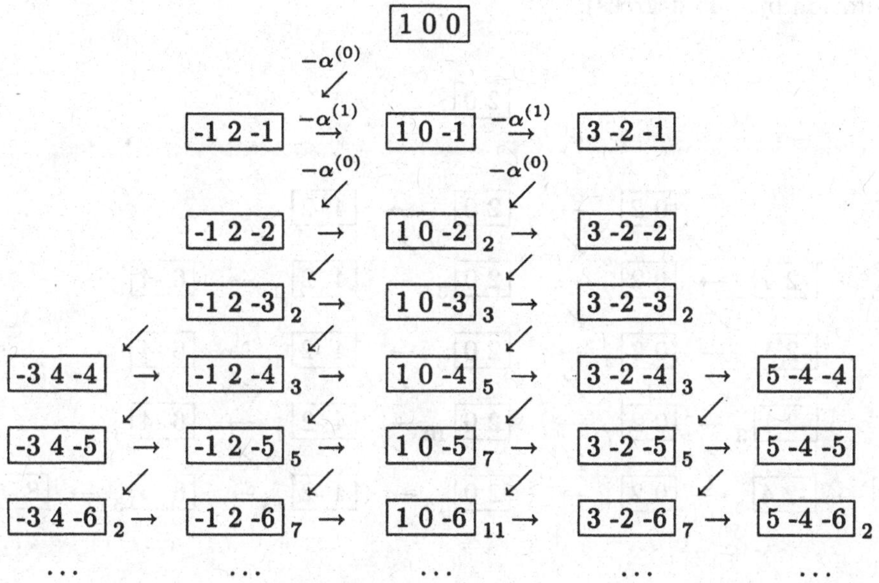

$$(2.5.13)$$

Let us identify the lowest grade null vector in the corresponding Verma module. As already said, the subtraction of $\alpha^{(1)}$ from Λ does not produce a weight, as is clear from $\Lambda^1 = 0$. We *can* subtract $\alpha^{(0)}$ from the highest weight (but not $2\alpha^{(0)}$). From the weight $(-1, 2, -1)$ obtained that way we can subtract $\alpha^{(1)}$ twice, thus obtaining the other two weights at grade one. So far everything holds both for the Verma module and its irreducible quotient; the three possibilities of obtaining vectors at grade one from the single vector at grade zero correspond to the application of H_{-1} and $E_{-1}^{\pm}$. Analogously, at grade two the Verma module contains nine vectors, whereas according to the above picture the irreducible quotient contains only four. Thus at grade two there are five null vectors, namely $(-3, 4, -2)$, $(-1, 2, -2)$, $(1, 0, -2)$, $(3, -2, -2)$, $(5, -4, -2)$; of these, obviously the first one is a primitive null vector, with the others being obtained from it by subsequent subtraction of $\alpha^{(1)}$.

The irreducible highest weight module with highest weight $\Lambda = \Lambda_{(0)}$ is (for any affine $\mathfrak{g}$) called the *basic* module of $\mathfrak{g}$. This is in a natural way the simplest non-trivial module of $\mathfrak{g}$. In the case of $\mathfrak{g} = A_1^{(1)}$, the structure of R_Λ with $\Lambda = \Lambda_{(1)} = (0, 1, 0)$ is just as simple (the same holds for $\mathfrak{g} = A_r^{(1)}$ and $\Lambda = \Lambda_{(i)}$, $i = 1, ..., r$); namely, its weight system is obtained from the array (2.5.13) by exchanging everywhere the first two entries of the weights and replacing the grade $-n$, which in (2.5.13) just counts the lines from top to bottom, by $-m$ where m counts the diagonal "lines" from top left to bottom right (i.e. the lines of the diagram obtained by a rotation by -45 degrees).

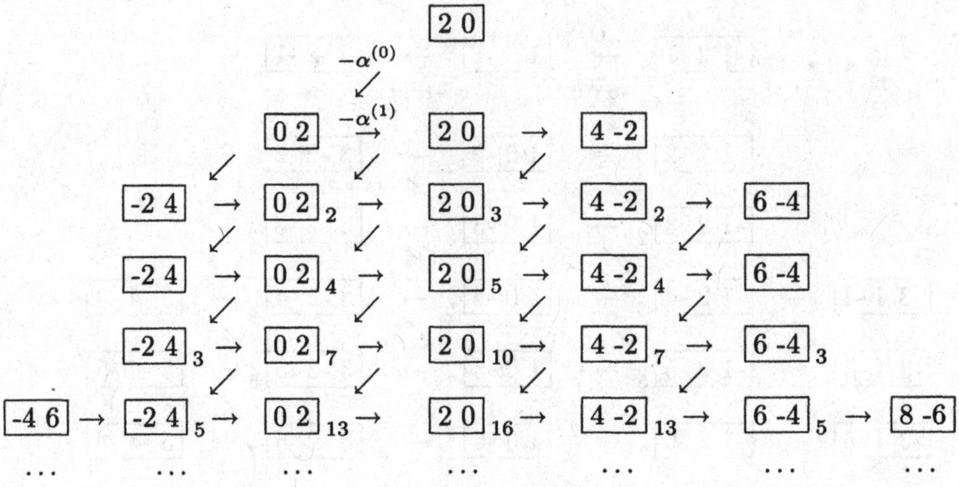

$$
\begin{array}{c}
\boxed{2\ 0} \\[4pt]
\quad {\scriptstyle -\alpha^{(0)}} \searrow \\[2pt]
\boxed{0\ 2} \xrightarrow{\ -\alpha^{(1)}\ } \boxed{2\ 0} \to \boxed{4\ \text{-}2} \\[4pt]
\boxed{\text{-}2\ 4} \to \boxed{0\ 2}_2 \to \boxed{2\ 0}_3 \to \boxed{4\ \text{-}2}_2 \to \boxed{6\ \text{-}4} \\[4pt]
\boxed{\text{-}2\ 4} \to \boxed{0\ 2}_4 \to \boxed{2\ 0}_5 \to \boxed{4\ \text{-}2}_4 \to \boxed{6\ \text{-}4} \\[4pt]
\boxed{\text{-}2\ 4}_3 \to \boxed{0\ 2}_7 \to \boxed{2\ 0}_{10} \to \boxed{4\ \text{-}2}_7 \to \boxed{6\ \text{-}4}_3 \\[4pt]
\boxed{\text{-}4\ 6} \to \boxed{\text{-}2\ 4}_5 \to \boxed{0\ 2}_{13} \to \boxed{2\ 0}_{16} \to \boxed{4\ \text{-}2}_{13} \to \boxed{6\ \text{-}4}_5 \to \boxed{8\ \text{-}6} \\[4pt]
\cdots \qquad \cdots \qquad \cdots \qquad \cdots \qquad \cdots \qquad \cdots \qquad \cdots
\end{array}
$$

$$\tag{2.5.14}$$

The second example we display is R_Λ with $\Lambda = 2\Lambda_{(0)} = (2,0,0)$; proceeding as above, one gets figure (2.5.14). The D-eigenvalue of the weights which is not displayed in figure (2.5.14) is just $-n$ with n numbering the lines from top to bottom; also, analogously to the first example, the weight system of R_Λ with $\Lambda = (0,2,0)$ is obtained from (2.5.14) by exchanging the first two entries in the weights and counting lines from top left to bottom right.

2.6   The Weyl group. Characters

In analogy to simple Lie algebras, one defines the *Weyl reflection* σ_α of the weight lattice of an affine algebra by

$$
\sigma_\alpha : \begin{array}{l} L_w \to L_w \\ \lambda \mapsto \sigma_\alpha(\lambda) := \lambda - (\lambda, \alpha^\vee)\,\alpha. \end{array} \tag{2.6.1}
$$

Here α is any real root (for imaginary roots α the definition would not make sense because then $\alpha^\vee$ is not defined). Because of

$$
(\sigma_\alpha(\lambda), \alpha^\vee) = -(\lambda, \alpha^\vee), \tag{2.6.2}
$$

this is indeed again the reflection at the hyperplane perpendicular to α, and hence together with the identity map the Weyl reflections form a group under composition, called the *affine Weyl group* W.

Many properties of the affine Weyl group are analogous to those of the Weyl group of a simple algebra, e.g. W is generated by the reflections $\sigma_{\alpha^{(i)}}$, $i = 0, 1, ..., r$ with respect to simple roots, and each of these elementary reflections permutes the set of positive roots. There are, however, also new features which are related to the existence of imaginary roots. In particular, since $(\alpha, \delta) = 0$ for any real root α, one has

$$\sigma_\alpha(\delta) \equiv \delta - (\delta, \alpha)\alpha^\vee = \delta, \tag{2.6.3}$$

and hence any Weyl reflection acts as the identity on the set $\Phi_{\rm i} = \{n\delta | n \neq 0\}$ of imaginary roots,

$$\sigma_\alpha|_{\Phi_{\rm i}} = id_{\Phi_{\rm i}}. \tag{2.6.4}$$

Since any reflection is an automorphism of the root lattice, this also means that the Weyl group maps the set $\Phi_{\rm r}$ of real roots onto itself,

$$W(\Phi_{\rm r}) = \Phi_{\rm r}. \tag{2.6.5}$$

This is easily verified: for any pair $\alpha = (\bar\alpha, 0, n)$, $\beta = (\bar\beta, 0, m)$ of real roots, one has

$$\sigma_\alpha(\beta) = \beta - \tfrac{2}{(\alpha,\alpha)}\,[(\bar\alpha, \bar\beta) + 0\cdot m + n\cdot 0]\,\alpha = (\sigma_{\bar\alpha}(\bar\beta), 0, m - (\bar\alpha^\vee, \bar\beta)n), \tag{2.6.6}$$

which is again a real root.

More generally, the Weyl reflection with respect to $\alpha = (\bar\alpha, 0, n)$ takes any weight $\lambda = (\bar\lambda, k, m)$ to

$$\begin{aligned}\sigma_\alpha(\lambda) &= \lambda - \tfrac{2}{(\alpha,\alpha)}((\bar\lambda, \bar\alpha) + kn)\,\alpha \\ &= (\bar\lambda - (\bar\lambda, \bar\alpha^\vee)\,\bar\alpha - kn\bar\alpha^\vee,\ k,\ m - (\bar\lambda, \bar\alpha^\vee)\,n + \tfrac{2}{(\alpha,\alpha)}n^2),\end{aligned} \tag{2.6.7}$$

which may be rewritten as

$$\sigma_\alpha(\lambda) = (\sigma_{\bar\alpha}(\bar\lambda + nk\bar\alpha^\vee),\ k,\ m + \frac{1}{2k}[(\bar\lambda, \bar\lambda) - (\bar\lambda + nk\bar\alpha^\vee, \bar\lambda + nk\bar\alpha^\vee)]). \tag{2.6.8}$$

Defining, for any root $\bar\beta \in \bar\Phi$, the translation $t_{\bar\beta}$ as

$$t_{\bar\beta} :\ \lambda = (\bar\lambda, k, m) \mapsto (\bar\lambda + k\bar\beta,\ k,\ m + \frac{1}{2k}[(\bar\lambda, \bar\lambda) - (\bar\lambda + k\bar\beta, \bar\lambda + k\bar\beta)]), \tag{2.6.9}$$

the expression (2.6.8) for σ_α becomes

$$\sigma_\alpha = \bar\sigma_{\bar\alpha} \circ (t_{\bar\alpha^\vee})^n, \tag{2.6.10}$$

where $\bar\sigma_{\bar\alpha}$ acts as the element $\sigma_{\bar\alpha}$ of the $\bar{\mathfrak g}$-Weyl group $\bar W$ on the components $\bar\lambda$ of λ, and as the identity on the last two components.

In short, any $\sigma_\alpha \in W$ is of the form

$$\sigma_\alpha = \bar\sigma_{\bar\alpha} \circ t_{\bar\beta} \tag{2.6.11}$$

for some $\bar\beta$ in the coroot lattice $L^\vee$ of $\bar{\mathfrak{g}}$. Moreover, a straightforward calculation shows that

$$t_{\sigma(\bar\alpha)} = \sigma \circ t_{\bar\alpha} \circ \sigma^{-1} \tag{2.6.12}$$

for all $\sigma \in W$. Finally, the abelian group T of translations of the coroot lattice $L^\vee(\bar{\mathfrak{g}})$ is a normal subgroup of W, and $\bar W \cap T = \{id\}$. Together it then follows that W is the semidirect product of the $\bar{\mathfrak{g}}$-Weyl group and T,

$$W = \bar W \ltimes T. \tag{2.6.13}$$

Also, since $t_{\bar\alpha} \circ t_{\bar\beta} = t_{\bar\alpha+\bar\beta}$, the translations T may be in fact identified with the coroot lattice.

An important property of the coroot lattice of $\bar{\mathfrak{g}}$ is that it is generated by the highest coroot $\bar\theta^\vee$ together with its Weyl images $\bar\sigma(\bar\theta^\vee)$, $\bar\sigma \in \bar W$ (thus, employing the conventional normalization $(\bar\theta, \bar\theta) = 2$, $L^\vee(\bar{\mathfrak{g}})$ can also be characterized as the lattice generated by the long roots of $\bar{\mathfrak{g}}$). As a consequence, all possible translations contained in the affine Weyl group W are the result of the reflection with respect to the zeroth root $\alpha^{(0)}$ and of combinations of this reflection with elements of $\bar W$. Using the fact that $\bar W$ is generated by the fundamental reflections (1.7.9), it then follows that the affine Weyl group is generated by a finite set of fundamental transformations as well (which now include the reflection with respect to $\alpha^{(0)}$),

$$\sigma_{(i)} \equiv \sigma_{\alpha^{(i)}}, \quad i = 0, 1, \ldots, r. \tag{2.6.14}$$

It then also follows that, analogously to the case of $\bar W$, the affine Weyl group permutes transitively and freely the affine Weyl chambers which are those open subsets of the weight space which are obtained by removing all hyperplanes which are left invariant by some Weyl reflection. Moreover, any integral weight can be obtained by applying a suitable Weyl reflection to an element of the closure P_k^+ of the *dominant affine Weyl chamber*,

$$P_k^+ := \left\{ \sum_{i=0}^{r} \Lambda^i \Lambda_{(i)} \mid \Lambda^i \geq 0 \right\}. \tag{2.6.15}$$

Explicitly, the Weyl reflection with respect to $\alpha^{(0)}$ acts on $\lambda = (\bar\lambda, k, m)$ as (compare (2.6.7))

$$\sigma_{(0)}(\lambda) = (\bar\sigma_{\bar\theta}(\bar\lambda) + k\,\bar\theta^\vee,\ k,\ m + (\bar\lambda, \bar\theta^\vee) - k^\vee). \tag{2.6.16}$$

In particular, the horizontal projection $\bar{\sigma}_{(0)}$ acts as

$$\bar{\sigma}_{(0)}(\bar{\lambda}) = \bar{\sigma}_{\bar{\theta}}(\bar{\lambda}) + k^\vee \bar{\theta}. \tag{2.6.17}$$

For the case of $A_1^{(1)}$ where according to (1.7.25) one has $\bar{\sigma}_{\bar{\theta}}(\bar{\lambda}) \equiv \bar{\sigma}_{(1)}(\bar{\lambda}) = -\bar{\lambda}$, this means that $\bar{\sigma}_{(0)}(\bar{\lambda}) = 2k^\vee - \bar{\lambda}$ and

$$\sigma_{(0)}(\bar{\lambda}, k, m) = (2k^\vee - \bar{\lambda}, k, m + \bar{\lambda} - k^\vee), \tag{2.6.18}$$

and the other simple reflection of the $A_1^{(1)}$ Weyl group acts as

$$\sigma_{(1)}(\bar{\lambda}, k, m) = (-\bar{\lambda}, k, m). \tag{2.6.19}$$

By inspection, one then finds that the translation subgroup of the Weyl group is generated by $t := \sigma_{(0)} \circ \sigma_{(1)}$, which acts as

$$t(\bar{\lambda}, k, m) = (\bar{\lambda} + 2k^\vee, k, m - \bar{\lambda} - k^\vee). \tag{2.6.20}$$

As a second example, consider the Weyl group of $A_2^{(1)}$. The simple reflections are

$$\begin{aligned}
&\sigma_{(1)}(\bar{\lambda}^1, \bar{\lambda}^2, k, m) = (-\bar{\lambda}^1, \bar{\lambda}^1 + \bar{\lambda}^2, k, m),\\
&\sigma_{(2)}(\bar{\lambda}^1, \bar{\lambda}^2, k, m) = (\bar{\lambda}^1 + \bar{\lambda}^2, -\bar{\lambda}^2, k, m),\\
&\sigma_{(0)}(\bar{\lambda}^1, \bar{\lambda}^2, k, m) = (k^\vee - \bar{\lambda}^2, k^\vee - \bar{\lambda}^1, k, m + \bar{\lambda}^1 + \bar{\lambda}^2 - k^\vee),
\end{aligned} \tag{2.6.21}$$

and the translation subgroup is generated by the two elements

$$\begin{aligned}
t_{(1)} &= \sigma_{(2)} \circ \sigma_{(0)} \circ \sigma_{(2)} \circ \sigma_{(1)},\\
t_{(2)} &= \sigma_{(1)} \circ \sigma_{(0)} \circ \sigma_{(1)} \circ \sigma_{(2)},
\end{aligned} \tag{2.6.22}$$

which act as

$$\begin{aligned}
t_{(1)}(\bar{\lambda}, k, m) &= (\bar{\lambda} + k^\vee \alpha^{(1)}, k, m - k^\vee - \bar{\lambda}^1),\\
t_{(2)}(\bar{\lambda}, k, m) &= (\bar{\lambda} + k^\vee \alpha^{(2)}, k, m - k^\vee - \bar{\lambda}^2).
\end{aligned} \tag{2.6.23}$$

A modified action of the affine Weyl group can be defined in analogy to the finite-dimensional case by

$$\sigma \star \lambda = \sigma(\lambda + \rho) - \rho \tag{2.6.24}$$

with $\rho = \sum_{i=0}^r \Lambda_{(i)}$ the affine Weyl vector. For the reflections $\sigma_{(i)}$ with $i = 1, ..., r$ this simply means

$$\sigma_{(i)} \star (\bar{\lambda}, k, m) = (\bar{\sigma}_{(i)} \star \bar{\lambda}, k, m), \tag{2.6.25}$$

while for $\sigma_{(0)}$ one must take into account that the weight $\lambda + \rho$ is at level $k^\vee + g^\vee$ if λ is at level $k^\vee$, so that together with (1.7.21) one finds

$$\sigma_{(0)} \star (\bar{\lambda}, k, m) = (\bar{\sigma}_{\bar{\theta}}(\bar{\lambda}) + (k^\vee + 1)\bar{\theta}, k, m + (\bar{\lambda}, \bar{\theta}^\vee) - k^\vee - 1). \tag{2.6.26}$$

Thus for $\mathfrak{g} = A_1^{(1)}$, the modified Weyl group action is generated by

$$\begin{aligned}\sigma_{(1)} \star (\bar\lambda, k, m) &= (-\bar\lambda - 2, k, m),\\ \sigma_{(0)} \star (\bar\lambda, k, m) &= (2k^\vee + 2 - \bar\lambda, k, m + \bar\lambda - k^\vee - 1).\end{aligned} \tag{2.6.27}$$

In the case of $A_1^{(1)}$ the weights that are left invariant by the fundamental reflections are easily identified. For the ordinary action, a weight is left invariant by $\sigma_{(1)}$ if $\bar\lambda = 0$, and by $\sigma_{(0)}$ if $\bar\lambda = k^\vee$, respectively; for the modified action, the corresponding requirements are $\bar\lambda = -1$ and $\bar\lambda = k^\vee + 1$, respectively. For arbitrary affine $\mathfrak{g}$, this generalizes as follows. A weight $\lambda = (\bar\lambda, k, m)$ is left invariant by the ordinary action of $\sigma_{(i)}$, $i = 1, ..., r$, if $\bar\lambda^i = 0$, and by the ordinary action of $\sigma_{(0)}$ if $(\bar\lambda, \bar\theta^\vee) = k^\vee$, i.e. if $\lambda^0 = 0$ (or in other words, if the bound for $(\bar\Lambda, \bar\theta^\vee)$ on integrable highest weights $\bar\Lambda$ is saturated); for the modified action, the requirements are, respectively, $\bar\lambda^i = -1$, and $(\bar\lambda, \bar\theta^\vee) = k^\vee + 1$ (i.e. $\lambda^0 = -1$).

As in the finite-dimensional case, one of the main applications of the Weyl group is in the calculation of characters. The characters χ_Λ of a Kac–Moody algebra are defined analogously to the case of simple Lie algebras. Thus (compare (1.7.33) and (1.7.35)),

$$\chi_\Lambda = \sum_\lambda \mathrm{mult}_\Lambda(\lambda)\, \mathrm{e}^\lambda \tag{2.6.28}$$

where e^λ is defined by a formal exponential power series, i.e. is the function

$$\mathrm{e}^\lambda : \begin{array}{l} \mathfrak{g}_0 \to F \\ \mu \mapsto \mathrm{e}^{(\lambda,\mu)} \end{array} \tag{2.6.29}$$

on the Cartan subalgebra (or, equivalently, on the weight space), with the right hand side denoting the ordinary exponential.

Since the nontrivial modules of a non-simple Kac–Moody algebra are infinite-dimensional, the sum in (2.6.28) is an infinite sum so that, in contrast to the case of simple algebras, it cannot be computed by hand. Nevertheless it is possible to determine the characters more explicitly. Namely, it turns out that the Weyl character formula (1.7.37) generalizes to the case of arbitrary Kac–Moody algebras: for any irreducible integrable highest weight module R_Λ the character obeys the *Weyl–Kac character formula*

$$\chi_\Lambda = \frac{\sum_{\sigma\in W} (\mathrm{sign}\,\sigma)\, \mathrm{e}^{\sigma(\Lambda+\rho)}}{\sum_{\sigma\in W} (\mathrm{sign}\,\sigma)\, \mathrm{e}^{\sigma(\rho)}}, \tag{2.6.30}$$

with $\rho = \sum_{i=0}^r \Lambda_{(i)}$ the Weyl vector of $\mathfrak{g}$. The infinite sums appearing in this formula converge at least for appropriately chosen highest weights.

There is also an analogue of the denominator identity (1.7.39):

$$\sum_{\sigma\in W} (\mathrm{sign}\,\sigma)\, \mathrm{e}^{\sigma(\rho)} = \mathrm{e}^{\rho} \prod_{\alpha>0} (1 - e^{-\alpha})^{\mathrm{mult}(\alpha)}; \tag{2.6.31}$$

the exponent mult(α) appearing here takes into account the fact that the multiplicity of a root can now be larger than one.

As an example for (2.6.31), consider the case $\mathfrak{g} = A_1^{(1)}$. Then $L^\vee$ equals the A_1-root lattice $\alpha^{(1)} \cdot \mathbb{Z}$, $\rho = \alpha^{(1)}/2$, $\mathrm{g}^\vee = 2$, and the finite-dimensional Weyl group is $\bar{W} \cong \{\pm 1\}$. Setting $u = \mathrm{e}^{-\alpha^{(0)}}$, $v = \mathrm{e}^{-\alpha^{(1)}}$, one has

$$\begin{aligned} \prod_{n\in\mathbb{Z}_{>0}} &(1 - u^n v^n)(1 - u^{n-1} v^n)(1 - u^{n+1} v^n) \\ &= \sum_{n\in\mathbb{Z}} (-1)^n\, u^{n(n+1)/2} v^{n(n-1)/2}. \end{aligned} \tag{2.6.32}$$

This is the so-called Jacobi triple product identity.

It should be noted that the Weyl-Kac formula is valid for arbitrary Kac–Moody algebras. However, only for affine algebras where the multiplicities of the roots and the structure of the Weyl group are known explicitly, can this formula be exploited further. Let us consider the case of untwisted affine algebras in some detail. With the help of the decomposition (2.6.13) of W into a semidirect product, one obtains

$$\sum_{\sigma\in W} (\mathrm{sign}\,\sigma)\, e^{\sigma(\mu)} = \sum_{\bar\sigma\in\bar W} (\mathrm{sign}\,\bar\sigma) \sum_{\bar\beta\in L^\vee} \mathrm{e}^{t_{\bar\beta}\bar\sigma(\mu)}. \tag{2.6.33}$$

Next define

$$\Theta_\mu := \mathrm{e}^{-\frac{1}{2k}(\mu,\mu)\delta} \sum_{t\in T} \mathrm{e}^{t(\mu)} \tag{2.6.34}$$

(recall that $\delta = (0,0,1)$), which with the explicit form (2.6.9) of the translations reads

$$\begin{aligned} \Theta_\mu &= \sum_{\bar\beta\in L^\vee} \exp\left[(\bar\mu + k\bar\beta,\, k,\, -\tfrac{1}{2k}(\bar\mu + k\bar\beta, \bar\mu + k\bar\beta)\right] \\ &= \mathrm{e}^{k^\vee \Lambda_{(0)}} \sum_{\bar\beta\in L^\vee + k^{-1}\bar\mu} \mathrm{e}^{k(\bar\beta - \frac{1}{2}(\bar\beta,\bar\beta)\delta)}. \end{aligned} \tag{2.6.35}$$

Here $\Lambda_{(0)}$ is as in (2.4.28), and in going to the final expression we have shifted the vectors $\bar\beta$ by $k^{-1}\bar\mu$. Comparing this result with (2.6.33) (and taking into account the form (2.4.39) of the Weyl vector), the denominator of the Weyl–Kac formula becomes

$$\sum_{\sigma\in W} (\mathrm{sign}\,\sigma)\, \mathrm{e}^{\sigma(\rho)} = \mathrm{e}^{\,(\bar\rho,\bar\rho)\delta/((\bar\theta,\bar\theta)\mathrm{g}^\vee)} \sum_{\bar\sigma\in\bar W} (\mathrm{sign}\,\bar\sigma)\, \Theta_{\bar\sigma(\rho)}. \tag{2.6.36}$$

Similarly, for the numerator of the Weyl–Kac formula one finds

$$\sum_{\sigma\in W} (\operatorname{sign}\sigma)\, \mathrm{e}^{\sigma(\Lambda+\rho)} = \mathrm{e}^{\,(\Lambda+\rho,\Lambda+\rho)\delta/((\bar\theta,\bar\theta)(k^\vee+g^\vee))} \sum_{\bar\sigma\in\bar W} (\operatorname{sign}\bar\sigma)\, \Theta_{\bar\sigma(\Lambda+\rho)}. \tag{2.6.37}$$

Thus for an untwisted affine algebra, the character formula (2.6.30) can be rewritten as

$$\chi_\Lambda = \mathrm{e}^{s_\Lambda\delta} \cdot \frac{\sum_{\bar\sigma\in\bar W}(\operatorname{sign}\bar\sigma)\,\Theta_{\bar\sigma(\Lambda+\rho)}}{\sum_{\bar\sigma\in\bar W}(\operatorname{sign}\bar\sigma)\,\Theta_{\bar\sigma(\rho)}}, \tag{2.6.38}$$

with

$$s_\Lambda = \frac{1}{(\bar\theta,\bar\theta)} \left(\frac{(\Lambda+\rho,\Lambda+\rho)}{k^\vee+g^\vee} - \frac{(\rho,\rho)}{g^\vee} \right). \tag{2.6.39}$$

The number s_Λ is called the *modular anomaly* of R_Λ; using $(\Lambda+\rho,\Lambda+\rho) = (\Lambda,\Lambda+2\rho)+(\rho,\rho)$ and the strange formula (1.4.71), it may also be written as

$$s_\Lambda = \frac{1}{(\bar\theta,\bar\theta)} \left(\frac{C_{\bar\Lambda}}{k^\vee+g^\vee} - \frac{1}{24}\frac{kd}{k^\vee+g^\vee} \right). \tag{2.6.40}$$

The sums in the above formulae become particularly simple if the character is evaluated on a weight proportional to $\Lambda_{(0)}$; one then speaks of the *basic specialization* of the character. This is because for $\mu = (0,\xi,0)$ and $\lambda = (\bar\lambda,k,n)$ the scalar product is $(\mu,\lambda) = n\xi$ so that

$$\chi_\Lambda(0,\xi,0) = \sum_{n\in\mathbb{Z}_{\geq 0}} d_n\, q^n, \quad q \equiv \mathrm{e}^{\xi}. \tag{2.6.41}$$

Here d_n is the dimension of the vector space that is obtained by considering all vectors of R_Λ with grade n. In other words, the basic specialization of the character is the generating function for the dimensions of the subvectorspaces of R_Λ having fixed values of the grade. It is conventional to introduce a new variable τ by writing $\xi = 4\pi i\tau/(\bar\theta,\bar\theta)$ and to normalize the highest $\bar{\mathfrak{g}}$-root as $(\bar\theta,\bar\theta) = 2$ so that

$$q \equiv \mathrm{e}^{2\pi i\tau}. \tag{2.6.42}$$

Employing the Weyl–Kac character formula and the denominator identity in the basic specialization, one can derive numerous combinatorial identities. Some of these, such as

$$\prod_{n=1}^{\infty}\left(1+q^{n-1/2}\right)^8 - \prod_{n=1}^{\infty}\left(1-q^{n-1/2}\right)^8 = 16\sqrt{q}\prod_{n=1}^{\infty}(1+q^n)^8, \tag{2.6.43}$$

were known to mathematicians already for a long time (in the case of (2.6.43), the identity was obtained in 1829 by Jacobi), but many others had not yet been derived by other methods. Below, we just list a few of them, all involving the infinite product

$$\eta(\tau) = q^{1/24} \prod_{n=1}^{\infty} (1-q^n)\,, \tag{2.6.44}$$

namely

$$\begin{aligned} \eta(8\tau)\eta(16\tau) &= \sum_{\substack{m,n\in\mathbb{Z}\\ m\le 3|n|}} (-1)^m\, q^{(2m+1)^2-32n^2}\,, \\ \eta(12\tau)\eta(12\tau) &= \sum_{\substack{m,n\in\mathbb{Z}\\ m\le 2|n|}} (-1)^{m+n}\, q^{(3(2m+1)^2-(6n+1)^2)/2}\,, \\ \eta(24\tau)\eta(96\tau) &= \sum_{\substack{m,n\in\mathbb{Z}\\ 2m\le n\le 0}} (-1)^{\frac12 n(n+1)}\, q^{8(3m+1)^2-3(2n+1)^2}(1-q^{24(2m+1)})\,, \end{aligned} \tag{2.6.45}$$

and also

$$\begin{aligned} \eta(\tau)^2 \prod_{m\le 0} &(1-q^m z)^{-1}(1-q^{m+1}z^{-1})^{-1} \\ &= q^{1/12} \sum_{m\in\mathbb{Z}} (-1)^m (1-q^m z)^{-1}\, q^{m(m+1)/2}\,. \end{aligned} \tag{2.6.46}$$

The function $\eta(\tau)$ is known as the *Dedekind eta function.* Its inverse plays the role of the character of the Heisenberg algebra $\hat{u}_1$.

2.7 Modular transformations

An important feature of the affine characters is that they possess a simple transformation property with respect to the *modular group* $PSL_2(\mathbb{Z})$. By definition, $SL_2(\mathbb{Z})$ is the group of 2×2-matrices with integer entries and determinant one; the "*P*" in $PSL_2(\mathbb{Z})$ means that in addition any such matrix is to be identified with its negative. To define the action of $PSL_2(\mathbb{Z})$ on the weight space, we first relabel the weights as

$$\lambda \equiv (\bar{\lambda}, \mu, \nu) = \frac{4\pi i}{(\bar\theta,\bar\theta)}\,(\zeta,\tau,t)\,. \tag{2.7.1}$$

An action of $SL_2(\mathbb{Z})$ on the coordinates (ζ,τ,t), which for $\zeta = 0$ restricts to an action of $PSL_2(\mathbb{Z})$, is defined as

$$\begin{pmatrix} a & b \\ c & d \end{pmatrix} (\zeta,\tau,t) = \left(\frac{\zeta}{c\tau+d}, \frac{a\tau+b}{c\tau+d}, t - \frac{c\,(\zeta,\zeta)}{2(c\tau+d)} \right). \tag{2.7.2}$$

The modular group is generated by two elements called S and T. More precisely, $PSL_2(\mathbb{Z})$ can be abstractly defined as the group generated by

$\mathcal{S}$ and $\mathcal{T}$ subject to the relations

$$\mathcal{S}^2 = 1, \quad (\mathcal{S}\mathcal{T})^3 = 1. \tag{2.7.3}$$

A realization of the generators is

$$\mathcal{S} = \begin{pmatrix} 0 & -1 \\ 1 & 0 \end{pmatrix}, \quad \mathcal{T} = \begin{pmatrix} 1 & 1 \\ 0 & 1 \end{pmatrix}. \tag{2.7.4}$$

Thus for a function which depends only on τ, $\mathcal{S}$ and $\mathcal{T}$ act as

$$f(\tau)\big|_{\mathcal{S}} = f\Big(-\frac{1}{\tau}\Big), \quad f(\tau)\big|_{\mathcal{T}} = f(\tau+1). \tag{2.7.5}$$

For example, for the Dedekind eta function (2.6.44), one finds

$$\begin{aligned} \eta(\tau)\ |_{\mathcal{S}} &= (-i\tau)^{1/2}\,\eta(\tau), \\ \eta(\tau)\ |_{\mathcal{T}} &= \mathrm{e}^{\pi i/12}\,\eta(\tau). \end{aligned} \tag{2.7.6}$$

When evaluated on (ζ,τ,t), the function Θ_μ of (2.6.35) reads, with $q = \mathrm{e}^{2\pi i\tau}$,

$$\Theta_\mu(\zeta,\tau,t) = \mathrm{e}^{2\pi i k^\vee t} \sum_{\bar\beta\in L^\vee + k^{-1}\bar\mu} \mathrm{e}^{2\pi i k^\vee(\bar\beta,\zeta)}\, q^{k^\vee(\bar\beta,\bar\beta)/2}, \tag{2.7.7}$$

which is a classical theta function (it converges absolutely on $\{(\zeta,\tau,t) \mid \tau,t\in\mathbb{C};\ \Im(\tau)>0;\ \zeta\in\mathfrak{g}_0\}$ to an analytic function). The transformation of this function under the modular group is found to be

$$\begin{aligned} \Theta_\mu(\zeta,\tau,t)|_{\mathcal{S}} &= (-i\tau)^{r/2}\,|L_w/L^\vee|^{-1/2}(k^\vee)^{-1/2} \\ &\qquad \sum_{\lambda\in P_k \bmod k^\vee L^\vee} \mathrm{e}^{-2\pi i(\bar\lambda,\bar\mu)/k}\,\Theta_\lambda(\zeta,\tau,t), \\ \Theta_\mu(\zeta,\tau,t)|_{\mathcal{T}} &= \mathrm{e}^{\pi i(\bar\mu,\bar\mu)/k}\,\Theta_\mu(\zeta,\tau,t). \end{aligned} \tag{2.7.8}$$

Here L_w and $L^\vee$ are the weight and coroot lattices, respectively, of the horizontal subalgebra (thus for simply laced $\bar{\mathfrak{g}}$, $|L_w/L^\vee| = |L_w/L|$ is the index of connection which equals the determinant of the Cartan matrix of $\bar{\mathfrak{g}}$). Also, P_k denotes the set of affine weights at level $k^\vee$, or in other words the set of weights which are Weyl group images of the integrable highest weights at level $k^\vee$. Thus (2.7.8) means in particular that the theta functions appearing at a given value of the level transform under the modular group into linear combinations of themselves. The coefficients depend in general explicitly on τ via the factor $\tau^{r/2}$ in the $\mathcal{S}$-transformation, but this dependence can be removed by considering $\Theta_\mu/(\eta(\tau))^r$ instead of Θ_μ.

As an example, consider the simplest case $\mathfrak{g} = A_1^{(1)}$. Then the theta functions appearing at level $k^\vee$ are

$$\Theta_{\bar\mu;k^\vee}(\zeta,\tau,t) = \mathrm{e}^{-2\pi i k t} \sum_{\bar\lambda\in 2\mathbb{Z}+\bar\mu/k^\vee} \exp\Big(\pi i k\,(\tfrac12\bar\lambda^2\tau - \bar\lambda\zeta)\Big), \tag{2.7.9}$$

and their transformation law under $\mathcal{S}$ reads

$$\Theta_{\bar\mu;k^\vee}(\tfrac{\zeta}{\tau},-\tfrac{1}{\tau},t-\tfrac{\zeta^2}{4\tau}) = \sqrt{\frac{-i\tau}{2k^\vee}} \sum_{\bar\lambda\in\mathbf{Z}\bmod 2k^\vee\mathbf{Z}} \mathrm{e}^{-\pi i\bar\lambda\bar\mu/k}\, \Theta_{\bar\lambda;k^\vee}(\zeta,\tau,t). \tag{2.7.10}$$

It turns out that the theta functions are actually invariant up to a phase under a large subgroup of the modular group. Namely, define

$$\Gamma(m) = \left\{ \begin{pmatrix} a & b \\ c & d \end{pmatrix} \mid a=d=1 \bmod m,\ b=c=0 \bmod m \right\} \tag{2.7.11}$$

for m any positive integer; then

$$\Theta_\mu|_\gamma = \mathrm{e}^{i\phi_\gamma}\Theta_\mu \qquad \text{for all} \quad \gamma\in\Gamma(k^\vee\ell), \tag{2.7.12}$$

where the phase ϕ depends on γ (and on $k^\vee$), but not on μ, and ℓ is the smallest positive integer such that $\ell L_w \subseteq L^\vee$.

Having established the transformation properties of the theta functions, it is straightforward to deduce how the characters transform. According to the character formula (2.6.38), one has

$$\chi_\Lambda = \mathrm{e}^{s_\Lambda\delta}\, \Xi_{\Lambda+\rho}/\Xi_\rho\,, \tag{2.7.13}$$

where s_Λ is the modular anomaly (2.6.39), and

$$\Xi_\Lambda := \sum_{\bar\sigma\in\bar W} (\operatorname{sign}\bar\sigma)\, \Theta_{\bar\sigma(\Lambda)}. \tag{2.7.14}$$

From the previous results it follows that the theta functions $\{\Xi_\Lambda \mid \Lambda\in P_k^{++}\}$ transform into each other under the modular group; here P_k^{++} denotes the set of strictly dominant integrable highest weights at level $k^\vee$, defined similarly as P_k^+ in (2.6.15),

$$P_k^{++} = \left\{ \sum_{i=0}^{r} \Lambda^i\,\Lambda_{(i)} \mid \Lambda^i > 0 \text{ for } i=0,1,...,r \right\}. \tag{2.7.15}$$

Using the fact that inner products are invariant under $\bar W$, the transformation formulae can be written as

$$\Xi_\Lambda|_{\mathcal{S}} = (-i\tau)^{r/2}\,|L_w/L^\vee|^{-1/2}(k^\vee)^{-1/2} \cdot \sum_{\Lambda'\in P_k^{++}} \left(\sum_{\bar\sigma\in\bar W} (\operatorname{sign}\bar\sigma)\, \exp\left[-\tfrac{2\pi i}{k}\,(\bar\sigma(\bar\Lambda),\bar\Lambda')\right] \right) \Xi_{\Lambda'} \tag{2.7.16}$$

and

$$\Xi_\Lambda|_{\mathcal{T}} = \exp\left[\tfrac{\pi i}{k}\,(\bar\Lambda,\bar\Lambda)\right]\,\Xi_\Lambda. \tag{2.7.17}$$

Now in particular at level $k^\vee = g^\vee$ there is only a single theta function of this type, namely Ξ_ρ. It follows that modular transformations only multiply Ξ_ρ by some constant; explicitly one finds, with the help of the denominator identity,

$$\Xi_\rho|_{\mathcal{S}} = (-i)^{d/2}\tau^{r/2}\,\Xi_\rho. \tag{2.7.18}$$

As a consequence, the characters at some given level also transform into each other. In addition, the prefactor $\tau^{r/2}$ in the $\mathcal{S}$-transformation (2.7.16) cancels between the numerator and denominator of (2.7.13). The only explicit τ-dependence in modular transformations is then due to the prefactor $\exp(s_\Lambda\delta)$ in (2.7.13). To get rid of this factor, one defines

$$\tilde\chi_\Lambda := \mathrm{e}^{2\pi i\tau\, s_\Lambda}\,\chi_\Lambda(\zeta,\tau,t), \tag{2.7.19}$$

which according to (2.7.13) means

$$\tilde\chi_\Lambda = (\Xi_{\Lambda+\rho}/\Xi_\rho)\,(\zeta,\tau,t). \tag{2.7.20}$$

These new objects are called *modified characters,* but often also just characters. Summarizing, the (modified) characters of the set of integrable modules at a given level transform under the modular group into each other without any explicit τ-dependence, or in other words they span a module of the modular group. Thus

$$\tilde\chi_\Lambda|_{\mathcal{S}} = \sum_{\Lambda'\in P_k^+} S_{\Lambda\Lambda'}\,\tilde\chi_{\Lambda'}, \tag{2.7.21}$$

and

$$\tilde\chi_\Lambda|_{\mathcal{T}} = \sum_{\Lambda'\in P_k^+} T_{\Lambda\Lambda'}\,\tilde\chi_{\Lambda'}, \tag{2.7.22}$$

where S is some matrix which can be deduced from the previous results and

$$T_{\Lambda\Lambda'} = \mathrm{e}^{2\pi i\, s_\Lambda}\,\delta_{\Lambda,\Lambda'}. \tag{2.7.23}$$

In fact, it turns out that in general the representation of the modular group generated by S and T is not a faithful one, but rather a representation of a twofold covering of the modular group so that one has $S^4 = \mathbf{1} = (ST)^6$ for the representation matrices instead of (2.7.3). The explicit form of the matrix $S_{\Lambda\Lambda'}$ can be read off (2.7.16):

$$\begin{aligned} S_{\Lambda\Lambda'} = (i)^{(d-r)/2}\,|L_w/L^\vee|^{-1/2}(k^\vee+g^\vee)^{-1/2} \\ \sum_{\bar\sigma\in\bar W}(\mathrm{sign}\,\bar\sigma)\,\exp\left[-\tfrac{2\pi i}{k+I_{\mathrm{ad}}/2}\,(\bar\sigma(\bar\Lambda+\bar\rho),\bar\Lambda'+\bar\rho)\right] \end{aligned} \tag{2.7.24}$$

(recall that if Λ has level $k^\vee$, then $\Lambda+\rho$ has level $k^\vee+g^\vee$). Note that the weights appearing on the right hand side are shifted by $\bar\rho$; this turns out to be very relevant in the proof that the matrix S has the important property of being unitary. Namely, it can be shown that the matrix representing the S-transformation on the theta functions is unitary; this implies that the matrix (2.7.24) satisfies

$$(S\,S^{\dagger})_{\Lambda+\rho,\Lambda'+\rho} \propto \sum_{\bar\sigma\in\bar W}\sum_{\bar\sigma'\in\bar W} (\mathrm{sign}\,\bar\sigma\bar\sigma')\,\delta_{\bar\sigma(\bar\Lambda+\bar\rho),\bar\sigma'(\bar\Lambda'+\bar\rho)}. \tag{2.7.25}$$

On the right hand side, only the terms with $\bar\Lambda = (\bar\sigma\bar\sigma') \star \bar\Lambda'$ give a nonvanishing contribution; but since both $\bar\Lambda$ and $\bar\Lambda'$ are dominant weights and $\bar W$ permutes the Weyl chambers freely, this is only possible if $\bar\Lambda = \bar\Lambda'$ so that

$$(S\,S^{\dagger})_{\Lambda+\rho,\Lambda'+\rho} \propto \sum_{\bar\sigma,\bar\sigma'\in\bar W} (\mathrm{sign}\,\bar\sigma\bar\sigma')\,\delta_{\bar\sigma(\bar\Lambda+\bar\rho),\bar\sigma'(\bar\Lambda+\bar\rho)}. \tag{2.7.26}$$

Now, again due to the basic properties of the Weyl group $\bar W$, this sum gets contributions only from the terms with $\bar\sigma = \bar\sigma'$, and the result is independent of $\bar\Lambda$. Thus one can conclude that $S\,S^{\dagger} \propto \mathbf{1}$. Moreover, because of $S^4 = 1$ the constant of proportionality must in fact be equal to one; thus the matrix S is unitary as promised. Another property of the matrix S is that it is symmetric. This follows immediately from (2.7.24) with the help of the $\bar W$-invariance of inner products. Together one then has

$$S^{-1} = S^{\dagger} = S^{*}. \tag{2.7.27}$$

The representation of the (covering of) the modular group on the set of (modified) characters of integrable modules is therefore a unitary one. A particularly interesting consequence of this is that the bilinear form

$$\sum_{\Lambda\in P_k^+} \chi_\Lambda\,\chi_\Lambda^{*} \tag{2.7.28}$$

is invariant under any modular transformation.

In the following we list some further properties of the modular matrix S. First, it can be shown that

$$\begin{aligned} S_{\Lambda\Lambda'} &= a_\Lambda\,\mathrm{tr}_{\bar\Lambda'}\,\exp\left(-2\pi i\,\frac{(\bar\Lambda+\bar\rho,\cdot\,)}{k+I_{\mathrm{ad}}/2}\right) \\ &\equiv a_\Lambda\,\bar\chi_{\bar\Lambda'}\,(-\ln p\cdot(\bar\Lambda+\bar\rho))\,. \end{aligned} \tag{2.7.29}$$

Here $\bar\chi$ denotes the character of the horizontal subalgebra $\bar{\mathfrak g}$ of $\mathfrak g$ and I_{ad} the quadratic Casimir eigenvalue of the adjoint of $\bar{\mathfrak g}$; also, we defined

$$a_\Lambda := |L_w/L^{\vee}|^{-1/2}\,(k^{\vee}+g^{\vee})^{-1/2}\prod_{\bar\alpha>0} 2\,\sin\left(\frac{(\bar\Lambda+\bar\rho,\bar\alpha)}{k+I_{\mathrm{ad}}/2}\,\pi\right), \tag{2.7.30}$$

the summation being over the positive roots of $\bar{\mathfrak g}$, and

$$p := \mathrm{e}^{2\pi i/(k+I_{\mathrm{ad}}/2)}. \tag{2.7.31}$$

To see that (2.7.29) is correct, just use the Weyl character formula to rewrite it as

$$S_{\Lambda\Lambda'} = i^{(d-r)/2}\, a_\Lambda \frac{\sum_{\bar\sigma\in\bar W} (\operatorname{sign}\bar\sigma)\, p^{(\bar\Lambda+\bar\rho,\bar\sigma(\bar\Lambda'+\bar\rho))}}{\sum_{\bar\sigma\in\bar W} (\operatorname{sign}\bar\sigma)\, p^{(\bar\Lambda+\bar\rho,\bar\sigma(\bar\rho))}}, \tag{2.7.32}$$

which is indeed nothing but the previous result (2.7.24) for S. These considerations also show that

$$a_\Lambda = S_{\Lambda,k^\vee\Lambda_{(0)}} > 0. \tag{2.7.33}$$

Similarly, using the Weyl denominator identity, the quantities a_Λ can also be written as

$$\begin{aligned} a_\Lambda &= a_{k^\vee\Lambda_{(0)}}\, \mathrm{tr}_{\bar\Lambda}\left(\exp\left(2\pi i \frac{\bar\rho}{k+I_{\mathrm{ad}}/2}\right)\right) \\ &\equiv a_{k^\vee\Lambda_{(0)}}\, \bar\chi_{\bar\Lambda}(\bar\rho\ln p)\,. \end{aligned} \tag{2.7.34}$$

Now considering the column corresponding to $\Lambda' = k^\vee\Lambda_{(0)}$ and using the unitarity of S, one obtains

$$\sum_{\Lambda\in P_k^+} (a_\Lambda)^2 = 1. \tag{2.7.35}$$

Combining this with (2.7.34), one has

$$a_{k^\vee\Lambda_{(0)}} = \left(\sum_{\Lambda\in P_k^+} |\bar\chi_{\bar\Lambda}(\bar\rho\ln p)|^2\right)^{-1/2} \tag{2.7.36}$$

Finally we note the following property: applying the S-transformation twice, one gets

$$\tilde\chi_\Lambda|_{S^2} = \tilde\chi_{\Lambda^+} \tag{2.7.37}$$

where $\Lambda^+ = (k^\vee - (\bar\Lambda,\bar\theta))\Lambda_{(0)} + \bar\Lambda^+$ with $\bar\Lambda^+$ the $\bar{\mathfrak{g}}$-weight conjugate to $\bar\Lambda$. In other words, the property $S^4 = \mathbf{1}$ can be strengthened to

$$S^2 = C, \qquad C_{\Lambda\Lambda'} = \delta_{\Lambda',\Lambda^+}, \tag{2.7.38}$$

and analogously one deduces that $(ST)^3 = C$. In particular, if all highest weight modules of $\bar{\mathfrak{g}}$ are selfconjugate (i.e. for all $\bar{\mathfrak{g}}$ other than A_r, $r \geq 2$, D_{2r+1} and E_6), then $S^2 = \mathbf{1} = (ST)^3$ so that the characters transform faithfully under the modular group.

As an example, consider again the simplest case $\mathfrak{g} = A_1^{(1)}$. The Weyl group of A_1 is isomorphic to $\{\pm 1\}$, and $\bar\rho = \frac{1}{2}\bar\theta = 1$. Hence in the notation of (2.7.9), one has

$$\Xi_{\bar\mu,k^\vee} = \Theta_{\bar\mu,k^\vee} - \Theta_{-\bar\mu,k^\vee}, \tag{2.7.39}$$

and

$$\tilde{\chi}_{\bar{\mu},k^\vee} = \frac{\Xi_{\bar{\mu}+1,k^\vee+2}}{\Xi_{1,2}}. \qquad (2.7.40)$$

Note that according to (2.7.7) one has, for $A_1^{(1)}$, $\Theta_{\bar{\mu},k^\vee} = \Theta_{-\bar{\mu},k^\vee}$ for $\zeta = 0$; thus in order for $\tilde{\chi}|_{\zeta=0}$ to be well defined, it is actually necessary to restore first the z-dependence and then take the limit $\zeta \to 0$ by application of l'Hospital's rule. Also, the matrices S and T are easily obtained as

$$S_{\lambda\mu} = \sqrt{\frac{2}{k^\vee+2}} \sin\left(\frac{(\bar{\lambda}+1)(\bar{\mu}+1)}{k^\vee+2}\,\pi\right) \qquad (2.7.41)$$

and

$$T_{\lambda\mu} = \exp\left(\pi i \left[\frac{(\bar{\lambda}+1)^2}{2(k^\vee+2)} - \frac{1}{4}\right]\right) \delta_{\lambda\mu}. \qquad (2.7.42)$$

It is straightforward to verify that these matrices are unitary and satisfy $S^2 = (ST)^3 = \mathbf{1}$.

2.8 Branching rules

As discussed in section 1.8, for embeddings $\bar{\mathfrak{h}} \subset \bar{\mathfrak{g}}$ of semisimple Lie algebras, the unitary highest weight modules of $\bar{\mathfrak{g}}$ decompose into a finite number of unitary highest weight modules of $\bar{\mathfrak{h}}$. One says that these embeddings are *finitely reducible*. For affine algebras, this is in general no longer the case because the nontrivial unitary highest weight modules are now infinite-dimensional, so that the highest weight modules of $\mathfrak{h}$ may appear with infinite multiplicity in those of $\mathfrak{g}$. Indeed, it turns out that this is the generic situation. To see this, first notice that as a trivial consequence of the definition of the characters, for any decomposition

$$R_\Lambda(\mathfrak{g}) \rightsquigarrow \bigoplus_i R_{\Lambda_i}(\mathfrak{h}), \qquad (2.8.1)$$

the characters obey the sum rule $\chi_\Lambda(\mathfrak{g}) = \sum_i \chi_{\Lambda_i}(\mathfrak{h})$, or equivalently (for a suitable choice of the D-eigenvalues of the highest weight vectors)

$$\tilde{\chi}_\Lambda(\mathfrak{g}) = \sum_i \tilde{\chi}_{\Lambda_i}(\mathfrak{h}). \qquad (2.8.2)$$

To investigate the implications of this sum rule, one considers the limit of the modified characters $\tilde{\chi}_\Lambda(\tau)$ for $\tau \to 0$. This limit is best evaluated by relating it to the limit $\tau \to i\infty$ (note that because of the convergence properties of the characters the limit $\tau \to 0$ is to be taken with a positive imaginary part of τ); this is done via the modular transformation S:

$$\lim_{\tau\to 0} \tilde\chi_\Lambda(\tau) = \lim_{\tau'\to i\infty} \tilde\chi_\Lambda(-\tfrac{1}{\tau'}) = \sum_{\Lambda'\in P_k^+} S_{\Lambda\Lambda'} \lim_{\tau'\to i\infty} \tilde\chi_{\Lambda'}(\tau'). \tag{2.8.3}$$

For $\tau \to i\infty$, clearly only those terms in the definition (2.6.28) of the character survive which are independent of τ, so that

$$\lim_{\tau\to i\infty} \chi_\Lambda(\tau) = d_{\bar\Lambda} \tag{2.8.4}$$

gives the dimensionality of the underlying module of the horizontal subalgebra. For the modified characters, this means

$$\lim_{\tau'\to i\infty} \tilde\chi_\Lambda(\tau') = \lim_{\tau'\to i\infty} (\mathrm{e}^{2\pi i s_\Lambda \tau'} \chi_\Lambda(\tau')) = \mathrm{e}^{-2\pi i s_\Lambda/\tau}\, d_{\bar\Lambda}. \tag{2.8.5}$$

In the sum over $\Lambda' \in P_k^+$, the leading term is then the one with minimal modular anomaly $s_{\Lambda'}$, i.e. $\Lambda' = k^\vee \Lambda_{(0)}$. Thus one finally obtains

$$\lim_{\tau\to 0} \tilde\chi_\Lambda(\tau) = S_{\Lambda,k^\vee\Lambda_{(0)}}\, d_{\bar\Lambda_{(0)}}\, \mathrm{e}^{-2\pi i s_0/\tau} \tag{2.8.6}$$

with $s_0 \equiv s_{k^\vee\Lambda_{(0)}}$, or shortly

$$\lim_{\tau\to 0} \tilde\chi_\Lambda(\tau) = a_\Lambda\, \mathrm{e}^{\pi i c/12\tau}, \tag{2.8.7}$$

with a_Λ as in (2.7.34) and

$$c = 48 s_0 = \frac{k^\vee d}{k^\vee + g^\vee}. \tag{2.8.8}$$

The number c is called the *conformal anomaly* of $\mathfrak{g}$ at level $k^\vee$.

Implementing now this asymptotic behavior into the character sum rule, one learns that the sum over i is finite if and only if the conformal anomalies of $\mathfrak{g}$ and $\mathfrak{h}$ are equal,

$$c(\mathfrak{g}) = c(\mathfrak{h}). \tag{2.8.9}$$

(In particular, one has to consider both $\mathfrak{g}$ and $\mathfrak{h}$ at fixed values of the respective levels.) An embedding fulfilling this condition is called a *conformal embedding.* In contrast, for non-conformal embeddings the sum is infinite; in fact the sum then makes sense only formally and should more carefully be written as

$$\tilde\chi_\Lambda(\mathfrak{g}) = \sum_j b_{\Lambda;\Lambda_j}\, \tilde\chi_{\Lambda_j}(\mathfrak{h}), \tag{2.8.10}$$

where the sum is now finite (because the number of integrable highest weight modules at a given level is finite), but the characters are multiplied with functions $b_{\Lambda;\Lambda_j}$, known as *branching functions,* that depend nontrivially on τ.

Another possibility of deducing the equality of conformal anomalies as the criterion for simple reducibility is to investigate the multiplicities of the weights $\lambda - n\delta$ for fixed λ; one finds the asymptotic behavior

$$\lim_{n\to\infty} \mathrm{mult}_\Lambda(\lambda - n\delta) \propto n^{-(r+3)/4} \exp(4\pi\sqrt{nc}), \tag{2.8.11}$$

from which the criterion is read off immediately.

Any embedding $\bar{\mathfrak{h}} \subset \bar{\mathfrak{g}}$ of semisimple (or possibly reductive) Lie algebras determines in a natural (and, in fact, unique) way an embedding $\mathfrak{h} \subset \mathfrak{g}$ of untwisted affine algebras, namely by constructing $\mathfrak{h}$ and $\mathfrak{g}$ from the horizontal subalgebras $\bar{\mathfrak{h}}$ and $\bar{\mathfrak{g}}$, respectively, as the central extension of the respective loop algebras. Since, as explained in section 1.8, the maximal subalgebras of semisimple Lie algebras are fully classified, it is thus also possible to obtain a complete list of all conformal embeddings; such lists are available in the literature (although they are not quite complete; see below). One finds in particular that for a maximal conformal embedding the level of $\bar{\mathfrak{g}}$ must be equal to one. Also, for any regular embedding $\bar{\mathfrak{h}} \subset \bar{\mathfrak{g}}$ with $\bar{\mathfrak{g}}$ simply laced, the corresponding embedding $\mathfrak{h} \subset \mathfrak{g}$ is a conformal one, while for a regular embedding of $\bar{\mathfrak{h}}$ in non-simply laced $\bar{\mathfrak{g}}$, the embedding $\mathfrak{h} \subset \mathfrak{g}$ is conformal iff all short roots of $\bar{\mathfrak{g}}$ are contained in the root system of $\bar{\mathfrak{h}}$.

The construction of the affine embedding in terms of the embedding of the horizontal subalgebras also shows that the central extension K has to be the same for $\mathfrak{g}$ and $\mathfrak{h}$ or in other words that the eigenvalues k and h of the central generators of $\mathfrak{g}$ and $\mathfrak{h}$ must coincide. For the levels $h^\vee$ and $k^\vee$ this implies that their ratio is equal to the relative length squared of the highest roots of $\bar{\mathfrak{h}}$ and $\bar{\mathfrak{g}}$, i.e. to the Dynkin index (1.8.8) of the embedding $\bar{\mathfrak{h}} \subset \bar{\mathfrak{g}}$,

$$h^\vee = I_{\bar{\mathfrak{h}}\subset\bar{\mathfrak{g}}}\, k^\vee . \tag{2.8.12}$$

In particular, for maximal conformal embeddings one has $h^\vee = I_{\bar{\mathfrak{h}}\subset\bar{\mathfrak{g}}}$.

Let us now list a few maximal conformal embeddings. For $\mathfrak{g}$ at level $k^\vee$ we write shortly $(\mathfrak{g})_{k^\vee}$. Then the simplest example is

$$(A_1^{(1)})_4 \subset (A_2^{(1)})_1. \tag{2.8.13}$$

Some others are

$$\begin{aligned}
&(E_6^{(1)})_6 \subset (A_{26}^{(1)})_1, \\
&(A_{n-1}^{(1)})_m \oplus (A_{m-1}^{(1)})_n \subset (A_{mn-1}^{(1)})_1, \\
&(\mathfrak{g})_{g^\vee} \subset ((so_{\dim\bar{\mathfrak{g}}})^{(1)})_1 \qquad \text{for any simple } \bar{\mathfrak{g}}, \\
&(so_m^{(1)})_n \oplus (so_n^{(1)})_m \subset (so_{mn}^{(1)})_1.
\end{aligned} \tag{2.8.14}$$

In the latter series, the labels m and n are restricted by $m, n \geq 3$; thus the first two members of the series would better be written as

$$\begin{aligned}(A_1^{(1)})_6 \oplus (A_1^{(1)})_6 &\subset (B_4^{(1)})_1, \\ (A_1^{(1)})_3 \oplus (A_1^{(1)})_3 \oplus (A_1^{(1)})_8 &\subset (D_6^{(1)})_1,\end{aligned} \tag{2.8.15}$$

owing to the identifications $(B_1^{(1)})_m \cong (A_1^{(1)})_{2m}$ and $(D_2^{(1)})_m \cong (A_1^{(1)})_m \oplus (A_1^{(1)})_m$ which follow from the corresponding isomorphisms of the horizontal subalgebras. The maximal conformal embeddings in untwisted affine algebras of type A, B and D will be discussed further in section 3.8.

Let us also present one example for the branching rules of a conformal embedding, namely for the embedding $(E_6^{(1)})_6 \subset (A_{26}^{(1)})_1$. In terms of the dimensionalities with respect to the horizontal subalgebras, the branching rules read

$$\begin{aligned}
1 &\rightsquigarrow 1 + 650 + 70\,070 + 1\,337\,050 \\
27 &\rightsquigarrow 27 + 7\,371 + 386\,100 \\
351 &\rightsquigarrow 351 + 51\,975 + 638\,820 \\
2\,925 &\rightsquigarrow 2\,925 + 43\,758 + 252\,252 \\
17\,550 &\rightsquigarrow 17\,550 + 412\,776 + 459\,459 \\
80\,730 &\rightsquigarrow 34\,398 + 46\,332 + 2\,088\,450 \\
296\,010 &\rightsquigarrow 43\,758 + 252\,252 + 3\,309\,696 \\
888\,030 &\rightsquigarrow 393\,822 + 494\,208 + 3\,281\,850 \\
2\,220\,075 &\rightsquigarrow 579\,150 + 1\,640\,925 + 1\,837\,836 \\
4\,686\,825 &\rightsquigarrow 442\,442 + 371\,800 + 1\,337\,050 + 2\,977\,975 \\
8\,436\,285 &\rightsquigarrow 100\,386 + 2\,559\,843 + 5\,776\,056 \\
13\,037\,895 &\rightsquigarrow 853\,281 + 4\,582\,656 + 7\,601\,958 \\
17\,383\,860 &\rightsquigarrow 1\,559\,376 + 3\,309\,696 + 12\,514\,788 \\
20\,058\,300 &\rightsquigarrow 5\,553\,900 + 6\,747\,300 + 7\,757\,100
\end{aligned} \tag{2.8.16}$$

(for brevity we do not display which one of two conjugate modules of E_6 is meant; this can be inferred from the consideration of conjugacy classes provided that they are nontrivial; also, the 351-dimensional module of E_6 is the one with highest weight $\Lambda_{(2)}$ (respectively $\Lambda_{(4)}$) rather than $2\Lambda_{(1)}$ (respectively $2\Lambda_{(5)}$)).

For conformal embeddings, the asymptotic form of the character sum rule reduces to

$$a_\Lambda(\mathfrak{g}) = \sum_i a_{\Lambda_i}(\mathfrak{h}), \tag{2.8.17}$$

where the sum is of course meant to count also multiplicities. Note the similarity of this sum rule to the corresponding sum rule of dimensions that holds for embeddings of semisimple Lie algebras. Indeed a_Λ can in a sense be considered as describing the size of an affine module; for a more detailed description of what is meant by this phrase, see sections 5.3 (p. 343) and 5.5.

In full generality one can compute branching rules only by matching the states of a given $\mathfrak{g}$-module, grade by grade, by the states of the various possible $\mathfrak{h}$-modules, which in particular requires the complete knowledge of the null vector structure of the modules. But since the numbers $a_\Lambda = S_{\Lambda,k^\vee\Lambda_{(0)}}$ are typically irrational, the sum rule (2.8.17) fortunately implies that these numbers can already determine a branching rule to a large extent. In fact, at low levels of $\mathfrak{h}$ the knowledge of these numbers for all integrable Λ is often sufficient to determine the branching rules completely. More precisely, one usually also has to implement some simple information about the conjugacy classes of the relevant modules of the horizontal subalgebras, owing to degeneracies of the numbers a_Λ. In particular, from the definition of a_Λ and the unitarity of the matrix S it follows immediately that conjugate modules possess the same values, $a_\Lambda = a_{\Lambda^+}$. More generally, one has $a_\Lambda = a_{\omega(\Lambda)}$ for any automorphism $\dot\omega$ of the Dynkin diagram of the affine algebra (for the precise definition of $\omega(\Lambda)$, see p. 286); for example, for $(A_r^{(1)})_1$ one finds $a_{\Lambda_{(i)}} = a_{\Lambda_{(0)}} = 1/\sqrt{r+1}$ for all $i = 1, ..., r$.

In the above example, the sum rule is sufficient to fix all branching rules uniquely, except for the one of the basic module and the one corresponding to the 2925-dimensional A_{26}-module for which a more detailed investigation is required.

2.9 Literature

The standard reference for affine Lie algebras is the textbook by Kac [B19, B20]. Other books which contain parts of the material of chapter 2 are [B21] and [B29]. Reviews are given in [R18, R15, R19, R5], and as a short introduction one can use [R28] or the appendix of [251]. An extensive list of original papers on Kac-Moody algebras is provided in [49], and a guide to the literature on vertex operators is [373]. The classification of affine algebras was obtained by Kac [307, 308, 310] and Moody [426, 427, 428], and from a different point of view by Kantor [324]; a complete list of affine root systems was also given in [398]. For the classification of inequivalent gradations of affine algebras, see [309]. For non-affine Kac–Moody algebras, see [242, B19], and also [70] and the references cited there. For Kac–Moody *groups*, see [317]. Affine algebras first appeared in the physics literature in the context of current algebras including the Schwinger [518] term. A review of four-dimensional current algebra is given in [B1].

In the book [B19], the nomenclature for the twisted affine algebras $X_r^{(\ell)}$, $\ell = 2, 3$, follows the original article by Kac [307]; here we have not used Kac's notation, but rather essentially that [127], which is close to the notation in the original paper by Moody [426] and to the one in [399]. For the relation between the two notations, see the remarks after (2.2.42).

The hyperbolic Dynkin diagrams displayed in section 2.1 are taken from [503]. For the relation between the automorphism groups of the Dynkin diagrams of $\mathfrak{g}$ and $\bar{\mathfrak{g}}$, see [458].

Vertex operator constructions of level one highest weight modules in the homogenous gradation were first obtained in the physics literature by Banks, Horn, and Neuberger [35], and in the mathematics literature by Frenkel and Kac [214] and by Segal [519]. In the principal gradation, vertex operators were constructed in [376, 315]. The link between these two types of vertex operators is described in [319, 374]; vertex operator constructions in other gradations are discussed in [162, 541]. Spinor type constructions of level one highest weight modules were discovered in [211, 212, 316, 198]. Further constructions of affine Lie algebra modules arose in the context of conformal field theory; for the corresponding literature, see section 3.9.

Verma modules were introduced by Verma [554] for semisimple Lie algebras; their reduction to irreducible modules was analyzed by Bernstein, Gelfand, and Gelfand [55]. The Weyl group of affine algebras was first considered in [311, 398, 429]. The Weyl-Kac character formula was derived by Kac [311], following the Bernstein–Gelfand–Gelfand proof [55] of the Weyl character formula. Combinatorial identities following from the character formula and their comparison with known [398, 555] ones are contained in [311, 375, 372, 312, 199, 377, 378]. Root multiplicities of non-affine Kac–Moody algebras are discussed in [52, 466, 365, 129].

The modular transformation properties of characters (section 2.7) which play an important role in conformal field theory are found in [318], with some new results added in [322].

The conformal embeddings of affine algebras were classified in [508, 32]. For a proof of the finite reducibility of these embeddings, see [84]. Branching rules for conformal embeddings have been computed in [320, 322, 566, 11, 284, 558, 559]; the branching rules given in (2.8.16) appear here for the first time.

3
WZW theories

We now apply what we have learned so far to some problems in two-dimensional conformal field theory. There exist several ways of deriving the basic concepts of conformal field theory, such as analysing the two-dimensional models of statistical mechanics at the critical point of a second order phase transition in terms of a euclidean field theory, or by describing the world sheet swept out by a relativistic string. These approaches differ from each other, among other things, by the precise meaning that is given to the coordinate z on which the fields $\phi(z)$ of the theory depend. Here we choose a presentation which shows how these concepts emerge rather naturally in a purely algebraic context. In this approach, the variable z is introduced as a formal parameter, which allows for rewriting the brackets of an affine Lie algebra in terms of generating functions called the current fields. This is done in section 3.1, where we also introduce the related concepts of operator products and normal ordering. In section 3.2, we describe how one of the main ingredients of conformal field theory, the energy–momentum tensor, arises in the context of affine Lie algebras, namely via the so-called Sugawara construction; we also briefly comment on the representation theory of the Virasoro algebra and on the Goddard–Kent–Olive coset construction. All these results are set into the context of two-dimensional conformal field theory in section 3.3. There also the basic principles of conformal field theory are summarized, such as the operator product algebra, the notion of primary and descendant fields and of correlation functions, and the various possible meanings of the two-dimensional world sheet on which the theory is defined.

In section 3.4 we explain how the operator product coefficients can be determined by analysing four-point functions of primary fields, and introduce the Knizhnik–Zamolodchikov equation which can be employed to compute these four-point functions for the special case of Wess–Zumino–Witten (WZW) theories, i.e. theories whose fields furnish irreducible highest weight modules of an affine Lie algebra and which possess an energy–momentum tensor of the Sugawara form (most WZW theories can be realized in a lagrangian setting as sigma models on group manifolds, supplemented by a Wess–Zumino term). An algebraic equation restricting the WZW correlation functions is analysed in section 3.5; when combined with the Knizhnik–Zamolodchikov equation, this allows us to identify the special class of primary fields known as cominimal fields.

In section 3.6 the conformal field theory of free fermions and the fermionic construction of affine Lie algebras is described. Section 3.7 discusses the general issue of quantum equivalence between conformal field theories; analysed in detail is the example of free fermion systems possessing a Sugawara-type energy–momentum tensor, which are classified by the symmetric space theorem of Goddard, Nahm, and Olive. In section 3.8 this example is investigated further, and the relation with conformal embeddings in level one WZW theories is presented.

3.1 Operator products

In this section an economic way of rewriting the affine Lie algebra relations

$$[T^a_m, T^b_n] = f^{ab}{}_c\, T^c_{m+n} + m\bar{\kappa}^{ab}\delta_{m+n,0}K \tag{3.1.1}$$

is considered. The idea is to introduce a generating function for the generators T^a_m, i.e. a single object depending on a formal parameter z such that the generators T^a_m are obtainable as Fourier–Laurent coefficients of this object. Thus we define

$$J^a(z) := \sum_{n\in\mathbf{Z}} T^a_n\, z^{-n-1}. \tag{3.1.2}$$

Any quantity depending on the parameter z will be called a *field* or *field operator*. In particular, the quantity $J^a(z)$ will be called the affine current field or affine current operator, or shortly the *current*. In the applications we have in mind, only modules of the affine algebra at some fixed value of the level occur so that K acts as a constant. Consequently, no field is introduced for the central generator K.

So far, the parameter z is just a formal variable. However, there exist various applications where z can be given a physical meaning, namely as a parameter of the unit circle $S^1 \subset \mathbb{C}$ on which some conformally invariant quantum field theory is defined. This will be explained in detail in section 3.3 (p. 172). If the variable z takes values in the unit circle, then in terms of z the expansion (3.1.2) is interpreted as the Laurent series obtained by expanding the operator-valued quantity $J^a(z)$ around zero, where it is assumed that this series converges on S^1. Thus the components T^a_n can be recovered from $J^a(z)$ as the contour integral

$$T^a_m = \frac{1}{2\pi i} \oint_0 \mathrm{d}z\, z^m J^a(z). \tag{3.1.3}$$

Alternatively one may parametrize the unit circle as $z = \exp(2\pi i x/L)$ with $x \in \mathbb{R}$ and fixed $L \in \mathbb{R}$; then the right hand side of (3.1.2) is just the Fourier series decomposition of J^a considered as a function of x with period L.

It is tempting to try to identify a Lie algebra structure in terms of the current fields by defining a bracket of the currents as

$$[J^a(z), J^b(w)] = \sum_{m,n\in\mathbb{Z}} z^{-m-1} w^{-n-1} [T^a_m, T^b_n], \tag{3.1.4}$$

but with this one runs immediately into problems. Namely, using the brackets (3.1.1) and introducing $\ell = m+n$ as a new summation variable, (3.1.4) can be rewritten as

$$[J^a(z), J^b(w)] = \sum_{n\in\mathbb{Z}} \left(\frac{z}{w}\right)^{n+1} \sum_{\ell\in\mathbb{Z}} z^{-\ell-2} \,(f^{ab}{}_c T^c_\ell + n\bar{\kappa}^{ab}\delta_{\ell,0}K). \tag{3.1.5}$$

The summation over $\ell \in \mathbb{Z}$ could now be performed, leading to the expression $z^{-1}T^c(z) + nz^{-2}\bar{\kappa}^{ab}K$, but then the n-summation does not converge for z, $w \in S^1$ (nor for any other complex values of z, w).

The key to the resolution of this problem lies in the observation that the n-summation in (3.1.5) can be split up into two parts, one of which converges for $|z| > |w|$, the other one for $|z| < |w|$. This motivates us to admit for z in the expansion (3.1.2) arbitrary nonzero complex values, and in addition to prescribe special integration contours in the relation

$$[T^a_m, T^b_n] = \frac{1}{2\pi i} \oint_0 \mathrm{d}z \,\frac{1}{2\pi i} \oint_0 \mathrm{d}w\, z^m w^n \,[J^a(z), J^b(w)] \tag{3.1.6}$$

which follows from the contour integral representation (3.1.3). Namely, while the bracket on the left hand side stands for the usual commutator, on the right hand side the bracket is meant as a kind of modified commutator that is defined as follows: one formally writes the Lie bracket $[J^a(z), J^b(w)]$ as $J^a(z)J^b(w) - J^b(w)J^a(z)$, but in addition for each individual term the integration contour is chosen such that the absolute value of the first argument is larger than that of the second one. Thus

$$\begin{aligned}[T^a_m, T^b_n] &= \frac{1}{2\pi i} \oint_0 \mathrm{d}z \,\Big(\frac{1}{2\pi i} \oint_{C_1} \mathrm{d}w\, z^m w^n J^a(z) J^b(w) \\ &\qquad\qquad - \frac{1}{2\pi i} \oint_{C_2} \mathrm{d}w\, z^m w^n J^b(w) J^a(z)\Big) \\ &= \frac{1}{2\pi i} \oint_0 \mathrm{d}z \,\frac{1}{2\pi i} \oint_{C_1\cup C_2} \mathrm{d}w\, z^m w^n \,\mathcal{R}(J^a(z)J^b(w)).\end{aligned} \tag{3.1.7}$$

Here for fixed z the w-integration contours are as depicted in figure (3.1.8).

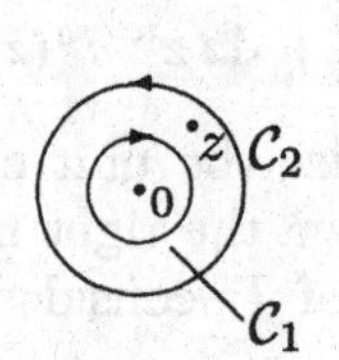

(3.1.8)

Also, we defined

$$\mathcal{R}(J^a(z)J^b(w)) := \begin{cases} J^a(z)J^b(w) & \text{for} \quad |z| > |w| \\ J^b(w)J^a(z) & \text{for} \quad |z| < |w|. \end{cases} \tag{3.1.9}$$

The expression (3.1.9) is called the *radially ordered product*, or shortly *operator product* of the two currents $J^a(z)$ and $J^b(w)$. Note that the radial ordering is forced on us by the aim to obtain convergent power series when z is interpreted as a complex number. In physical applications, the radial direction corresponds to the direction of proper time, and hence the radial ordering requirement reproduces the familiar time ordering prescription of quantum field theory.

The radial ordering prescription effectively splits the n-summation into pieces which converge (namely, to $(z-w)^{-1}$ and $(z-w)^{-2}$ for the terms proportional to J^c_ℓ and to K, respectively). In fact one gets

$$\mathcal{R}(J^a(z)J^b(w)) = (z-w)^{-2}\bar{\kappa}^{ab}K - (z-w)^{-1}f^{ab}{}_c J^c(w) + \mathcal{O}(z-w)^0, \tag{3.1.10}$$

as is most easily proven by showing that this reproduces correctly the affine Lie brackets (3.1.1). To see this, one deforms the integration contour $C_1 \cup C_2$ into a contour encircling w (in the negative sense) according to the figure (3.1.11).

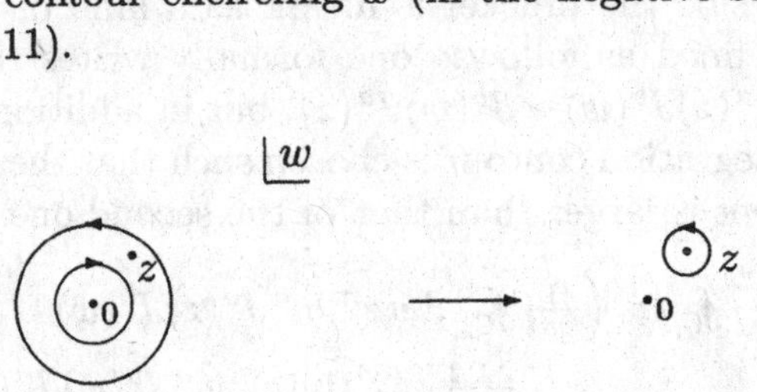

(3.1.11)

The w-integration can now be performed with the help of Cauchy's theorem:

$$\begin{aligned}[T^a_m, T^b_n] &= \tfrac{1}{2\pi i}\oint_0 \mathrm{d}z \, \tfrac{1}{2\pi i}\oint_z \mathrm{d}w \; z^m w^n \, \mathcal{R}(J^a(z)J^b(w)) \\ &= \tfrac{1}{2\pi i}\oint_0 \mathrm{d}z \; z^m \tfrac{1}{2\pi i}\oint_z \mathrm{d}w \; w^n \, [(z-w)^{-2}\bar\kappa^{ab} K \\ &\qquad - (z-w)^{-1} f^{ab}{}_c T^c(w) + \mathcal{O}(z-w)^0] \qquad (3.1.12) \\ &= \tfrac{1}{2\pi i}\oint_0 \mathrm{d}z \; z^m \, [-(\tfrac{\mathrm{d}}{\mathrm{d}z}\, z^n)\, \bar\kappa^{ab} K + z^n f^{ab}{}_c T^c(z) + 0] \\ &= -n\bar\kappa^{ab}\delta_{m+n,0} K + f^{ab}{}_c T^c_{m+n} .\end{aligned}$$

Note that the terms of order $(z-w)^0$ and higher which where left unspecified in (3.1.10) do not contribute to the final result. Thus in particular we could have written $J^c(\tilde w)$ in (3.1.10) instead of $J^c(w)$ with any $\tilde w$ fulfilling $\tilde w \to w$ for $z \to w$, e.g. $\tilde w = \frac{1}{2}(z+w)$. In terms of the n-summation in (3.1.5), this freedom corresponds to the fact that one is free to require $|w| > |z|$ or $|w| < |z|$ for an arbitrary finite number of terms since this does not influence the convergence of the series.

Let us also write down the operator product (3.1.10) in terms of the variable x related to z via the exponential map $z = \exp(2\pi i x/L)$. This relation looks nicest if one defines

$$\tilde J^a(x) := \frac{z}{L}\, J^a(z). \qquad (3.1.13)$$

With this definition, the sums over n corresponding to those in (3.1.5) are of the form

$$\begin{aligned}\sum_{n\in\mathbb{Z}} \mathrm{e}^{2\pi i n x/L} &= L\,\delta(x), \\ \sum_{n\in\mathbb{Z}} n\, \mathrm{e}^{2\pi i n x/L} &= \frac{L^2}{2\pi i}\,\delta'(x)\end{aligned} \qquad (3.1.14)$$

where δ and δ' are the δ-distribution with period L and its derivative (these are just the analogues of z^{-1} and z^{-2}, respectively). Thus one gets

$$[\tilde J^a(x), \tilde J^b(y)] = \tfrac{1}{2\pi i}\,\bar\kappa^{ab} K \delta'(x-y) + f^{ab}{}_c \tilde J^c(x)\delta(x-y). \qquad (3.1.15)$$

A relation of this type was encountered in physics as commutation relation of local fields in quantum field theory. In that context, the fields $\tilde J$ had the meaning of currents in the sense of being fields which transform according to the adjoint module of a finite-dimensional internal symmetry algebra, and as a vector with respect to the space–time Lorentz algebra. Therefore the relation (3.1.15) was named a *current algebra*. (Alternatively, owing to the non-ultralocal term involving δ' (called the *Schwinger term* in that context), the term *anomalous* commutation relation was also used.) Nowadays, the name current algebra has been taken over to describe also the operator product version (3.1.10) of the commutator algebra (3.1.15), and often also the affine Lie algebra generated by the Fourier–Laurent coefficients T^a_m.

We will soon see that many calculations can be performed with more ease when formulated in terms of the current fields rather than in terms of the generators T^a_m. This is also true for relations involving other z-dependent quantities encountered below. Thus we define the operator product of arbitrary *fields* $A(z)$, $B(z)$ in analogy to (3.1.9):

$$\mathcal{R}(A(z)B(w)) := \begin{cases} A(z)B(w) & \text{for} \quad |z| > |w| \\ B(w)A(z) & \text{for} \quad |z| < |w|. \end{cases} \tag{3.1.16}$$

When deforming the integration contours according to figure (3.1.11), we have implicitly assumed that $J^b(w)$ is analytic in $\mathbb{C} \setminus \{w = 0\}$ except for $w = z$. In fact, we will from now on usually assume that any fields we introduce are analytic in $\mathbb{C} \setminus \{0\}$ except at the points of coincidence with other fields, or in other words that all operator products $\mathcal{R}(\prod_i A_i(z_i))$ are analytic in $(\mathbb{C} \setminus \{0\})^n \setminus \{z_i = z_j \mid i,j = 1,\dots,n,\ i \neq j\}$; more precisely, this must hold inside any correlation function, as will be explained in more detail in section 3.4. In particular the contour deformation in figure (3.1.11) is strictly correct only if no other fields besides $J^a(z)$ and $J^b(w)$ are present.

One can expand the operator product (3.1.16) in a Laurent series with the coefficients being fields $C(z)$ (this is possible in all applications, i.e. there never arises an essential singularity at $z = w$),

$$\mathcal{R}(A(z)B(w)) = \sum_{n=-n_0}^{\infty} (z-w)^n \, C_{(n)}(w). \tag{3.1.17}$$

After introducing the *contraction*

$$\underbrace{A(z)\,B(w)} := \sum_{n=-n_0}^{-1} (z-w)^n \, C_{(n)}(w) \tag{3.1.18}$$

and the *normal ordered product*

$$:A(w)B(w): \; := C_{(0)}(w) \tag{3.1.19}$$

of $A(z)$ and $B(w)$, (3.1.16) is written as

$$\mathcal{R}(A(z)B(w)) = \underbrace{A(z)\,B(w)} + :A(w)B(w): + \mathcal{O}(z-w). \tag{3.1.20}$$

For example, according to (3.1.10) the contraction of the current fields reads

$$\underbrace{J^a(z)J^b(w)} = (z-w)^{-2}\bar{\kappa}^{ab}K - (z-w)^{-1} f^{ab}{}_c J^c(w). \tag{3.1.21}$$

In the following the regular contributions $\mathcal{O}(z-w)$ to operator products will usually be suppressed so that the equality sign in formulae containing contractions will mean equality up to terms which are regular in the relevant limit.

It is easily checked that the normal ordered product defined this way obeys

$$:A(z)B(z):=\frac{1}{2\pi i}\oint_z \mathrm{d}w\,(w-z)^{-1}\mathcal{R}(A(w)B(z)). \qquad (3.1.22)$$

From this one can see that the operation of building the normal ordered product is neither associative nor commutative; e.g. the normal ordered product of a commutator is found to be

$$\begin{aligned} :[A(z),B(z)]: &\equiv\ :A(z)B(z):-:B(z)A(z): \\ &= \sum_{n=1}^{n_0}\frac{(-1)^{n+1}}{n!}\,\partial^n C_{(-n)}(z), \end{aligned} \qquad (3.1.23)$$

where the fields $C_{(m)}(z)$ and the number n_0 are defined via the expansion (3.1.17). (Commutativity could be obtained by using a more symmetric definition, taking e.g. $\frac{1}{2}(z+w)$ rather than w as argument in the definition of the coefficient fields $C_{(n)}$ in (3.1.17), but further inspections shows that this would be no real advantage.) For example, for the current fields one obtains

$$:[J^a(z),J^b(z)]:=f^{ab}{}_c\,\partial J^c(z). \qquad (3.1.24)$$

It will also sometimes be necessary to consider normal ordered products of more than two factors; these can be defined by multiple use of (3.1.22). The result of course depends on the order in which the individual normal orderings are performed. For example, one has

$$\begin{aligned} :(:AB::CD:):(z) =&\ :ABCD:(z)+:[:AB:,:CD:]:(z) \\ &+:(:[:CD:,A]:B):(z)+:(A:[:CD:,B]:):(z), \end{aligned} \qquad (3.1.25)$$

where the first term on the right hand side denotes the *fully normal ordered* product defined as

$$:ABC...GH:(z)=:A:(B:(C:(\ldots(:GH:):):\ldots):):\,. \qquad (3.1.26)$$

If the fields A, B and $:AB:$ possess Laurent expansions of the form

$$A(z)=\sum_n z^{-n-\Delta_A}\,a_n,\quad B(z)=\sum_n z^{-n-\Delta_B}\,b_n, \qquad (3.1.27)$$

and

$$:A(z)B(z):=\sum_n z^{-n-\Delta_A-\Delta_B}\,c_n \qquad (3.1.28)$$

(here $n\in\mathbb{Z}-\Delta_A$ etc. where Δ_A is some non-negative real number which will play the role of the conformal dimension of the field A, see section 3.3), we can obtain a connection between the Laurent modes a_n, b_n, c_n

by interpreting the integration contour in (3.1.22) as the difference of two contours around zero, one for $|w| > |z|$ and one for $|w| < |z|$, and expanding $(w-z)^{-1}$ in the respective regions in a Taylor series. The integration then yields

$$
\begin{aligned}
:A(z)B(z): = \sum_{m,n}\sum_{p=0}^{\infty}\Big\{ & \frac{1}{2\pi i}\oint_{0\atop |z|<|w|} \mathrm{d}w\; w^{-1}\left(\frac{z}{w}\right)^{p} w^{-n-\Delta_A} a_n\, z^{-m-\Delta_B} b_m \\
& -\frac{1}{2\pi i}\oint_{0\atop |z|>|w|} \mathrm{d}w\; (-z^{-1})\left(\frac{w}{z}\right)^{p} z^{-m-\Delta_B} b_m\, w^{-n-\Delta_A} a_n \Big\} \\
= \sum_{m,n} z^{-m-n-\Delta_A-\Delta_B}\, & \{\Theta(-n-\Delta_A)\, a_n b_m \\
& + \Theta(n+\Delta_A-1)\, b_m a_n\}\,,
\end{aligned}
\tag{3.1.29}
$$

where

$$
\Theta(x) := \begin{cases} 1 & \text{for } x \ge 0 \\ 0 & \text{for } x < 0. \end{cases} \tag{3.1.30}
$$

In terms of the modes, this means

$$
c_m = \sum_{n \le -\Delta_A} a_n b_{m-n} + \sum_{n > -\Delta_A} b_{m-n} a_n , \tag{3.1.31}
$$

or shortly

$$
c_m = \sum_{n\in\mathbb{Z}} : a_n b_{m-n} : , \tag{3.1.32}
$$

with the normal ordering of Laurent modes defined by

$$
: a_m b_n : \; := \begin{cases} a_m b_n & \text{for } m \le -\Delta_A \\ b_n a_m & \text{for } m > -\Delta_A . \end{cases} \tag{3.1.33}
$$

In the literature, various other prescriptions for normal ordering are used (in particular (3.1.19) does not coincide with the normal ordering which is usually chosen in the description of free field theories). They all differ from our prescription (3.1.19) only insofar as in the relation (3.1.31) between Laurent modes the order of the factors a_n, b_{m-n} is changed (or some linear combination of the two possible orders is chosen) for at most a finite number of terms. For example, a common choice is

$$
(: a_m b_n :)' \; := \begin{cases} a_m b_n & \text{for } m < n \\ b_n a_m & \text{for } m > n \\ \frac{1}{2}\{a_m, b_n\} & \text{for } m = n \end{cases} \tag{3.1.34}
$$

(the curly brackets denote the anticommutator); it can be seen that this ordering reproduces the normal ordering prescription (3.1.19) provided that the fields $C_{(n)}$ on the right hand side of the operator product expansion (3.1.17) are defined as being evaluated at $\tilde{w} = \sqrt{zw}$ rather than at w.

Considerations similar to those leading to (3.1.31) can be used to show that for the contraction of a field with a normal ordered product the *Wick rule*

$$\underbrace{A(y)\,:B(z)C(z):} = \frac{1}{2\pi i}\oint_z \mathrm{d}w\,(w-z)^{-1} \Big[\mathcal{R}(\underbrace{A(y)B(w)}\,C(z)) + \mathcal{R}(B(w)\,\underbrace{A(y)C(z)})\Big] \tag{3.1.35}$$

is valid. In the case of free field theories, this rule is equivalent to the usual quantum field theoretic Wick theorem.

By iterating the Wick rule, one obtains formulae for evaluating contractions of multiply normal ordered products. A particularly simple situation arises if the contractions of the elementary fields are $\mathbb{C}$-numbers rather then fields themselves, as is the case in free field theories, but also e.g. for those current fields $J^a(z)$ such that the corresponding generator T^a belongs to the Cartan subalgebra $\mathfrak{g}_0 \subset \mathfrak{g}$. In particular, it is not difficult to show by induction that, if both $\underbrace{A(z)B(w)}$ and $\underbrace{B(z)B(w)}$ are $\mathbb{C}$-numbers, then

$$\underbrace{A(z)\,:B^n(w):} = n\,(\underbrace{A(z)B(w)})\,:B^{n-1}(w):, \tag{3.1.36}$$

and hence

$$\underbrace{A(z)\,:\mathrm{e}^{B(w)}:} = (\underbrace{A(z)B(w)})\,:\mathrm{e}^{B(w)}:. \tag{3.1.37}$$

If both factors of the contraction are normal ordered, one has to be a bit more careful because the contraction pieces in the radially ordered products that arise in the integrands typically give nonvanishing contributions (i.e. effectively one considers multiple contractions, just as in the ordinary Wick theorem); for example, if $\underbrace{A(z)A(w)}$, $\underbrace{A(z)B(w)}$ and $\underbrace{B(z)B(w)}$ are all $\mathbb{C}$-numbers, one gets

$$\begin{aligned}\underbrace{:\mathrm{e}^{A(z)}::\mathrm{e}^{B(w)}:} &:= \sum_{m,n} \frac{1}{m!n!}\,\underbrace{:A^m(z)::B^n(w):}\\ &= \sum_{m,n,\ell} \frac{\ell!}{m!n!}\binom{m}{\ell}\binom{n}{\ell}[\underbrace{A(z)B(w)}]^\ell\;:A^{m-\ell}(w)B^{n-\ell}(w):\\ &= \exp\Big(\underbrace{A(z)B(w)}\Big)\;:\mathrm{e}^{A(w)}\,\mathrm{e}^{B(w)}:.\end{aligned} \tag{3.1.38}$$

If $A(z)$ is the field describing a free boson, then the normal ordered exponential $:\exp(\xi\, A(z)):$ is called a *vertex operator.* Vertex operators play a prominent role in the construction of explicit realizations of irreducible highest weight modules of affine Lie algebras.

In section 3.6 we will also encounter fields which do not possess a proper Laurent expansion, but do so when divided by $\sqrt{z}$; for such fields one could still use the normal ordering prescription (3.1.22), but it would not be possible to deform the contours as in (3.1.29) in order to get a relation for the (pseudo-) Laurent modes. This can be remedied by changing (3.1.22) in a way which is irrelevant for w close to z, namely

$$\begin{aligned} :A(z)B(z):\, = \tfrac{1}{2\pi i} \oint_z \mathrm{d}w\, (w-z)^{-1} \\ \cdot \tfrac{1}{2}\left(\sqrt{\tfrac{z}{w}}+\sqrt{\tfrac{w}{z}}\right) \mathcal{R}(A(w)B(z)) . \end{aligned} \tag{3.1.39}$$

Proceeding as in (3.1.29) then gives

$$\begin{aligned} :A(z)B(z):\, = \sum_{m,n} z^{-m-n-\Delta_A-\Delta_B} \\ \cdot\tfrac{1}{2}\left\{[\Theta(-n-\Delta_A-\tfrac{1}{2})+\Theta(-n-\Delta_A+\tfrac{1}{2})]\, a_n b_m \right. \\ \left. +[\Theta(n+\Delta_A-\tfrac{3}{2})+\Theta(n+\Delta_A-\tfrac{1}{2})]\, b_m a_n \right\} . \end{aligned} \tag{3.1.40}$$

In terms of modes (defined as in (3.1.28)), this means

$$\begin{aligned} c_m = \sum_{n\le -\Delta_A-1/2} a_n b_{m-n} + \sum_{n\ge -\Delta_A+3/2} b_{m-n} a_n \\ +\tfrac{1}{2}\,(a_{-\Delta_A+1/2}\, b_{m+\Delta_A-1/2} + b_{m+\Delta_A-1/2}\, a_{-\Delta_A+1/2}) . \end{aligned} \tag{3.1.41}$$

Actually, for the fields we will be really interested in we also have to change the definition (3.1.16) of radial ordering into

$$\mathcal{R}(A(z)B(w)) := \begin{cases} A(z)B(w) & \text{for} \quad |z|>|w| \\ -B(w)A(z) & \text{for} \quad |z|<|w|, \end{cases} \tag{3.1.42}$$

because those fields are *fermionic* whereas the fields obeying (3.1.16) are *bosonic.* As a consequence, the formula (3.1.41) for the modes c_m gets replaced by

$$\begin{aligned} c_m = \sum_{n\le -\Delta_A-1/2} a_n b_{m-n} - \sum_{n\ge -\Delta_A+3/2} b_{m-n} a_n \\ +\tfrac{1}{2}\,(a_{-\Delta_A+1/2}\, b_{m+\Delta_A-1/2} - b_{m+\Delta_A-1/2}\, a_{-\Delta+1/2}) . \end{aligned} \tag{3.1.43}$$

Similar changes arise in the other formulae derived above; e.g. if both A and B are fermionic, the second term in the Wick rule (3.1.35) gets a minus sign, and the commutation rule (3.1.23) is replaced by an analogous formula for the anticommutator,

$$
\begin{aligned}
:\{A(z),B(z)\}: &= \;:A(z)B(z):+:B(z)A(z): \\
&= \sum_{n=1}^{n_0} \frac{(-1)^{n+1}}{n!}\, \partial^n C_{(-n)}(z).
\end{aligned} \tag{3.1.44}
$$

In the following we will usually encounter only operator products which are either normal ordered or else radially ordered. We will therefore save space by suppressing the symbol $\mathcal{R}$ for radial ordering from now on, even in the case of fermionic fields where the radial ordering prescription involves a minus sign.

3.2 The Sugawara construction

A natural generalization of the analysis of an affine algebra g is to study also its universal enveloping algebra $\mathsf{U}(\mathsf{g})$. It turns out that the subspace of quadratic elements, generated by $T^a_m T^b_n$, is of particular interest. In terms of current fields, one should thus consider the normal ordered product $:J^a(z)J^b(z):$ of two currents. This can be split into an antisymmetric part proportional to the $\bar{\mathsf{g}}$-structure constants $f^{ab}{}_c$, and a symmetric part proportional to the $\bar{\mathsf{g}}$-Killing form $\bar\kappa^{ab}$. The antisymmetric part is actually fixed uniquely by the result (3.1.24); thus upon introduction of a new field denoted by T one may write

$$
:J^a(z)J^b(z): \;= \tfrac{1}{2}\, f^{ab}{}_c\, J^c(z) + \frac{\bar\kappa^{ab}}{\xi\, d}\, T(z). \tag{3.2.1}
$$

Here d denotes the dimension of $\bar{\mathsf{g}}$, and the constant ξ is incorporated to tune the normalization of the field $T(z)$ in a convenient way (the value of ξ will be fixed in (3.2.6) below). When the above result is contracted with the Killing form of $\bar{\mathsf{g}}$, the antisymmetric part drops out so that one is left with

$$
T(z) = \xi\, \bar\kappa_{ab} \;:J^a(z)J^b(z):\, . \tag{3.2.2}
$$

We will see below that the zero mode of the field $T(z)$ can be identified with minus the derivation of g. But even the whole field T, not only its zero mode, has a concise meaning in physical applications, namely as the energy-momentum tensor of a two-dimensional conformal field theory; this will be discussed in detail in the following section. Also note that one way of thinking of (3.2.2) is as a generalization of the quadratic Casimir operator. Since unlike in the case of simple Lie algebras one now has an infinite number of generators, the summation over the generators has to be regulated, and this is implemented by the normal ordering prescription.

Next, with the help of the Wick rule (3.1.35) and the current operator product (3.1.10), one deduces

$$\begin{aligned}\underbrace{J^a(z)\,T(w)} &= \xi\,\tfrac{1}{2\pi i}\oint_w \mathrm{d}y\,(y-w)^{-1}\\ &\quad\cdot\Big\{[(z-y)^{-2}K\,J^a(w) - f^{ab}{}_c(z-y)^{-1}J^c(y)J_b(w)]\\ &\quad+[(z-w)^{-2}K\,J^a(y) - f^{ab}{}_c(z-w)^{-1}J^c(w)J_b(y)]\Big\}.\end{aligned} \tag{3.2.3}$$

Using again (3.1.10) (which produces a term $\propto (y-w)^{-1}$ in $J^c(y)J_b(w)$) and performing the integration, the right hand side of (3.2.3) becomes

$$\begin{aligned}\xi\Big\{&[(z-w)^{-2}K\,J^a(w) - (z-w)^{-1}f^{ab}{}_c\,:J^c(w)J_b(w):\\ &\qquad+(z-w)^{-2}f^{ab}{}_c f^c{}_{bd}\,J^d(w)]\\ &+[(z-w)^{-2}K\,J^a(w) - (z-w)^{-1}f^{ab}{}_c\,:J_b(w)J^c(w):]\Big\}.\end{aligned} \tag{3.2.4}$$

Owing to the antisymmetry of f^{abc}, the two terms involving normal ordered products cancel; using also (1.6.40), we finally get

$$\underbrace{J^a(z)\,T(w)} = \xi\,(2K+C_\theta)\,(z-w)^{-2}\,J^a(w). \tag{3.2.5}$$

From now on we will consider the generators of the affine Lie algebra as given in some definite representation, but suppress the corresponding symbol R throughout. Then in particular the central generator K can be replaced by its eigenvalue k on the corresponding module. Thus, choosing the normalization

$$\xi = \frac{1}{2k+C_\theta} \equiv \frac{1}{(\bar\theta,\bar\theta)(k^\vee+g^\vee)}, \tag{3.2.6}$$

the contraction (3.2.5) of J^a and T becomes

$$\underbrace{J^a(z)\,T(w)} = (z-w)^{-2}\,J^a(w). \tag{3.2.7}$$

Calculations similar to those above show that we have the further operator products

$$\underbrace{T(z)\,J^a(w)} = (z-w)^{-2}J^a(w) + (z-w)^{-1}\partial J^a(w) \tag{3.2.8}$$

and

$$\underbrace{T(z)T(w)} = \frac{c}{2}\,(z-w)^{-4} + 2(z-w)^{-2}\,T(w) + (z-w)^{-1}\partial T(w). \tag{3.2.9}$$

Here $\partial \equiv \frac{\mathrm{d}}{\mathrm{d}z}$, and we defined

$$c = \frac{k^\vee d}{k^\vee+g^\vee} \tag{3.2.10}$$

with d the dimension of $\mathfrak{g}$. (Recall that the operator products are always radially ordered so that trivially $A(z)B(w) = B(w)A(z)$ for bosonic fields. This has to be taken into account when checking for example that the results (3.2.8) and (3.2.7) are compatible, which one verifies by expanding $J^a(z)$ as $J^a(z) = J^a(w) + (z-w)\partial J^a(w) + \mathcal{O}(z-w)^2$.) These operator products can of course also be reformulated in terms of the variable x used in (3.1.15); one then has

$$[\tilde{T}(x), \tilde{T}(y)] = \frac{c}{6\pi i}\,\delta'''(x-y) + 2\delta'(x-y)\,\tilde{T}(y) \tag{3.2.11}$$

etc., with

$$\tilde{T}(x) = 4\pi i\left(\frac{z}{L}\right)^2 T(z). \tag{3.2.12}$$

Also, introducing Laurent coefficients of the field $T(z)$,

$$T(z) = \sum_{n\in\mathbb{Z}} z^{-n-2} L_n, \tag{3.2.13}$$

the above operator products are equivalent to

$$[L_m, T^a_n] = -n\, T^a_{m+n}, \tag{3.2.14}$$

$$[L_m, L_n] = \frac{c}{12}\, m(m^2-1)\delta_{m+n,0} + (m-n)L_{m+n}. \tag{3.2.15}$$

It is easily checked that the bracket (3.2.15) is not only antisymmetric, but also obeys the Jacobi identity. Thus, interpreting c as the eigenvalue of some operator C, and taking C and the components L_m as the basis vectors of an infinite-dimensional vector space, (3.2.15) endows this space with a Lie algebra structure. This infinite-dimensional Lie algebra is called the *Virasoro algebra* $\mathcal{V}$; it can be realized as the (non-trivial) extension of the Lie algebra of $\mathbb{C}^\infty$ maps of the unit circle to itself by the one-dimensional center C. Similarly, the brackets (3.2.15) and (3.2.14) together with (3.1.1) (and with the trivial brackets $[C,x] = [K,x] = 0$ for all generators x) constitute the defining relations of a Lie algebra which is the semidirect sum $\mathcal{V} \oplus \mathfrak{g}$ of the Virasoro algebra with an affine Lie algebra $\mathfrak{g}$. Also note that for $n = 0$, (3.2.14) reads

$$[L_0, T^a_m] = -m\, T^a_m, \tag{3.2.16}$$

i.e. the generator L_0 can be identified with minus the derivation D of the affine algebra.

The eigenvalue c of the central generator of the Virasoro algebra is known as the *conformal central charge* or also (as already introduced in section 2.8) the *conformal anomaly.*

The particular construction (3.2.2) of $T(z)$ in terms of the currents

$J^a(z)$ is called the *Sugawara construction.* Expressed in Laurent components, it reads

$$L_m = \frac{1}{(\theta,\theta)(k^\vee+g^\vee)}\,\kappa_{ab}\sum_{n\in\mathbb{Z}} :T^a_n T^b_{m-n}: \tag{3.2.17}$$

with the normal ordering of the current modes T^a_m defined by

$$:T^a_n T^b_m: \;:= \begin{cases} T^a_n T^b_m & \text{for} \quad n<0 \\ T^b_m T^a_n & \text{for} \quad n\ge 0 \end{cases} \tag{3.2.18}$$

(this normal ordering corresponds to the formula (3.1.33) with $\Delta_A = 1$; recall that other normal ordering prescriptions may be chosen which differ from the one used here in a finite number of terms).

One may also perform the Sugawara construction for an affine Lie algebra $\mathfrak{g}^\omega$ which is twisted by an inner $\mathfrak{g}$-automorphism ω and hence isomorphic to the corresponding untwisted algebra $\tilde{\mathfrak{g}}$. Recalling the relation (2.2.41) between these algebras, it follows that the current fields of $\mathfrak{g}^\omega$ and $\tilde{\mathfrak{g}}$ are related by

$$\begin{aligned} E^\alpha(z) &= z^{(\zeta_\omega,\alpha)}\,\tilde{E}^\alpha(z), \\ H^i(z) &= \tilde{H}^i(z) - z^{-1}\,\zeta^i_\omega K, \end{aligned} \tag{3.2.19}$$

together with $K = \tilde{K}$ and $D = \tilde{D} - (\zeta_\omega)_i H^i_0$. As a consequence, one has

$$T(z) = \tilde{T}(z) + \frac{1}{(\theta,\theta)(k^\vee+g^\vee)}\left(-2z^{-1}k\,(\zeta_\omega)_i\tilde{H}^i(z) + z^{-2}\,(\zeta_\omega,\zeta_\omega)\,k^2\right) \tag{3.2.20}$$

if T is defined as in (3.2.2). In terms of Laurent modes, this is

$$L_n = \tilde{L}_n + \frac{1}{(\theta,\theta)(k^\vee+g^\vee)}\left(-2\,k\,(\zeta_\omega)_i\tilde{H}^i_n + (\zeta_\omega,\zeta_\omega)\,k^2\,\delta_{n,0}\right). \tag{3.2.21}$$

(Note that for the Virasoro algebra the generator K gets replaced by its eigenvalue k; thus the contribution proportional to $(\zeta_\omega,\zeta_\omega)$ to L_0 need not have a counterpart for the derivation D for which a constant term is irrelevant. However, in order to have $D = -L_0 + \text{const}$, one must also add the term $-g^\vee(\zeta_\omega)_i\tilde{H}^i_n/(k^\vee+g^\vee)$ to L_n, i.e. use a normal ordering prescription different from (3.2.2).)

The representation theory of the Virasoro algebra runs closely parallel to that of simple or affine Lie algebras. We will not discuss it in detail, but only summarize a few basic facts. $\mathcal{V}$ can be decomposed as

$$\mathcal{V}_+ \oplus \mathcal{V}_0 \oplus \mathcal{V}_- \tag{3.2.22}$$

into the subalgebras generated by the positive, zero, and negative modes $\mathcal{V}_+ = \{L_n \,|\, n>0\}$, $\mathcal{V}_0 = \{L_0, C\}$, $\mathcal{V}_- = \{L_n \,|\, n<0\}$, respectively. $\mathcal{V}_0$ is a Cartan subalgebra of $\mathcal{V}$, and $\mathcal{V}_\pm \oplus \mathcal{V}_0$ are Borel subalgebras. Verma modules of the Virasoro algebra contain a highest weight vector $v_{c,\Delta}$ which is annihilated by the raising operators $L_n \in \mathcal{V}_+$,

$$R(L_n) \cdot v_{c,\Delta} = 0 \quad \text{for} \quad n > 0. \tag{3.2.23}$$

All other vectors in the module are obtained by applying an appropriate number of lowering operators to $v_{c,\Delta}$,

$$v = R(L_{-n_1})R(L_{-n_2}) \ldots R(L_{-n_p}) \cdot v_{c,\Delta}, \tag{3.2.24}$$

with $n_i > 0$ (and $n_i \geq n_j$ for $i < j$). The labelling of the highest weight vectors corresponds to the eigenvalues with respect to the generators C and L_0 of $\mathcal{V}_0$,

$$R(C) \cdot v_{c,\Delta} = c\, v_{c,\Delta}, \qquad R(L_0) \cdot v_{c,\Delta} = \Delta\, v_{c,\Delta}. \tag{3.2.25}$$

Since C is a central generator, all vectors in the Verma module obtained from $v_{c,\Delta}$ possess the same eigenvalue c; also, from (3.2.16) it follows immediately that the L_0-eigenvalue of v is given by Δ augmented by the grade $n = \sum_{i=1}^{p} n_i$ (for v as given in (3.2.24)),

$$R(L_0) \cdot v = (\Delta + n)\, v. \tag{3.2.26}$$

Thus among all vectors in the Verma module, $v_{c,\Delta}$ has the lowest L_0-eigenvalue, i.e. it is in fact a highest weight vector with respect to $-L_0$ rather than L_0. Analogously to the case of affine Lie algebras, the Verma representation can also be described in a more compact way as the quotient of the universal enveloping algebra $\mathsf{U}(\mathcal{V})$ of the Virasoro algebra by an ideal which is now the space spanned by $\{L_n \mid n > 0\} \cup \{L_0 - \Delta\} \cup \{C - c\}$.

At this point one should recall that the D-eigenvalue of the highest weight of a module of an affine algebra could be chosen at will. This is no longer the case for the Virasoro zero mode L_0 because for the semidirect sum $\mathfrak{g} \oplus \mathcal{V}$, L_0 does appear on the right hand side of the bracket relations. This means in particular that the fundamental weights $\Lambda_{(i)}$ of $\mathfrak{g}$ now should not be chosen precisely as in section 2.4 (where the D-eigenvalues were set to zero by convention), but rather a term $-\Delta_{(i)}\delta$ ($\delta \equiv (0,0,1)$) with appropriate constants $\Delta_{(i)}$ must be added. As we will learn in the following section, the eigenvalues Δ are in fact determined entirely by the level $k^\vee$ of $\mathfrak{g}$ and by the highest weight $\bar{\Lambda}$ with respect to the zero mode subalgebra of $\mathfrak{g}$ (see (3.4.20)).

The Virasoro generators can be defined similarly as in (3.2.2) or (3.2.20) for the gradation of $\mathfrak{g}$ that corresponds to an arbitrary automorphism of an underlying simple Lie algebra $\bar{\mathfrak{g}}$. Since the level $k^\vee$ of a $\mathfrak{g}$-module is intrinsically defined, a change in the gradation does not alter the value (3.2.10) of the Virasoro central charge. It does, however, change the value of Δ. For brevity we report only the result: if we denote by Δ the L_0-eigenvalue in the homogeneous gradation that gives $\mathfrak{g}_{\rm hor}$ as the zero mode subalgebra, and by ω an automorphism of order ℓ that results in a semisimple Lie algebra $\bar{\mathfrak{h}}$ as zero mode subalgebra, then the L_0 eigenvalue in the gradation defined by ω is

$$\Delta^{(\omega)} = (I_{\bar{\mathfrak{h}}\subset\mathfrak{g}_{\rm hor}})^{-1}\,\Delta + \Delta_0^{(\omega)} \tag{3.2.27}$$

where $I_{\bar{\mathfrak{h}}\subset\mathfrak{g}_{\rm hor}}$ is the Dynkin index of the embedding $\bar{\mathfrak{h}} \subset \mathfrak{g}_{\rm hor}$, and

$$\Delta_0^{(\omega)} = \frac{c}{4\ell^2} \sum_{j=1}^{\ell-1} j(\ell-j)\, \frac{\dim \bar{\mathfrak{g}}_{[j]}}{\dim \bar{\mathfrak{g}}} \tag{3.2.28}$$

with $\bar{\mathfrak{g}}_{[j]}$, $j = 0, 1, \dots, \ell-1$, the eigenspaces of ω corresponding to the eigenvalue $\exp(2\pi i j/\ell)$ and $\bar{\mathfrak{g}}$ the simple Lie algebra on which the automorphism ω is acting. Note that for $\ell = 2$ and $\ell = 3$ (and hence in particular for the twisted affine algebras in their homogeneous gradations) the formula for $\Delta_0^{(\omega)}$ can also be written as

$$\Delta_0^{(\omega)} = \frac{c}{4\ell^2}\,(\ell-1)\left(1 - \frac{\dim \bar{\mathfrak{g}}_{[0]}}{\dim \bar{\mathfrak{g}}}\right). \tag{3.2.29}$$

For generic values of c and Δ, the Verma module having $v_{c,\Delta}$ as its highest weight vector is irreducible. It becomes reducible (i.e. contains null vectors) iff $c < 1$ and Δ take values in the discrete sets

$$\begin{aligned} c &= 1 - \frac{6}{(\ell+2)(\ell+3)}, && \ell \in \mathbb{Z}_{\geq 0}, \\ \Delta &= \frac{((\ell+3)r - (\ell+2)s)^2 - 1}{4(\ell+2)(\ell+3)}, && r = 1, 2, \dots, \ell+1, \\ & && s = 1, 2, \dots, r. \end{aligned} \tag{3.2.30}$$

For the values in this so-called *Kac table* the Verma module contains a unique proper maximal submodule so that a unique irreducible module is obtained by dividing out this submodule. The values (3.2.30) also show up if one investigates whether the irreducible highest weight module based on the Verma module is unitarizable (namely, such that $(L_n)^\dagger = L_{-n}$). As is easily checked, the norm of the vector $R(L_{-n})\cdot v_{c,\Delta}$ is $2n\Delta + cn(n^2-1)/12$, and hence unitarizability requires that $c \geq 0$ and $\Delta \geq 0$. A careful study of the inner products of all other states in the module shows that in the region $0 \leq c \leq 1$, the only allowed values of c and Δ are those of (3.2.30) (this restriction on c and Δ is not only necessary, but also proves to be sufficient; see p. 167 below). In contrast, for $c \geq 1$, unitarity does not place any further constraint on the value of Δ.

Also note that for $c = 0$ (i.e. $\ell = 0$ in (3.2.30)), there is a single unitary module, having $\Delta = 0$. This is in fact the trivial one-dimensional module because in the corresponding Verma module any vector other than the highest weight vector is a null vector; compare e.g. the above result for the inner product of $R(L_{-n}) \cdot v_{c,\Delta}$ with itself. The simplest way to prove this is to investigate the matrix of inner products of $R(L_{-2n}) \cdot v_{c,\Delta}$ and $(R(L_{-n}))^2 \cdot v_{c,\Delta}$. For $c = 0$, the determinant of this matrix is found to be $4n^3\Delta^2(4\Delta - 5n)$, and thus is negative for sufficiently large n unless $\Delta = 0$. Thus unitarity at $c = 0$ implies $\Delta = 0$; but then the norm of *any* vector in the Verma module except the highest weight vector is zero. Summarizing, $c = 0$ together with unitarity requires that $R(L_n) = 0$ for all n, or symbolically

$$c = 0 \Longleftrightarrow T \equiv 0. \tag{3.2.31}$$

Analogous considerations hold for the semidirect sum of the Virasoro algebra with an affine algebra $\mathfrak{g}$. The highest weight vectors are now labelled by a highest weight Λ of $\mathfrak{g}$; there is no need to specify the values of c and Δ because c is completely fixed by (3.2.10) and Δ is determined by a similar formula (that such a relation should exist is clear from the identification of L_0 with the derivation D; its explicit form which involves the quadratic Casimir eigenvalue of $R_{\bar{\Lambda}}$ will be given in (3.4.20)). Any other vector in the Verma module has the form

$$v = R(J^{a_1}_{-n_1})...R(J^{a_p}_{-n_p})\, R(L_{-m_1})...R(L_{-m_q}) \cdot v_\Lambda \tag{3.2.32}$$

with n_i, $m_i > 0$ and a_i as in (2.4.2) (and the appropriate restrictions on the relative values of the n_i, n_j for $i \neq j$, etc.). The possible null vectors of such a Verma module are either purely affine null vectors of the form discussed in section 2.5, or they involve lowering operators of both the affine and the Virasoro algebra; the latter type of null vectors will be discussed in detail in section 3.4. In the Sugawara case there are no null vectors involving only the Virasoro generators: as just mentioned, these would require $c < 1$, whereas from (3.2.10) it is clear that the Sugawara construction leads to $c \geq 1$. More precisely, (3.2.10) implies that

$$r \leq c < d \tag{3.2.33}$$

where r and d are the rank and dimension of $\bar{\mathfrak{g}}$, respectively. Clearly, c tends to the upper bound in the limit $k \to \infty$; on the other hand, inspection shows that the lower bound $c = r$ is obtained iff $k^\vee = 1$ and $\bar{\mathfrak{g}}$ is simply laced.

Finally we note that the central term on the right hand side of (3.2.15) vanishes if $m \in \{0, \pm 1\}$. Thus the modes L_0, L_1 and L_{-1} play a special role. In fact, they generate a finite-dimensional subalgebra of the Virasoro algebra which is nothing but the simple Lie algebra $sl_2(\mathbb{C})$ (respectively, if the base field is taken to be $\mathbb{R}$, its real form $sl_2(\mathbb{R}) \cong su_{1,1}$).

Let us return now to the Sugawara construction. We have presented the construction for affine Lie algebras, but this can immediately be generalized to direct sums of such algebras by defining T as a sum of contributions $T^{(i)}$ of the form (3.2.2), each correponding to one of the various affine $\mathsf{g}_{(i)}$ in the direct sum $\mathsf{g} = \bigoplus_i \mathsf{g}_{(i)}$. Similarly one can treat the case of a $\hat{u}_1$ Kac–Moody algebra, and hence of any Kac–Moody algebra with reductive horizontal subalgebra: the $\hat{u}_1$-contribution to the energy–momentum tensor is

$$T(z) = -\tfrac{1}{2} \,:j(z)j(z):\tag{3.2.34}$$

where $j(z)$ is the $\hat{u}_1$ Kac–Moody current, which corresponds to an eigenvalue $c = 1$ of the central generator. Note that (3.2.34) and the relation $c = 1$ can be obtained from (3.2.2) and (3.2.10), respectively, via the substitutions $\bar{\kappa} \mapsto -1$, $g^\vee \mapsto 0$, $k^\vee \mapsto 1$.

When adding up contributions $T^{(i)}$ to the energy-momentum tensor, it is implicitly assumed that

$$\underbrace{T^{(i)}(z)\,T^{(j)}(w)} = 0 \quad \text{for} \quad i \neq j,\tag{3.2.35}$$

so that in particular the C-eigenvalues add up, $c = \sum_i c^{(i)}$. Because of the inequalities (3.2.33) it is thus impossible to obtain the interesting values in the region $c < 1$ this way. A way to obtain such values is of course to *subtract* two appropriate $T^{(i)}$ obeying (3.2.35). There is no obvious reason why such a construction could be used to obtain also the unitary modules corresponding to the values (3.2.30) of c and Δ. Nevertheless it turns out that this indeed can happen, namely if one takes the Sugawara energy-momentum tensors $T_{(\mathsf{g})}$ and $T_{(\mathsf{h})}$ for two affine Lie algebras g and h such that $\bar{\mathsf{h}}$ is a subalgebra of $\bar{\mathsf{g}}$. According to (3.2.8) one has (choosing the generators of $\bar{\mathsf{h}}$ and $\bar{\mathsf{g}}$ to be T^a with $a = 1, \dots, \dim(\mathsf{h})$, and with $a = 1, \dots, \dim(\mathsf{g})$, respectively)

$$\begin{aligned}\underbrace{T_{(\mathsf{g})}(z)\, J^a(w)} &= (z-w)^{-2} J^a(w) + (z-w)^{-1} \partial J^a(w) \\ &\qquad \text{for} \quad a = 1, \dots, \dim(\mathsf{g}), \\ \underbrace{T_{(\mathsf{h})}(z)\, J^a(w)} &= (z-w)^{-2} J^a(w) + (z-w)^{-1} \partial J^a(w) \\ &\qquad \text{for} \quad a = 1, \dots, \dim(\mathsf{h}).\end{aligned}\tag{3.2.36}$$

Thus the difference

$$T_{(\mathsf{g}/\mathsf{h})}(z) := T_{(\mathsf{g})}(z) - T_{(\mathsf{h})}(z)\tag{3.2.37}$$

obeys

$$\underbrace{T_{(\mathsf{g}/\mathsf{h})}(z)\, J^a(w)} = 0 \quad \text{for} \quad a = 1, \dots, \dim(\mathsf{h}),\tag{3.2.38}$$

and hence

$$\underbrace{T_{(\mathsf{g}/\mathsf{h})}(z)\,T_{(\mathsf{h})}(w)} = 0. \tag{3.2.39}$$

This also implies that

$$\underbrace{T_{(\mathsf{g})}(z)\,T_{(\mathsf{g})}(w)} = \underbrace{T_{(\mathsf{h})}(z)\,T_{(\mathsf{h})}(w)} + \underbrace{T_{(\mathsf{g}/\mathsf{h})}(z)\,T_{(\mathsf{g}/\mathsf{h})}(w)}. \tag{3.2.40}$$

Thus we are indeed in a situation where (3.2.35) holds. Moreover, as remarked in section 2.8, the embedding $\bar{\mathsf{h}} \subset \bar{\mathsf{g}}$ extends in a natural way to an embedding $\mathsf{h} \subset \mathsf{g}$ of the affine algebras; the level $k^\vee$ of g may be chosen arbitrarily, but the level $h^\vee$ of h is then fixed because g and h must have the same central extension K, so that $h^\vee/k^\vee$ is given by the Dynkin index (1.8.8) of the embedding $\bar{\mathsf{h}} \subset \bar{\mathsf{g}}$, $h^\vee = I_{\bar{\mathsf{h}}\subset\bar{\mathsf{g}}}\, k^\vee$. So it is in fact possible to decompose the modules of g with respect to $\mathsf{h} \oplus \mathcal{V}_{(\mathsf{g}/\mathsf{h})}$ where $\mathcal{V}_{(\mathsf{g}/\mathsf{h})}$ is the Virasoro algebra generated by $T_{(\mathsf{g}/\mathsf{h})}$. This is the *coset construction* of Goddard, Kent, and Olive. The characters of $\mathcal{V}_{(\mathsf{g}/\mathsf{h})}$ are essentially the branching functions $b_{\Lambda;\lambda}$ that were introduced in section 1.8. It can be proven that the unitary highest weight modules of g decompose into a *finite* number of highest weight modules of $\mathsf{h} \oplus \mathcal{V}_{(\mathsf{g}/\mathsf{h})}$ iff

$$c_{(\mathsf{g}/\mathsf{h})} = c(\mathsf{g}) - c(\mathsf{h}) \tag{3.2.41}$$

is less than one.

Furthermore, for appropriate choices of g and h one obtains indeed the c-and Δ-values of (3.2.30), namely e.g. with

$$\mathsf{g} = (C^{(1)}_{\ell+1})_1, \quad \mathsf{h} = (C^{(1)}_{\ell})_1 \oplus (A^{(1)}_1)_1, \tag{3.2.42}$$

and also with

$$\mathsf{g} = (A^{(1)}_1)_\ell \oplus (A^{(1)}_1)_1, \quad \mathsf{h} = (A^{(1)}_1)_{\ell+1}; \tag{3.2.43}$$

here the lower indices denote the levels of the respective affine algebras. The Virasoro modules obtained this way are in particular unitary because they inherit this structure from the affine modules. This proves that the restriction to the values (3.2.30) is not only necessary, but also sufficient for unitarity.

3.3 Conformal field theory

The constructions described in the two previous sections are tailored to application in two-dimensional conformal field theory. A conformal field theory is a quantum field theory which behaves covariantly under the conformal group. Conformal transformations are by definition those general coordinate transformations of space–time which preserve the angles between any two vectors, or, equivalently, change the space–time metric only by a local scale factor, $g_{\mu\nu}(x) \mapsto \sigma(x) g_{\mu\nu}(x)$. For an infinitesimal reparametrization

$$x^\mu \mapsto x^\mu + \varepsilon f^\mu(x), \tag{3.3.1}$$

the conservation of angles means that

$$d\,(\partial_\mu f_\nu + \partial_\nu f_\mu) - 2\delta_{\mu\nu} \textstyle\sum_{\rho=1}^{d} \partial_\rho f_\rho = 0. \tag{3.3.2}$$

Here the space–time is taken to be euclidean, and d is its dimension. For $d \neq 2$, (3.3.2) implies that $\partial_\mu \partial_\nu f = 0$ so that f is at most quadratic in x. The independent conformal transformations are then the

- translations (f constant),
- dilatations ($f_\mu = x_\mu$),
- rotations ($f_\mu = \sum_\nu m_{\mu\nu} x_\nu$ with $m_{\mu\nu} = -m_{\nu\mu}$), and the
- special conformal transformations ($f_\mu = c_\mu |x|^2 - 2 \sum_\nu c_\nu x_\mu x_\nu$ with c_μ constant).

These transformations generate a $(d{+}1)(d{+}2)/2$-dimensional (non-semisimple) Lie algebra. The corresponding group of finite transformations does not act properly on the d-dimensional space–time because finite special conformal transformations can map finite points to infinity; if the point at infinity is appropriately included in the space–time, the conformal group acts as a covering of the group $SO(d-2,2)$. For $d = 2$, the situation is rather different; the constraints (3.3.2) are now just the Cauchy–Riemann equations

$$\partial_0 f_0 = \partial_1 f_1, \qquad \partial_0 f_1 = -\partial_1 f_0. \tag{3.3.3}$$

In complex notation ($z, \bar z = x_0 \pm i x_1$; $\partial, \bar\partial = (\partial_0 \mp i\partial_1)/2$; $f, \bar f = f_0 \pm i f_1$), this means

$$\bar\partial f(z, \bar z) = 0 = \partial \bar f(z, \bar z), \tag{3.3.4}$$

i.e. $f = f(z)$ and $\bar f = \bar f(\bar z)$ are analytic functions of z and $\bar z$, respectively. The algebra generated by these transformations is clearly infinite-dimensional.

From now on we will take the dimensionality of space–time to be equal to two; the two-dimensional space–time will be called the *world sheet.* Then, in terms of complex coordinates on the world sheet, the finite conformal transformations form an infinite-dimensional Lie group which is the direct product of the group of holomorphic coordinate transformations with that of antiholomorphic coordinate transformations, i.e.

$$z \mapsto w = f(z), \quad \bar{z} \mapsto \bar{w} = \tilde{f}(\bar{z}) \tag{3.3.5}$$

with f and $\tilde{f}$ independent functions. The Lie algebra of each factor of this group turns out to be isomorphic to the so-called Witt algebra. This is the Lie algebra of smooth vector fields on the unit circle S^1, with generators $L_n^{(c)} = -z^{n+1}\frac{\mathrm{d}}{\mathrm{d}z}$ and brackets defined as commutators, which yields

$$[L_m^{(c)}, L_n^{(c)}] = (m-n)\, L_{m+n}^{(c)}. \tag{3.3.6}$$

In terms of the associated fields $T^{(c)}(x)$ related to the generators $L_m^{(c)}$ as in (3.2.13), (3.2.12), this reads

$$[T^{(c)}(x), T^{(c)}(y)] = \delta'(x-y)\, T^{(c)}(y). \tag{3.3.7}$$

In quantum field theory, the conformal symmetry develops an anomaly, which means that at the quantum level (i.e., with the generators acting as operators on some Hilbert space) the symmetry can not be maintained in the form given above. However, the following assertion, known as the *Lüscher–Mack theorem*, holds under very general circumstances: the anomaly just consists in the addition of a $\mathbb{C}$-number term on the right hand side of the bracket (3.3.7); from dimensional analysis this term must be of the form appearing in (3.2.11). The assumptions necessary for the theorem to hold are: validity of the Wightman axioms (i.e. locality of the interactions, positivity, spectrum condition, Lorentz invariance, and uniqueness of the vacuum state), dilatation invariance, and the existence of a conserved symmetric energy-momentum tensor. The coefficient of the $\mathbb{C}$-number is restricted to be positive; its explicit value depends on the particular field theory chosen. We are free to interprete this coefficient as the eigenvalue of an operator which furnishes a central extension of the Witt algebra; then at the quantum level, the Witt algebra just gets replaced by the Virasoro algebra.

When the conformal field theory is coupled to two-dimensional gravity, consistency requires that the combined system does not possess a conformal anomaly. This is so because the conformal transformations are then a subgroup of the general coordinate transformations, or more precisely, of their semidirect product with Weyl transformations which scale the two-dimensional metric but leave matter fields invariant (namely, the conformal transformations constitute the residual invariance in the so-called conformal gauge which is obtained by fixing the two-dimensional metric to the form of a scalar factor times some fixed reference metric); this means that a conformal anomaly would translate into an anomaly in the general coordinate transformations. While two-dimensional gravity does not possess propagating degrees of freedom, it nevertheless contributes to the conformal anomaly through the ghost system introduced by gauge fixing; this contribution turns out to be $c_{\rm ghost} = -26$. Thus consistent coupling to gravity requires the number c in the Virasoro algebra of the (matter) conformal field theory to be

$$c = 26. \tag{3.3.8}$$

In contrast, if we are just interested in the behavior of the conformal field theory as such, there is no reason to constrain the value of c.

In the following we will use the field $T(z)$ rather than its Laurent components L_n, and analogously the field $\bar{T}(\bar{z})$ generating the antiholomorphic Virasoro algebra. The fields T and $\bar{T}$ are introduced here abstractly as the generators of (quantum) conformal transformations. In the following we will call T and $\bar{T}$ the holomorphic and antiholomorphic *energy-momentum tensor*, respectively. The reason for this terminology is as follows. If a lagrangian formulation of the quantum field theory in question is available, then T and $\bar{T}$ can in fact be obtained as components of the energy-momentum tensor of the theory. The energy-momentum tensor is the field $T_{\mu\nu}(x)$ obtained by varying the action with respect to the world-sheet metric, $T_{\mu\nu} \propto \delta S/\delta g^{\mu\nu}$; in a generic two-dimensional quantum field theory, the energy-momentum tensor is symmetric, $T_{\mu\nu} = T_{\nu\mu}$, owing to the symmetry of the metric $g_{\mu\nu}$, and it is conserved, $\sum_\mu \partial T_{\mu\nu}(x)/\partial x^\mu = 0$, owing to translation invariance. In conformal field theory, owing to dilatation invariance it has the additional property of being traceless, $T_\mu^{\ \mu} = 0$. As a consequence, the energy-momentum tensor of a two-dimensional conformal field theory has only two independent components, say $T = T_{00} - T_{11} + 2iT_{01}$ and $\bar{T} = T_{00} - T_{11} - 2iT_{01}$; in terms of these components, the conservation of $T_{\mu\nu}$ means that T is purely holomorphic, $T = T(z)$, and $\bar{T}$ is purely antiholomorphic, $\bar{T} = \bar{T}(\bar{z})$, with $z = x_0 + ix_1$, $\bar{z} = x_0 - ix_1$.

Covariance of a quantum field theory under conformal transformations thus means that an action of T and $\bar{T}$ is defined on its Hilbert space of states, and hence on the *fields* (i.e. operators on the Hilbert space which create the various states by acting on some vacuum state) of the theory. In short, we can define a conformal field theory abstractly by requiring that it is a field theory such that

- the fields $\varphi(z, \bar{z})$ of the theory build up modules of the direct sum $\mathcal{V} \oplus \bar{\mathcal{V}}$ of two copies of the Virasoro algebra which are generated by the holomorphic and antiholomorphic energy-momentum tensors $T(z)$ and $\bar{T}(\bar{z})$, respectively.

If we want to describe the conformal field theory completely in terms of the fields φ (rather than, say, in a perturbative expansion around a classical theory defined by some lagrangian that contains only a small set of *elementary* fields), it is also necessary that the set of fields φ_i we consider is complete in the sense that the (radially ordered) operator product of fields can be defined and expanded in the given set of fields. Moreover, if multiple operator products are considered, the result should not depend on the order of the multiplications. Thus we also require that

- the fields φ_i span an associative algebra over the ring of functions in two complex arguments,

$$\varphi_i(z,\bar{z})\,\varphi_j(w,\bar{w}) = \sum_k d_{ij}{}^k(z,\bar{z},w,\bar{w})\,\varphi_k(w,\bar{w}). \tag{3.3.9}$$

This requirement (which is logically independent of the presence of conformal invariance) is called the conformal *bootstrap.* (3.3.9) is called the *operator product algebra* of the conformal field theory. Of course, in reality one deals with operator products where the arguments of the fields are considered as free parameters so that in fact one works with an algebra over a ring of complex-valued functions. The relation (3.3.9) then makes sense only if the right hand side possesses a nonzero radius of convergence; in fact if $\varphi(z,\bar{z})\varphi'(w,\bar{w})$ is part of an operator product which includes also other factors, then the radius of convergence can be assumed to be the smallest distance between any of the points z and w and any of the (holomorphic) arguments of the rest of the fields, and an analogous statement holds for the antiholomorphic arguments. Also note that (for $\bar{z}$, $\bar{w}$ complex conjugate to z, w) the structure "constants" $d_{ij}{}^k(z,\bar{z},w,\bar{w})$ are required to be single-valued functions rather than multivalued "functions" (i.e. sections of some vector bundle); it is common to call this requirement the *locality* of the operator product algebra (this must be distinguished from the concept of the locality of interactions in conventional lagrangian quantum field theory, and is also quite distinct from the concept of causality in (Minkowski space) field theory that requires the quantum fields to commute for spacelike distances).

Note that the operator product does not change the eigenvalues of the central operators of the symmetry algebra (because these values are the same for all fields of a given conformal field theory); as a consequence, the operator product *cannot* be interpreted in terms of some tensor product of the symmetry algebra.

Before we proceed to discuss the structure of a conformal field theory in more detail, a remark on the meaning of the variables z and $\bar{z}$ is in order. The notations employed here are adapted to the description of critical phenomena in statistical mechanics. More precisely, statistical systems that are close to a second order phase transition are characterized by long range fluctuations of order parameters which lead to singularities in the thermodynamic functions; this can be described by an effective two-dimensional euclidean field theory. At the critical point, the correlation lengths diverge and the effective field theory becomes massless and hence scale invariant; together with the assumption that the interactions of the theory are local, this implies conformal invariance. The space on which the effective field theory lives is the complex plane as described above.

Another realization of the complex plane is encountered in the theory of closed relativistic strings; the world sheet wrapped out by a free closed string has the topology of a cylinder, and the cylinder described by the compact space coordinate σ and the time coordinate τ can be mapped to the complex plane via

$$z = \mathrm{e}^{\tau+i\sigma}, \qquad \bar{z} = z^* = \mathrm{e}^{\tau-i\sigma}. \tag{3.3.10}$$

In particular, the time direction on the cylinder is the radial direction on the plane so that the radial ordering of fields means time ordering in the string picture.

In contrast, the relation to conventional quantum field theory in two-dimensional Minkowski space is less direct. Namely, the variables z and $\bar{z}$ are obtained from the light cones described by coordinates $x_1 \pm x_0$; each of these two light cones must be compactified (say, by stereographic projection to the circle S^1), and afterwards the coordinates are extended to the whole complex plane, with the original compactified light cone embedded as the unit circle. The origin of this construction is the fact that in order to reconcile conformal invariance with causality, the theory must be defined on an infinite covering $S^{d-1} \times \mathbb{R}$ of d-dimensional Minkowski space $\mathbb{R}^{d-1,1}$. For $d = 2$, the relation between the Minkowski space coordinates x_0, x_1 and the coordinates ϑ on S^1 ($\vartheta \in [0, 2\pi)$) and τ (the "conformal time") on $\mathbb{R}$ reads

$$x_\pm \equiv x_0 \pm x_1 = \tan(\tfrac{1}{2}(\tau \pm \vartheta)). \tag{3.3.11}$$

The relation to the variables z, $\bar{z}$ is then

$$\begin{aligned} z &= \frac{1+ix_+}{1-ix_+} = \mathrm{e}^{i(\tau+\sigma)}, \\ \bar{z} &= \frac{1+ix_-}{1-ix_-} = \mathrm{e}^{i(\tau-\sigma)}. \end{aligned} \tag{3.3.12}$$

In particular, after continuation of z and $\bar{z}$ to the whole complex plane, they have to be considered as independent complex variables. In order to identify $\bar{z} = z^*$ as the complex conjugate to z, one must analytically continue the coordinate τ to imaginary values, or in other words compactify the covering $S^1 \times \mathbb{R}$ of two-dimensional Minkowski space to $S^1 \times S^1$, with periodic conformal time. The theory obtained this way is called the *compact picture* or the *euclidean section* of the quantum field theory; the first of these names is the more appropriate one because what one considers is *not* the conventional euclidean version of a Minkowski space field theory, but rather a Minkowskian field theory which is analytically continued in a very special way.

In the string picture, z and $\bar{z}$ are automatically complex conjugate to each other, but owing to the factorization of conformal transformations into holomorphic and antiholomorphic ones, for many purposes z and $\bar{z}$ can formally be treated as independent. At the end one must however always set $\bar{z} = z^*$; the theory obtained this way is then also often called the *euclidean section*, but from the remarks above it should be clear that this terminology is somewhat misleading.

The Virasoro modules relevant to any given conformal field theory must all have the same eigenvalue of the central generator. A further restriction on the modules is obtained by considering the particular mode $L_0 + \bar{L}_0$ of the energy-momentum tensors; this mode generates dilatations and hence, upon identifying radial ordering as time ordering, plays the role of the energy operator of the theory so that its spectrum must be bounded from below. This implies that the Virasoro modules corresponding to the fields φ must in fact be highest weight modules of both the holomorphic and the antiholomorphic Virasoro algebras. The fields corresponding to the highest weight vectors of these modules are called *primary fields*; they will be denoted by $\phi(z, \bar{z}) \equiv \phi_{c,\Delta;\bar{c},\bar{\Delta}}(z, \bar{z})$. In the Hilbert space language of quantum field theory, the fields ϕ are in one to one correspondence with states $|\phi\rangle$ which are obtained by applying $\lim_{z\to 0} \phi(z)$ to a conformally invariant *vacuum state* $|0\rangle$ which corresponds to the identity primary field. The non-highest weight states in the Hilbert space are then obtained by applying the lowering operators of the Virasoro algebra to the highest weight states $|\phi\rangle$.

The highest weight property of primary fields reads in terms of the Laurent components of T

$$\begin{aligned} L_n\,\phi &= 0 \quad \text{for} \quad n > 0, \\ L_0\,\phi &= \Delta\phi. \end{aligned} \tag{3.3.13}$$

Also, one has

$$L_{-1}\,\phi = -\partial\phi. \tag{3.3.14}$$

Analogous relations are valid for the corresponding antiholomorphic quantities. The origin of (3.3.14) is the fact that for L_{-1} the central term in the Virasoro bracket always vanishes so that it can be identified with its classical analogue $L_{-1}^{(c)} = -\frac{\mathrm{d}}{\mathrm{d}z}$. Likewise, L_0 can be identified with its classical counterpart $-z\frac{\mathrm{d}}{\mathrm{d}z}$ which implies that it measures the scaling behavior of the fields; therefore Δ and $\bar{\Delta}$ are called the scaling dimensions or *conformal dimensions* of ϕ. Under a finite conformal transformation (3.3.5), primary fields transform as

$$\phi(z,\bar{z}) \mapsto \phi(w,\bar{w}) \cdot \left(\frac{\mathrm{d}w}{\mathrm{d}z}\right)^{\Delta} \left(\frac{\mathrm{d}\bar{w}}{\mathrm{d}\bar{z}}\right)^{\bar{\Delta}} . \tag{3.3.15}$$

The fields corresponding to non-highest weight vectors are called *secondary fields* or *descendants* of the corresponding primary field. From the commutation relations of the Virasoro algebra it follows that they still possess definite L_0-eigenvalues, namely $\Delta+m$ where $m \in \mathbb{Z}_{>0}$ is the grade (cf. (3.2.26)). However, for descendant fields the formula (3.3.15) is no longer generally valid, but only for the so-called *restricted* conformal transformations which are generated by L_0, L_{-1}, and $L_{+1} = -z^2 \frac{\mathrm{d}}{\mathrm{d}z}$ (recall that these generate a subalgebra of $\mathcal{V}$). For example, for the energy–momentum tensor one finds the transformation law

$$T(z) \mapsto T(w) \left(\frac{\mathrm{d}w}{\mathrm{d}z}\right)^2 + \frac{c}{12}\, s(z,w), \tag{3.3.16}$$

where

$$s(z,w) := \frac{\mathrm{d}^3 w}{\mathrm{d}z^3} \Big/ \frac{\mathrm{d}w}{\mathrm{d}z} - \frac{3}{2}\left(\frac{\mathrm{d}^2 w}{\mathrm{d}z} \Big/ \frac{\mathrm{d}w}{\mathrm{d}z}\right)^2 \tag{3.3.17}$$

is the so-called *Schwarzian derivative.*

The relations (3.3.13), (3.3.14) imply that the operator product of T with a primary field ϕ is such that

$$\underbrace{T(z)\,\phi(w,\bar{w})} = (z-w)^{-2}\,\Delta\,\phi(w,\bar{w}) + (z-w)^{-1}\,\partial\phi(w,\bar{w}), \tag{3.3.18}$$

while for a secondary field of grade m, more singular terms appear (namely, up to terms proportional to $(z-w)^{-m-2}$). Recalling now the operator products (3.2.8) and (3.2.9) of T and J, we see that the current $J(z)$ is a primary field of dimension $\Delta = 1$, while according to (3.3.16) $T(z)$ is a secondary field of dimension $\Delta + m = 2$. (Also, since these fields do not depend on the coordinate $\bar{z}$, they are both primary with respect to the antiholomorphic Virasoro algebra, with $\bar{\Delta} = 0$.) In fact, we are free to write $T(z) = T(z)\,\mathbf{1}$ where $\mathbf{1}$ denotes the *identity field* which acts as the unit operator in the Hilbert space (in particular, it does not depend on the coordinates at all and hence has $\Delta = \bar{\Delta} = 0$; the corresponding irreducible $\mathcal{V}$-representation with highest weight $\Delta = 0$ is isomorphic to the simple quotient of $\mathsf{U}(\mathcal{V}_-)$). From this it is clear that $T(z)$ is the descendant at grade two of the identity primary field (this descendant is unique because $L_{-1}\mathbf{1} = -\partial\mathbf{1} = 0$). Also note that as a consequence of (3.3.15), the conformal dimensions of T and J result in a factor of z^2 and z, respectively, if the particular finite transformation $z = \mathrm{e}^x \mapsto x$ is considered; this explains the presence of such factors in the definitions (3.1.13) and (3.2.12) that relate the fields on the real line to fields on the circle.

The fields which depend only on the coordinate z but not on $\bar{z}$ are obviously closed under the operator algebra, and analogously those depending only on $\bar{z}$. The subalgebra of the operator algebra formed by these fields is called the *symmetry algebra* of the theory; it has the structure of a direct sum of two (holomorphic respectively antiholomorphic) parts; these are called the chiral halves of the symmetry algebra, or shortly the *chiral algebras* of the theory. Until further notice we will concentrate on the holomorphic chiral half; analogous considerations will apply to the antiholomorphic part. It is natural to try to classify the fields of the conformal field theory not only in terms of the Virasoro algebra, but in terms of the whole chiral algebra. For most chiral algebras, however, the explicit form of the operator products of the chiral fields among themselves is not known completely, and even less is known about the representation theory of these algebras. The most notable exception to this is the case where the chiral algebra is the semidirect sum $\mathcal{V}\oplus\mathfrak{g}$ of the Virasoro algebra with an affine Lie algebra, the chiral fields being just $T(z)$ and $J(z)$. In this case the primary fields must also correspond to highest weight vectors of the affine algebra (recall that $-L_0$ has the meaning of the derivation D of the affine algebra). Thus (in the case of untwisted algebras) primary fields $\phi \equiv \phi_\Lambda$ obey

$$T^a_n\,\phi = 0 \quad \text{for} \quad n > 0 \tag{3.3.19}$$

and

$$T^a_0\,\phi = T^a_{(\bar{\Lambda})}\,\phi. \tag{3.3.20}$$

Here we use the shorthand notation

$$T^a_{(\bar{\Lambda})} := R_{\bar{\Lambda}}(T^a). \tag{3.3.21}$$

Also, the eigenvalue of the central generator K is again fixed to a given value k for all fields of a given theory. In terms of operator products, (3.3.19) and (3.3.20) mean that

$$\underline{J^a(z)\,\phi_\Lambda(w,\bar{w})} = (z-w)^{-1}\,T^a_{(\bar{\Lambda})}\phi_\Lambda(w,\bar{w}). \tag{3.3.22}$$

In view of the current operator product (3.1.10), this means in particular that while J^a is a primary field of the Virasoro algebra, it is not primary with respect to the affine algebra; rather, it is the descendant of the identity field **1** at grade one.

The conformal field theories which we will discuss in considerable detail in the following sections are the so-called *Wess–Zumino–Witten theories*

(*WZW theories*, for short). These are by definition those conformal field theories for which the chiral symmetry algebra is generated by (at least) $T(z)$ and $J^a(z)$ and, in addition, the energy-momentum tensor is of the Sugawara form (3.2.2). The name WZW theory has the following origin. Most of these theories can be realized in a lagrangian formulation as two-dimensional nonlinear sigma models for which the target space (i.e. the space in which the elementary quantum fields γ of the theory take values) is a compact Lie group manifold and which are supplemented by a Wess–Zumino term which, as realized by Witten, guarantees conformal invariance of the theory at the quantum level. The action of such a WZW sigma model reads

$$S_{\rm WZW} = 2\pi\,\tilde{k}\,(S_\sigma + S_{\rm WZ})\,, \tag{3.3.23}$$

where

$$S_\sigma = \frac{1}{32\pi^2}\int \mathrm{d}^2 z\,\mathrm{tr}(\partial_\mu\gamma\,\partial^\mu\gamma^{-1}) \tag{3.3.24}$$

and

$$S_{\rm WZ} = \frac{1}{48\pi^2}\int \mathrm{d}^3 y\,\epsilon^{\mu\nu\tau}\mathrm{tr}(\dot\gamma^{-1}\partial_\mu\dot\gamma\,\dot\gamma^{-1}\partial_\nu\dot\gamma\,\dot\gamma^{-1}\partial_\tau\dot\gamma). \tag{3.3.25}$$

Here γ takes values in some Lie group G, and the integral in the *Wess–Zumino action* $S_{\rm WZ}$ extends over a three-dimensional ball $\mathcal{B}$ whose surface $\partial\mathcal{B}$ is to be identified with the compactified two-dimensional world sheet such that on the surface, $\dot\gamma$ coincides with γ, and is smooth throughout the interior of the ball.

The quantum field theory defined by the action $S = \lambda S_\sigma$, with λ a coupling constant, is called the *principal* sigma model associated with the group G. Its field equation has the form of a conservation equation, $\partial^\mu J_\mu = 0$, for the current $J_\mu := \gamma^{-1}\partial_\mu\gamma$. Conformal invariance would require that the dual current $J^\star_\mu \equiv \epsilon_{\mu\nu}J^\nu$ must be conserved as well; for the principal sigma model this is not the case so that it is not conformally invariant. If, however, the Wess–Zumino action $S_{\rm WZ}$ is added to S_σ with a definite relative normalization, namely the one chosen in (3.3.23), then the field equation (obtained in the standard quantum field theoretic way as the Euler-Lagrange equation for the action $S_{\rm WZW}$) indeed can be written as $\partial_\mu J^\mu = 0 = \epsilon^{\mu\nu}\partial_\mu J_\nu$, or in complex notation

$$\bar\partial\,J = 0 = \partial\,\bar J. \tag{3.3.26}$$

More precisely, this is true iff the definition of J^μ is chosen a bit differently from what it was in the case of S_σ, namely, in complex notation

$$\begin{aligned} J &= \gamma^{-1}\partial\gamma, \\ \bar J &= (\bar\partial\gamma)\,\gamma^{-1} \equiv \gamma\,(\gamma^{-1}\bar\partial\gamma)\,\gamma^{-1} \end{aligned} \tag{3.3.27}$$

(respectively, if the sign of the coupling constant $\tilde{k}$ is negative, the other way round, $J = (\partial\gamma)\gamma^{-1}$, $\bar{J} = \gamma^{-1}\bar{\partial}\gamma$). In addition, it is straightforward to show that (for G a simple Lie group) the currents J and $\bar{J}$ defined this way generate two commuting copies of the untwisted affine Lie algebra $\mathfrak{g}$ whose horizontal subalgebra $\mathfrak{g}_{\text{hor}}$ is the Lie algebra of the group G, with the central generator K at a fixed value, namely such that the level of $\mathfrak{g}$ is equal to the absolute value of the coupling constant $\tilde{k}$ that appears in (3.3.23),

$$k^{\vee} = |\, \tilde{k}\, | \,. \tag{3.3.28}$$

Finally, one finds that the canonical energy-momentum tensor (which splits into T and $\bar{T}$ in complex notation) of the theory is of the Sugawara form; at the classical level, the prefactor in the Sugawara formula for T contains however the factor $1/k^{\vee}$ instead of $1/(k^{\vee}+g^{\vee})$. In other words, this prefactor receives a finite multiplicative renormalization from quantum effects (this finite renormalization can be calculated in perturbation theory with standard Feynman diagrammatic techniques, yielding the perturbation expansion $(1/k^{\vee})\,(1-g^{\vee}/k^{\vee}+\ldots)$ of the exact value $1/(k^{\vee}+g^{\vee})$).

Thus, in short, the WZW sigma model is a conformal field theory with the symmetry algebra given by the direct sum of a holomorphic and an antiholomorphic copy of $\mathfrak{g} \oplus \mathcal{V}$, and with the energy-momentum tensor being of the Sugawara form.

The existence of a smooth continuation of a given configuration $\gamma(x)$ on the compactified world sheet to the interior of the ball $\mathcal{B}$ is not a trivial issue, because usually there exist topological obstructions to such continuations. Indeed, in the case of a nonlinear sigma model with target space being an arbitrary manifold $\mathcal{M}$, the addition of a Wess–Zumino term is possible iff the homology group $H_d(\mathcal{M},\mathbb{Z})$ is trivial, with d the dimensionality of the world "sheet". The Wess–Zumino term is then a topological invariant so that its prefactor in the action must be quantized; this quantization is described by the homology group $H_{d+1}(\mathcal{M},\mathbb{Z})$. In the case of our interest one has

$$H_2(\mathsf{G}) = 1, \qquad H_3(\mathsf{G}) \cong \mathbb{Z}, \tag{3.3.29}$$

i.e. a Wess–Zumino term exists, and in appropriate normalization its prefactor is an integer. This normalization is actually related to the one chosen in the form (3.3.23) of S_{WZW} above in such a way that $\tilde{k}$, and hence the level $k^{\vee}$ of the affine algebra $\mathfrak{g}$, is always an integer in these theories. (In quantum field theory language, the integrality requirement can be interpreted as follows. The fact that S_{WZW} is defined only up to a multiplicative ambiguity does not cause any problems as far as the classical field equation is concerned, because each of the actions labelled by $\tilde{k}$ is stationary for the same set of paths. Quantum mechanically, one must however require that $\exp(iS_{\rm WZW})$ is single-valued, and this is only the case if $\tilde{k}$ is an integer.)

Not for all WZW theories, i.e. conformal field theories with chiral symmetry algebra $\mathfrak{g}\oplus\mathcal{V}$ and Sugawara type energy-momentum tensor, is a langrangian description known. However, the integrality property of the level also turns out to be fulfilled for those WZW theories of our interest for which a lagrangian description is lacking. Also, in all these WZW theories the fields must be unitary. Thus we have to consider WZW theories for which the fields correspond to irreducible integrable highest weight modules. The existence of null vectors in the associated Verma modules then leads to non-trivial relations among the fields of the WZW theory. These will be the subject of the two following sections. The primary fields of the WZW theories will be denoted by $\phi\equiv\phi_{\Lambda,\tilde{\Lambda}}$ with Λ and $\tilde{\Lambda}$ being highest weights of the holomorphic and antiholomorphic affine algebra, respectively (we do not need any indices referring to the Virasoro algebra because via the Sugawara construction all conformal properties of a primary field follow from its current algebra properties; also, we write $\tilde{\Lambda}$ rather than $\bar{\Lambda}$ in order to avoid confusion with the horizontal part of Λ).

The correspondence between primary WZW fields $\phi_{\Lambda,\tilde{\Lambda}}$ and highest weight vectors $v_{\Lambda,\tilde{\Lambda}}$ of integrable $\mathfrak{g}\oplus\mathfrak{g}$-modules is however not one to one. Some highest weights may not be realized in a given WZW theory at all while other weights may appear more than once (the only combination which must be present precisely once is $\Lambda=\tilde{\Lambda}=0$ because this corresponds to the identity primary field, which must be unique). The combinations $(\Lambda,\tilde{\Lambda})$ which do appear constitute the *spectrum* of primary fields of the WZW theory. For any compact Lie group G with Lie algebra $\mathfrak{g}$, there exists a WZW theory whose spectrum corresponds to the action (3.3.23). In the case of the simply connected covering group, each integrable highest weight module appears precisely once, in a left–right (i.e. holomorphic–antiholomorphic) symmetric way, i.e. the spectrum is

$$\{(\Lambda,\Lambda)\mid R_\Lambda \text{ integrable}\}. \tag{3.3.30}$$

Such a theory exists at any level of $\mathfrak{g}$. We will call it the *diagonal* WZW theory associated to the untwisted affine Lie algebra $\mathfrak{g}$; often this particular theory is also simply referred to as *the* WZW theory based on $\mathfrak{g}$.

For many values of the level of $\mathfrak{g}$, there are also other possible spectra; some of them will be discussed in sections 5.2 and 5.4.

3.4 The Knizhnik–Zamolodchikov equation

All properties of a (two-dimensional) conformal field theory are encoded in its operator product algebra. The operator product algebra naturally splits into three pieces of information, namely the symmetry algebra, its spectrum of primary fields, and the structure constants involving only primary fields. Indeed, once the symmetry algebra has been identified, all structure constants involving at least one secondary field can be expressed uniquely through the structure constants of the symmetry algebra and the structure constants for primary fields. Here we use the term structure *constants* for the coefficients of (3.3.9) although they depend on the world-sheet variables, because the world-sheet dependence is in fact rather trivial. Namely, combining the associativity property of the operator product algebra with the T-ϕ operator product (3.3.18), one finds that the structure constants involving three primary fields ϕ_i, ϕ_j, ϕ_k are

$$d_{ij}{}^{k}(z,\bar z,w,\bar w) = (z-w)^{\Delta_k-\Delta_i-\Delta_j}(\bar z-\bar w)^{\bar\Delta_k-\bar\Delta_i-\bar\Delta_j}\,C_{ij}{}^{k} \tag{3.4.1}$$

where $C_{ij}{}^{k}$ are complex constants, called the *operator product coefficients*. From this result and the locality of the operator product algebra it follows that $C_{ij}{}^{k}$ can be nonvanishing only if the conformal dimensions of the involved fields obey

$$\Delta_i+\Delta_j-\Delta_k = \bar\Delta_i+\bar\Delta_j-\bar\Delta_k \mod \mathbb{Z}. \tag{3.4.2}$$

A formula analogous to (3.4.1), with Δ replaced by $\Delta+m$, holds if secondary fields (at grade m) are involved; more precisely, if the secondary field φ_k is obtained by applying generators W_{-m_i}, $\bar W_{-\bar m_i}$ of the symmetry algebra $\mathcal{W}\oplus\bar{\mathcal{W}}$ to the primary field ϕ_k, then the operator product coefficient relating φ_k to the primaries ϕ_i and ϕ_j is

$$d_{ij}{}^{(k,\{m_i\},\{\bar m_i\})} = (z-w)^{m}(\bar z-\bar w)^{\bar m}\, d_{ij}{}^{k}\, c_{ij}{}^{(k,\{m_i\})}\,\bar c_{ij}{}^{(k,\{\bar m_i\})}. \tag{3.4.3}$$

Here the last two factors are completely determined by the holomorphic and antiholomorphic part of the symmetry algbera, respectively; also, the grades m and $\bar m$ are $m\equiv\sum_i m_i$, $\bar m\equiv\sum_i \bar m_i$, as a consequence of the relation $[L_0,W_m]=-m\,W_m$ which generalizes (3.2.16) to the case of arbitrary extended algebras.

The operator product of two primary fields thus reads

$$\phi_i(z,\bar z)\,\phi_j(w,\bar w) = \textstyle\sum_k (z-w)^{\Delta_k-\Delta_i-\Delta_j}(\bar z-\bar w)^{\bar\Delta_k-\bar\Delta_i-\bar\Delta_j} \cdot \Big(C_{ij}{}^{k}\,\phi_k(w,\bar w)+\ldots\Big) \tag{3.4.4}$$

where the ellipsis stands for a power series in $(z-w)$ and $(\bar z-\bar w)$ coming from the contributions involving descendants of ϕ_k.

Assuming that the symmetry algebra and its spectrum of primary fields are known for some given conformal field theory, the task of solving the conformal field theory then consists in determining the operator product coefficients ${C_{ij}}^k$ of the theory. Some of these coefficients are very simple, namely those involving the identity primary field which is present in any conformal field theory (because its descendants provide the fields in the symmetry algebra): writing $\mathbf{1} \equiv \phi_0$, one has

$$ {C_{i0}}^j = \delta_i^{\,j} . \tag{3.4.5} $$

Also, for fixed index i, ${C_{ij}}^0$ is nonzero for precisely one primary field ϕ_j, which is called the field *conjugate* to ϕ_i; we will denote the index j which is conjugate to i by i^+, and the conjugate field by $\phi_i^+ \equiv \phi_{i^+}$. The dimensions of conjugate fields are equal,

$$ \Delta_{i^+} = \Delta_i . \tag{3.4.6} $$

Owing to the uniqueness of i^+ for given i, the matrix $C_{ij} := {C_{ij}}^0$ of operator product coefficients and its inverse C^{ij} (defined by the requirement $C^{ij}C_{jl} = \delta^i_l$) can be used as metric tensors which lower and raise indices, so that in particular one can define operator product coefficients with lower indices only,

$$ C_{ijk} := {C_{ij}}^{k^+} \, {C_{k^+k}}^0 . \tag{3.4.7} $$

From now on it will be assumed that the primary fields are normalized approriately such that

$$ {C_{ij}}^0 = \delta_{i\,j^+} . \tag{3.4.8} $$

To be able to determine the operator product coefficients for generic primary fields, one has to introduce the concept of *correlation functions*. These are by definition vacuum expectation values of (radially ordered) products of fields, i.e.

$$ \langle \Pi(z_1, ..., \bar z_N) \rangle \equiv \langle 0 \,|\, \Pi(z_1, \dots, \bar z_N) \,|\, 0 \rangle . \tag{3.4.9} $$

Here

$$ \Pi(z_1, \dots, \bar z_N) \equiv \prod_{i=1}^{N} \varphi_i(z_i, \bar z_i) , \tag{3.4.10} $$

and the in-vacuum state $|0\rangle$ and out-vacuum state $\langle 0|$ are defined by the requirements that

$$ \begin{aligned} W_n \,|\, 0 \rangle = 0 \quad \text{for} \quad n \geq 0 , \\ \langle 0 \,|\, W_n = 0 \quad \text{for} \quad n \leq 0 , \end{aligned} \tag{3.4.11} $$

and that one point functions of primary fields other than the identity field vanish,

$$\langle \phi_i \rangle = \delta_{i0}. \tag{3.4.12}$$

In (3.4.11), W_n stands for the Laurent modes of the chiral symmetry algebra $\mathcal{W}$; the modes with $n > 0$, $n = 0$, and $n < 0$ correspond to the raising, Cartan subalgebra, and lowering operators, respectively, i.e. to the subalgebras in a Cartan-like decomposition $\mathcal{W} = \mathcal{W}_+ \oplus \mathcal{W}_0 \oplus \mathcal{W}_-$. Also, for unitary highest weight modules, the modes W_n behave under hermitian conjugation as $(W_n)^\dagger = W_{-n}$. In the case of the Virasoro algebra, one has in addition

$$L_{-1} \mid 0 \rangle = 0 = \langle 0 \mid L_1, \tag{3.4.13}$$

because $|0\rangle \equiv \mathbf{1}\,|0\rangle$ and $L_{-1}\mathbf{1} = \partial\mathbf{1} = 0$. Note that an n-point correlation function is defined only on that subset of $\mathbb{C}^N$ for which the positions of the fields are radially ordered; however, by analytic continuation in the coordinates z_i (and also, independently, in the antianalytic coordinates $\bar{z}_i$) one can in fact extend the domain of definition to $\mathbb{C}^N \backslash \Sigma$ with Σ consisting of those points for which at least two of the positions of the fields coincide (compare the remarks in section 3.1, p. 154). It must emphasized that, strictly speaking, all manipulations involving the deformation of integration contours (such as (3.1.7) above or (3.4.15) below) are valid only inside correlation functions, because the positions of additional fields constitute singular points through which the contours cannot be deformed.

The two- and three-point correlation functions of primary fields have a very simple form. First, using the operator product expansion (3.4.4) together with (3.4.11) and (3.4.12), it follows immediately that

$$\begin{aligned} \langle \phi_i(z,\bar{z})\phi_j(w,\bar{w})\rangle &= \textstyle\sum_k (z-w)^{\Delta_k-\Delta_i-\Delta_j}(\bar{z}-\bar{w})^{\bar{\Delta}_k-\bar{\Delta}_i-\bar{\Delta}_j}\langle \phi_k(w,\bar{w})\rangle \\ &= C_{ij}{}^0\,(z-w)^{-\Delta_i-\Delta_j}(\bar{z}-\bar{w})^{-\bar{\Delta}_i-\bar{\Delta}_j} \\ &= (z-w)^{-2\Delta_i}(\bar{z}-\bar{w})^{-2\bar{\Delta}_i}\,\delta_{ij^+}. \end{aligned} \tag{3.4.14}$$

This result can also be obtained as follows. By contour deformation and implementation of the T-ϕ operator product (3.3.18) one finds that in general

$$\begin{aligned} &\tfrac{1}{2\pi i}\oint_0 \mathrm{d}z\, z^n\, T(z) \textstyle\prod_{i=1}^N \phi_i(z_i,\bar{z}_i) \\ &\quad = \sum_{j=1}^n \Big(-\tfrac{1}{2\pi i}\oint_{z_j} \mathrm{d}z\, z^n\, T(z)\phi_j(z_j,\bar{z}_j)\Big) \prod_{\substack{i=1\\ i\neq j}}^N \phi_i(z_i,\bar{z}_i) \end{aligned} \tag{3.4.15}$$

for any integer n; here all points z_i are to be exterior to the contour around zero which appears on the left hand side. These relations are called the *conformal Ward identities*; analogous identities are obtained when the energy–momentum tensor is replaced by any other field belonging to the chiral symmetry algebra. The Ward identities of the symmetry algebra can in principle be used to calculate (inductively grade by grade) the coefficients $c_{ij}^{(k,\{m_i\})}$ in the operator algebra (3.4.3). For $n = 1, 0, -1$ (corresponding to the sl_2 subalgebra of the Virasoro algebra that is generated by L_0, $L_{\pm 1}$, or in terms of finite transformations, to the subgroup of the conformal group that consists of special conformal transformations, dilatations and translations), the relations (3.4.15) are called the *projective Ward identities.* Their general solution reads

$$\langle \phi_1(z_1, \bar{z}_1) \ldots \phi_N(z_N, \bar{z}_N) \rangle = \prod_{\substack{i,j=1 \\ i<j}}^{N} (z_i - z_j)^{\Delta_{ij}} (\bar{z}_i - \bar{z}_j)^{\bar{\Delta}_{ij}} F(z_{ijkl}, \bar{z}_{ijkl}) ; \tag{3.4.16}$$

here the numbers Δ_{ij} are constants satisfying $\Delta_{ij} = \Delta_{ji}$ and $\sum_{i \neq j} \Delta_{ij} = 2\Delta_j$ (and analogously for the $\bar{\Delta}_{ij}$). Also, the function F is required to be sl_2 invariant, but is not constrained otherwise; the sl_2 invariance is made manifest by writing F as a function of the anharmonic ratios $z_{ijkl} = (z_i - z_j)(z_k - z_l)/(z_j - z_k)(z_l - z_i)$ (precisely $N - 3$ of these are independent) and their antiholomorphic analogues. In the case of the two point function, this reproduces (3.4.14). Similarly, the three point function becomes

$$\begin{aligned} \langle \phi_i(z_1, \bar{z}_1) \phi_j(z_2, \bar{z}_2) \phi_k(z_3, \bar{z}_3) \rangle &= (z_1 - z_2)^{\Delta_k - \Delta_i - \Delta_j} (\bar{z}_1 - \bar{z}_2)^{\bar{\Delta}_k - \bar{\Delta}_i - \bar{\Delta}_j} (z_1 - z_3)^{\Delta_j - \Delta_i - \Delta_k} \\ &\cdot (\bar{z}_1 - \bar{z}_3)^{\bar{\Delta}_j - \bar{\Delta}_i - \bar{\Delta}_k} (z_2 - z_3)^{\Delta_i - \Delta_j - \Delta_k} (\bar{z}_2 - \bar{z}_3)^{\bar{\Delta}_i - \bar{\Delta}_j - \bar{\Delta}_k} C_{ijk} . \end{aligned} \tag{3.4.17}$$

(To be precise, (3.4.14) and (3.4.17) are literally true if there is at most one independent possibility of coupling the relevant primary fields.) As a consequence, the determination of operator product coefficients is equivalent to the calculation of the three-point correlation functions of primary fields. One may hope to be able to compute the three point functions through solving appropriate constraints following from the symmetry algebra of the theory. However, it turns out that any such constraint provides a linear equation and hence can determine the three point function only up to an overall factor. The same is true for all higher-point correlation functions. However, considering the limit of a four-point function where two arguments approach each other, it effectively reduces to a sum of three point functions (with coefficients given by operator product coefficients) so that the normalization of four-point functions can be expressed uniquely in terms of the normalizations of three point functions. Moreover, considering various limits of various four-point functions and using the associativity of the operator algebra, it is in fact possible to relate the normalizations to the fixed normalizations of the operator product coefficients ${C_{ij}}^0$. In short, the knowledge of all four-point functions of a conformal field theory up to their overall normalization is sufficient to calculate the operator product coefficients for primary fields and hence (provided the symmetry algebra and the spectrum of primary fields are known) to solve the theory completely. Therefore the computation of four-point functions is one of the main tasks in conformal field theory.

In the following we derive an equation which allows us to compute any four-point function in a WZW theory. This equation is an example for a *null vector equation.* Such equations are obtained from the fact that any correlation function involving a *null field* (i.e. a descendant field corresponding to a null vector in the Verma module headed by the corresponding primary field) must vanish identically if the fields of the theory are required to belong to irreducible modules of the symmetry algebra. This is so because in terms of the irreducible modules, the null vectors are just zero. In the WZW case, a null vector of the combined Virasoro and current algebra is obtained as a consequence of the Sugawara form (3.2.2) of the energy–momentum tensor. Namely, by employing (3.3.22) and the Wick rule (3.1.35), one obtains

$$\begin{aligned}\underline{:J^a(z)J^b(z):\phi_\Lambda(w,\bar{w})} &= \tfrac{1}{2\pi i}\oint_z \mathrm{d}y\,(y-z)^{-1}\\ &\Big[(w-y)^{-1}T^a_{(\bar\Lambda)}\phi_\Lambda(y,\bar{w})J^b(z)+(w-z)^{-1}J^a(y)T^b_{(\bar\Lambda)}\phi_\Lambda(z,\bar{w})\Big]\\ &=(w-z)^{-2}T^a_{(\bar\Lambda)}T^b_{(\bar\Lambda)}\phi_\Lambda(z,\bar{w})+2(w-z)^{-1}:J^a(z)T^b_{(\bar\Lambda)}\phi_\Lambda(z,\bar{w}):\,.\end{aligned} \tag{3.4.18}$$

Combining this with the T-ϕ operator product (3.3.18) and the Sugawara relation, one obtains

$$\begin{aligned}&(z-w)^{-2}\,\Delta_\Lambda\,\phi_\Lambda(w,\bar{w})+(z-w)^{-1}\partial\phi_\Lambda(w,\bar{w})\\ &\quad=\frac{1}{(\theta,\theta)(k^\vee+g^\vee)}\,\bar\kappa_{ab}\,\Big[(z-w)^{-2}R_{\bar\Lambda}(T^aT^b)\phi_\Lambda(w,\bar{w})\\ &\qquad\qquad -2(z-w)^{-1}\,:J^a(w)R_{\bar\Lambda}(T^b)\phi_\Lambda(w,\bar{w}):\Big]\,.\end{aligned} \tag{3.4.19}$$

Comparing the $(z-w)^{-2}$ terms, we see that the conformal dimension of

ϕ_Λ is related via

$$\Delta_\Lambda = \frac{1}{(\bar\theta,\bar\theta)\,(k^\vee + g^\vee)} \cdot C_{\bar\Lambda} = \frac{(\bar\Lambda + 2\bar\rho, \bar\Lambda)}{(\bar\theta,\bar\theta)\,(k^\vee + g^\vee)} \tag{3.4.20}$$

to the quadratic Casimir eigenvalue of $R_{\bar\Lambda}$. Note that this fixes the eigenvalue of the derivation D on the highest weight vector v_Λ. In other words, unlike in the case where the module R_Λ is considered in its own right (see p. 117), it is now not allowed to redefine this eigenvalue to zero. Also, the formula holds literally only if the Lie algebra $\mathfrak{g}$ is simple; if $\mathfrak{g}$ is semisimple, then one has a sum of terms of the given type, while any u_1-factor of $\mathfrak{g}$ provides a contribution

$$\Delta = \tfrac{1}{2}\,q^2 \tag{3.4.21}$$

where q is the u_1-charge of the primary field. Also, combining the result (3.4.20) for the conformal dimension with the formula (3.2.10) for the conformal central charge, one can write

$$\Delta_\Lambda - \frac{1}{24}\,c_\Lambda = s_\Lambda , \tag{3.4.22}$$

where s_Λ is the modular anomaly (2.6.40) which plays an important role in the modular transformation properties of the affine characters.

Let us also note that the term linear in m on the right hand side of the Virasoro brackets (3.2.15) is not intrinsic: it may be changed to any desired value by replacing L_n by $L_n + \xi\,\delta_{n,0}$ which results in a central term $(c/12)\,(m^3 - (1+24\xi/c)m)\,\delta_{m+n,0}$. In particular, for the choice $\xi = -c/24$ the linear term disappears completely. Thus the modular anomaly may be understood as a result of a modification of the naive Lie algebra conventions that is motivated by conformal field theory: the characters are multiplied by a factor $\exp(-2\pi i\,\tau\Delta_\Lambda)$ because the eigenvalue of $D = -L_0$ is fixed to $-\Delta_\Lambda$ rather than to zero, and in addition the value of Δ that is the preferred choice for the modular transformation properties of the characters is the one for which the Virasoro algebra looks most simple, and hence differs by $-c/24$ from the conventional one. In string theory, the addition of the term $-c/24$ can be traced back to the contribution of the Schwarzian derivative to the conformal transformation property (3.3.16) of the energy-momentum tensor that arises in the map from the cylindrical string world-sheet to the plane.

A further observation can be made by comparison of the $(z-w)^{-1}$-terms in (3.4.19); namely, one finds

$$\tfrac{1}{2}\,\partial\phi_\Lambda = \frac{1}{(\theta,\theta)(k^\vee + g^\vee)}\,\bar\kappa_{ab}\ :J^a R_{\bar\Lambda}(T^b)\phi_\Lambda : . \tag{3.4.23}$$

In terms of the Laurent modes of the symmetry algebra, this reads

$$\left(\tfrac{1}{2}L_{-1}-\frac{1}{(\theta,\theta)(k^\vee+g^\vee)}\,\bar{\kappa}_{ab}\,J^a_{-1}R_{\bar\Lambda}(T^b)\right)\phi_\Lambda=0, \tag{3.4.24}$$

which just says the the left hand side (with ϕ_Λ replaced by the corresponding highest weight vector v_Λ) is a null vector of the combined current and Virasoro algebra; in fact this is the unique primitive null vector of this kind.

After inserting (3.4.23) into an n-point correlation function, one may deform the integration contour which is implicit in the normal ordered product (cf. (3.1.22)) into a sum of contours encircling the other fields in the correlator,

$$\begin{aligned}&\tfrac{1}{2\pi i}\oint_z dw\,(w-z)^{-1}J^a(w)\phi_\Lambda(z,\bar z)\prod_{i=1}^n\phi_{\Lambda_i}(z_i,\bar z_i)\\ &\quad=\phi_\Lambda(z,\bar z)\sum_{j=1}^n\left(-\tfrac{1}{2\pi i}\oint_{z_j}dw\,(w-z)^{-1}J^a(w)\phi_{\Lambda_j}(z_j,\bar z_j)\right)\\ &\qquad\qquad\cdot\prod_{\substack{i=1\\ i\neq j}}^n\phi_{\Lambda_i}(z_i,\bar z_i).\end{aligned} \tag{3.4.25}$$

Making use of the J-ϕ operator product (3.3.22), the w-integration is easily carried out, and one gets

$$\begin{aligned}&\left[\frac{1}{2}\frac{\partial}{\partial z}-\frac{\bar\kappa_{ab}}{(\theta,\theta)(k^\vee+g^\vee)}\sum_{j=1}^n\frac{1}{z-z_j}R_{\bar\Lambda}(T^a)R_{\bar\Lambda_j}(T^b)\right]\\ &\qquad\left\langle\phi_\Lambda(z,\bar z)\prod_{i=1}^n\phi_{\Lambda_i}(z_i,\bar z_i)\right\rangle=0.\end{aligned} \tag{3.4.26}$$

This differential equation expresses the fact that any correlation function containing a null vector vanishes; it is called the *Knizhnik–Zamolodchikov* equation. Of course there is also an analogous equation involving the antiholomorphic current and Virasoro algebras (the corresponding indices $\tilde\Lambda$ of $\phi\equiv\phi_{\Lambda,\tilde\Lambda}$ etc. have been suppressed in the above discussion). Also, it has again been assumed that $\mathfrak{g}$ is simple. For semisimple $\mathfrak{g}$, the second term in the square bracket is replaced by a sum of such terms, while any u_1-piece of $\mathfrak{g}$ adds a term

$$-\sum_{j=1}^n\frac{1}{z-z_j}\,q_\Lambda q_{\Lambda_j} \tag{3.4.27}$$

with q_{Λ_i} the u_1-charge carried by ϕ_{Λ_i}.

For correlation functions of more than four fields, the Knizhnik–Zamolodchikov equation is a partial differential equation. In contrast, because of (3.4.16), for the four-point function one has a linear ordinary differential equation. E.g. for

$$\mathcal{F}(z_1, z_2, z_3, z_4) = \langle \phi_{\Lambda_1}(z_1)\phi_{\Lambda_2}(z_2)\phi_{\Lambda_3}(z_3)\phi_{\Lambda_4}(z_4)\rangle \tag{3.4.28}$$

(from now on we suppress both the antiholomorphic indices and coordinates), we can write

$$\begin{aligned}\mathcal{F}(z_1, z_2, z_3, z_4) = {} & (z_1 - z_4)^{-2\Delta_1}(z_2 - z_3)^{-\Delta_1-\Delta_2-\Delta_3+\Delta_4} \\ & (z_2 - z_4)^{\Delta_1-\Delta_2+\Delta_3-\Delta_4}(z_3 - z_4)^{\Delta_1+\Delta_2-\Delta_3-\Delta_4} \cdot F(z).\end{aligned} \tag{3.4.29}$$

Here z is the sl_2 invariant anharmonic ratio

$$z = \frac{(z_1 - z_2)(z_3 - z_4)}{(z_2 - z_3)(z_4 - z_1)} \tag{3.4.30}$$

and $F(z)$ is defined by

$$F(z) = \langle \phi_{\Lambda_1}(z)\phi_{\Lambda_2}(0)\phi_{\Lambda_3}(1)\phi_{\Lambda_4}(\infty)\rangle, \tag{3.4.31}$$

with the argument "∞" of the fourth field meant in the sense that

$$\langle \textstyle\prod \phi_i(z_i)\ \phi(\infty)\rangle := \lim_{z_\infty \to \infty} (z_\infty)^{2\Delta(\phi)} \langle \textstyle\prod \phi_i(z_i)\ \phi(z_\infty)\rangle. \tag{3.4.32}$$

In terms of $F(z)$, the Knizhnik–Zamolodchikov equation reads

$$\frac{1}{2}\,\partial F(z) = \frac{1}{(\bar\theta,\bar\theta)(k^\vee + g^\vee)}\left(-\frac{1}{z}\,P + \frac{1}{1-z}\,Q\right)F(z) \tag{3.4.33}$$

with

$$P \equiv \bar\kappa_{ab}\, R_{\bar\Lambda_1}(T^a) \otimes R_{\bar\Lambda_2}(T^b), \qquad Q \equiv \bar\kappa_{ab}\, R_{\bar\Lambda_1}(T^a) \otimes R_{\bar\Lambda_3}(T^b). \tag{3.4.34}$$

The notation in (3.4.31) is adapted to the case where all fields in the correlation function are strictly primary in the sense that they correspond to highest weight vectors of the affine algebra. However, the generators appearing in the Knizhnik–Zamolodchikov equation generically change a highest weight vector into a weight vector for a weight λ which differs from Λ in its $\bar{\mathfrak{g}}$-part, or in other words map the primary field to a descendant with respect to the horizontal subalgebra, which we denote as $\phi_\Lambda^{\bar\lambda}$. Thus the Knizhnik–Zamolodchikov equation is in fact a matrix differential equation which connects various components

$$F^{\bar\lambda_1\bar\lambda_2\bar\lambda_3\bar\lambda_4}(z) = \langle \phi_{\Lambda_1}^{\bar\lambda_1}(z)\phi_{\Lambda_2}^{\bar\lambda_2}(0)\phi_{\Lambda_3}^{\bar\lambda_3}(1)\phi_{\Lambda_4}^{\bar\lambda_4}(\infty)\rangle \tag{3.4.35}$$

of the four-point function. Now inserting $T_0^a = \frac{1}{2\pi i}\oint_0 \mathrm{d}z\, J^a(z)$ into a correlation function, and deforming the integration contour as in (3.4.25), the fact that T_0^a annihilates the vacuum results in

$$\sum_{i=1}^{n} R_{\bar{\Lambda}_i}(T^a)\langle\Pi(z_1,\dots,z_n)\rangle = 0. \tag{3.4.36}$$

With this relation one can express all components of the correlation function through a number m of (a priori) independent ones; the Knizhnik–Zamolodchikov equation is then an $m\times m$-matrix differential equation connecting these basic components. Note that (3.4.36) implies in particular that

$$F^{\bar{\lambda}_1\bar{\lambda}_2\bar{\lambda}_3\bar{\lambda}_4}(z) \propto \delta_{\bar{\lambda}_1+\bar{\lambda}_2+\bar{\lambda}_3+\bar{\lambda}_4} \tag{3.4.37}$$

for the four-point function (and analogously for correlation functions of an arbitrary number of fields).

If $R_1,\dots,R_n$ are the $\mathfrak{g}$-modules corresponding to the fields in a correlation function, then the number m defined above is nothing but the number of one-dimensional modules in the decomposition of the Kronecker product $R_1\times\dots\times R_n$ into its irreducible parts. This can be made explicit by using the language of invariant tensors rather than that of weights. The correlation function then has n tensor indices. The invariance property (3.4.36) translates into the fact that F can be decomposed into a set of m invariant amplitudes F_A which are multiplied with the independent invariant tensors having the index structure of F,

$$F_{i_1\dots i_n} = \sum_{A=1}^{m} F_A\,(I_A)_{i_1\dots i_n}, \tag{3.4.38}$$

with $(I_A)_{i_1\dots i_n}$ denoting the invariant tensors. The Knizhnik–Zamolodchikov equation (3.4.33) for the four-point function reads

$$\tfrac{1}{2}\,\partial F_A(z) = \frac{1}{(\bar{\theta},\bar{\theta})(k^\vee+g^\vee)} \textstyle\sum_{B=1}^{m} \left(-\tfrac{1}{z}P_{AB}+\tfrac{1}{1-z}Q_{AB}\right)F_B(z). \tag{3.4.39}$$

Even for the four-point function the number m is typically rather large; as a consequence, the explicit form of the Knizhnik–Zamolodchikov equation is only known for a relatively small number of cases. Among these, there are all four-point functions for $\mathfrak{g}=A_1$, while for other choices of $\mathfrak{g}$ results are known only for a few four-point functions involving low-dimensional modules.

In the following we present two specific examples which should give a flavor of how such computations work. First take the four-point function of two fields carry ing the defining module $R_{\bar{\Lambda}_{(1)}}$ of $\mathfrak{g}=A_r$ and of two fields carrying the conjugate module $R_{\bar{\Lambda}_{(r)}}$. The decompositions of the relevant Kronecker products read

$$\begin{aligned} R_{\bar\Lambda_{(1)}} \times R_{\bar\Lambda_{(r)}} &= R_{\bar\Lambda_{(0)}} \oplus R_{\bar\theta}, \\ R_{\bar\Lambda_{(1)}} \times R_{\bar\Lambda_{(1)}} &= R_{2\bar\Lambda_{(1)}} \oplus R_{\bar\Lambda_{(2)}} \end{aligned} \tag{3.4.40}$$

(recall that $R_{\bar\Lambda_{(0)}}$ denotes the singlet, and $R_{\bar\theta}$ the adjoint module). From this it follows that the product $R_{\bar\Lambda_{(1)}} \times R_{\bar\Lambda_{(1)}} \times R_{\bar\Lambda_{(r)}} \times R_{\bar\Lambda_{(r)}}$ contains just two singlets. We will use the tensor notation; in this language, $\phi \equiv \phi_{\bar\Lambda_{(1)}}$ carries a lower index $i \in \{1,2,...,r+1\}$ while the conjugate field $\phi^+ \equiv \phi_{\bar\Lambda_{(r)}}$ has a corresponding upper index. Since there are two possibilities of combining the fields in the four-point function

$$F_{il}^{jk}(z) \equiv \langle \phi_i(z) \phi^{+\,j}(0) \phi^{+\,k}(1) \phi_l(\infty) \rangle \tag{3.4.41}$$

into a singlet, it decomposes into two invariant amplitudes according to

$$F_{il}^{jk}(z) = \sum_{A=1}^{2} (I_A)_{il}^{jk}\, F_A(z). \tag{3.4.42}$$

The two independent tensors I_A may obviously be chosen as

$$(I_1)_{il}^{jk} = \delta_i^{\,j} \delta_l^{\,k}, \quad (I_2)_{il}^{jk} = \delta_i^{\,k} \delta_l^{\,j}, \tag{3.4.43}$$

but of course also any independent set of linear combinations of these tensors could be used. The matrices P and Q appearing in (3.4.39) can be computed with the help of the decomposition (1.6.80) of $\bar\kappa_{ab} T^a \otimes T^b$ into invariant tensors. For example, with $t_{ij}^{\;\;kl} := \bar\kappa_{ab}\, (R_{\bar\Lambda_{(1)}}(T^a))_i^{\;k} (R_{\bar\Lambda_{(1)}}(T^b))_j^{\;l}$ one gets with the help of (1.6.80)

$$\begin{aligned} t_{ij'}^{\;\;i'j}\, F_{i'l}^{j'k} &= t_{ip}^{\;\;pj}\, \delta_l^{\,k}\, F_1 + t_{il}^{\;\;kj}\, F_2 \\ &= -\tfrac{r(r+2)}{2(r-1)}\, I_1\, F_1 - I_1\, F_2 + \tfrac{1}{r+1}\, I_2\, F_2. \end{aligned} \tag{3.4.44}$$

Together with $R_{\bar\Lambda_{(r)}}(T^b) = -(R_{\bar\Lambda_{(1)}}(T^b))^t$ we can therefore deduce that

$$P = \frac{1}{2(r+1)} \begin{pmatrix} r(r+2) & r+1 \\ 0 & -1 \end{pmatrix}. \tag{3.4.45}$$

Similarly, one finds

$$Q = \frac{1}{2(r+1)} \begin{pmatrix} -1 & 0 \\ r+1 & r(r+2) \end{pmatrix}. \tag{3.4.46}$$

We have thus found the explicit form of the Knizhnik-Zamolodchikov equation. This system of two first order differential equations for two functions F_1 and F_2 is easily decoupled. The decoupling leads to second order differential equations for the amplitudes F_1 and F_2. These equations are of the so-called Riemann form, and the solutions of such equations are well known: they are combinations of powers of z and $1-z$ with hypergeometric functions.

In short, once the matrices P and Q are known, for the example at hand it is a simple exercise to give the complete solution to the Knizhnik-Zamolodchikov equation. We will not bother to write this solution out, but rather discuss a more complicated example which leads to special functions that are not as well known as the hypergeometric ones. To do so, we have to look for a set of four fields for which the multiple $\mathfrak{g}$-Kronecker product contains more than two singlets. Let us choose $\mathfrak{g}=E_6$ and two fields $\phi \equiv \phi_{\bar\Lambda_{(1)}}$ together with two conjugate fields $\phi^+ \equiv \phi_{\bar\Lambda_{(5)}}$; thus according to the list (1.6.78) the fields ϕ and ϕ^+ carry the defining 27-dimensional defining module of E_6 and its conjugate module, respectively. Again we use the tensor notation so that ϕ carries a lower index $i \in \{1,2,...,27\}$ while ϕ^+ has a corresponding upper index. The relevant E_6 Kronecker products are easily calculated with the Racah–Speiser algorithm; for example, from (1.7.57) we learn that

$$R_{\bar\Lambda_{(1)}} \times R_{\bar\Lambda_{(5)}} = R_{\bar\Lambda_{(0)}} \oplus R_{\bar\Lambda_{(6)}} \oplus R_{\bar\Lambda_{(1)}+\bar\Lambda_{(5)}}, \tag{3.4.47}$$

which in terms of dimensions reads $27\times 27^+ = 1+78+650$, and similarly one finds $27\times 27 = 27^+ + 351 + 351'$. Accordingly, there are now three singlets in $R_{\bar\Lambda_{(1)}} \times R_{\bar\Lambda_{(1)}} \times R_{\bar\Lambda_{(5)}} \times R_{\bar\Lambda_{(5)}}$. Thus the four-point function $F^{jk}_{il}(z) \equiv \langle \phi_i(z)\phi^{+\,j}(0)\phi^{+\,k}(1)\phi_l(\infty)\rangle$ decomposes into invariant components as

$$F^{jk}_{il}(z) = \sum_{A=1}^{3} (I_A)^{jk}_{il}\, F_A(z). \tag{3.4.48}$$

The three independent tensors I_A may be chosen as

$$(I_1)^{jk}_{il} = \delta^j_i \delta^k_l, \quad (I_2)^{jk}_{il} = \delta^k_i \delta^j_l, \quad (I_3)^{jk}_{il} = d_{ilp}\, d^{jkp} \tag{3.4.49}$$

(or as an independent set of linear combinations of these). Here d_{ijk} stands for the totally symmetric traceless three-index tensor mentioned in (1.6.80), normalized such that $d_{ipq}\, d^{jpq} = \delta^j_i$ (and summation on $p = 1, \ldots, 27$ etc. is implicit).

The matrices P and Q of (3.4.39) are again computed with the help of the decomposition (1.6.80) of $\bar\kappa_{ab}\, T^a \otimes T^b$ into invariant tensors; in addition, one has to use the identity

$$d_{ipq} d_{lp'q'} d^{jp'q} d^{kpq'} = \tfrac{1}{20}\,(\delta^j_i \delta^k_l + \delta^k_i \delta^j_l) - \tfrac{2}{5}\, d_{ilp} d^{jkp}, \tag{3.4.50}$$

which can be obtained with the help of the so-called Springer relation

$$\begin{aligned} &10\,(d_{ijk}d_{lmn} + d_{ilk}d_{jmn} + d_{imk}d_{jln})\, d^{kpn} \\ &\quad = \delta^p_i d_{jlm} + \delta^p_j d_{ilm} + \delta^p_l d_{ijm} + \delta^p_m d_{ijl}. \end{aligned} \tag{3.4.51}$$

Using the invariants (3.4.49), the result for P and Q is

$$\begin{aligned} P &= \tfrac{1}{12}\begin{pmatrix} 104 & 6 & 3 \\ 0 & 2 & -3 \\ 0 & -60 & 26 \end{pmatrix} \\ Q &= \tfrac{1}{12}\begin{pmatrix} 2 & 0 & -3 \\ 6 & 104 & 3 \\ -60 & 0 & 26 \end{pmatrix}. \end{aligned} \tag{3.4.52}$$

Actually, it is convenient to choose a basis of invariants in which P is diagonal. In such a basis each invariant corresponds precisely to one of the irreducible modules in the decomposition of $R_{\bar\Lambda_{(1)}} \times R_{\bar\Lambda_{(5)}}$ (rather than to some non-trivial linear combination); it is given by

$$\tilde I_A = \sum_{B=1}^{3} I_B\, M_{BA}, \quad M \equiv \begin{pmatrix} 1 & 1 & 1 \\ 0 & 3 & -9 \\ 0 & -30 & -18 \end{pmatrix} \tag{3.4.53}$$

This leads to

$$\begin{aligned} \tilde P &= \tfrac{1}{3}\begin{pmatrix} 26 & 0 & 0 \\ 0 & 8 & 0 \\ 0 & 0 & -1 \end{pmatrix} \\ \tilde Q &= \tfrac{1}{18}\begin{pmatrix} 0 & 104 & 0 \\ 3 & 54 & -60 \\ 0 & -20 & 144 \end{pmatrix}. \end{aligned} \tag{3.4.54}$$

Thus we have again derived the explicit form of the Knizhnik–Zamolodchikov equation. To find also its solutions is a more complicated task. There is a general algorithm to decouple the $m \times m$-matrix differential equation ($m = 3$ in the case of our interest) leading to mth order differential equations for the components F_A. It is not difficult to describe the analytic properties of the solutions to these equations, but usually not much is known about their explicit form, except for the case $m = 2$, where as mentioned above they are essentially hypergeometric functions. In the present case, however, the (third order) differential equations belong to a special class for which the solutions are known in more detail, and they take a particularly simple form in the basis (3.4.53). Namely, they can be expressed through certain contour integrals, with the ends of the integration contours at the singular points $0, 1, \infty$ of the differential equation. For brevity, we just present the result for $\tilde F_1$ (the solutions for $\tilde F_2$ and $\tilde F_3$ may then be found by inserting this result into the Knizhnik–Zamolodchikov equation): three independent solutions are given by

$$\tilde{\mathcal{F}}_1^{(\nu)}(z) = (z(1-z))^{-52\eta/3} \int_{C_\nu}\int \mathrm{d}s\,\mathrm{d}t\,(st)^{-\eta}\,(s-t)^{-6\eta} \cdot ((s-1)(t-1))^{-1-\eta}((s-z)(t-z))^{13\eta}. \tag{3.4.55}$$

Here we defined

$$\eta = \frac{1}{k^\vee + 12} \tag{3.4.56}$$

with $k^\vee$ the level of $E_6^{(1)}$, and the integration contours C_ν, $\nu = 1,2,3$ are

$$\begin{aligned} \textstyle\int_{C_1}\int \mathrm{d}s\,\mathrm{d}t &= \textstyle\int_0^1 \mathrm{d}s \int_0^1 \mathrm{d}t, \\ \textstyle\int_{C_2}\int \mathrm{d}s\,\mathrm{d}t &= \textstyle\int_0^1 \mathrm{d}s \int_z^\infty \mathrm{d}t, \\ \textstyle\int_{C_3}\int \mathrm{d}s\,\mathrm{d}t &= \textstyle\int_z^\infty \mathrm{d}s \int_z^\infty \mathrm{d}t. \end{aligned} \tag{3.4.57}$$

The functions $\mathcal{F}$ in (3.4.55) are called the *holomorphic blocks* of the correlation function F; they are not single-valued in the complex z-plane. However, assuming that the fields ϕ and ϕ^+ transform under the antiholomorphic affine algebra in the same way as they do under the holomorphic one, a dependence of F on the antiholomorphic coordinate $\bar{z}$ arises so that F decomposes into antiholomorphic blocks $\overline{\mathcal{F}}$ that depend on $\bar{z}$ analogously to the z-dependence in (3.4.55). As a consequence, it is possible to construct a single-valued linear combination of the solutions. Namely, the most general solution of both the holomorphic and antiholomorphic Knizhnik–Zamolodchikov equations for F then reads

$$F_{A\bar{A}}(z,\bar{z}) = \sum_{\nu,\tau=1}^{3} a_{\nu\tau}\,\mathcal{F}_A^{(\nu)}(z)\,\mathcal{F}_{\bar{A}}^{(\tau)}(\bar{z}) \tag{3.4.58}$$

with arbitrary complex coefficients $a_{\nu\tau}$. It can then be shown that, for generic values of $k^\vee$, there exists a unique (up to normalization) choice of these coefficients such that for $\bar{z} = z^*$ the function (3.4.58) is single-valued in the whole complex plane: single-valuedness around $z = \bar{z} = 0$ requires

$$a_{\nu\tau} = a_\nu\,\delta_{\nu\tau}, \tag{3.4.59}$$

and single-valuedness around $z = \bar{z} = 1$ then fixes the relative values of the a_ν to

$$\begin{aligned} \frac{a_1}{a_3} &= \frac{\sin(5\pi\eta)\sin(6\pi\eta)\sin(8\pi\eta)}{\sin(10\pi\eta)\sin(12\pi\eta)\sin(13\pi\eta)}, \\ \frac{a_2}{a_3} &= \frac{\sin(\pi\eta)\sin(3\pi\eta)\sin(5\pi\eta)\sin(9\pi\eta)}{\sin(4\pi\eta)\sin(6\pi\eta)\sin(10\pi\eta)\sin(12\pi\eta)}. \end{aligned} \tag{3.4.60}$$

The single-valuedness of a four-point function which in the whole (compactified) complex plane is equivalent to the *crossing symmetry* of the four-point function. Crossing symmetry means that the set of functions defined by $F_{ijkl}(z,\bar{z}) \equiv \langle\phi_i(z,\bar{z})\phi_j(0,0)\phi_k(1,1)\phi_l(\infty,\infty)\rangle$ obeys

$$F_{ijkl}(z,\bar{z}) = F_{ikjl}(1-z,1-\bar{z}) = z^{-2\Delta_i}\bar{z}^{-2\bar{\Delta}_i}\, F_{ilkj}\left(\frac{1}{z},\frac{1}{\bar{z}}\right). \qquad (3.4.61)$$

Analogous crossing symmetry relations hold for arbitrary n-point correlation functions; they are nothing else than the translation of the locality and associativity of the operator product algebra into the language of correlation functions. Since locality and associativity of the operator product algebra are fundamental properties of any conformal field theory, it is clear that any differential equation for a (sensible) correlation function of a conformal field theory admits a single valued solution eventhough its generic solution is multivalued. These matters will be discussed in much more detail in section 5.6.

3.5 The Gepner–Witten equation

In the previous section we have seen how the presence of null vectors leads to constraints on the correlation functions of conformal field theories. Namely, the null vector (3.4.24) of the combined current and Virasoro algebra gives rise to the Knizhnik-Zamolodchikov equation (3.4.26). It is natural to apply these results also to the null vectors of the affine Lie algebra alone. The relevant null vector is the one given by (2.5.6) which arises at integer values of the level. (In contrast, the null vectors (2.5.4) which are present for simple Lie algebras as well do not give rise to new constraints; they are already implemented through the sum rule (3.4.36).) From (2.5.6) we get

$$0 = \Big\langle \Big((R_\Lambda(E^{\bar{\theta}}_{-1}))^{k^\vee-(\bar{\theta}^\vee,\bar{\Lambda})+1}\phi_\Lambda(z,\bar{z})\Big) \textstyle\prod_i \phi_{\Lambda_i}(z_i,\bar{z}_i)\Big\rangle . \qquad (3.5.1)$$

Now by a contour deformation argument as in (3.4.25), it is easy to see that, for any (not necessarily primary) field φ belonging to the family of ϕ_Λ,

$$\begin{aligned}&\Big\langle\Big(R_\Lambda(E^{\bar{\theta}}_{-1})\varphi(z,\bar{z})\Big)\textstyle\prod_i \phi_{\Lambda_i}(z_i,\bar{z}_i)\Big\rangle \\ &\qquad = \sum_j \frac{1}{z-z_j}\, R_{\bar{\Lambda}_j}(E^{\bar{\theta}})\,\langle\varphi(z,\bar{z})\textstyle\prod_i \phi_{\Lambda_i}(z_i,\bar{z}_i)\rangle .\end{aligned} \qquad (3.5.2)$$

Applying this result repeatedly, the identity (3.5.1) for an $n+1$-point correlation function can be rewritten as

$$\sum_{\substack{l_1,\ldots,l_n=1\\ l_1+\ldots+l_n\\ =k^\vee-(\bar\theta^\vee,\bar\Lambda)+1}}^{k^\vee+1} \left\{ \prod_{j=1}^{n} \frac{\left(R_{\bar\Lambda_i}(E^{\bar\theta})\right)^{l_j}}{(l_j)!\,(z-z_j)^{l_j}} \right\} \left\langle \phi_\Lambda(z,\bar z) \prod_{i=1}^{n} \phi_{\Lambda_i}(z_i,\bar z_i) \right\rangle = 0. \tag{3.5.3}$$

This equation is due to Gepner and Witten, and we will call it the *Gepner–Witten equation.*

With the particular choice $\phi_\Lambda = \mathbf{1}$ (so that $\bar\Lambda = 0$), the correlator in (3.5.3) becomes independent of the variable z so that (3.5.3) can be easily expanded in a power series in z. Each term in this series must vanish individually. In particular it follows that

$$\left(R_{\bar\Lambda_i}(E^{\bar\theta})\right)^{k^\vee+1} \left\langle \prod_{j=1}^{n} \phi_{\Lambda_j}(z_j,\bar z_j) \right\rangle = 0 \tag{3.5.4}$$

for any $i = 1,\ldots,n$, or in other words, that inside correlation functions the primary fields obey

$$\left(R_{\bar\Lambda}(E^{\bar\theta})\right)^{k^\vee+1} \phi_\Lambda = 0. \tag{3.5.5}$$

(Owing to the relation (3.4.36) expressing the invariance under the horizontal subalgebra, this result remains true if $\phi_\Lambda \equiv \phi_\Lambda^{\bar\Lambda}$ is replaced by any of its horizontal descendants $\phi_\Lambda^{\bar\lambda}$.) This shows that the highest weight modules carried by the primary fields indeed have to be the irreducible modules R_Λ rather than the Verma modules V_Λ. Since the irreducible integrable highest weight modules are unitarizable, this means in particular that we do not have to *assume* (as we have done on p. 178) that WZW theories are unitary quantum field theories, but that in fact this can be *shown* by simple algebraic considerations.

The Gepner–Witten equation strongly constrains the correlation functions of primary WZW fields. The analysis of these constraints is particularly straightforward if the sum in (3.5.3) contains only terms with one generator $R(E^{\bar\theta})$, i.e. if

$$(\bar\Lambda, \bar\theta) = k. \tag{3.5.6}$$

We will soon (p. 196) discuss this simple case, but first consider another special case, namely the three-point function. For $n = 3$, the Gepner–Witten equation becomes simple because then according to (3.4.17) the correlation function is known explicitly up to an overall constant. In particular, it follows that

$$(R_{\bar\Lambda_1}(E^{\bar\theta}))^{l_1} (R_{\bar\Lambda_2}(E^{\bar\theta}))^{l_2} \langle \phi_\Lambda(z,\bar z)\phi_{\Lambda_1}(z_1,\bar z_1)\phi_{\Lambda_2}(z_2,\bar z_2)\rangle = 0 \tag{3.5.7}$$

for $\ell_1+\ell_2 = k^\vee - (\bar\Lambda,\bar\theta^\vee)+1$, and in fact (using also (3.4.36)) for all for $\ell_1+\ell_2 \ge k^\vee - (\bar\Lambda,\bar\theta^\vee)+1$. Furthermore, according to (3.4.3) there must exist a simple relation between the operator product coefficients for primary fields $\phi^{\bar\Lambda}_{\Lambda}$ and those for their horizontal descendants $\phi^{\bar\lambda}_{\Lambda}$; with the help of (3.4.37) this can be found to be

$$C_{\phi^{\bar\lambda}_{\Lambda}\phi^{\bar\lambda'}_{\Lambda'}}^{\phi^{\bar\lambda''}_{\Lambda''}} = C^{\bar\lambda\bar\lambda';\bar\lambda''}_{\bar\Lambda\bar\Lambda';\bar\Lambda''}\cdot C_{\Lambda\Lambda'}^{\;\;\Lambda''}. \tag{3.5.8}$$

Here the constants $C^{\bar\lambda\bar\lambda_;'\bar\lambda''}_{\bar\Lambda\bar\Lambda';\bar\Lambda''}$ are the Clebsch–Gordan coefficients of the horizontal subalgebra $\bar{\mathfrak g}$ (see p. 62), and for the operator product coefficients of primaries we use the notation

$$C_{\Lambda\Lambda'}^{\;\;\Lambda''} \equiv C(\phi^{\bar\Lambda}_{\Lambda},\phi^{\bar\Lambda'}_{\Lambda'};\phi^{\bar\Lambda''}_{\Lambda''}) = C_{\phi^{\bar\Lambda}_{\Lambda}\phi^{\bar\Lambda'}_{\Lambda'}}^{\phi^{\bar\Lambda''}_{\Lambda''}}. \tag{3.5.9}$$

Next define the number

$$\ell = \mathrm{depth}_{\bar\Lambda}(\bar\lambda) \tag{3.5.10}$$

as the largest integer such that $\bar\lambda - \ell\bar\theta$ is a weight of the $\bar{\mathfrak g}$-module $R_{\bar\Lambda}$. For $\ell_1 = \mathrm{depth}_{\bar\Lambda_1}(\bar\lambda_1)$ and $\ell_2 = \mathrm{depth}_{\bar\Lambda_2}(\bar\lambda_2)$, one then has

$$\langle\phi^{\bar\Lambda}_{\Lambda}\phi^{\bar\lambda_1}_{\Lambda_1}\phi^{\bar\lambda_2}_{\Lambda_2}\rangle \propto (R_{\bar\Lambda_1}(E^{\bar\theta}))^{\ell_1}(R_{\bar\Lambda_2}(E^{\bar\theta}))^{\ell_2}\,\langle\phi^{\bar\Lambda}_{\Lambda}\phi^{\bar\lambda_1-\ell_1\bar\theta}_{\Lambda_1}\phi^{\bar\lambda_2-\ell_2\bar\theta}_{\Lambda_2}\rangle, \tag{3.5.11}$$

where for notational convenience we suppress the dependence of the fields on the world-sheet coordinates. According to (3.5.7), the right hand side of this equation must vanish whenever $\ell_1+\ell_2 > k^\vee - (\bar\Lambda,\bar\theta^\vee)$. On the other hand, because of (3.5.8) the left hand side of (3.5.11) is zero only if the Clebsch–Gordan coefficient $C^{\bar\lambda\bar\lambda_1;\bar\lambda_2}_{\bar\Lambda\bar\Lambda_1;\bar\Lambda_2}$ or the operator product coefficient $C_{\Lambda\Lambda_1}^{\;\;\Lambda_2}$ vanishes. Thus one arrives at the following theorem: the operator product coefficient $C_{\Lambda\Lambda_1}^{\;\;\Lambda_2}$ is zero unless for any pair of weights $\bar\lambda_1$, $\bar\lambda_2$ of $R_{\bar\Lambda_1}$, $R_{\bar\Lambda_2}$ either of the conditions

$$\begin{aligned} &(3.5.12a)\quad C^{\bar\lambda,\bar\lambda_1;\bar\lambda_2}_{\bar\Lambda,\bar\Lambda_1;\bar\Lambda_2} = 0 \\ \text{or}\quad &(3.5.12b)\quad \mathrm{depth}_{\bar\Lambda_1}(\bar\lambda_1) + \mathrm{depth}_{\bar\Lambda_2}(\bar\lambda_2) \le k^\vee - (\bar\Lambda,\bar\theta^\vee), \end{aligned} \tag{3.5.12}$$

is fulfilled. The selection rule (3.5.12) is called the *depth rule.*

The depth rule is most easily applied to the case $\mathfrak g = A_1^{(1)}$, but we will defer the discussion of this case until section 5.1. In the general case, the implementation of (3.5.12) is less straightforward, partly because of the lack of a general formula for the Clebsch–Gordan coefficients of $\bar{\mathfrak g}$. To compute the depth of a weight, one has to consider the sl_2-subalgebra of

$\mathfrak{g}$ corresponding to the highest root, $sl_2^{(\bar\theta)}$. It is not difficult to see that

$$\text{depth}_{\bar\Lambda}(\bar\lambda) = \tfrac{1}{2}\,(\bar\lambda^{(\bar\theta)} + \bar\Lambda^{(\bar\theta)}), \tag{3.5.13}$$

where $\bar\lambda^{(\bar\theta)}$ is the weight of $v_{\bar\lambda}$ with respect to $sl_2^{(\bar\theta)}$, and $\bar\Lambda^{(\bar\theta)}$ the highest weight of the irreducible module of $sl_2^{(\bar\theta)}$ to which $\bar\lambda^{(\bar\theta)}$ belongs. Let us consider just one specific example of how the selection rule works, namely the case $g = E_8^{(1)}$ at level 2. There are three integrable modules of $E_8^{(1)}$ at level two; with respect to the horizontal subalgebra E_8, their highest weights correspond to the singlet ($R_{\bar\Lambda_{(0)}}$), adjoint ($R_{\bar\Lambda_{(1)}}$) and the 3875-dimensional ($R_{\bar\Lambda_{(7)}}$) module, respectively. Denoting the irreducible E_8-modules by their dimensionality, the relevant Kronecker products are

$$\begin{aligned} &248 \times 248 = 1 + 248 + 3\,875 + 27\,000 + 30\,380, \\ &248 \times 3\,875 = 248 + 3\,875 + 30\,380 + 147\,250 + 779\,247, \\ &3\,875 \times 3\,875 = 1 + 248 + 3\,875 + 27\,000 + 30\,380 + 147\,250 \\ &\qquad\qquad + 779\,247 + 2\,450\,240 + 4\,881\,384 + 6\,696\,000 \end{aligned} \tag{3.5.14}$$

(together with the trivial products involving the singlet). The centralizer of $sl_2^{(\bar\theta)}$ in E_8 turns out to be E_7; with respect to the embedding $sl_2^{(\bar\theta)} \oplus E_7 \subset E_8$, the E_8-modules decompose as

$$\begin{aligned} &248 \rightsquigarrow 3 \otimes 1 + 1 \otimes 133 + 2 \otimes 56, \\ &3\,875 \rightsquigarrow 1 \otimes 1 + 2 \otimes 56 + 3 \otimes 133 + 1 \otimes 1\,539 + 2 \otimes 1\,912, \end{aligned} \tag{3.5.15}$$

where we again denote the modules by their dimensionality (e.g. 3 stands for the adjoint of sl_2 and 133 for the adjoint of E_7).

Now consider first the three-point function $\langle \phi_1 \phi_{248} \phi_{248} \rangle$; since the singlet only has the zero weight, the weights of the two adjoints must add up to zero in order to give rise to a nonvanishing Clebsch–Gordan coefficient. Thus we have to look at the combination $\text{depth}_{248}(\bar\lambda) + \text{depth}_{248}(-\bar\lambda)$; according to the branching rules (3.5.15), the adjoint contains the modules with highest weights $\bar\Lambda^{(\bar\theta)} = 0, 1, 2$ with respect to $sl_2^{(\bar\theta)}$, so that the possible values for the depth are

$$\text{depth} = \begin{cases} 0 & \text{for } (\bar\Lambda^{(\bar\theta)}; \bar\lambda^{(\bar\theta)}) = (0;0),\ (1;-1),\ (2;-2) \\ 1 & \text{for } (\bar\Lambda^{(\bar\theta)}; \bar\lambda^{(\bar\theta)}) = (1;1),\ (2;0) \\ 2 & \text{for } (\bar\Lambda^{(\bar\theta)}; \bar\lambda^{(\bar\theta)}) = (2;2). \end{cases} \tag{3.5.16}$$

Thus in particular $\text{depth}_{248}(\bar\lambda) + \text{depth}_{248}(-\bar\lambda) \le 2$ for all weights of the adjoint; but this is just the condition (3.5.12b) (with $k^\vee = 2$ and $\bar\Lambda = 0$). Hence we conclude that the operator product coefficient $C_{1,248,248}$ is non-zero. This could in fact have been concluded much more directly from the general property (3.4.5) of the operator product coefficients. Similarly, this general relation implies that $C_{1,3875,3875}$ is non-zero, while $C_{1,248,3875}$=0. In contrast, in the case of operator product coefficients which do not involve the singlet field, we really must go through the analysis of the depth values. First, one can see that $C_{248,248,3875}$ is nonvanishing, since for combinations of weights which give non-zero Clebsch–Gordans among the corresponding modules, the sum of depths appearing in the expression (3.5.12b) always turns out to be zero. Next consider the three point function of three fields carrying the adjoint module. The Clebsch–Gordan coefficients involving $-\bar{\theta}$ are then just the structure constants of the horizontal algebra, so that for the triple $-\bar{\theta}$, 0, $-\bar{\theta}$ the Clebsch–Gordan coefficient is non-zero. But the depth of the zero weight in the adjoint module, and hence also the sum of depths appearing in (3.5.12b) for this particular choice of weights, is clearly equal to one. On the other hand, the right hand side of (3.5.12b) is now zero because $(\bar{\theta},\bar{\theta}^{\vee}) = 2$. Hence we have to conclude that $C_{248,248,248} = 0$. Similarly one finds that $C_{248,3875,3875} = 0 = C_{3875,3875,3875}$ because in both cases a triple of weights exists which gives a nonvanishing Clebsch–Gordan coefficient and a sum of depths which is larger than zero. In the former case, the weight of the adjoint module is $(\Lambda^{(\bar{\theta})};\lambda^{(\bar{\theta})}) \otimes (R^{(E_7)};\lambda^{(E_7)}) = (2;-2) \otimes (1;0)$, and the weights of the 3875-dimensional modules are $(2;0)\otimes(133;\bar{\alpha})$ and $(2;-2)\otimes(133;\bar{\alpha})$ (that these weights are contained in the respective E_8-modules follows from the decomposition (3.5.15)); here $\bar{\alpha}$ is any weight of the 133-dimensional adjoint module of E_7 (i.e. any root of E_7). In the latter case, the three weights are $(2;-2)\otimes(133;\bar{\alpha})$, $(2;0)\otimes(133;\bar{\beta})$ and $(2;-2)\otimes(133;\bar{\alpha}+\bar{\beta})$, with the E_7-roots $\bar{\alpha}$ and $\bar{\beta}$ chosen such that $\bar{\alpha}+\bar{\beta}$ is an E_7-root, too.

Summarizing, the only non-zero operator product coefficients of the level two $E_8^{(1)}$ WZW theory are $C_{1,\phi,\phi}$ and $C_{248,248,3875}$.

We now return to the case where $\bar{\Lambda}$ obeys $(\bar{\Lambda},\bar{\theta}) = k$. Then the Gepner–Witten equation takes the form

$$\sum_{j=1}^{n} \frac{R_{\bar{\Lambda}_j}(E^{\bar{\theta}})}{z-z_j} \left\langle \phi_\Lambda(z,\bar{z}) \prod_{i=1}^{n} \phi_{\Lambda_i}(z_i,\bar{z}_i) \right\rangle = 0. \tag{3.5.17}$$

As an application of this formula, we consider the case $n = 3$, i.e. the four-point function, and for notational convenience specialize to $\bar{\Lambda}_3 = \bar{\Lambda}^+$, $\bar{\Lambda}_2 = \bar{\Lambda}_1^+$. Fixing three of the world-sheet coordinates as in (3.4.35) and using the $\mathfrak{g}$-invariance (3.4.36), every component $F^{\bar{\lambda}\bar{\lambda}_1\bar{\lambda}_2\bar{\lambda}_3}$ of the four point function is then a linear combination of the particular components

$$F^{\bar{\lambda}}(z) = \langle \phi_\Lambda^{\bar{\Lambda}}(z)\phi_{\Lambda_1}^{\bar{\lambda}}(0)\phi_{\Lambda_1^+}^{-\bar{\lambda}}(1)\phi_{\Lambda^+}^{-\bar{\Lambda}}(\infty)\rangle. \tag{3.5.18}$$

Not all of these components are independent, because for any positive $\mathfrak{g}$-root $\bar\alpha$ with the property $(\bar\alpha,\bar\Lambda)=0$, (3.4.36) implies that

$$\begin{aligned} 0 &= \textstyle\sum_{i=1}^4 R_i(E^{-\bar\alpha})\,\langle\phi_\Lambda^{\bar\lambda}(z)\phi_{\Lambda_1}^{\bar\lambda+\bar\alpha}(0)\phi_{\Lambda_1^+}^{-\bar\lambda}(1)\phi_{\Lambda^+}^{-\bar\Lambda}(\infty)\rangle \\ &= \textstyle\sum_{i=2}^3 R_i(E^{-\bar\alpha})\,\langle\phi_\Lambda^{\bar\lambda}(z)\phi_{\Lambda_1}^{\bar\lambda+\bar\alpha}(0)\phi_{\Lambda_1^+}^{-\bar\lambda}(1)\phi_{\Lambda^+}^{-\bar\Lambda}(\infty)\rangle \\ &= c_{\bar\lambda}\,F^{\bar\lambda}+c_{\bar\lambda+\bar\alpha}\,F^{\bar\lambda+\bar\alpha} \end{aligned} \tag{3.5.19}$$

with numerical constants $c_{\bar\lambda}$, $c_{\bar\lambda+\bar\alpha}$ (here $R_1=R_{\bar\Lambda}$, $R_2=R_{\bar\Lambda_1}$, $R_3=R_{\bar\Lambda_1^+}$, $R_4=R_{\bar\Lambda^+}$). In other words,

$$F^{\bar\lambda+\bar\alpha}\propto F^{\bar\lambda} \tag{3.5.20}$$

for any such root $\bar\alpha$. In fact, the number of components which remains independent after the implementation of (3.5.20) for all allowed roots $\bar\alpha$ is nothing but the number of singlets contained in the $\mathfrak{g}$-Kronecker product $R_{\bar\Lambda}\times R_{\bar\Lambda_1}\times R_{\bar\Lambda_1^+}\times R_{\bar\Lambda^+}$, and hence the dimension of the matrices in the Knizhnik–Zamolodchikov equation obeyed by $F(z)$, as discussed in the previous section (p. 187).

In the present case, the Gepner–Witten equation (3.5.17) reads

$$\left(\frac{R_{\bar\Lambda_1}(E^{\bar\theta})}{z}+\frac{R_{\bar\Lambda_1^+}(E^{\bar\theta})}{z-1}\right)\langle\phi_\Lambda^{\bar\lambda}(z)\phi_{\Lambda_1}^{\bar\lambda}(0)\phi_{\Lambda_1^+}^{-\bar\lambda-\bar\theta}(1)\phi_{\Lambda^+}^{-\bar\Lambda}(\infty)\rangle=0. \tag{3.5.21}$$

Evaluating the action of the generators, this yields

$$F^{\bar\lambda+\bar\theta}=\frac{z}{1-z}\,F^{\bar\lambda}. \tag{3.5.22}$$

(In the general case (3.5.3) of the Gepner–Witten equation, the corresponding identity for the four-point function reads

$$\sum_{j=0}^{m} c_j\,z^{-j}(1-z)^{j-m}\,F^{\bar\lambda+j\bar\theta}(z)=0, \tag{3.5.23}$$

where $m\equiv k^\vee-(\bar\Lambda,\bar\theta^\vee)+1$.) Actually, the formula (3.5.22) remains valid if $\bar\theta$ is replaced by any other positive root $\bar\alpha$ satisfying $(\bar\Lambda,\bar\alpha)=k$ because for any such root we can use the root $\bar\beta=\bar\theta-\bar\alpha$ in (3.5.20) (recall from section 1.4 (p. 35) that $\bar\beta$ is indeed a positive root) to obtain $F^{\bar\lambda+\bar\alpha}=F^{\bar\lambda+\bar\alpha+\bar\beta}=F^{\bar\lambda+\bar\theta}$, i.e.

$$F^{\bar\lambda+\bar\alpha}=\frac{z}{1-z}\,F^{\bar\lambda}. \tag{3.5.24}$$

Implementing (3.5.24), the number of independent components $F^{\bar\lambda}$ can be reduced considerably. An obvious question is whether it is possible to reduce the number of independent components this way to only one which then may be taken to be $F^{\bar\Lambda_1}$. Clearly this is possible iff any weight of the module $R_{\bar\Lambda_1}$ can be reached from the highest weight $\bar\Lambda_1$ by subtracting positive roots $\bar\alpha$ which can be used in (3.5.24). A necessary condition for this is that their span

$$\mathcal{M} := \mathrm{span}_F \{\bar\alpha > 0 \mid (\bar\Lambda, \bar\alpha) = k\} \tag{3.5.25}$$

is already the whole weight space. Since the weight space of $\bar{\mathfrak{g}}$ is r-dimensional, this means that we need

$$\dim \mathcal{M} = r. \tag{3.5.26}$$

To see when this is possible, we observe that the Dynkin labels $\bar\Lambda^i$ of $\bar\Lambda$ are non-negative (since $\bar\Lambda$ is a highest weight of a finite-dimensional $\bar{\mathfrak{g}}$-module), and that the same is true for the coefficients $\bar\beta_i$ of the expansion of $\bar\beta = \bar\theta - \bar\alpha$ in the basis of simple coroots; therefore $0 = (\bar\Lambda, \bar\beta) = \sum_{i=1}^r \bar\Lambda^i \bar\beta_i$ implies that $\bar\Lambda^i \bar\beta_i$ (no sum on i) must vanish for any $i = 1, \dots, r$. Now, owing to the defining property (1.4.22) of the fundamental $\bar{\mathfrak{g}}$-weights, the number $\tilde r$ of vanishing coefficients $\bar\Lambda^i$ is equal to the dimension of $\mathrm{span}\{\bar\beta > 0 | (\bar\Lambda, \bar\beta) = 0\}$, and because of $\bar\alpha = \bar\theta - \bar\beta$ (where also $\bar\beta = 0$ is allowed) the dimension of the latter space is one less than the dimension of $\mathcal{M}$. So we conclude

$$\dim \mathcal{M} = \tilde r + 1. \tag{3.5.27}$$

Hence the requirement (3.5.26) is fulfilled iff $\tilde r = r - 1$, i.e. iff only one Dynkin label of $\bar\Lambda$ does not vanish, or in other words, iff $\bar\Lambda$ is a multiple of a fundamental $\bar{\mathfrak{g}}$-weight,

$$\bar\Lambda = s\, \bar\Lambda_{(j)} \tag{3.5.28}$$

for some positive integer s and some $j \in \{1, \dots, r\}$. Inserting (3.5.28) into $(\bar\Lambda, \bar\alpha) = k$, we see that the jth component of $\bar\alpha$ in the basis of simple coroots has the value $\bar\alpha_j = k/\bar\Lambda^j = k/s$ (in particular $(k/s)(2/(\bar\alpha^{(j)}, \bar\alpha^{(j)}))$ must be an integer; also note that $s \le k^\vee$, due to $(\bar\Lambda_{(j)}, \bar\theta) \ge 1$). This is valid for all $\bar\alpha$ which can appear in (3.5.24), and so we learn that $F^{\bar\lambda}$ can be related to $F^{\bar\Lambda_1}$ via (3.5.24) and (3.5.20) iff

$$\bar\Lambda_1 - \bar\lambda = n_j \, \frac{k}{s} \, \frac{2}{(\bar\alpha^{(j)}, \bar\alpha^{(j)})} \, \bar\alpha^{(j)} + \sum_{\substack{i=1 \\ i \ne j}}^{r} n_i \, \bar\alpha^{(i)} \tag{3.5.29}$$

for some nonnegative integers n_i, $i = 1, \dots, r$. Since in particular n_j is an integer, there will be $(k/s)(2/(\bar\alpha^{(j)}, \bar\alpha^{(j)}))$ distinct classes of weights for which the components $F^{\bar\lambda}$ cannot be related through (3.5.24) and (3.5.20). Hence it is possible to express all components of F through the one for the highest component iff

$$\frac{2}{(\bar\alpha^{(j)},\bar\alpha^{(j)})}\,\frac{k}{s}\equiv\frac{(\bar\theta,\bar\theta)}{(\bar\alpha^{(j)},\bar\alpha^{(j)})}\,\frac{k^\vee}{s}=1. \tag{3.5.30}$$

Because of $s\le k^\vee$ and $(\bar\alpha,\bar\alpha)\le(\bar\theta,\bar\theta)$, this means that we need both

$$(\bar\alpha^{(j)},\bar\alpha^{(j)})=(\bar\theta,\bar\theta), \tag{3.5.31}$$

i.e. $\bar\alpha^{(j)}$ must be a long root, and $s=k^\vee$, i.e.

$$\bar\Lambda=k^\vee\,\bar\Lambda_{(j)}. \tag{3.5.32}$$

Note that (3.5.31) and the relation $(\bar\Lambda_{(j)},\bar\theta^\vee)=1$ which follows from $(\bar\Lambda,\bar\theta)=k$ and (3.5.32) together imply that the jth Coxeter label is one, $a_j=1$. Thus the fundamental $\bar{\mathfrak g}$-weights which can appear in (3.5.32) are precisely the cominimal fundamental weights of $\bar{\mathfrak g}$. Consequently, the fields ϕ_Λ which carry the highest weight module R_Λ with $\bar\Lambda$ obeying (3.5.31) and (3.5.32) are called *cominimal fields.*

We have thus derived a necessary condition for being able to relate all components of F to a single one through the identities (3.5.17) and (3.4.36). It is easy to see that this condition is also sufficient. Namely, for a cominimal fundamental weight the inner product with any positive root $\bar\alpha$ is either $(\bar\Lambda_{(j)},\bar\alpha)=0$ or $(\bar\Lambda_{(j)},\bar\alpha)=1$, and hence the construction discussed above can be used. This property of $(\bar\Lambda_{(j)},\bar\alpha)$ is valid because $(\bar\Lambda_{(j)},\bar\theta)=1$ and $\bar\theta-\bar\alpha$ is a positive root, so that $b_j=0$ or $b_j=1$ in the expansion $\alpha=\sum_{i=1}^r b_i\alpha^{(i)}$ of α with respect to the basis of simple roots.

The four-point functions for cominimal fields turn out to be remarkably simple. Namely, any component (3.4.35) of a four-point function containing at least one cominimal field is of the form

$$F^{\bar\lambda_1\bar\lambda_2\bar\lambda_3\bar\lambda_4}\propto z^a(1-z)^b \tag{3.5.33}$$

with appropriate real numbers a and b. This is seen most easily in the special case $\bar\Lambda_3=\bar\Lambda^+,\ \bar\Lambda_2=\bar\Lambda_1^+$ (compare (3.5.18); the general case works analogously, but complicates the notation considerably). Combining (3.5.20) and (3.5.24), it follows that any component (3.5.18) can be expressed through the highest one as

$$F^{\bar\lambda}(z)=\left(\frac{z}{1-z}\right)^{(\bar\Lambda,\bar\lambda-\bar\Lambda_1)/k}F^{\bar\Lambda_1}(z). \tag{3.5.34}$$

To prove the statement, it is therefore sufficient to establish it for the highest component

$$F^{\bar\Lambda_1}(z)=\langle\phi_\Lambda^{\bar\Lambda}(z)\phi_{\Lambda_1}^{\bar\Lambda_1}(0)\phi_{\Lambda_1^+}^{-\bar\Lambda_1}(1)\phi_{\Lambda^+}^{-\bar\Lambda}(\infty)\rangle. \tag{3.5.35}$$

Now owing to (3.5.24), the Knizhnik–Zamolodchikov equation for any correlation function containing cominimal fields becomes very simple. In the case of the highest component, it even simplifies further. Using a Cartan-Weyl basis (so that the Killing form $\bar\kappa$ takes the form (1.4.10)) and employing the identities (1.6.6) and (1.6.25) for the action of the $\bar{\mathfrak g}$-generators, the Knizhnik–Zamolodchikov equation (3.4.26) for the component $F^{\bar\Lambda_1}$ reads

$$\tfrac12\,(\bar\theta,\bar\theta)(k^\vee+g^\vee)\,\partial F^{\bar\Lambda_1}(z)=\left(\frac1z-\frac1{z-1}\right)(\bar\Lambda,\bar\Lambda_1)\,F^{\bar\Lambda_1}(z) + \frac{1}{z-1}\sum_{\bar\alpha>0}\sqrt{(\bar\Lambda,\bar\alpha)}\sqrt{(\bar\Lambda_1,\bar\alpha)}\,\langle\phi_\Lambda^{\bar\Lambda-\bar\alpha}(z)\phi_{\Lambda_1}^{\bar\Lambda_1}(0)\phi_{\Lambda_1^+}^{\bar\alpha-\bar\Lambda_1}(1)\phi_{\Lambda^+}^{-\bar\Lambda}(\infty)\rangle. \tag{3.5.36}$$

Using also the $\bar{\mathfrak g}$-invariance (3.4.36), this may be rewritten as

$$\tfrac12\,(\bar\theta,\bar\theta)(k^\vee+g^\vee)\,\partial F^{\bar\Lambda_1}(z) = \left[\frac1z\,(\bar\Lambda,\bar\Lambda_1)+\frac{1}{1-z}\,(\bar\Lambda+2\bar\rho,\bar\Lambda_1)\right]F^{\bar\Lambda_1}(z) + \frac{1}{1-z}\sum_{\bar\alpha>0}(\bar\Lambda_1,\bar\alpha)\,F^{\bar\Lambda_1-\bar\alpha}(z)\,. \tag{3.5.37}$$

This result is valid also in the non-cominimal case. Specializing now to the case where ϕ_Λ is a cominimal field, one can employ (3.5.24) to reduce this to an equation involving only the component $F^{\bar\Lambda_1}$:

$$\tfrac12\,(\bar\theta,\bar\theta)(k^\vee+g^\vee)\,\partial F^{\bar\Lambda_1}(z) = \left[\frac1z\,(\bar\Lambda+2\bar\rho-2\tilde\rho,\bar\Lambda_1)+\frac{1}{1-z}\,(\bar\Lambda+2\bar\rho,\bar\Lambda_1)\right]F^{\bar\Lambda_1}(z). \tag{3.5.38}$$

Here $\tilde\rho$ is defined as

$$\tilde\rho=\tfrac12\sum_{\substack{\bar\alpha>0\\(\bar\Lambda,\bar\alpha)=0}}\bar\alpha. \tag{3.5.39}$$

(Thus $\tilde\rho$ is the Weyl vector of the regular subalgebra $\tilde{\mathfrak g}\subset\bar{\mathfrak g}$ that is obtained by removing the node corresponding to the relevant cominimal fundamental weight $\bar\Lambda_{(i)}=\bar\Lambda/k^\vee$ from the Dynkin diagram of $\bar{\mathfrak g}$.) The solution of (3.5.38) has indeed the promised form:

$$F^{\bar\Lambda_1}(z)\propto z^{(\bar\Lambda+2\bar\rho-2\tilde\rho,\bar\Lambda_1)/(k+C_{\bar\theta})}(1-z)^{-(\bar\Lambda+2\bar\rho,\bar\Lambda_1)/(k+C_{\bar\theta})}. \tag{3.5.40}$$

With the help of the formula (3.4.20) for the conformal dimensions of primary WZW fields, this may also be written as

$$F^{\bar\Lambda_1}(z)\propto z^{-(2\tilde\rho,\bar\Lambda_1)/(k+C_{\bar\theta})}\left(\frac{z}{1-z}\right)^{\Delta_\Lambda+\Delta_{\Lambda_1}-\Delta_{\Lambda'}} \tag{3.5.41}$$

with $\bar{\Lambda}' := \bar{\Lambda} - \bar{\Lambda}_1$ (if $\bar{\Lambda}'$ is not a dominant $\bar{\mathfrak{g}}$-weight, then this is only a formal definition, i.e. there is no corresponding primary field $\phi_{\Lambda'}$).

The four-point function looks particularly simple if $\Lambda_1 = \Lambda$. Then the highest component is

$$F^{\bar{\Lambda}}(z) \propto \left(\frac{z}{1-z}\right)^{2\Delta_\Lambda} \tag{3.5.42}$$

because $(\bar{\Lambda}, \bar{\rho}) = 0$ and because in this case $\bar{\Lambda}' = 0$ so that also $\Delta_{\Lambda'} = 0$.

A further remarkable property of cominimal fields is that their conformal dimensions are proportional to the level of the affine algebra. More precisely,

$$\Delta_{k^\vee \Lambda_{(j)}} = k^\vee \Delta_{\Lambda_{(j)}} \tag{3.5.43}$$

holds for cominimal fundamental weights $\Lambda_{(j)}$, but not for any non-cominimal fundamental weight. This follows with the help of the formula (3.4.20) for the conformal dimension from the corresponding property of quadratic Casimir eigenvalues,

$$C_{k^\vee \Lambda_{(j)}} = k^\vee C_{\Lambda_{(j)}} \cdot \frac{C_\theta + k}{C_\theta + (\bar{\theta}, \bar{\theta})/2}, \tag{3.5.44}$$

which may be checked case by case.

The above relations suggest that the level one cominimal fields are of fundamental importance. Indeed, it turns out that all other cominimal fields can be viewed as normal ordered products of level one cominimal fields. Namely, consider the $k^\vee$-fold direct product of a level one WZW theory with itself. The currents of the product theory are by definition the sum of the currents $J^a_{(\nu)}$ of the subtheories,

$$J^a(z) = \sum_{\nu=1}^{k^\vee} J^a_{(\nu)}(z), \tag{3.5.45}$$

and hence generate a level $k^\vee$ current algebra. The primary fields of the product theory are of the form $\Phi(z) = \phi_{(1)}(z) \otimes \phi_{(2)}(z) \otimes \ldots \otimes \phi_{(k^\vee)}(z)$ with $\phi_{(i)}$ denoting generically the primary fields of the ith level one WZW theory. Thus for every cominimal fundamental highest weight $\bar{\Lambda}$ of $\bar{\mathfrak{g}}$, the product theory contains in particular a set of $k^\vee$ distinct primary fields $\Phi_{(\nu)} \equiv (\phi_\Lambda)_{(\nu)}$, $\nu = 1, \ldots, k^\vee$, each carrying the module $R_{\tilde{\Lambda}}$ with highest weight $\Lambda = (\bar{\Lambda}, k, 0)$, namely

$$\Phi_{(\nu)}(z) = \mathbf{1} \otimes \ldots \otimes \mathbf{1} \otimes \phi_{\tilde{\Lambda}} \otimes \mathbf{1} \otimes \ldots \otimes \mathbf{1} \tag{3.5.46}$$

($\phi_{\tilde{\Lambda}}$ at the νth place) with $\tilde{\Lambda} = (\bar{\Lambda}, \frac{1}{2}(\bar{\theta}, \bar{\theta}), 0)$. A primary field carrying the module with highest weight $\Lambda' := k^\vee \tilde{\Lambda} = (k^\vee \bar{\Lambda}, k, 0)$ can then be obtained by forming the $k^\vee$-fold normal ordered product of the fields

$\Phi_{(\nu)}$ and contracting with an appropriate Clebsch–Gordan coefficient of $\bar{\mathfrak{g}}$:

$$\Phi_{\Lambda'}^{\bar\lambda'}(z) := \sum_{\bar\lambda_1,\dots,\bar\lambda_{k^\vee}} C^{\bar\lambda_1\bar\lambda_2\dots\bar\lambda_{k^\vee};\bar\lambda'} :\Phi_{(1)}^{\bar\lambda_1}(z)\Phi_{(2)}^{\bar\lambda_2}(z)\dots\Phi_{(k^\vee)}^{\bar\lambda_{k^\vee}}(z): \qquad (3.5.47)$$

with $C^{\bar\lambda_1\bar\lambda_2\dots\bar\lambda_{k^\vee};\bar\lambda'} \equiv C^{\bar\lambda_1\bar\lambda_2\dots\bar\lambda_{k^\vee};\bar\lambda'}_{\bar\Lambda\bar\Lambda\dots\bar\Lambda;k^\vee\bar\Lambda}$ the Clebsch–Gordan coefficient describing how the module $R_{\bar\Lambda'}$ is obtained in the $k^\vee$-fold Kronecker product of the module $R_{\bar\Lambda}$. (The normal ordering in (3.5.47) is rather trivial because the corresponding radially ordered operator product does not contain any singular terms.)

By construction, the primary field (3.5.47) is a cominimal field. Its correlation functions must therefore be of the simple form of (3.5.40) etc. Indeed, rewriting (3.5.47) as

$$\Phi_{\Lambda'}^{\bar\lambda'}(z) := \sum C^{\bar\lambda_1\bar\lambda_2\dots\bar\lambda_{k^\vee};\bar\lambda'}\, \phi_{\bar\Lambda}^{\bar\lambda_1}(z)\otimes\phi_{\bar\Lambda}^{\bar\lambda_2}(z)\otimes\dots\otimes\phi_{\bar\Lambda}^{\bar\lambda_{k^\vee}}(z), \qquad (3.5.48)$$

it is clear that the correlation functions of $\Phi_{\Lambda'}$ factorize into $k^\vee$-fold products of correlation functions of the level one cominimal field $\phi_{\bar\Lambda}$. For example, for the highest component $\Phi_{\Lambda'}^{\bar\Lambda'}$ (for which $C^{\bar\Lambda\bar\Lambda\dots\bar\Lambda;\bar\Lambda'}=1$ is the only Clebsch–Gordan coefficient contributing to the sum (3.5.47)), one has

$$\begin{aligned} F^{\bar\Lambda'}(z) &\equiv \langle\Phi_{\Lambda'}^{\bar\Lambda'}(z)\Phi_{\Lambda'^+}^{-\bar\Lambda'}(0)\Phi_{\Lambda'^+}^{-\bar\Lambda'}(1)\Phi_{\Lambda'}^{\bar\Lambda'}(\infty)\rangle \\ &= \Big(\langle\phi_{\bar\Lambda}^{\bar\Lambda}(z)\phi_{\bar\Lambda^+}^{-\bar\Lambda}(0)\phi_{\bar\Lambda^+}^{-\bar\Lambda}(1)\phi_{\bar\Lambda}^{\bar\Lambda}(\infty)\rangle\Big)^{k^\vee}. \end{aligned} \qquad (3.5.49)$$

Using the results (3.5.33) and (3.5.43), this becomes

$$F^{\bar\Lambda'}(z) \propto \left[\left(\frac{z}{1-z}\right)^{2\Delta_{\bar\Lambda}}\right]^{k^\vee} = \left(\frac{z}{1-z}\right)^{2\Delta_{\Lambda'}}. \qquad (3.5.50)$$

This shows that the highest component of the four-point function is correctly reproduced; for the other components $F^{\bar\lambda'}$, the same statement follows if one uses in addition that $(\bar\Lambda',\bar\Lambda'-\bar\lambda') = k^\vee\sum_{i=1}^{k^\vee}(\bar\Lambda,\bar\Lambda-\bar\lambda_i)$ (this is a consequence of the fact that $C^{\bar\lambda_1\dots\bar\lambda_{k^\vee};\bar\lambda'}\neq 0$ requires that $\bar\lambda_1+\dots+\bar\lambda_{k^\vee}=\bar\lambda'$).

For the case of selfconjugate modules, the results (3.5.34) and (3.5.42) also imply that $F^{-\bar\Lambda}\propto(F^{\bar\Lambda})^{-1}$, and from this one concludes that

$$\Delta_\Lambda = \frac{1}{2k}\,(\bar\Lambda,\bar\Lambda). \qquad (3.5.51)$$

Thus in particular

$$\Delta_{\Lambda_{(j)}} = \frac{(\bar\Lambda_{(j)},\bar\Lambda_{(j)})}{(\bar\theta,\bar\theta)}, \qquad (3.5.52)$$

and hence the relation $\Delta_{k^\vee\Lambda_{(j)}} = k^\vee \Delta_{\Lambda_{(j)}}$ (see (3.5.43)) is reobtained.

The level one cominimal fields and their conformal dimensions are listed in table (3.5.53). Apart from the weights listed there, also of course in all cases the zero weight corresponding to the identity primary field is allowed.

$\mathfrak{g}$	cominimal $\bar\Lambda_{(i)}$	Δ
A_r	$\bar\Lambda_{(i)},\ i=1,\ldots,r$	$\frac{j(r+1-j)}{2(r+1)}$
B_r	$\bar\Lambda_{(1)}$	$\frac{1}{2}$
C_r	$\bar\Lambda_{(r)}$	$\frac{r}{4}$
D_r	$\bar\Lambda_{(1)}$	$\frac{1}{2}$
	$\bar\Lambda_{(r-1)},\ \bar\Lambda_{(r)}$	$\frac{r}{8}$
E_6	$\bar\Lambda_{(1)},\ \bar\Lambda_{(5)}$	$\frac{2}{3}$
E_7	$\bar\Lambda_{(6)}$	$\frac{3}{4}$
$E_8,\ F_4,\ G_2$	–	–

(3.5.53)

So far we have used only the identities (3.5.17) and (3.4.36) in order to relate all components of the four point function to a single one. If there are further identities that can be used to this end, of course further fields in addition to the cominimal ones may be allowed. Two types of such further identities are conceivable. First, there may be further null vectors in the theory. This is only possible if there is an enlarged symmetry algebra which extends the affine algebra of the WZW theory; examples of this possibility will be encountered in the sections 3.7 and 3.8. Second, it may happen that even without a higher symmetry, the several classes $\{F^{\bar\lambda}\}$ of components which are not related by (3.5.20) or (3.5.24) nevertheless are accidentally equal to each other up to powers $z^a(1-z)^b$ (e.g. all but one class may vanish). This seems not very plausible because it violates an intuitively reasonable naturality property, but in fact a single example of this phenomenon is known; the corresponding primary field possessing simple power-like four-point functions is the field ϕ_{3875} of the level two $E_8^{(1)}$ WZW theory which was discussed above.

3.6 Free fermions

While the abstract definition of irreducible integrable highest weight modules as the irreducible quotients of Verma modules can be employed to deduce many important results, such as the Knizhnik–Zamolodchikov equation and the Gepner–Witten equation, there are also situations in which a more concrete realization of these modules is desirable. It turns out that one can indeed find such realizations, and that in terms of two-dimensional field theory they correspond to conformal field theories describing free fields. Currently three distinct types of such realizations are known, corresponding to the conformal field theory of free bosons, of free bosons and so-called superconformal ghosts, and of free fermions, respectively. The realization in terms of free bosons is known as the *vertex operator construction* and describes the level one cominimal modules. The realization in terms of free bosons and ghosts, the so-called *Wakimoto construction*, is much more general and in fact describes all integrable modules at arbitrary level as well as a very large class of non-integrable modules. For lack of space we do not describe here these two types of realizations, but only the one involving free fermions; the free boson construction will however be mentioned briefly in section 4.8 (p. 317), and the Wakimoto construction in section 5.8 (p. 386).

Consider a free, massless Majorana fermion in two space–time dimensions. This field has only two independent degrees of freedom, which can be chosen as Ψ_+ and Ψ_- such that in light-cone coordinates $x_\pm = x_0 \pm x_1$ the action of the field theory is

$$S = \tfrac{1}{2} \int \mathrm{d}^2 x \, (i\Psi_+ \tfrac{\partial}{\partial x_-} \Psi_+ + i\Psi_- \tfrac{\partial}{\partial x_+} \Psi_-). \tag{3.6.1}$$

The corresponding field equations read

$$\frac{\partial}{\partial x_-} \Psi_+ = 0 = \frac{\partial}{\partial x_+} \Psi_-, \tag{3.6.2}$$

with the general solution

$$\Psi_+ = \Psi_+(x_+), \quad \Psi_- = \Psi_-(x_-); \tag{3.6.3}$$

As a consequence, Ψ_+ and Ψ_- are completely decoupled degrees of freedom. Therefore we can in the following restrict our attention to a single component which we denote by $\Psi(x)$. Let us assume that there are several such fields Ψ^i, and that they transform into each other according to a finite-dimensional orthogonal representation $R_{\bar\Lambda}$ of a simple Lie algebra $\mathfrak{g}$. (In particular, the index i runs from 1 to $d_{\bar\Lambda}$; in the following for notational convenience we use upper indices for both the vectors of $R_{\bar\Lambda}$ and those of the conjugate module which is isomorphic to $R_{\bar\Lambda}$, and also write δ^{ij} for $\delta_i^{\ j}$.) It is then not difficult to see that the canonical anticommutation relations

$$\{\Psi^i(x), \Psi^j(y)\} = \delta^{ij} \delta(x-y) \tag{3.6.4}$$

imply that the composite fields

$$\tilde{J}^a(x) := \tfrac{1}{2} \sum_{i,j=1}^{d_\Lambda} :\Psi^i(x)(T^a_{(\Lambda)})_{ij}\Psi^j(x): \tag{3.6.5}$$

obey the current algebra commutation relations (3.1.15). (Here normal ordering is defined e.g. by a suitable point-splitting procedure; we do not discuss this in any detail because the corresponding calculations in the compact picture will be displayed explicitly below.) In order to go to the compact picture, we now require that the currents $\tilde{J}^a$ are periodic,

$$\tilde{J}^a(x+L) = \tilde{J}^a(x), \tag{3.6.6}$$

and go from $[0, L) \subset \mathbb{R}$ to the unit circle by defining $J^a(z) := (L/z)\tilde{J}^a(x)$ with $z = \exp(2\pi i x/L)$ (compare (3.1.13)). In order to have an expression analogous to (3.6.5) also on the circle, fermions on the circle have to be introduced via

$$\psi(z) := \sqrt{\frac{L}{z}}\,\Psi(x). \tag{3.6.7}$$

For compatibility of (3.6.5) with (3.6.6), two possible boundary conditions on the fermions Ψ can be imposed. Namely, they can be either periodic or antiperiodic; in the former case they are called *Ramond fermions*, and in the latter case *Neveu–Schwarz fermions*, shortly denoted as *R*- and *NS*-fermions, respectively (in the language of ordinary field theory, the two possible boundary conditions are due to the fact that only local *observables* – such as bilinears in the fermions, but not the fermions themselves – are expected to be well defined on S^1). Thus

$$\Psi(x+L) = \begin{cases} -\Psi(x) & (NS) \\ +\Psi(x) & (R). \end{cases} \tag{3.6.8}$$

In terms of the circle, these boundary conditions translate into

$$\psi(\mathrm{e}^{2\pi i}z) = \begin{cases} +\psi(z) & (NS) \\ -\psi(z) & (R) \end{cases} \tag{3.6.9}$$

(notice the additional minus sign coming from $\sqrt{z}$ in (3.6.7) which affects a change in the periodicity properties). Expanding the fermions into a Fourier–Laurent series, we have

$$\begin{aligned} \Psi(x) &= L^{-1/2} \sum_s \mathrm{e}^{-2\pi i s x/L}\,\psi_s, \\ \psi(z) &= \sum_s z^{-s-1/2}\,\psi_s, \end{aligned} \tag{3.6.10}$$

where

$$s \in \begin{cases} \mathbb{Z}+\frac{1}{2} & (NS) \\ \mathbb{Z} & (R). \end{cases} \tag{3.6.11}$$

The canonical anticommutation relations (3.6.4) imply that the Fourier-Laurent coefficients ψ_s obey

$$\{\psi^i_s, \psi^j_t\} = \delta^{ij}\delta_{s+t,0}. \tag{3.6.12}$$

In terms of the fields $\psi(z)$ (with z extended from the circle to $\mathbb{C}\setminus\{0\}$), the anticommutation relations translate into radially ordered operator products

$$\mathcal{R}(\psi^i(z)\psi^j(w)) = \begin{cases} \psi^i(z)\psi^j(w) & \text{for } |z|>|w| \\ -\psi^i(w)\psi^j(z) & \text{for } |w|>|z|. \end{cases} \tag{3.6.13}$$

Their explicit form is obtained by appropriate deformations of integration contours just as e.g. in (3.1.7), now with an additional minus sign for $|z|<|w|$ owing to the use of anticommutators rather than commutators (compare p. 158; in the following we will again suppress the radial ordering symbol $\mathcal{R}$ just as in the case of bosonic fields). One finds

$$\underbrace{\psi^i(z)\,\psi^j(w)} = \begin{cases} -\dfrac{1}{z-w}\,\delta^{ij} & (NS) \\ -\dfrac{1}{z-w}\,\delta^{ij}\cdot\dfrac{1}{2}\left(\sqrt{\dfrac{z}{w}}+\sqrt{\dfrac{w}{z}}\right) & (R). \end{cases} \tag{3.6.14}$$

Here we display only the contraction, i.e. the singular terms in the operator product. However, whereas in the Neveu–Schwarz case the singular terms can be identified unambigously (compare the definition (3.1.18) of the contraction) in the Ramond case there is an ambiguity due to the branch cuts of $\sqrt{z}$ and $\sqrt{w}$ which must be present as a consequence of the branch cut in $\psi(z)$. The result (3.6.14) arises from summations of the form

$$\frac{1}{w}\sum_{s>0}\left(\frac{z}{w}\right)^{s-1/2} = \begin{cases} -\frac{1}{z-w} & \text{for } s\in\mathbb{Z}+\frac{1}{2} \\ -\frac{1}{z-w}\sqrt{\frac{z}{w}} & \text{for } s\in\mathbb{Z}. \end{cases} \tag{3.6.15}$$

The validity of the result (3.6.14) is easily checked by inserting it into

$$\psi_s = \tfrac{1}{2\pi i}\oint_0 \mathrm{d}z\, z^{s-1/2}\,\psi(z), \tag{3.6.16}$$

which leads to

$$\{\psi^i_s, \psi^j_t\} = \tfrac{1}{2\pi i}\oint_0 \mathrm{d}z\,\tfrac{1}{2\pi i}\oint_z \mathrm{d}w\, z^{s-1/2}w^{t-1/2}\,(-\delta^{ij}\,(z-w)^{-1}) \tag{3.6.17}$$

for Neveu–Schwarz fermions, and

$$\{\psi^i_s, \psi^j_t\} = \frac{1}{2\pi i} \oint_0 \mathrm{d}z \, \frac{1}{2\pi i} \oint_z \mathrm{d}w \qquad \frac{1}{2}\,(z^s w^{t-1} + z^{s-1} w^t)\,(-\delta^{ij}\,(z-w)^{-1}) \tag{3.6.18}$$

for Ramond fermions. Performing the integrations then indeed yields the anticommutation relations (3.6.12). Note that in the Ramond case, the correct result is also obtained if in the bracket on the right hand side of (3.6.14) any other (suitably normalized) linear combination of $\sqrt{z/w}$ and $\sqrt{w/z}$ is used; the symmetric linear combination chosen there has the advantage that, just as in the Neveu–Schwarz case, the contraction simply changes sign upon interchanging z and w. Moreover, this choice leads to a simple formula for the square of (3.6.14):

$$\Big(\underbrace{\psi(z)\,\psi(w)}\Big)^2 = (z-w)^{-2} + \begin{cases} 0 & (NS) \\ (4zw)^{-1} & (R). \end{cases} \tag{3.6.19}$$

Next we want to show that the currents

$$J^a(z) = \tfrac{1}{2} \textstyle\sum_{i,j} :\psi^i(z) (T^a_{(\bar\Lambda)})_{ij} \psi^j(z): \tag{3.6.20}$$

possess the operator product (3.1.10) that identifies them as the fields generating an affine Lie algebra. First consider Neveu–Schwarz fermions. We use (3.1.22) as the definition of the normal ordering in (3.6.20) and then employ the Wick rule (3.1.35) which for this free field theory is completely equivalent to the ordinary quantum field theoretical Wick theorem. Then straightforward manipulations show that (3.6.14) implies

$$\underbrace{J^a(z)\,\psi^i(w)} = (z-w)^{-1} \textstyle\sum_j (T^a_{(\bar\Lambda)})_{ij}\, \psi^j(w), \tag{3.6.21}$$

and

$$\underbrace{J^a(z)\,J^b(w)} = \tfrac{1}{2} \textstyle\sum_{i,j,l} (T^a_{(\bar\Lambda)})_{li} (T^b_{(\bar\Lambda)})_{jl} \cdot \big[(z-w)^{-1} (:\psi^j(w)\psi^i(w): - :\psi^i(w)\psi^j(w):) + (z-w)^{-2}\,\delta^{ij}\big]. \tag{3.6.22}$$

The latter formula can be rewritten as

$$\underbrace{J^a(z)\,J^b(w)} = (z-w)^{-2} \bar\kappa^{ab}\, k - (z-w)^{-1} f^{ab}{}_c\, J^c(w). \tag{3.6.23}$$

Here the number k is defined through $k\,\bar\kappa^{ab} = \frac{1}{2}\,\mathrm{tr}(R_{\bar\Lambda}(T^a) R_{\bar\Lambda}(T^b))$, i.e. (compare (1.6.44)) $k = \frac{1}{2} I_{\bar\Lambda}$ with $I_{\bar\Lambda}$ the Dynkin index of $R_{\bar\Lambda}$, or in other words

$$k^\vee = \frac{1}{(\bar\theta, \bar\theta)}\, I_{\bar\Lambda}. \tag{3.6.24}$$

For Ramond fermions, the calculation runs completely parallel. More precisely, one first gets (3.6.21) multiplied by $\frac{1}{2}(\sqrt{z/w}+\sqrt{w/z})$, but expanding the square roots then gives back (3.6.21) up to irrelevant regular terms. Also note that the Wick rule (3.1.35) is still valid because the integration contour does not meet the square root branch cuts which arise according to (3.6.14) (similarly, we may use either of the normal ordering prescriptions (3.1.22) or (3.1.39)).

Thus what we have obtained are indeed the current algebra relations (3.1.10), albeit specialized to representations with level equal to $1/(\bar{\theta},\bar{\theta})$ times the Dynkin index of $R_{\bar{\Lambda}}$. Of course we want to act with these currents on fields corresponding to a highest weight module with highest weight $\Lambda = (\bar{\Lambda}, k, 0)$. Since the state space of the theory of free fermions (the usual Fock space of a free field theory) is, without any truncation, a Hilbert space, this highest weight module must in fact be unitary. Now recall from section 1.6 (p. 59) that $I_{\bar{\Lambda}}/(\bar{\theta},\bar{\theta})$ is an integer for any orthogonal module $R_{\bar{\Lambda}}$; hence we are dealing with a unitary highest weight module at positive integer level, which means that it is the irreducible integrable highest weight module R_Λ. The fermions $\psi^i(z)$ correspond to the highest weight vector (and its horizontal descendants) of this module, while the (not purely horizontal) descendants are obtained as suitable normal ordered products of ψ^i with the currents.

Thus we have found a construction realizing irreducible integrable highest weight modules for arbitrary untwisted affine Lie algebras, at any level which is equal to the Dynkin index of a finite-dimensional module of the horizontal subalgebra. The above discussion was for the case of irreducible orthogonal $\bar{\mathfrak{g}}$-modules with $\bar{\mathfrak{g}}$ simple; but direct sums of irreducible modules and the case of semisimple $\bar{\mathfrak{g}}$ can be treated analogously. Now recall that the Dynkin index is additive when forming direct sums of modules. As a consequence, the identification of the level with the Dynkin index means that most, albeit not all, integer levels can be reached via the above construction. E.g. the defining module of so_n has $I_{\bar{\Lambda}}/(\bar{\theta},\bar{\theta}) = 1$ so that for so_n all integer levels can be obtained. In contrast, for $\bar{\mathfrak{g}} = E_8$ the lowest possible level is $k^\vee = 30$; this value is obtained with fermions in the adjoint module (for arbitrary simple $\bar{\mathfrak{g}}$, fermions in the adjoint module give a level equal to the dual Coxeter number of $\bar{\mathfrak{g}}$).

Free fermions carrying non-selfconjugate $\bar{\mathfrak{g}}$-modules can be treated by considering Dirac fermions rather than Majorana fermions. By definition, Dirac fermions possess twice the number of degrees of freedom as Majorana fermions, i.e. in addition to the field ψ there is an independent primary field ψ^+, and the elementary operator products are

$$\underbrace{\psi(z)\,\psi(w)} = 0 = \underbrace{\psi^+(z)\,\psi^+(w)} \tag{3.6.25}$$

and

$$\underbrace{\psi(z)\,\psi^+(w)} = -(z-w)^{-1} \tag{3.6.26}$$

(in the Neveu–Schwarz case, and analogously in the Ramond case). In the case of our interest, $\psi = \psi_i$ carries an index of some (non-selfconjugate) $\mathfrak{g}$-module $R_{\bar\Lambda}$, and $\psi^+ = \psi^{+i}$ a corresponding index of the conjugate module $R_{\bar\Lambda^+}$. Thus the contraction (3.6.26) reads more explicitly

$$\underbrace{\psi_i(z)\,\psi^{+j}(w)} = -(z-w)^{-1}\,\delta_i^{\;j}\,, \tag{3.6.27}$$

while (3.6.25) may be viewed as a consequence of the absence of an invariant tensor with two upper (or two lower) indices for $R_{\bar\Lambda}$. The currents $J^a(z)$ are now obtained as

$$\begin{aligned} J^a(z) &= \tfrac12\left(:\psi^{+i}(z)(R_{\bar\Lambda}(T^a))_i^{\;j}\psi_j(z) + \psi_i(z)(R_{\bar\Lambda^+}(T^a))^i{}_j\psi^{+j}(z):\right) \\ &= :\psi^{+i}(z)(T^a_{(\bar\Lambda)})_{ij}\psi_j(z): \end{aligned} \tag{3.6.28}$$

Again it is straightforward to check that these fields generate an affine Lie algebra, now with the level appropriate to the reducible module $R_{\bar\Lambda}\oplus R_{\bar\Lambda^+}$,

$$k^\vee = \frac{1}{(\bar\theta,\bar\theta)}\,(I_{\bar\Lambda}+I_{\bar\Lambda^+}) = \frac{2}{(\bar\theta,\bar\theta)}\,I_{\bar\Lambda}. \tag{3.6.29}$$

Again the level is always an integer. In addition, however, we can now also define the composite field

$$J^0(z) = :\psi_j(z)\psi^{+j}(z): \tag{3.6.30}$$

(the corresponding field in the Majorana case vanishes identically owing to the antisymmetry of the operator product of fermions). This field obeys

$$\underbrace{J^0(z)\,J^0(w)} = d_{\bar\Lambda}\,(z-w)^{-2} \tag{3.6.31}$$

as well as

$$\underbrace{J^0(z)\,J^a(w)} = 0. \tag{3.6.32}$$

Thus the currents $J^a(z)$ and $J^0(z)$ generate a Kac-Moody algebra which is the direct sum $\mathfrak{g}\oplus\hat u_1$ of the affine algebra $\mathfrak{g}$ and the Heisenberg algebra $\hat u_1$.

The same construction works for Dirac fermions carrying a symplectic representation of $\bar{\mathfrak{g}}$. In this case, however, it is even possible to define an additional $A_1^{(1)}$ current algebra (containing the above $\hat u_1$). This is related to the fact that the representation matrices for a symplectic module $R_{\bar\Lambda}$ can be written as $(d_{\bar\Lambda}/2\times d_{\bar\Lambda}/2)$-matrices with quaternionic entries (recall that symplectic modules are even-dimensional); these quaternions can be represented as real 4×4-matrices, and the construction then follows with the help of the isomorphism $so_4 \cong so_3 \oplus so_3$.

As a free field theory, the theory of free fermions is of course conformally invariant. For Neveu–Schwarz-type Majorana fermions carrying an irreducible $\mathfrak{g}$-module, the energy momentum tensor, obtained according to the standard field theory prescription (compare p. 170), reads

$$T(z) = \tfrac{1}{2} \textstyle\sum_i : \psi^i(z) \partial \psi^i(z) : . \tag{3.6.33}$$

Using the Wick rule (3.1.35), it then follows immediately that

$$\begin{aligned} \underbrace{\psi^j(z)\, T(w)} &= \tfrac{1}{2} \textstyle\sum_i [\underbrace{\psi^j(z) \psi^i(w)}\, \partial \psi^i(w) + \underbrace{(\partial \psi^i(w)) \psi^j(z)}\, \psi^i(w) \\ &= \tfrac{1}{2} [-(z-w)^{-1} \partial \psi^j(w) + (z-w)^{-2} \psi^j(w)] \\ &= \tfrac{1}{2} [(z-w)^{-2} \psi^j(w) - (z-w)^{-1} \partial \psi^j(w)], \end{aligned} \tag{3.6.34}$$

or, what is the same,

$$\underbrace{T(z)\, \psi^j(w)} = \tfrac{1}{2} (z-w)^{-2} \psi^j(w) + (z-w)^{-1} \partial \psi^j(w). \tag{3.6.35}$$

Comparing this with the general formula (3.3.18), we see that ψ is a primary conformal field, with conformal dimension

$$\Delta(\psi) = \tfrac{1}{2}. \tag{3.6.36}$$

Note that this equals the canonical dimension of ψ; also, this value could have been expected from the expressions (3.6.20) and (3.6.33) of J and T through ψ because the currents and the energy–momentum tensor have conformal dimension $\Delta(J) = 1$, $\Delta(T) = 2$ (and because $\Delta(\partial \phi) = \Delta(\phi) + 1$). It must be noted, however, that generically the conformal dimension of a normal ordered product is different from the sum of the dimensions of its factors; that the dimensions just add up is a special situation which is typically encountered only for free fields.

It is also straightforward to compute the operator product of the energy-momentum tensor with itself; using the Wick rule together with (3.6.34), one has

$$\begin{aligned} \underbrace{T(z)\, T(w)} = \tfrac{1}{2} \tfrac{1}{2\pi i} \oint_w \mathrm{d}y\ (y-w)^{-1} & \\ \cdot \Big\{ \tfrac{1}{2} (z-y)^{-2} \psi^i(y) \partial \psi^i(w) &+ (z-y)^{-1} \partial \psi^i(y) \partial \psi^i(w) \\ + \partial_w [\tfrac{1}{2} (z-w)^{-2} \psi^i(y) \psi^i(w)] &+ \partial_w [(z-w)^{-1} \psi^i(y) \partial \psi^i(w)] \Big\} . \end{aligned} \tag{3.6.37}$$

Upon integration, the terms in the second line lead directly to normal ordered products, while for the terms in the first line one has to insert the ψ-ψ operator product (3.6.14) and expand $(z-y)^{-1}$ around $z=w$; this yields

$$\begin{aligned}\underbrace{T(z)\,T(w)} = \tfrac{1}{2}\textstyle\sum_i \Big\{ & \tfrac{1}{2}\,(z-w)^{-2} : \psi^i(w)\partial\psi^i(w) : -\tfrac{3}{2}\,d_{\bar\Lambda}\,(z-w)^{-4} \\ & + (z-w)^{-1} : \partial\psi^i(w)\partial\psi^i(w) : +2\,d_{\bar\Lambda}\,(z-w)^{-4} \\ & - (z-w)^{-3} : \psi^i(w)\psi^i(w) : +\tfrac{1}{2}\,(z-w)^{-2} : \psi^i(w)\partial\psi^i(w) : \\ & +(z-w)^{-2} : \psi^i(w)\partial\psi^i(w) : +(z-w)^{-1} : \psi^i(w)\partial^2\psi^i(w) : \Big\}\end{aligned} \tag{3.6.38}$$

(the four lines correspond to the four terms in the curly bracket in (3.6.37)). Now it follows directly from (3.6.14) and (3.1.44) that

$$: \{\psi^i(z), \psi^j(z)\} : = 0, \tag{3.6.39}$$

and hence

$$: \psi^i(z)\psi^i(z) : = 0 = : \partial\psi^i(z)\partial\psi^i(z) :, \tag{3.6.40}$$

and so we finally get

$$\underbrace{T(z)\,T(w)} = \tfrac{1}{4}\,d_{\bar\Lambda}\,(z-w)^{-4} + 2\,(z-w)^{-2}\,T(w) + (z-w)^{-1}\,\partial T(w). \tag{3.6.41}$$

Thus comparison with (3.2.9) shows that $T(z)$ generates the Virasoro algebra (i.e. indeed is the energy-momentum tensor of a conformal field theory), with the central charge fixed to half the number of fermions,

$$c = \tfrac{1}{2}\,d_{\bar\Lambda}. \tag{3.6.42}$$

The generalization to the case of reducible and non-selfconjugate $\mathfrak{g}$-modules and to non-simple $\mathfrak{g}$ is also straightforward. The energy-momentum tensor is a sum with each term being of the form discussed above. In particular the c-values just add up; thus in the case of direct sums of orthogonal modules, and in the case of semisimple $\mathfrak{g}$, c is equal to the total number of Majorana fermions; similarly in the case of symplectic or non-selfconjugate modules, c is given by twice the number of Dirac fermions.

Let us also describe, for the case of orthogonal $R_{\bar\Lambda}$, the energy-momentum tensor of Ramond fermions. If we define $T(z)$ as in (3.6.33), we can again immediately derive the operator product (3.6.34), using now (3.1.39) as the normal ordering prescription. Owing to this new prescription, we then get the relation (3.6.37) with an additional factor of

$$\frac{1}{2}\left(\sqrt{\frac{y}{w}} + \sqrt{\frac{w}{y}}\right) = 1 + \frac{1}{8}\left(\frac{y-w}{w}\right)^2 - \frac{1}{8}\left(\frac{y-w}{w}\right)^3 + \dots, \tag{3.6.43}$$

and with analogous factors arising in the expansion of the terms in its second line. Upon integration, the only effect of these factors is to change the term $2d_{\bar{\Lambda}}(z-w)^{-4}$ in the curly bracket of (3.6.38) to $d_{\bar{\Lambda}}[2(z-w)^{-4}+\frac{1}{8}w^{-2}(z-w)^{-2}-\frac{1}{8}w^{-3}(z-w)^{-1}]$ so that the right hand side of (3.6.41) gets replaced by

$$\begin{aligned}&\tfrac{1}{4}\,d_{\bar{\Lambda}}\,(z-w)^{-4}+2\,(z-w)^{-2}\,[\tfrac{1}{2}\,:\psi^i(w)\partial\psi^i(w):+\tfrac{1}{16}\,d_{\bar{\Lambda}}\,w^{-2}]\\ &\qquad+(z-w)^{-1}\,\partial_w\,[\tfrac{1}{2}\,:\psi^i(w)\partial\psi^i(w):+\tfrac{1}{16}\,d_{\bar{\Lambda}}\,w^{-2}].\end{aligned} \tag{3.6.44}$$

Thus in order to get the correct operator product of the energy-momentum tensor with itself, it is no longer possible to use the definition (3.6.33); rather, T must be defined as

$$T(z)=\tfrac{1}{2}\,:\psi^i(z)\partial\psi^i(z):+\tfrac{1}{16}\,d_{\bar{\Lambda}}\,z^{-2}, \tag{3.6.45}$$

i.e. as compared with Neveu–Schwarz fermions, the different normal ordering results in an additional term which changes the zero mode L_0 of $T(z)$.

We note that for Ramond fermions it is not possible to derive the correct operator product $T(z)T(w)$ with the normal ordering (3.1.22); in the formula analogous to (3.4.58) one would then get a contribution proportional to $(z-w)^{-2}w^{-2}$ (with a coefficient different from that in (3.6.44)), but no term proportional to $(z-w)^{-1}w^{-3}$. It should also be noted that in the literature (3.6.44) is usually derived by a different method. Namely, one uses the ordinary Wick theorem (which is possible because one is dealing with a free field theory), so that (3.6.33) implies

$$\begin{aligned}\underbrace{T(z)\,T(w)}=\tfrac{1}{4}\textstyle\sum_{i,j}\Big[&-\underbrace{\psi^i(z)\psi^j(w)}\;\underbrace{\partial\psi^i(z)\partial\psi^j(w)}\\ &-\underbrace{\psi^i(z)\psi^j(w)}\;:\partial\psi^i(z)\partial\psi^j(w):-\,:\psi^i(z)\psi^j(w):\;\underbrace{\partial\psi^i(z)\partial\psi^j(w)}\\ &+\underbrace{\psi^i(z)\partial\psi^j(w)}\;\underbrace{\partial\psi^i(z)\psi^j(w)}+\underbrace{\psi^i(z)\partial\psi^j(w)}\;:\partial\psi^i(z)\psi^j(w):\\ &+:\psi^i(z)\partial\psi^j(w):\;\underbrace{\partial\psi^i(z)\psi^j(w)}\Big].\end{aligned} \tag{3.6.46}$$

In the Neveu–Schwarz sector, this is evaluated as

$$\begin{aligned}\underbrace{T(z)\,T(w)}&=\tfrac{1}{4}\,[(z-w)^{-4}(2-1)\delta^{ii}-2(z-w)^{-3}:\psi^i(z)\psi^i(w):\\ &\qquad+(z-w)^{-2}\,(:\psi(z)\partial\psi(w):-:\partial\psi(z)\psi(w):)]\\ &=\tfrac{1}{4}\,d_{\bar{\Lambda}}\,(z-w)^{-4}+2\,(z-w)^{-2}\,T(w)+(z-w)^{-1}\,\partial T(w).\end{aligned} \tag{3.6.47}$$

(Here we have expanded $\psi(z)$ in a Taylor series around $z = w$ in normal ordered products like $:\psi(z)\psi(w):$, and used the identity (3.6.39).) The calculation in the Ramond sector runs completely analogous; the additional contributions which enforce the identification (3.6.45) then arise from the terms involving double contractions (which differ from the analogous terms in the NS-case, as described by (3.6.19)).

In terms of Laurent modes, one has for the case of Ramond fermions (using the general expression (3.1.41) for the modes of normal ordered products, and $(\partial\psi^i)_n = -(n+\frac{1}{2})\psi^i_n$)

$$(:\psi^i\psi^j:)_m = \sum_{n<0}(n-m-\tfrac{1}{2})\,\psi^i_n\psi^j_{m-n} - \sum_{n>0}(n-m-\tfrac{1}{2})\,\psi^j_{m-n}\psi^i_n - (m+\tfrac{1}{2})[\psi^i_n,\psi^j_{m-n}]. \tag{3.6.48}$$

In particular, the zero mode L_0 of the energy-momentum tensor reads

$$L_0 = -\sum_{n>0} n\,\psi^i_{-n}\psi^i_n + \tfrac{1}{16}\,d_{\bar\Lambda}. \tag{3.6.49}$$

This shows that the ground state of the theory of Ramond fermions, which is the Fock vacuum state $|0\rangle_R$ defined by $\psi^i_n|0\rangle_R = 0$ for $n > 0$, has a positive L_0-eigenvalue,

$$L_0\,|\,0\,\rangle_R = \tfrac{1}{16}\,d_{\bar\Lambda}\,|\,0\,\rangle_R. \tag{3.6.50}$$

In field theory, the distinctive feature of free theories is that one can use Wick's theorem to deduce arbitrary correlation functions from the one- and two-point correlation functions of the elementary fields. In the case of free Neveu–Schwarz-type Majorana fermions, the one point functions $\langle\psi\rangle = \langle 0\,|\,\psi\,|\,0\rangle$ vanish (because any mode ψ^i_n annihilates either $|0\rangle$ or $\langle 0|$); the operator product (3.6.14) then shows that the two point functions are

$$\langle\psi^i(w)\psi^j(z)\rangle = (z-w)^{-1}\delta^{ij}. \tag{3.6.51}$$

All other correlation functions then follow with the help of Wick's theorem. In particular all correlators of an odd number of fermions vanish, and the four-point function reads

$$\langle\psi^i(z)\psi^j(0)\psi^p(1)\psi^q(\infty)\rangle = -z^{-1}\delta^{ij}\delta^{pq} - (1-z)^{-1}\delta^{ip}\delta^{jq} + \delta^{iq}\delta^{jp}. \tag{3.6.52}$$

(Here the correlator with one field at infinity is to be interpreted in the sense of the limiting procedure (3.4.32); since the conformal dimension of the fermions is $\Delta(\psi) = \frac{1}{2}$, the infinite factor $(z_\infty)^{2\Delta}$ precisely cancels out as it must be.) Let us also write the latter formulae in the language of weights rather than of invariant tensors of $R_{\bar\Lambda}$. The fermions are then labelled as $\psi^{\bar\lambda} \equiv \psi^{\bar\lambda}_{\bar\Lambda}$, and the two point function reads

$$\langle \psi^{\bar\lambda}(z)\psi^{\bar\mu}(w)\rangle = -(z-w)^{-1}\,\delta^{\bar\lambda+\bar\mu,0}. \tag{3.6.53}$$

Using Wick's theorem, this leads to the four-point function

$$\langle \psi^{\bar\Lambda}(z)\psi^{\bar\lambda}(0)\psi^{-\bar\lambda}(1)\psi^{-\bar\Lambda}(\infty)\rangle = 1 - z^{-1}\delta^{\bar\lambda,-\bar\Lambda} - (1-z)^{-1}\delta^{\bar\lambda,\bar\Lambda}. \tag{3.6.54}$$

(Recall from the previous section (p. 196) that any component $F^{\bar\lambda_1\bar\lambda_2\bar\lambda_3\bar\lambda_4}$ is equal to a component of the form (3.6.54) up to a constant.)

For Dirac fermions, the correlation functions are obtained analogously from the two-point functions

$$\langle \psi_i(w)\psi^{+j}(z)\rangle = (z-w)^{-1}\,\delta_i^{\,j} \tag{3.6.55}$$

and $\langle \psi_i(w)\psi_j(z)\rangle = 0 = \langle \psi^{+i}(w)\psi^{+j}(z)\rangle$. This yields the four-point function

$$\langle \psi_i(z)\psi_j(0)\psi^{+p}(1)\psi^{+q}(\infty)\rangle = -(1-z)^{-1}\delta_i^p\delta_j^q + \delta_i^q\delta_j^p, \tag{3.6.56}$$

while all correlators involving unequal numbers of ψs and of ψ^+s (e.g. an odd number of fermions) vanish. In the language of weights, (3.6.55) becomes analogous to (3.6.53), and the four-point function reads

$$\langle \psi^{\bar\Lambda}(z)\psi^{\bar\lambda}(0)(\psi^+)^{-\bar\lambda}(1)(\psi^+)^{-\bar\Lambda}(\infty)\rangle = 1 - (1-z)^{-1}\delta^{\bar\lambda,\bar\Lambda}. \tag{3.6.57}$$

3.7 Quantum equivalence

A surprising feature of two-dimensional quantum field theory is that theories which are defined at the classical level by totally different actions can in fact be equivalent to each other at the quantum level. For example, the WZW theories on a simply connected simply laced group manifold at level one of the underlying affine algebra $\mathfrak{g}$ are quantum equivalent to a system of rank($\bar{\mathfrak{g}}$) free bosons compactified on the root lattice of $\bar{\mathfrak{g}}$. The translation between these two types of theories is provided by the Frenkel–Kac–Segal vertex operator construction of level one highest weight modules of simply laced affine Lie algebras. Another interesting class of quantum equivalences is between WZW theories and free fermion theories. This will be the subject of the present section.

In conventional quantum field theory, it is extremely difficult to decide whether two given theories are fully equivalent, because generically many of their properties are only accessible in some appropriate type of perturbation theory. In contrast, in the case of two-dimensional conformally invariant field theories, the notion of quantum equivalence can be given a precise nonperturbative meaning, and the calculations necessary to decide the question of equivalence are, at least in principle, straightforward. Namely, complete information about a conformal field theory is contained in its operator product algebra; this is the conformal bootstrap hypothesis which was already mentioned in section 3.3 (p. 171). Accordingly, two conformal field theories are quantum equivalent iff they possess the same operator algebra. In practice, it is convenient to separate this necessary and sufficient condition for quantum equivalence into two parts, namely to require that

- the two theories possess the same symmetry algebra,

and

- the spectra of primary fields are in one to one correspondence, and the operator product coefficients involving corresponding primary fields coincide.

Clearly, each of these two requirements is necessary for the coincidence of the whole operator product algebras, and hence for quantum equivalence. Together, they are also sufficient, because as we learnt in section 3.4, the symmetry algebra can be employed to express any operator product coefficient in terms of those involving only primary fields. In the second condition, we may equivalently require coincidence of the four-point function of primary fields instead of coincidence of their operator product coefficients, because the latter quantities are determined uniquely in terms of the former ones.

Also note that the two requirements are not logically independent because the term "primary" in the second condition of course refers to the particular symmetry algebra given by the first condition. However, it is important to realize that the first condition does not imply the second condition at all. We emphasize this because sometimes the first condition is erroneously taken to be the *the* condition for quantum equivalence. For example, it has been argued that a sufficient condition for quantum equivalence is that the two energy-momentum tensors coincide (which follows a fortiori from the first condition) because then the two systems possess the same Hamiltonian. However, even if two Hamiltonians are formally identical, the relevant Hilbert spaces of physical states on which they act may be different, and hence the systems described by this Hamiltonian can still be inequivalent. In path integral language, coincidence of the Hamiltonians corresponds to equality of the classical actions, but for equivalence at the quantum level the path integral measures must coincide as well. As an example, consider the WZW sigma models on two group manifolds G and G' such that G is a finite covering of G'; these theories possess the same Lagrangians, but different path integral measures.

Let us now discuss the first of the above conditions for the case of our interest, i.e. the comparison of WZW theories and of free fermion theories. We already know that for both types of theories the symmetry algebra is the semidirect sum $\mathcal{V}\oplus\mathfrak{g}$ of the Virasoro algebra and an untwisted affine Lie algebra (or more precisely, the direct sum of two such algebras, one holomorphic and one antiholomorphic; as in the previous sections, we will concentrate on the holomorphic part). Of course, also the central generators of $\mathcal{V}$ and $\mathfrak{g}$ must coincide; since for the free fermion theory the eigenvalues of these generators are fixed by (3.6.42) and (3.6.24) while for the WZW theory one has the relation (3.2.10) between the corresponding eigenvalues, we must consider a WZW theory at level $k^\vee = I_{\bar\Lambda}/(\bar\theta,\bar\theta)$, and the relation

$$\frac{k^\vee d}{k^\vee + g^\vee} = \tfrac{1}{2}\, d_{\bar\Lambda} \tag{3.7.1}$$

must be fulfilled. Moreover, for a WZW theory the energy-momentum tensor is of the Sugawara form (3.2.2) while for free fermions the canonical energy-momentum tensor is given by (3.6.33) (for Majorana fermions, and analogously for Dirac fermions). Thus in order to have quantum equivalence between free fermions and WZW theories, the two types of energy-momentum tensors must coincide.

At first sight, it seems rather implausible that this can happen, because for fermions the affine currents are quadratic in the fermions so that the Sugawara energy-momentum tensor becomes quartic in the fermions, whereas the canonical free fermion energy-momentum tensor is only quadratic. Nevertheless it turns out that this is possible. To see this, one has to express the canonical energy-momentum tensor

$$T_{\rm can}(z) = \tfrac{1}{2} :\psi^i(z)\partial\psi^i(z): \tag{3.7.2}$$

of free fermions through their Sugawara energy-momentum tensor, which reads

$$T_{\rm sug}(z) = \frac{1}{(\theta,\theta)(k^\vee + g^\vee)}\,\bar\kappa_{ab} :J^a(z)J^b(z): \tag{3.7.3}$$

with

$$J^a(z) = \tfrac{1}{2}\,(T^a_{(\bar\Lambda)})_{ij} :\psi^i(z)\psi^j(z): . \tag{3.7.4}$$

(The formulae are adopted to the case of Majorana fermions carrying an irreducible module $R_{\bar\Lambda}$ of a simple Lie algebra $\bar{\mathfrak{g}}$; also, in the following we will consider only Neveu–Schwarz type fermions. All other cases are treated analogously; compare the calculations in the previous section.)

To obtain the desired relation, we rewrite $T_{\rm sug}$ in terms of fully normal ordered products of fermions, using the formula (3.1.25) that relates different orders of normal orderings. From the elementary contraction

(3.6.14) it follows that

$$\underbrace{\psi^i(z) :\psi^j(w)\psi^k(w)}: = (z-w)^{-1} [\delta^{ik}\psi^j(w) - \delta^{ij}\psi^k(w)],$$

$$\begin{aligned} \underbrace{:\psi^i(z)\psi^j(z)::\psi^k(w)\psi^l(w)}: &= (z-w)^{-2} (\delta^{jk}\delta^{il} - \delta^{ik}\delta^{jl}) \\ &+ (z-w)^{-1}[\delta^{ik} :\psi^j(w)\psi^l(w): + \delta^{jl} :\psi^i(w)\psi^k(w): \\ &- \delta^{il} :\psi^j(w)\psi^k(w): - \delta^{jk} :\psi^i(w)\psi^l(w):], \end{aligned} \tag{3.7.5}$$

and hence, according to the rule (3.1.23),

$$\begin{aligned} &: [\psi^i(z), :\psi^j(z)\psi^k(z):] : = \delta^{ik}\,\partial\psi^j(z) - \delta^{ij}\,\partial\psi^k(z), \\ &: [:\psi^i(z)\psi^j(z):, :\psi^k(z)\psi^l(z):] : \\ &\qquad = \partial_z\, [\delta^{ik} :\psi^j(z)\psi^l(z): + \delta^{jl} :\psi^i(z)\psi^k(z): \\ &\qquad\quad - \delta^{il} :\psi^j(z)\psi^k(z): - \delta^{jk} :\psi^i(z)\psi^l(z):] . \end{aligned} \tag{3.7.6}$$

With the help of the results (3.1.25) and (3.6.39) for the normal ordering, one then obtains

$$\begin{aligned} : (:\psi^i\psi^j::\psi^k\psi^l:) : = &:\psi^i\psi^j\psi^k\psi^l: + \delta^{ik} :\psi^j\partial\psi^l: \\ &+ \delta^{jl} :\psi^i\partial\psi^k: - \delta^{il} :\psi^k\partial\psi^j: - \delta^{jk} :\psi^l\partial\psi^i:, \end{aligned} \tag{3.7.7}$$

where the first term on the right hand side denotes the fully normal ordered product of the four fermions. This implies

$$\begin{aligned} \bar\kappa_{ab}\,(T^a_{(\bar\Lambda)})_{ij}(T^b_{(\bar\Lambda)})_{kl} &\left\{: (:\psi^i\psi^j::\psi^k\psi^l:) : - :\psi^i\psi^j\psi^k\psi^l :\right\} \\ &= 4\,\bar\kappa_{ab}\,(T^a_{(\bar\Lambda)}T^b_{(\bar\Lambda)})_{ij} :\psi^i\partial\psi^j: \\ &= 4\,C_{\bar\Lambda} :\psi^i\partial\psi^i: . \end{aligned} \tag{3.7.8}$$

Combining all these informations, one arrives at the relation

$$T_{\rm sug} = T_{4\psi} + \eta\, T_{\rm can} \tag{3.7.9}$$

between the Sugawara and canonical energy-momentum tensors. Here we defined

$$T_{4\psi}(z) = \frac{\bar\kappa_{ab}}{4(I_{\bar\Lambda} + C_{\bar\theta})}\,(T^a_{(\bar\Lambda)})_{ij}(T^b_{(\bar\Lambda)})_{kl} :\psi^i(z)\psi^j(z)\psi^k(z)\psi^l(z): \tag{3.7.10}$$

and

$$\eta = \frac{2C_{\bar\Lambda}}{I_{\bar\Lambda} + C_{\bar\theta}} = \frac{2d}{d_{\bar\Lambda}(1 + I_{\bar\theta}/I_{\bar\Lambda})} . \tag{3.7.11}$$

Thus in order to have equality between $T_{\rm sug}$ and $T_{\rm can}$, one needs $T_{4\psi}=0$ and $\eta=1$. The latter condition is equivalent to (3.7.1) and hence adds nothing new. Moreover, the former condition is in fact not independent, but is a direct consequence of $\eta=1$. Namely, for $\eta=1$ one has $T_{4\psi}=T_{\rm sug}-T_{\rm can}$, and hence also the central generator $C_{4\psi}$ of the Virasoro algebra generated by $T_{4\psi}$ is the difference of the central generators of the Virasoro algebras generated by $T_{\rm sug}$ and $T_{\rm can}$; validity of (3.7.1) then means that $C_{4\psi}$ acts as zero, which in turn implies (see section 3.2, p. 165) that also $T_{4\psi}=0$. It turns out that it is convenient to look first for the solution to the requirement $T_{4\psi}=0$.

To get a handle on this condition, we first observe that, as a consequence of (3.6.39), the fully normal ordered product $:\psi^i\psi^j\psi^k\psi^l:$ is totally antisymmetric in the indices i,j,k,l, so that the requirement may be rewritten as

$$\bar\kappa_{ab}\left\{(T^a_{(\bar\Lambda)})_{ij}(T^b_{(\bar\Lambda)})_{kl}+(T^a_{(\bar\Lambda)})_{jk}(T^b_{(\bar\Lambda)})_{il}+(T^a_{(\bar\Lambda)})_{ik}(T^b_{(\bar\Lambda)})_{lj}\right\}=0 \tag{3.7.12}$$

(to see that this is fully antisymmetric, one must use that $T^a_{(\bar\Lambda)}$ is antisymmetric for orthogonal $R_{\bar\Lambda}$). This condition can indeed be met; e.g. for the adjoint module for which the representation matrices $(T^a_{(\bar\theta)})_{bc}$ are given by the structure constants, (3.7.12) is nothing but the Jacobi identity of $\bar{\mathfrak g}$. This suggests that in order to find the most general solution, one has to look for an interpretation of (3.7.12) as a sort of Jacobi identity. A more careful analysis shows that to implement this idea, one has to add further generators t^i to the $\bar{\mathfrak g}$-generators T^a so that $\bar{\mathfrak g}$ is extended to a larger algebra

$$\bar{\mathfrak h}=\bar{\mathfrak g}\oplus\bar{\mathfrak t}, \tag{3.7.13}$$

and require that $\bar{\mathfrak h}$ possesses a $\mathbb Z_2$-grading with the subalgebras $\bar{\mathfrak g}$ and $\bar{\mathfrak t}$ being the even and odd parts of $\bar{\mathfrak h}$, respectively. This means that the Lie brackets of $\bar{\mathfrak g}$ are of the form

$$\begin{aligned}&[\bar{\mathfrak g},\bar{\mathfrak g}]\subseteq\bar{\mathfrak g},\\&[\bar{\mathfrak h},\bar{\mathfrak g}]\subseteq\bar{\mathfrak h},\\&[\bar{\mathfrak h},\bar{\mathfrak h}]\subseteq\bar{\mathfrak g},\end{aligned} \tag{3.7.14}$$

or, more explicitly,

$$\begin{aligned}&[T^a,T^b]=f^{ab}{}_c\,T^c,\\&[T^a,t^i]=-x^{ai}{}_j\,t^j,\\&[t^i,t^j]=-y^{ij}{}_a\,T^a.\end{aligned} \tag{3.7.15}$$

Note that the Jacobi identity for the triple T^a, T^b, t^i of generators can be written as

$$[x^a, x^b]^i{}_j = f^{ab}{}_c\,(x^c)^i{}_j, \tag{3.7.16}$$

where the structure constants $x^{ai}_{\;j}$ are interpreted as the elements of some matrices x^a; thus these structure constants provide a matrix representation $R_{[t]}$ of g. Moreover, using the cyclic invariance of the trace operation, one has $\mathrm{tr}([t^i, t^j]\,T^a) = \mathrm{tr}(t^i\,[t^j, T^a])$, or in other words

$$y^{ija} = x^{aij}, \tag{3.7.17}$$

where indices are raised and lowered with the Killing matrix $\bar\kappa^{ab}(\bar{\mathsf{h}})$ of $\bar{\mathsf{h}}$ and its inverse. The Jacobi identity for the triple t^i, t^j, t^k then yields

$$0 = \bar\kappa_{ab}(\bar{\mathsf{h}}) \cdot \left\{ x^a{}_{ij} x^b{}_{kl} + x^a{}_{jk} x^b{}_{il} + x^a{}_{ki} x^b{}_{jl} \right\}. \tag{3.7.18}$$

Thus by setting

$$(T^a_{(\bar\Lambda)})_{ij} = x^a{}_{ij}, \tag{3.7.19}$$

i.e. $R_{\bar\Lambda} = R_{[t]}$, the equation (3.7.12) is fulfilled identically. Conversely, given (3.7.12), it is consistent with the Jacobi identities to construct $\bar{\mathsf{h}}$ from g via (3.7.15) (with $x = y$), and hence (3.7.12) is equivalent to the existence of a Lie algebra fulfilling (3.7.13) and (3.7.15). The virtue of the latter presentation of the problem is that the classification of the Lie algebras of the required form is well known. Namely, in terms of the compact Lie groups G and H which are related to g and $\bar{\mathsf{h}}$ via the exponential map, the quotient manifold H/G must be a so-called compact *symmetric space*, and the quantities

$$\mathcal{R}_{ijkl} = \bar\kappa_{ab}\, x^a{}_{ij}\, x^b{}_{kl} \tag{3.7.20}$$

are the components of the Riemann tensor of H/G. The representation $R_{\bar\Lambda} = R_{[t]}$ is called the *tangent space* representation of H/G. The classification shows that even in the irreducible cases (i.e. with H/G not a direct product of other symmetric spaces) the algebra g need not be simple nor even semisimple.

Thus the free fermion theories having an energy–momentum tensor of the Sugawara form are in one to one correspondence with the compact symmetric spaces. This assertion is known as the *symmetric space theorem* or as the Goddard–Nahm–Olive classification. The table (3.7.21) contains the classification of (compact, irreducible) symmetric spaces; the columns list the name of the symmetric space, the algebras g and $\bar{\mathsf{h}}$, the tangent space module (denoted either by its dimension or, in the case of infinite series of symmetric spaces, by its highest g-weight $\bar\Lambda_{[t]}$; if the tangent space module is of the form $R \oplus R^+$ with R non-selfconjugate, only R is displayed), and finally the level of g.

name	$\bar{\mathfrak{g}}$	$\bar{\mathfrak{h}}$	$\bar{\Lambda}_{[t]}$ or $d_{\bar{\Lambda}_{[t]}}$	level
S^n	so_n	so_{n+1}	$\bar{\Lambda}_{(1)}$	1
CP^n	u_n	su_{n+1}	$\bar{\Lambda}_{(1)}$	1
AI	so_n $(n{\geq}4)$	su_n	$2\bar{\Lambda}_{(1)}$	$n+2$
AII	C_n $(n{\geq}3)$	su_{2n}	$\bar{\Lambda}_{(2)}$	$n-1$
AIII	$su_m \oplus su_n \oplus u_1$	su_{m+n}	$\bar{\Lambda}_{(1)} \otimes \bar{\Lambda}_{(1)}$	(n,m)
BDI	$so_m \oplus so_n$	so_{m+n}	$\bar{\Lambda}_{(1)} \otimes \bar{\Lambda}_{(1)}$	(n,m)
CI	u_n	C_n	$2\bar{\Lambda}_{(1)}$	$n+2$
CII	$C_m \oplus C_n$	C_{m+n}	$\bar{\Lambda}_{(1)} \otimes \bar{\Lambda}_{(1)}$	(n,m)
DIII	u_n $(n{\geq}4)$	so_{2n}	$\bar{\Lambda}_{(2)}$	$n-2$
AI	su_2	su_3	5	10
EI	C_4	E_6	42	7
EIII	$so_{10} \oplus u_1$	E_6	16	4
EIV	F_4	E_6	26	3
EV	su_8	E_7	70	10
EVII	$E_6 \oplus u_1$	E_7	27	6
EVIII	so_{16}	E_8	128	16
FII	so_9	F_4	16	2
EII	$su_6 \oplus su_2$	E_6	$20 \otimes 2$	(6,10)
EVI	$so_{12} \oplus su_2$	E_7	$32 \otimes 2$	(8,16)
EIX	$E_7 \oplus su_2$	E_8	$56 \otimes 2$	(12,28)
FI	$C_3 \oplus su_2$	F_4	$14 \otimes 2$	(5,7)
G	$su_2 \oplus su_2$	G_2	$4 \otimes 2$	(10,2)
	su_n $(n{\geq}3)$	$su_n \oplus su_n$	$\bar{\Lambda}_{(1)} + \bar{\Lambda}_{(n-1)}$	n
	so_n	$so_n \oplus so_n$	$\bar{\Lambda}_{(2)}$	$n-2$
	C_n	$C_n \oplus C_n$	$2\bar{\Lambda}_{(1)}$	$n+1$
type II	E_6	$E_6 \oplus E_6$	78	12
	E_7	$E_7 \oplus E_7$	133	18
	E_8	$E_8 \oplus E_8$	248	30
	F_4	$F_4 \oplus F_4$	52	9
	G_2	$G_2 \oplus G_2$	14	4

(3.7.21)

The table (3.7.21) has been divided into four parts. The first three parts contain the (irreducible) symmetric spaces of type I. These spaces are either S^n (the n-dimensional sphere) or CP^n (the n-dimensional complex projective space), or else carry the names AI to G. The fourth part of the table contains the (irreducible) type II symmetric spaces. For the type II spaces the Lie algebra $\mathfrak{g}$ is the diagonal subalgebra of the semisimple algebra $\bar{\mathfrak{h}} = \mathfrak{g} \oplus \mathfrak{g}$, where $\mathfrak{g}$ is any of the simple Lie algebras, and the tangent space representation is the adjoint representation of $\mathfrak{g}$. Among the type I symmetric spaces, we have listed the infinite series first, afterwards the exceptional cases for which $\mathfrak{g}$ is either simple or the direct sum of a simple algebra with u_1 (the u_1-part of the modules is suppressed in the table; the u_1-charges are given by $q = 1/\sqrt{2c}$; recall that $2c$ is the total number of fermionic degrees of freedom), and finally the exceptional cases for which $\mathfrak{g}$ is the direct sum of a simple algebra with su_2. Also, we used the notations so_n instead of $B_{(n-1)/2}$ (for n odd) or $D_{n/2}$ (for n even) because this makes it easier to write down the corresponding infinite series, and also su_n instead of A_{n-1} as well as $u_n \equiv su_n \oplus u_1$ (in the remaining cases, the symbol $\mathfrak{g}$ is understood to stand for the compact real form as well).

Inspection shows that the u_1- and su_2-parts of $\mathfrak{g}$ are present precisely if the tangent space representation is non-selfconjugate respectively symplectic. In terms of the free fermion theory, the presence of these algebras is understood as describing the u_1- and su_2-currents which exist automatically (see p. 209) for a system of fermions in a non-selfconjugate and a symplectic module, respectively. These contributions to $\mathfrak{g}$ are indeed needed in order that the conformal central charge of the theory satisfies the relation (3.7.1). Consider e.g. the simplest case of a non-selfconjugate module, namely the defining module $R_{\Lambda_{(1)}}$ of su_n. Then the dimension, dual Coxeter number and level are given by $d = n^2 - 1$, $g^\vee = n$, $k^\vee = 1$, respectively, so that the Sugawara construction for su_n yields a conformal central charge $c_{\text{sug}} = n - 1$ (this is a special case of the lower limit $c_{\text{sug}} = \text{rank}\,\mathfrak{g}$ for the conformal central charge; compare p. 165). The central charge of the Virasoro algebra generated by the canonical energy–momentum tensor T_{can} however equals half the total (i.e., including the conjugate module) number of fermions, i.e. $c_{\text{can}} = n$; the missing contribution $\delta c = 1$ that is needed to obtain equality is provided when su_n is extended to $su_n \oplus u_1$ so that $T_{\text{sug}} = T_{\text{sug}}(su_n) + T_{\text{sug}}(u_1)$, with the second part given by the formula (3.2.34). Similarly, the contribution of the u_1 part to the conformal dimension of the fermions is $\Delta = \frac{1}{2}q^2 = \frac{1}{2n}$ (see also (3.8.6) below).

The condition (3.7.1) can also be obtained purely in terms of the symmetric space description. Namely, the Lie brackets (3.7.15) imply that the Killing matrix of $\bar{\mathfrak{h}}$ reads

$$\begin{aligned}\bar\kappa^{ab}(\bar{\mathfrak{h}}) = \mathrm{tr}\,(T^a_{(\bar\theta)}T^b_{(\bar\theta)}) &= \frac{1}{I_{\rm ad}(\bar{\mathfrak{h}})}\,(f^{ad}{}_c f^{bc}{}_d + x^{ai}{}_j x^{bj}{}_i) \\ &= \frac{I_{\rm ad}(\bar{\mathfrak{g}})}{I_{\rm ad}(\bar{\mathfrak{h}})}\,[\bar\kappa^{ab}(\bar{\mathfrak{g}}) + \bar\kappa^{ab}_{[t]}(\bar{\mathfrak{g}})],\end{aligned} \tag{3.7.22}$$

i.e. (using the relation (1.6.43) between $\bar\kappa$ and $\bar\kappa_{[t]}$) one has

$$I_{\rm ad}(\bar{\mathfrak{h}})\,\bar\kappa^{ab}(\bar{\mathfrak{h}}) = [I_{\rm ad}(\bar{\mathfrak{g}}) + I_{[t]}(\bar{\mathfrak{g}})]\,\bar\kappa^{ab}(\bar{\mathfrak{g}}). \tag{3.7.23}$$

On the other hand, specializing to elements of this matrix which correspond to the subalgebra $\bar{\mathfrak{t}} \subset \bar{\mathfrak{h}}$, one can also write

$$\bar\kappa^{ij}(\bar{\mathfrak{h}}) = \mathrm{tr}(T^i_{(\bar\theta)}T^j_{(\bar\theta)}) = \frac{1}{I_{\rm ad}(\bar{\mathfrak{h}})}\cdot 2\,y^{ik}{}_a y^{j}{}_k{}^a = \frac{2}{I_{\rm ad}(\bar{\mathfrak{h}})}\,\bar\kappa^{ab}(\bar{\mathfrak{g}})\,y^{ik}{}_a y^j{}_{kb}, \tag{3.7.24}$$

or (in a rather symbolic notation, and using the definition (1.6.34) of the quadratic Casimir operator), $I_{\rm ad}(\bar{\mathfrak{h}})\,\bar\kappa^{ij}(\bar{\mathfrak{h}}) = 2\,C_{[t]}\,\bar\kappa^{ij}(\bar{\mathfrak{g}})$. Comparing this with (3.7.23), it follows that one must have

$$I_{\rm ad}(\bar{\mathfrak{g}}) + I_{\bar\Lambda} = 2\,C_{\bar\Lambda} \tag{3.7.25}$$

with $\bar\Lambda \equiv \bar\Lambda_{[t]}$. Using the relation (1.6.45) between the Dynkin index and the quadratic Casimir eigenvalue, and the fact that $I_{\bar\Lambda} = k$, this is indeed just a rephrasing of (3.7.1).

Summarizing, the table (3.7.21) gives the complete solution to the requirement of having identical symmetry algebras for free fermion systems and WZW theories, i.e. to the first of the two conditions of p. 215 for quantum equivalence. The rest of this section is devoted to the discussion of the second of those conditions. The first part of that condition, namely the coincidence of the spectra of primary fields, is almost tautological. Namely, one may just *define* the WZW theory as possessing precisely those primary fields which are present in the corresponding free fermion theory. There is no obstacle to impose this spectrum; notice, however, that there is also no reason whatsoever to expect that the WZW theory based on some affine Lie algebra $\mathfrak{g}$ obtained this way is the WZW theory which possesses a langrangian formulation as a nonlinear sigma model with target space given by the (simply connected) covering group G associated to the horizontal subalgebra $\bar{\mathfrak{g}}$ of $\mathfrak{g}$.

The remaining task is thus to check whether four-point functions of free fermion theories possessing an energy–momentum tensor of the Sugawara form are identical to the four-point functions of the WZW primary fields (so that the respective operator product coefficients are identical as well). Technically, this means that it must be analysed whether the free fermion correlators obey the relevant Knizhnik–Zamolodchikov equations of the WZW theory.

This check has not yet been carried out for all cases in the list (3.7.21), simply because it is a very lengthy exercise to write down the relevant Knizhnik–Zamolodchikov equations in an explicit form. However, in all cases where Knizhnik–Zamolodchikov equations *have* been worked out, the analysis can be carried through, with the result that the four-point functions indeed agree. The cases in question are the infinite series corresponding to the symmetric spaces S^n, CP^n, AIII, BDI, CI, CII and DIII. Below we will also check the case of EVII, which is accessible owing to the results obtained at the end of section 3.4. Before doing so, it must however be pointed out that there is a different line of reasoning which implies that the four point functions in fact agree in all cases; this is based on the classification of so-called conformal embeddings and will be presented in the following section.

It should also be mentioned that there is a single class of WZW theories that are quantum equivalent both to free fermions and to free bosons, namely those for $D_r^{(1)}$ at level one. The equivalence to such WZW theories provides a way of understanding the equivalence of free bosons and free fermions that had been observed in a direct manner many years before the connection with affine Lie algebras could be uncovered.

Consider now the case of the symmetric space EVII, i.e. fermions carrying the defining module of E_6 (the u_1-part modifies the four-point function in a trivial manner and can therefore be suppressed). This module is not selfconjugate so that we have to consider Dirac fermions. This then gives rise to the level $k^\vee = 6$ of the affine algebra $E_6^{(1)}$. The only non-trivial four-point function of the elementary fermions is the one that contains two fermions carrying the defining module of E_6 and two fermions carrying its conjugate module. The corresponding correlator of the WZW theory has already been considered in section 3.4, the result being given in (3.4.55). It turns out that for $k^\vee = 6$, the functions obtained there simplify considerably. To be specific, one finds

$$\begin{aligned} F_1^{(1)} + F_1^{(3)}(z) &= \frac{1}{z}\,(z(1-z))^{1/27}, \\ F_2^{(1)} + F_2^{(3)}(z) &= \frac{1}{1-z}\,(z(1-z))^{1/27}, \\ F_3^{(1)} + F_3^{(3)} &= 0, \end{aligned} \tag{3.7.26}$$

while the combinations $F_A^{(1)} - F_A^{(3)}$ and $F_A^{(2)}$, $A = 1, 2, 3$, are combinations of powers $z^a(1-z)^b$ with certain hypergeometric functions (namely, with $\mathrm{F}(\frac{2}{9}, \frac{7}{9}; \frac{5}{3}; z)$, $\mathrm{F}(-\frac{4}{9}, \frac{1}{9}; \frac{1}{3}; z)$ and their derivatives). Here the components corresponding to the invariants (3.4.49) rather than to (3.4.53) are given. These results can be obtained by using various special properties of the contour integrals (3.4.55); alternatively, they may be checked by comparing the (pseudo) Taylor series of the functions (3.4.55) with those of the functions presented here.

The factor of $(z(1-z))^{1/27}$ in (3.7.26) gets compensated if the nontrivial transformation of the fermions with respect to the additional u_1-part of the algbra $\mathfrak{g}$ is taken into account. Thus it follows that from the general solution

$$F(z) = \sum_{\nu=1}^{3} a_\nu \, F^{(\nu)}(z) \tag{3.7.27}$$

of the Knizhnik–Zamolodchikov equation, one obtains a solution which is single valued as a function of z by choosing the particular linear combination

$$a_1 = a_3 = 1, \quad a_2 = 0. \tag{3.7.28}$$

Using the explicit form (3.4.49) of the invariant tensors, this solution then reads

$$F_{il}^{jk}(z) = \frac{1}{z}\, \delta_i^{\,j} \delta_l^{\,k} + \frac{1}{1-z}\, \delta_i^{\,k} \delta_l^{\,j}. \tag{3.7.29}$$

Hence we have indeed reproduced the four-point function of Dirac fermions as given by (3.6.56). Besides the elementary fermions, the free fermion theory contains further primary fields obtained as normal ordered products of the fermions. Their correlation functions are easily calculated with the help of Wick's theorem, and so in principle it is straightforward to check that these correlation functions again fulfill the relevant Knizhnik–Zamolodchikov equations of the WZW theory. However, the calculation of the explicit form of the Knizhnik–Zamolodchikov equations is a rather tedious exercise, and in view of the proof of equivalence that will be obtained in the next section it is not necessary to present them here.

Let us, however, consider the four-point function of WZW fields $\phi_{27,27}(z, \bar{z})$ carrying the defining E_6-module (respectively, its conjugate) with respect both to the holomorphic and to the antiholomorphic symmetry algebra. According to (3.4.58), the most general solution to the holomorphic and antiholomorphic Knizhnik–Zamolodchikov equations is then given by

$$F(z, \bar{z}) = \sum_{\nu,\tau=1}^{3} a_{\nu\tau} \, F^{(\nu)}(z) F^{(\tau)}(\bar{z}). \tag{3.7.30}$$

It turns out that this is single-valued for $\bar{z} = z^*$ iff the coefficients $a_{\nu\tau}$ are given (up to overall normalization) by

$$(a_{\nu\tau}) = \begin{pmatrix} 1 & 0 & 1-\xi \\ 0 & \gamma\xi & 0 \\ 1-\xi & 0 & 1 \end{pmatrix}, \tag{3.7.31}$$

where

$$\gamma = 2\,\frac{\Gamma^2(\frac{1}{3})\Gamma(\frac{8}{9})\Gamma(\frac{13}{9})}{\Gamma^2(\frac{2}{3})\Gamma(\frac{1}{9})\Gamma(\frac{5}{9})} \tag{3.7.32}$$

and where ξ is an arbitrary constant. Thus at level $k^\vee = 6$, the single-valued solution involves a free parameter and is thus more general than the solution (3.4.59), (3.4.60) obtained for generic level. Technically, this arises because in the limit $z \to 0$, the two solutions $F^{(1)}(z)$ and $F^{(3)}(z)$ of the Knizhnik–Zamolodchikov equation differ only by an integer power of z if $k^\vee = 6$. In terms of conformal field theory, the underlying reason is that the conformal dimension of the WZW primary field carrying the 650-dimensional module of E_6 is an integer at level 6 (this primary field is relevant to the four-point function in question owing to the Kronecker product (1.7.58)); namely, the general formula (3.4.20) for the conformal dimension of a WZW primary yields

$$\Delta(\phi_{650}) = \frac{18}{k^\vee + 12}, \tag{3.7.33}$$

and hence $\Delta(\phi_{650}) = 1$ for $k^\vee = 6$.

The presence of a free parameter ξ in (3.7.31) is to be interpreted as follows. At level 6 there exist two distinct WZW theories which possess primary fields $\phi_{27,27}$. The Knizhnik–Zamolodchikov equations for the four-point functions of these fields are the same in both theories; the four-point functions themselves, however, are not identical, but rather provide two independent solutions to the Knizhnik–Zamolodchikov equation each of which is single valued. As a consequence, the space of single valued solutions to the Knizhnik–Zamolodchikov equation is two-dimensional, i.e. in addition to the overall normalization there is a further free parameter.

It is not difficult to identify the two WZW theories in question. One of them is just the WZW theory already discussed, i.e. the one which is quantum equivalent to the free fermion system. The relevant fields are then the bilinears

$$\phi_{27,27}(z,\bar{z}) = \psi(z)\,\bar{\psi}(\bar{z}). \tag{3.7.34}$$

The four-point function factorizes as

$$F(z,\bar{z}) = F(z)\,F(\bar{z}) \tag{3.7.35}$$

with $F(z)$ given by (3.7.29); this is reproduced by (3.7.30) if the matrix $a_{\nu\tau}$ factorizes as

$$a_{\nu\tau} = a_\nu \, a_\tau , \tag{3.7.36}$$

with a_ν chosen as in (3.7.28), and hence corresponds to the value $\xi = 0$ of the free parameter. The second WZW theory is the one realized as a sigma model with Wess–Zumino term, with target space given by the simply connected group manifold G whose Lie algebra is E_6. This theory contains only left–right symmetric primary fields; as a consequence, the matrix $a_{\nu\tau}$ relevant to the four-point function of this theory must be diagonal; thus it corresponds to the value $\xi = 1$ of the free parameter in (3.7.31). (The result obtained for $\xi = 1$ also reproduces the formula (3.4.60) for a_ν at level 6; to see this, one has to use the functional identities obeyed by the gamma function.)

3.8 Conformal embeddings

In section 3.5 a classification was obtained of those primary WZW fields which possess simple power-like four-point functions (see (3.5.53)). According to section 3.6, the four-point functions of free fermions are of this simple form, too, and as a consequence this is also the case for the relevant primary fields of those WZW theories which in the previous section were argued to be quantum equivalent to free fermion systems. However, with the exception of the first two cases in the list (3.7.21) of the latter WZW theories, these primary fields are *not* contained in the classification of section 3.5. As already mentioned at the end of section 3.5, such a situation can be explained through the presence of an enlarged symmetry algebra. It turns out that for the theories listed in the table (3.7.21) an enlarged symmetry algebra does exist, and that it is again given by the semidirect sum $\mathcal{V} \oplus \mathsf{f}$ of the Virasoro algebra and an untwisted affine Kac–Moody algebra; the original affine Lie algebra g is a subalgebra of the affine algebra f appearing here.

We learned in section 2.8 that for an embedding $\mathsf{g} \subset \mathsf{f}$ of affine Lie algebras, the irreducible highest weight modules of f generically decompose into an infinite number of irreducible highest weight modules of g (which is possible because they are infinite-dimensional), but that there exist special cases where this unpleasant feature is circumvented. In the latter situation the embedding is a called a conformal embedding. To be more precise, recall from section 2.8 (p. 144) that an embedding $\mathsf{g} \subset \mathsf{f}$ is by definition a conformal embedding iff the central terms c of the Virasoro algebras that are associated to them by the Sugawara construction have the same values, i.e.

$$\frac{k^\vee(\mathfrak{g})\,d(\bar{\mathfrak{g}})}{k^\vee(\mathfrak{g})+g^\vee(\bar{\mathfrak{g}})}=\frac{k^\vee(\mathfrak{f})\,d(\bar{\mathfrak{f}})}{k^\vee(\mathfrak{f})+g^\vee(\bar{\mathfrak{f}})}. \tag{3.8.1}$$

This was derived in section 2.8 with the help of the asymptotic behavior of the affine characters. In terms of conformal field theory, one can rephrase this condition by saying that the embedding $\mathfrak{g}\subset\mathfrak{f}$ of affine Lie algebras is a conformal one iff the algebras $\mathfrak{g}$ and $\mathfrak{f}$ give rise, via the Sugawara construction, to the same Virasoro algebra, i.e. (in the notation of section 3.2)

$$T_{(\mathfrak{g})}(z)=T_{(\mathfrak{f})}(z). \tag{3.8.2}$$

In other words, the difference of these two energy–momentum tensors, i.e. the coset energy–momentum tensor $T_{(\mathfrak{f}/\mathfrak{g})}$, must vanish.

Let us also explain in conformal field theory terms why this characterization of conformal embeddings is correct. To verify that the condition (3.8.2) is necessary, recall that for any embedding $\mathfrak{g}\subset\mathfrak{f}$, the unitary highest weight modules of $\mathfrak{f}$ decompose into a direct sum of highest weight modules of $\mathfrak{g}\oplus\mathcal{V}_{(\mathfrak{f}/\mathfrak{g})}$. Since the non-trivial modules of $\mathcal{V}_{(\mathfrak{f}/\mathfrak{g})}$ are infinite-dimensional, the infinite reducibility follows unless only the trivial $\mathcal{V}_{(\mathfrak{f}/\mathfrak{g})}$-module is allowed to occur, that is (compare p. 165), unless $\mathcal{V}_{(\mathfrak{f}/\mathfrak{g})}$ vanishes. Conversely, if (3.8.2) is fulfilled, then in particular the zero modes $L_0(\mathfrak{g})$ and $L_0(\mathfrak{f})$ of the two Virasoro algebras are identical. But for a unitary highest weight module of $\mathfrak{f}$, any eigenvalue of $L_0(\mathfrak{f})$ occurs with finite multiplicity (because L_0 equals minus the derivation D, and the subspaces of a unitary highest weight module at a fixed grade are finite-dimensional). The same property must then hold for the zero mode $L_0(\mathfrak{g})$; this is clearly incompatible with an infinite multiplicity of any given highest weight module of $\mathfrak{g}$. Moreover, only a finite number of distinct highest weight modules of $\mathfrak{g}$ can appear because they must be unitary (they inherit unitarizability from the $\mathfrak{f}$-module), or, what is the same, integrable. Putting this information together, the finite reducibility follows.

It has been seen on p. 167 that the relative value $k^\vee(\bar{\mathfrak{f}})/k^\vee(\bar{\mathfrak{g}})$ is given by the Dynkin index $I\equiv I_{\bar{\mathfrak{g}}\subset\bar{\mathfrak{f}}}$ of the embedding $\bar{\mathfrak{g}}\subset\bar{\mathfrak{f}}$ of the horizontal subalgebras which is induced by the imbedding $\mathfrak{g}\subset\mathfrak{f}$. Thus (3.8.1) can be rewritten as

$$k^\vee(\mathfrak{f})=\frac{1}{d(\bar{\mathfrak{g}})-d(\bar{\mathfrak{f}})}\left[I^{-1}d(\bar{\mathfrak{f}})g^\vee(\bar{\mathfrak{g}})-d(\bar{\mathfrak{g}})g^\vee(\bar{\mathfrak{f}})\right]. \tag{3.8.3}$$

This can be interpreted as follows: whenever for an embedding $\bar{\mathfrak{g}} \subset \bar{\mathfrak{f}}$ of semisimple Lie algebras the right hand side of (3.8.3) is an integer, the embedding extends to a conformal embedding of the associated untwisted affine algebras. It is then straightforward to classify all conformal embeddings $\mathfrak{g} \subset \mathfrak{f}$ by looking at the known (see section 1.8) list of embeddings $\bar{\mathfrak{g}} \subset \bar{\mathfrak{f}}$. This way one obtains the classification that has been mentioned in section 2.8.

By an inspection of the list of all conformal embeddings one readily verifies the claim that for all entries in (3.7.21) a conformal embedding of $\mathfrak{g}$ exists. In the first two cases (for which the relevant WZW primary field is cominimal so that no extended symmetry algebra is required), the "embedding" is just the trivial embedding of the algebra $\mathfrak{g}$ into itself. In the rest of the cases, the embedding is a maximal embedding into a level one affine algebra provided that the tangent space module $R_{[t]}$ is not symplectic, namely such that $\bar{\mathfrak{f}} = so_d$ and $\bar{\mathfrak{f}} = su_d \oplus u_1$ with $d = \dim R_{[t]}$ if $R_{[t]}$ is orthogonal and non-selfconjugate, respectively. (To be precise, there are two exceptional cases where $R_{[t]}$ is selfconjugate and nevertheless $\mathfrak{f} = u_d$; these are the lowest members of the infinite series CI and DIII, namely su_2 at level 4 and su_4 at level 2, respectively. The existence of these exceptions can be understood through the fact that for all other members of these series the tangent space module is not selfconjugate.) If $R_{[t]}$ is symplectic, the situation is a bit more complicated; in these cases the Lie algebra $\mathfrak{g}$ is a semisimple one containing an su_2-piece, i.e. $\mathfrak{g} = \mathfrak{g}_\circ \oplus su_2$, and the embedding can be performed in two steps: first, there is a maximal embedding $\mathfrak{g}_\circ \subset \mathfrak{f}_\circ$ into the symplectic affine algebra $\mathfrak{f}_\circ = C^{(1)}_{d/4}$ at level one, and then $\mathfrak{f}_\circ \oplus A^{(1)}_1$ is maximally embedded in $\mathfrak{f}$ at level one, where $\bar{\mathfrak{f}} = so_d$.

The table (3.8.4) lists all cases in the same sequence as in (3.7.21). The first two columns give again the name of the corresponding symmetric space and $\mathfrak{g}$, while the next two columns show $\bar{\mathfrak{f}}$ and the Dynkin index $I = I_{\bar{\mathfrak{g}}\subset\bar{\mathfrak{f}}}$; the contents of the last column will be explained below. In the case of symplectic tangent space modules, the third and fourth columns of the table (3.8.4) have been split into two pieces, the first piece consisting of the algebra $\bar{\mathfrak{f}}_\circ$ respectively the Dynkin index $I_\circ$ of $\bar{\mathfrak{g}}_\circ \subset \bar{\mathfrak{f}}_\circ$, and the second piece being $\bar{\mathfrak{f}}$ respectively the number $\tilde{I}$ such that the pair $(1, \tilde{I})$ is the Dynkin index for the embedding $\bar{\mathfrak{f}}_\circ \oplus su_2 \subset \bar{\mathfrak{f}}$ (the Dynkin index of $\bar{\mathfrak{g}}_\circ \oplus su_2 \subset \bar{\mathfrak{f}}$ is then the pair $I = (I_\circ, \tilde{I})$). Note that although the classification (3.7.21) was obtained in a constructive manner, the construction did not provide us with an explanation for the encountered values of the levels of the affine algebras. From table (3.8.4) we now learn that these levels are in fact just the Dynkin indices of the associated conformal embeddings.

name	$\bar{\mathfrak{g}}$	$\bar{\mathfrak{f}}$	I	$\bar{\Lambda}_{[f]}$
S^n	so_n	so_n	1	–
CP^n	u_n	u_n	1	–
AI	so_n $(n{\geq}4)$	$so_{\frac{1}{2}(n-1)(n+2)}$	$n+2$	$2\bar{\Lambda}_{(1)}+\bar{\Lambda}_{(2)}$
AII	C_n $(n{\geq}3)$	$so_{(n-1)(2n+1)}$	$n-1$	$\bar{\Lambda}_{(1)}+\bar{\Lambda}_{(3)}$
AIII	$su_m \oplus su_n \oplus u_1$	u_{mn}	(n,m)	$(\bar{\Lambda}_{(1)}+\bar{\Lambda}_{(m-1)}) \otimes(\bar{\Lambda}_{(1)}+\bar{\Lambda}_{(n-1)})$
BDI	$so_m \oplus so_n$	so_{mn}	(n,m)	$(2\bar{\Lambda}_{(1)} \otimes \bar{\Lambda}_{(2)}) \oplus(\bar{\Lambda}_{(2)} \otimes 2\bar{\Lambda}_{(1)})$
CI	u_n	$u_{n(n+1)/2}$	$n+2$	$2\bar{\Lambda}_{(1)}+2\bar{\Lambda}_{(n-1)}$
CII	$C_m \oplus C_n$	so_{4mn}	(n,m)	$(2\bar{\Lambda}_{(1)} \otimes \bar{\Lambda}_{(2)}) \oplus(\bar{\Lambda}_{(2)} \otimes 2\bar{\Lambda}_{(1)})$
DIII	u_n $(n{\geq}4)$	$u_{n(n-1)/2}$	$n-2$	$\bar{\Lambda}_{(2)}+\bar{\Lambda}_{(n-2)}$
AI	su_2	C_2	10	7
EI	C_4	so_{42}	7	825
EIII	$so_{10} \oplus u_1$	u_{16}	4	210
EIV	F_4	so_{26}	3	273
EV	su_8	so_{70}	10	2352
EVII	$E_6 \oplus u_1$	u_{27}	6	650
EVIII	so_{16}	so_{128}	16	8008
FII	so_9	so_{16}	2	84
EII	$su_6 \oplus su_2$	C_{10} so_{40}	6 10	$175 \otimes 1 \oplus 189 \otimes 3$
EVI	$so_{12} \oplus su_2$	C_{16} so_{64}	8 16	$462 \otimes 1 \oplus 495 \otimes 3$
EIX	$E_7 \oplus su_2$	C_{28} so_{112}	12 28	$1463 \otimes 1 \oplus 1539 \otimes 3$
FI	$C_3 \oplus su_2$	C_7 so_{28}	5 7	$84 \otimes 1 \oplus 90 \otimes 3$
G	$su_2 \oplus su_2$	C_2 so_8	10 2	$7 \otimes 1 \oplus 5 \otimes 3$

(3.8.4)

name	$\bar{\mathfrak{g}}$	$\bar{\mathfrak{f}}$	I	$\bar{\Lambda}_{[f]}$
type II	su_n $(n\geq 3)$	so_{n^2-1}	n	$(2\bar{\Lambda}_{(1)}+\bar{\Lambda}_{(n-2)})$ $\oplus(\bar{\Lambda}_{(2)}+2\bar{\Lambda}_{(n-1)})$
	so_n	$so_{n(n-1)/2}$	$n-2$	$n{=}7$: $\bar{\Lambda}_{(1)}+2\bar{\Lambda}_{(3)}$ $n{=}8$: $\bar{\Lambda}_{(1)}+\bar{\Lambda}_{(3)}+\bar{\Lambda}_{(4)}$ $n{\geq}9$: $\bar{\Lambda}_{(1)}+\bar{\Lambda}_{(3)}$
	C_n	$so_{n(2n+1)}$	$n+1$	$2\bar{\Lambda}_{(1)}+\bar{\Lambda}_{(2)}$
	E_6	so_{78}	12	2 925
	E_7	so_{133}	18	8 645
	E_8	so_{248}	30	30 380
	F_4	so_{52}	9	1 274
	G_2	so_{14}	4	77

(3.8.4)
(continued)

By inspecting the tables, further observations can be made concerning the branching rules of the embedding $\mathfrak{g} \subset \mathfrak{f}$. We write these branching rules in the form

$$(\bar{\Lambda}(\bar{\mathfrak{f}}), k^\vee(\mathfrak{f}), \Delta(\mathcal{V}_{(\mathfrak{f})})) \rightsquigarrow \bigoplus_i (\bar{\Lambda}_i(\bar{\mathfrak{g}}), k^\vee(\mathfrak{g}), \Delta_i(\mathcal{V}_{(\mathfrak{g})})), \qquad (3.8.5)$$

i.e. instead of using affine weights $\Lambda = (\bar{\Lambda}, k, n)$ we write $(\bar{\Lambda}, k^\vee, -n)$ which in the context of WZW theories seems to be a more natural notation (also, we suppress any u_1 charges). We have already learned that the levels of $\mathfrak{f}$ and $\mathfrak{g}$ are $k^\vee(\mathfrak{f}) = 1$ and $k^\vee(\mathfrak{g}) = I$, and it is also clear that $\Delta(\mathcal{V}_{(\mathfrak{f})}) - \Delta_i(\mathcal{V}_{(\mathfrak{g})}) \in \mathbb{Z}_{\geq 0}$. The rest of the information in (3.8.4) must be computed case by case.

Consider first the $\mathfrak{f}$-module corresponding to the defining module $R_{\bar{\Lambda}_{(1)}}$ of $\bar{\mathfrak{f}}$. One might expect that in this case the direct sum on the right hand side of (3.8.5) contains a single term corresponding to the elementary free fermions so that $R_{\bar{\Lambda}}(\bar{\mathfrak{g}}) = R_{[t]}$ is as listed in (3.7.21) and $\Delta(\mathcal{V}_{(\mathfrak{g})}) = \frac{1}{2}$ (on the left hand side one also has $\Delta(\mathcal{V}_{(\mathfrak{f})}) = \frac{1}{2}$ because (compare (3.4.20) and (3.4.21))

$$\Delta_{(1)} \equiv \frac{C_{\bar\Lambda_{(1)}}(\bar{\mathsf f})}{1+C_{\bar\theta}(\bar{\mathsf f})} = \frac{(d-1)/2}{1+(d-2)} = \frac{1}{2} \qquad \text{for } \bar{\mathsf f} = so_d,$$

$$\Delta_{(1)} \equiv \frac{C_{\bar\Lambda_{(1)}}(\bar{\mathsf f})}{1+C_{\bar\theta}(\bar{\mathsf f})} + \frac{1}{2\dim R_{\bar\Lambda_{(1)}}} = \frac{(d^2-1)/2d}{1+d} + \frac{1}{2d} = \frac{1}{2} \qquad \text{for } \bar{\mathsf f} = u_d. \tag{3.8.6}$$

Note that the contribution to the conformal dimension coming from the u_1 part of $u_d \cong su_d \oplus u_1$ satisfies the requirement that the contributions from the total number of elementary fermions add up to the conformal dimension $\Delta = 1$ of a free boson). However, with this contribution alone the sum rule (2.8.17) is (generically) not saturated. Thus there must be further contributions with $\Delta(\mathcal{V}_{(\mathfrak g)}) = \frac{1}{2}+n$ with $n \in \mathbb{Z}_{>0}$, i.e. the descendants of the free fermions of the f-theory not only describe the descendants of the free fermions of the $\mathfrak g$-theory, but also appropriate normal ordered products of the latter fermions (and their descendants) which are primary fields of the $\mathfrak g$-theory. The actual computation of the branching rule (3.8.5) is in many cases quite tedious. The simplest examples are:

$$\begin{array}{llll}
5 & \rightsquigarrow & 5+11 & \text{for AI,} \\
16 & \rightsquigarrow & 16+1\,200 & \text{for EIII,} \\
26 & \rightsquigarrow & 26+1\,274 & \text{for EIV,} \\
27 & \rightsquigarrow & 27+7\,371+386\,100 & \text{for EVII,} \\
16 & \rightsquigarrow & 16 & \text{for FII,} \\
14 & \rightsquigarrow & 14+182 & \text{for type II with } \mathfrak g = G_2.
\end{array} \tag{3.8.7}$$

Here for brevity only the dimensionalities with respect to the horizontal subalgebras are displayed.

Let us nevertheless for the moment pretend that on the right hand side of (3.8.5) there is only the contribution involving the free fermions of the $\mathfrak g$-theory and analyse the implications of such an assumption. Thus we define a restricted branching rule that reads

$$(\bar\Lambda_{(1)}, 1, \tfrac{1}{2}) \overset{(\psi)}{\rightsquigarrow} (\bar\Lambda_{[t]}, I, \tfrac{1}{2}). \tag{3.8.8}$$

Then, analogously, the conjugate module which is present if $\mathsf f = u_d$ is decomposed as

$$(\bar\Lambda_{(d-1)}, 1, \tfrac{1}{2}) \overset{(\psi)}{\rightsquigarrow} ((\bar\Lambda_{[t]})^+, I, \tfrac{1}{2}). \tag{3.8.9}$$

Next consider the basic module of f. Concerning the branching rule of this module, one should first recall that in terms of conformal field theory, the module corresponds to the identity primary field so that the descendants at grade one are just the currents of $\mathfrak{f}$. These must contain in particular the currents of $\mathfrak{g}$, while the rest of the $\mathfrak{f}$-currents must correspond to primary fields of the WZW theory based on $\mathfrak{g}$. These latter fields can be identified (in a case by case analysis) unambiguously by implementing the information that the dimensions of the associated $\bar{\mathfrak{g}}$-modules must add up to $d(\bar{\mathfrak{f}}) - d(\bar{\mathfrak{g}})$, and that their conformal dimension must be equal to one (and, of course, that the $\mathfrak{g}$-modules must be integrable). The resulting $\bar{\mathfrak{g}}$-modules, denoted by $R_{[f]}$, are displayed in the last column of table (3.8.4) above. In principle, further $\mathfrak{g}$-modules can come from higher grades of the basic $\mathfrak{f}$-module; these correspond to primary WZW fields of integer conformal dimension larger than one. Inspection shows that such primary fields do indeed exist in appropriate WZW theories based on $\mathfrak{g}$ at the relevant level; however, it can be seen that nevertheless these fields cannot appear in the restricted branching rule. Namely, because the elementary fermions ψ in the $\mathfrak{f}$- and $\mathfrak{g}$-theory are identical, only such modules of the horizontal subalgebra $\bar{\mathfrak{g}}$ of $\mathfrak{g}$ can be carried by the fields appearing in the operator product $\psi(z)\psi^+(w)$ of the $\mathfrak{g}$-theory which are contained via the branching rules of $\bar{\mathfrak{g}} \subset \bar{\mathfrak{f}}$ in the irreducible $\bar{\mathfrak{f}}$-modules contained in the Kronecker product $R_{\bar{\Lambda}_{(1)}} \times R_{(\bar{\Lambda}_{(1)})^+}$ of $\bar{\mathfrak{f}}$. This Kronecker product is known explicitly; for su_d, one gets the singlet and the adjoint module,

$$R_{\bar{\Lambda}_{(1)}} \times R_{\bar{\Lambda}_{(d-1)}} = R_{\bar{\Lambda}_{(0)}} \oplus R_{\bar{\Lambda}_{(1)}+\bar{\Lambda}_{(d-1)}}, \tag{3.8.10}$$

while for so_d (where $\psi^+ \equiv \psi$) one has

$$R_{\bar{\Lambda}_{(1)}} \times R_{\bar{\Lambda}_{(1)}} = R_{\bar{\Lambda}_{(0)}} \oplus R_{\bar{\Lambda}_{(2)}} \oplus R_{2\bar{\Lambda}_{(1)}}. \tag{3.8.11}$$

Inspection shows that in the branching rules of the $\bar{\mathfrak{f}}$-modules on the right hand side of these equations, no $\bar{\mathfrak{g}}$-modules appear which would lead to primary fields with integer conformal dimension $\Delta > 1$. Summarizing, one arrives at the following restricted branching rule:

$$(0,1,0) \overset{(\psi)}{\rightsquigarrow} (0,I,0) \oplus (\bar{\Lambda}_{[f]},I,1). \tag{3.8.12}$$

For $\mathfrak{f} = u_d$, the restricted branching rules of the remaining level one integrable modules of $\mathfrak{f}$ can be obtained by noticing that, in terms of the free fermion system, they correspond to normal ordered products of the elementary fermions. The same must then hold for the corresponding $\mathfrak{g}$-modules; as a consequence, it is again not difficult to determine the branching rules explicitly (the above argument concerning the compatibility with the branching rules of the horizontal subalgebra can again be used to exclude primary fields of higher conformal dimension). The computation has been carried to the end only for a few cases; here we just list the results for the embedding $E_6^{(1)} \subset A_{26}^{(1)}$ corresponding to the symmetric space EVII for which the four-point function of elementary fermions was obtained in the previous section. First one verifies (3.3.8): from the Kronecker product $27 \times 27^+ = 1 + 78 + 650$ of E_6 (cf. (1.7.58)) one learns that the primary fields appearing in the operator product $\psi(z)\psi^+(w)$ can only carry the singlet and the 650-dimensional module of E_6 (the 78-dimensional adjoint module is not allowed because the corresponding field would have non-integer conformal dimension $\Delta = \frac{3}{4}$), while the E_6-modules with dimensions 70 070, 371 800, 442 442, 1 337 050 and 2 977 975 which also give rise to integrable $E_6^{(1)}$-modules with integer conformal dimension (namely, respectively, with $\Delta = 2, 3, 4, 3, 3$) cannot appear. (Also, both the singlet and the 650-dimensional module *must* appear, the first because the identity primary field occurs in any operator product of conjugate primary fields, and the second because it is needed to fill up the states of grade one in the basic module of $A_{26}^{(1)}$.) Similarly, for the rest of the integrable modules of $A_1^{(26)}$ one finds the following restricted branching rules:

$$
\begin{aligned}
(\bar\Lambda_{(1)}, 1, \tfrac{13}{27}) &\overset{(\psi)}{\rightsquigarrow} (\bar\Lambda_{(1)}, 6, \tfrac{13}{27}) \\
(\bar\Lambda_{(2)}, 1, \tfrac{25}{27}) &\overset{(\psi)}{\rightsquigarrow} (\bar\Lambda_{(2)}, 6, \tfrac{25}{27}) \\
(\bar\Lambda_{(3)}, 1, \tfrac{4}{3}) &\overset{(\psi)}{\rightsquigarrow} (\bar\Lambda_{(3)}, 6, \tfrac{4}{3}) \\
(\bar\Lambda_{(4)}, 1, \tfrac{46}{27}) &\overset{(\psi)}{\rightsquigarrow} (\bar\Lambda_{(4)} + \bar\Lambda_{(6)}, 6, \tfrac{46}{27}) \\
(\bar\Lambda_{(5)}, 1, \tfrac{55}{27}) &\overset{(\psi)}{\rightsquigarrow} (2\bar\Lambda_{(4)}, 6, \tfrac{55}{27}) \oplus (\bar\Lambda_{(5)} + 2\bar\Lambda_{(6)}, 6, \tfrac{55}{27}) \\
(\bar\Lambda_{(6)}, 1, \tfrac{7}{3}) &\overset{(\psi)}{\rightsquigarrow} (3\bar\Lambda_{(6)}, 6, \tfrac{7}{3}) \oplus (\bar\Lambda_{(4)} + \bar\Lambda_{(5)} + \bar\Lambda_{(6)}, 6, \tfrac{7}{3}) \\
(\bar\Lambda_{(7)}, 1, \tfrac{70}{27}) &\overset{(\psi)}{\rightsquigarrow} (\bar\Lambda_{(4)} + 2\bar\Lambda_{(6)}, 6, \tfrac{70}{27}) \oplus (\bar\Lambda_{(3)} + 2\bar\Lambda_{(5)}, 6, \tfrac{70}{27}) \\
(\bar\Lambda_{(8)}, 1, \tfrac{76}{27}) &\overset{(\psi)}{\rightsquigarrow} (\bar\Lambda_{(2)} + 3\bar\Lambda_{(5)}, 6, \tfrac{76}{27}) \oplus (\bar\Lambda_{(3)} + \bar\Lambda_{(5)} + \bar\Lambda_{(6)}, 6, \tfrac{76}{27}) \\
(\bar\Lambda_{(9)}, 1, 3) &\overset{(\psi)}{\rightsquigarrow} (\bar\Lambda_{(1)} + 4\bar\Lambda_{(5)}, 6, 3) \oplus (2\bar\Lambda_{(3)}, 6, 3) \\
&\qquad\qquad \oplus (\bar\Lambda_{(2)} + 2\bar\Lambda_{(5)} + \bar\Lambda_{(6)}, 6, 3) \\
(\bar\Lambda_{(10)}, 1, \tfrac{85}{27}) &\overset{(\psi)}{\rightsquigarrow} (5\bar\Lambda_{(5)}, 6, \tfrac{85}{27}) \oplus (\bar\Lambda_{(1)} + 3\bar\Lambda_{(5)} + \bar\Lambda_{(6)}, 6, \tfrac{85}{27}) \\
&\qquad\qquad \oplus (\bar\Lambda_{(2)} + \bar\Lambda_{(3)} + \bar\Lambda_{(5)}, 6, \tfrac{85}{27}) \\
(\bar\Lambda_{(11)}, 1, \tfrac{88}{27}) &\overset{(\psi)}{\rightsquigarrow} (4\bar\Lambda_{(5)} + \bar\Lambda_{(6)}, 6, \tfrac{88}{27}) \oplus (2\bar\Lambda_{(2)} + \bar\Lambda_{(4)}, 6, \tfrac{88}{27}) \\
&\qquad\qquad \oplus (\bar\Lambda_{(1)} + \bar\Lambda_{(3)} + 2\bar\Lambda_{(5)}, 6, \tfrac{88}{27})
\end{aligned}
$$

$$
\begin{aligned}
(\bar\Lambda_{(12)},1,\tfrac{10}{3}) &\overset{(\psi)}{\leadsto} (3\bar\Lambda_{(2)},6,\tfrac{10}{3}) \oplus (\bar\Lambda_{(3)}+3\bar\Lambda_{(5)},6,\tfrac{10}{3}) \\
&\qquad \oplus (\bar\Lambda_{(1)}+\bar\Lambda_{(2)}+\bar\Lambda_{(4)}+\bar\Lambda_{(5)},6,\tfrac{10}{3})) \\
(\bar\Lambda_{(13)},1,\tfrac{91}{27}) &\overset{(\psi)}{\leadsto} (2\bar\Lambda_{(1)}+2\bar\Lambda_{(4)},6,\tfrac{91}{27}) \oplus (\bar\Lambda_{(1)}+2\bar\Lambda_{(2)}+\bar\Lambda_{(5)},6,\tfrac{91}{27}) \\
&\qquad \oplus (\bar\Lambda_{(2)}+\bar\Lambda_{(4)}+2\bar\Lambda_{(5)},6,\tfrac{91}{27})
\end{aligned}
\tag{3.8.13}
$$

(for the modules with highest weights $\bar\Lambda_{(14)},\ldots,\bar\Lambda_{(26)}$, the restricted branching rules are completely determined by the ones given above because of $\bar\Lambda_{(i)} = (\bar\Lambda_{(27-i)})^+$). In terms of the dimensionalities of the $\bar{\mathfrak{g}}$-modules, these restricted branching rules read

$$
\begin{aligned}
27 &\overset{(\psi)}{\leadsto} 27 \\
351 &\overset{(\psi)}{\leadsto} 351 \\
2\,925 &\overset{(\psi)}{\leadsto} 2\,925 \\
17\,550 &\overset{(\psi)}{\leadsto} 17\,550 \\
80\,730 &\overset{(\psi)}{\leadsto} 34\,398 + 46\,332 \\
296\,010 &\overset{(\psi)}{\leadsto} 43\,758 + 252\,252 \\
888\,030 &\overset{(\psi)}{\leadsto} 393\,822 + 494\,208 \\
2\,220\,075 &\overset{(\psi)}{\leadsto} 579\,150 + 1\,640\,925 \\
4\,686\,825 &\overset{(\psi)}{\leadsto} 371\,800 + 1\,337\,050 + 2\,977\,975 \\
8\,436\,285 &\overset{(\psi)}{\leadsto} 100\,386 + 2\,559\,843 + 5\,776\,056 \\
13\,037\,895 &\overset{(\psi)}{\leadsto} 853\,281 + 4\,582\,656 + 7\,601\,958 \\
17\,383\,860 &\overset{(\psi)}{\leadsto} 1\,559\,376 + 3\,309\,696 + 12\,514\,788 \\
20\,058\,300 &\overset{(\psi)}{\leadsto} 5\,553\,900 + 6\,747\,300 + 7\,757\,100.
\end{aligned}
\tag{3.8.14}
$$

Note that the numbers on the right hand side of (3.8.14) just add up to those of the left hand side. This means that for this particular affine embedding the restricted affine branching rules are isomorphic to the branching rules of the horizontal subalgebras. In all other cases with $\bar{\mathfrak{f}} = u_d$ that have been checked, this turns out to be true as well. Thus one arrives at the following conjecture: For a conformal embedding $\mathfrak{g} \subset \mathfrak{f}$ with $\bar{\mathfrak{f}} = u_d$, all restricted branching rules can be uniquely determined from information about $\bar{\mathfrak{g}}$ and $\bar{\mathfrak{f}}$ (namely the $\bar{\mathfrak{g}}$-Kronecker products and the branching rules for $\bar{\mathfrak{g}} \subset \bar{\mathfrak{f}}$) together with the identification of the elementary free fermions which fixes the restricted branching rules (3.8.8),

(3.8.9); moreover, except for the basic module the restricted branching rules are isomorphic to the branching rules for the embedding of the horizontal subalgebras.

Just as one obtains restricted branching rules by considering only the term in (3.8.5) corresponding to the $\mathfrak{g}$-fermions, one may equally well define other restrictions of branching rules resulting from the consideration of any other term in that sum. The full branching rule of any field of the $\mathfrak{f}$-theory is then obtainable as the direct sum of its various restrictions. In other words, at least in principle it is possible to compute all branching rules for a conformal embedding $\mathfrak{g} \subset \mathfrak{f}$ with $\bar{\mathfrak{f}} = u_d$ and $k^\vee(\mathfrak{f}) = 1$ from the branching rule (3.8.5) of the defining module by employing only properties of $\bar{\mathfrak{f}}$ and $\bar{\mathfrak{g}}$.

In the case of the embedding $E_6^{(1)} \subset A_{26}^{(1)}$, one arrives this way at the branching rules that have been listed in (2.8.16).

For $\bar{\mathfrak{f}} = so_d$, the situation is completely analogous as far as the basic module and the module $R_{\Lambda_{(1)}}$ carried by the fermions are concerned. There is now however one additional integrable level one module if d is odd, while for d even there are two. In the language of free fermions, these correspond to the spin field(s) of the conformal field theory, i.e. to the primary fields that are associated with the ground state of Ramond fermions. Again we will only present the resulting branching rule for a special example, namely the type II symmetric spaces. Thus the tangent space module is the adjoint so that $d \equiv d_{[t]} = d(\bar{\mathfrak{g}}) \equiv \dim \bar{\mathfrak{g}}$. One finds

$$(\bar{\Lambda}_\sigma, 1, \tfrac{d}{16}) \rightsquigarrow (\textstyle\bigoplus \bar{\Lambda}_{\bar{\rho}}, g^\vee(\mathfrak{g}), \tfrac{d}{16}) . \tag{3.8.15}$$

Here $\bar{\Lambda}_\sigma$ stands for the highest weight of a spinor module of $\mathfrak{f} = so_d$, i.e. for $\bar{\Lambda}_{((d-1)/2)}$ if d is odd, and either for $\bar{\Lambda}_{(d/2)}$ or for $\bar{\Lambda}_{(d/2-1)}$ if d is even, while $\bigoplus \bar{\Lambda}_{\bar{\rho}}$ stands for the highest weights of the reducible $\bar{\mathfrak{g}}$-module which is the direct sum of 2^m copies of the irreducible module with highest weight equal to the Weyl vector $\bar{\rho}$ of $\bar{\mathfrak{g}}$. Here $m \equiv [\frac{1}{2}r]$ for d odd, and $m \equiv [\frac{1}{2}r] - 1$ for d even, with $r = \text{rank}\,(\bar{\mathfrak{g}})$ and $[x]$ denoting the integer part of x (thus, to be precise, $R_{\oplus\bar{\Lambda}_{\bar{\rho}}}$ is reducible only if the rank r is larger than two, while for $r = 1$ or $r = 2$ it is irreducible). Note that despite its reducibility, the latter module possesses a well-defined Casimir eigenvalue so that the associated primary field has a well-defined conformal dimension; the numerical value of the conformal dimension is obtained according to (3.4.20) as

$$\Delta_\rho = \frac{C_{\bar{\rho}}}{2C_{\bar{\theta}}(\bar{\mathfrak{g}})} = \frac{d}{16} \tag{3.8.16}$$

(note that $C_{\bar{\rho}} = 3(\bar{\rho}, \bar{\rho})$ and that $(\bar{\rho}, \bar{\rho})$ is given by the strange formula

(1.4.71)). In terms of $\bar{\mathfrak{f}}$, the same number arises as

$$\Delta_\sigma = \frac{C_\sigma}{1 + C_{\bar{\mathfrak{g}}}(so_d)} = \frac{d(d-1)/16}{1+(d-2)}. \tag{3.8.17}$$

According to the above considerations, it is consistent to identify the fermion system based on $\mathfrak{g}$ with the fermion system based on $\mathfrak{f}$. This is of course not a big surprise because when looked upon without prejudice, the two fermion systems are really one and the same thing, namely a system of d free (real respectively complex) fermions. Classically, the maximal symmetry of this set of free fermions is given by $\bar{\mathfrak{f}}$, but of course one is also allowed to employ any subalgebra of $\bar{\mathfrak{f}}$ to describe the theory, although this is certainly somewhat unnatural. In the quantum theory, the natural choice of taking $\bar{\mathfrak{f}}$ as the classical symmetry leads to the operator product

$$\psi(z)\,\psi^+(w) = \mathbf{1} + \ldots\,, \tag{3.8.18}$$

with the ellipsis standing for descendants, i.e. only a single primary field appears on the right hand side. This follows because the additional modules appearing in the corresponding $\bar{\mathfrak{f}}$-Kronecker product (3.8.10) respectively (3.8.11) do not give rise to integrable modules of $\mathfrak{f}$ at level one so that no primary fields carrying those $\bar{\mathfrak{f}}$-modules are allowed. The same reasoning also shows that for any other operator product in the fermion theory based on $\bar{\mathfrak{f}}$ only a single primary field appears on the right hand side. Free fermion theories with these properties are sometimes called *minimal* free fermion theories. Any unnatural choice of the symmetry algebra, on the other hand, leads to non-minimal free fermion theories; in particular, from the branching rule (3.8.12) one has

$$\psi(z)\,\psi^+(w) = \mathbf{1} + \phi_{[f]} + \ldots \tag{3.8.19}$$

(with $\phi_{[f]}$ denoting a set of three primary fields transforming as $1 \otimes R_{[f]}$, $R_{[f]} \otimes 1$ and $R_{[f]} \otimes R_{[f]}$ with respect to the left and right horizontal subalgebras, respectively) for the fermion theory based on $\mathfrak{g}$.

There is a unique WZW theory based on $\bar{\mathfrak{f}} = su_n$ or $\bar{\mathfrak{f}} = so_n$ at level one of the corresponding affine algebras. Therefore the WZW sigma model on the simply connected group manifold associated to $\bar{\mathfrak{f}}$ is, at level one, quantum equivalent to the free fermion system based on $\mathfrak{f}$. Analogously, the free fermion system based on $\mathfrak{g}$ must be quantum equivalent to a WZW theory based on $\mathfrak{g}$ because both of them are obtainable via the conformal embedding $\mathfrak{g} \subset \mathfrak{f}$ from the fermion and WZW theory based on $\mathfrak{f}$, respectively. However, the WZW theory based on $\mathfrak{g}$ defined this way is *not* the WZW sigma model on the simply connected group manifold having $\bar{\mathfrak{g}}$ as its Lie algebra; in particular, its spectrum is not of the diagonal form (3.3.30). Stated differently, the quantum equivalence of the fermion theories related by the conformal embedding $\mathfrak{g} \subset \mathfrak{f}$ implies that the corresponding WZW theories are quantum equivalent, too. While this equivalence is a rather trivial fact when formulated in terms of the fermion theories, it is still an important observation when expressed in terms of the WZW theories without referring to free fermions. In particular, it implies that the two WZW theories possess the same four-point functions, and as we have seen in the previous section it requires some work to establish this by direct computation.

When looking for the "natural" classical symmetry algebra of the free fermion system, the question also arises whether it is appropriate to distinguish between real and complex fermions. Clearly, the maximal symmetry of the classical Lagrangian for d complex free fermions is so_{2d} rather than u_d. In terms of affine algebras, this finds its counterpart in the existence of a conformal embedding $A^{(1)}_{d-1} \oplus \hat{u}_1 \subset D^{(1)}_d$. However, this embedding differs significantly from those considered so far. First, the Dynkin index of the associated horizontal embedding is equal to one. Moreover, the branching rule for the basic module of $D^{(1)}_d$ contains the basic module of $A^{(1)}_{d-1}$ twice, namely

$$(0,1,0) \rightsquigarrow 2\times(0,1,0) \oplus (\bar{\Lambda}_{(2)},1,1) \oplus (\bar{\Lambda}_{(d-2)},1,1), \qquad (3.8.20)$$

and the module carried by the fundamental fermions branches to the direct sum of two modules which are conjugate to each other rather than to a single module,

$$(\bar{\Lambda}_{(1)},1,\tfrac{1}{2}) \rightsquigarrow (\bar{\Lambda}_{(1)},1,\tfrac{1}{2}) \oplus (\bar{\Lambda}_{(d-1)},1,\tfrac{1}{2}). \qquad (3.8.21)$$

(One may also note that the embedding $u_d \subset so_{2d}$ is a (maximal) regular embedding whereas all embeddings $\bar{\mathfrak{g}} \subset \bar{\mathfrak{f}}$ encountered before are special embeddings (namely, the composition of two maximal special embeddings for the cases EII, EVI, EIX, FI and G, and a maximal special embedding for all other cases); also, so_{2d}/u_d is a symmetric space, namely DIII, whereas none of the coset spaces $\bar{\mathfrak{f}}/\bar{\mathfrak{g}}$ is symmetric.) In view of these differences, and also of the fact that the fermion theory based on u_d is already minimal in the sense introduced above, it seems that it does make sense to distinguish between real and complex free fermions.

In contrast, the notion of quaternionic fermions, carrying symplectic modules of a finite-dimensional Lie algebra, is much less natural because the corresponding theories based on $C_{d/4} \oplus su_2$ (i.e. $\bar{\mathfrak{g}}$ for the case of CII with $n = 1$, and $\bar{\mathfrak{f}}_\circ \oplus su_2$ for the cases with non-maximal embeddings) are not minimal fermion theories, but become minimal only when interpreted as theories based on so_d.

3.9 Literature

General information about WZW theories is contained in the various reviews of conformal field theory, e.g. [R15, R20, R2, R3, R14, R30, R25, R6, R13, R1, R23]. All these review papers also treat in detail various aspects of two-dimensional conformal field theory, including many issues which for lack of space could not be reproduced or even mentioned here. A topic which is dealt with in [R20, R23] (see also [94]), but not in most of the other reviews, is the meaning of the compact picture (see section 3.3); the paper [R20] also contains a comprehensive list of references to the older literature about conformal invariance in field theory. In [221], a different interpretation of conformal field theory is put forward, namely as the analytic geometry of the universal moduli space of Riemann surfaces. For the relevance of conformal field theory to string theory, see [R26, 216, B14, R24, B26, 359], and for applications in statistical mechanics consult the reprint volume [R16].

The Sugawara form of the energy–momentum tensor was first given in [535, 529], although with an incorrect value of the numerical prefactor; the correct prefactor was found for $A_r^{(1)}$ at level one in [143], and for the general case in [349]. In the mathematical literature, the Sugawara structure first showed up in [519, 212, 269]. Higher order Sugawara-like operators have been discussed in [33, 285, 270].

The most important original paper about conformal field theory is certainly the one by Belavin, Polyakov, and Zamolodchikov [47] in which the concept of the operator product algebra and the conformal bootstrap program were successfully applied to nontrivial theories, which was possible by implementing the Ward identities of the Virasoro algebra. However, the underlying ideas had been present long before: The relevance of the Virasoro algebra to conformal field theory was first pointed out in [392, 393]. The field theoretic approach to second order phase transitions was introduced in [274, 471]; the presence of conformal (rather than only scale) invariance in such systems was realized by Polyakov [472]; see also [421, 403]. Polyakov [473] and others [367, 391, 401] also combined conformal invariance with the concept [323, 579] of a closed operator algebra to arrive at the conformal bootstrap hypothesis. Early investigations of conformal transformations [516] and operator product expansions [517] were also undertaken in the framework of the algebraic approach to quantum field theory. (The algebraic approach has been developed e.g. in [278, 277, 73, 164, 165, 93]; for further references, see [B22, B16, R20, R25, R17].)

The determination of the operator product coefficients from four-point functions of primary fields was described in [166, 167, 230]; it was applied to minimal conformal models in [168, 467, 468], to diagonal $A_1^{(1)}$ WZW theories in [597, 124], and to nondiagonal $A_1^{(1)}$ WZW theories in [152, 331, 236, 231, 170, 133, 237].

The Virasoro algebra was discovered in [247], and independently in [91]. The latter reference was based on [560, 561] where the centerless version of the algebra, i.e. the Witt algebra, was considered (the Witt algebra was known much earlier; see for instance [600, 117, 483]). An analysis of the associated Virasoro *group* has been undertaken in [74]; see also [519, 582]. Verma modules for the Virasoro algebra were discussed in [195, 196]. The unitarity constraints for $c < 1$ were found in [217, 218, 220], based on a formula for the matrix of inner products of Verma module vectors at fixed grade that was conjectured by Kac [313] and proved by Feigin and Fuks [195]; the characters for these unitary modules were computed in [492]. Explicit formulae for the null vectors of Verma modules of the Virasoro algebra were obtained in [542, B21, 240, 200, 50, 39]. For a self-contained proof that unitarity of a highest weight Virasoro module with $c = 0$ implies $L_n = 0$, see [265].

The best known extended symmetry algebras for conformal field theories besides the affine algebras are the N=1 (see [453, 479] and [219, 181, 57]) and N=2 (see [158, 75, 449, 593]) superconformal algebras. Supersymmetric generalizations of affine algebras were considered in [321, 159, 224, 348, 450, 229, 416]. More recently, the extended algebras known as W-algebras have been the subject of intensive research; see e.g. [192, 33, 189, 68, 84, 263, 477, 63, 64, 65, 288]. General considerations about extended algebras are contained in [595, 77, 205].

WZW theories were introduced, as sigma models on group manifolds with Wess–Zumino [573, 456, 580, 146] term, by Witten [581]. In the language of conformal field theory, they were discussed by Knizhnik and Zamolodchikov [349], and by Gepner and Witten [251] who derived in particular the null vector equations described in sections 3.4 and 3.5, respectively. In the path integral framework, WZW theories were treated in [474, 481, 202, 203, 112, 326]. The coset construction of Goddard, Kent, and Olive was derived in [258, 259] following earlier work in [407, 37, 281]; it is also discussed in [262, 169]. A path integral realization of the construction in terms of gauged WZW theories is described in [246, 446, 10, 83, 514, 327]. Various series of mimimal conformal field theories have been identified with coset theories in [9, 57, 34, 30, 330, 480].

In [349], the explicit form of the Knizhnik–Zamolodchikov equation was worked out only for the four-point function involving the defining module of $\mathfrak{g} = A_r$ (for this special case this differential equation was in fact derived much earlier [143]). For some other four point functions, the Knizhnik–Zamolodchikov equation has meanwhile been worked out in the context of various applications: for arbitrary modules of $\mathfrak{g} = A_1$ [124, 597], for the defining modules of B_r, D_r [224], C_r [225], G_2, F_4 [232] and E_6 (see (3.4.52)) [227], for the spinor module of B_r [232], and for the symmetric and antisymmetric tensor modules of A_r [226]. The Gepner–Witten equation was employed in [251] to find the explicit form of the depth rule and the WZW operator products for the case of $\mathfrak{g} = A_1$; the example of E_8 at level two was discussed in [208]. The special case (3.5.17) of the Gepner–Witten equation, the derivation of the criterion (3.5.32) for cominimal fields, and the calculation of their four-point functions, were presented in [235].

The method of contour deformation to manipulate products of operators was first used in [90]. The particular normal ordering prescription used here (see e.g. (3.1.22) and (3.1.33)) was first described in [56, 33]; its extension to fields with non-integer moding (see (3.1.39)) is taken from unpublished notes by the present author.

Representations of affine Lie algebras expressed in terms of free fermions were first discussed in [212], and are reviewed e.g. in [R15]. However, the conformal field theory of free fermions is usually treated only with methods tailored to free field theories, whereas the discussion of section 3.6 emphasizes methods (such as the Wick rule (3.1.35)) which are in the spirit of generic conformal field theories. The concept of quantum equivalence is implicit already in [47]; it has been discussed in section 3.7 in detail because sometimes (see e.g. [R15]) the necessary conditions for quantum equivalence are not stated properly. The symmetric space theorem, classifying the free fermion theories with Sugawara type energy–momentum tensor, was derived by Goddard, Nahm, and Olive [260]. The quantum equivalence between free fermions based on $\mathfrak{g} = so_n$ and u_n and the corresponding (diagonal) WZW theories was found in [581] and [349], respectively; the other cases of the Goddard-Nahm-Olive list were discussed in [225, 226] (in those papers, the term "WZW theory" was reserved for *diagonal* WZW theories). The connection with conformal embeddings was pointed out in [21, 226]. The implications of the quantum equivalence theorems to affine branching rules (e.g. the results (3.8.13)) are stated here for the first time; the branching rule (3.8.15) was found in [316] (see also [229]).

The vertex operator construction of level one highest weight modules by which WZW theories become equivalent to free bosons was derived for $\mathfrak{g} = A_r$ in [280], and more generally for $\mathfrak{g}$ simply laced by Frenkel and Kac [214] and by Segal [519]; see also the references given in section 2.9. Vertex operator constructions for non-simply laced $\mathfrak{g}$ are discussed in [261, R5, 23, 17], and for the twisted affine algebras in [R5]. The equivalence of free bosons and free fermions was first observed by Skyrme [526] and others [37, 534, 128, 408], and was made precise in [280, 35, 212, 581]. An equivalence between WZW theories and theories of *constrained* fermions has been discussed in [20, 481, 482, 445]. WZW theories at arbitrary level can also be identified with the combination of rank($\mathfrak{g}$) free bosons and so-called parafermions (in a sense, this relation defines the parafermions, namely by the coset construction $\mathfrak{g}/\mathfrak{h}$ with $\mathfrak{h}$ the Cartan subalgebra of $\mathfrak{g}$); this has been shown in [191, 249]; for the mathematical background see [377, 379, 422].

The Coulomb gas construction which expresses other conformal field theories in terms of free bosons coupled to a background charge is due to Feigin and Fuks [197], and to Dotsenko and Fateev [166, 167, 168] who applied it to correlation functions of the minimal $c < 1$ conformal field theories on the sphere. Applications to higher genus Riemann surfaces are contained in [31, 200, 204], and the discussion of other minimal series can be found in [192, 189, 190, 442, 443, 289]. For the Coulomb gas picture of two-dimensional statistical mechanics models, see also [454].

The realization of (integrable, but also of non-integrable) modules of affine algebras through free bosons and ghosts is known as the Wakimoto construction; for the integrable modules it requires a proper resolution of the Fock spaces of the free fields analogous to the Bernstein–Gelfand–Gelfand resolution [55] of Verma modules of simple Lie algebras. This realization was obtained for $\mathfrak{g} = A_1$ in [565], and in more generality in [193, 194, 7, 54, 79, 80, 501, 370] (see also [157, 452] for similar constructions, for which however the resolution procedure is more problematic). As an important application of the Wakimoto construction, one can obtain a Coulomb gas representation of WZW theories [252, 253] and of coset theories [368].

4
Quantum groups

One of the fundamental tasks in quantum field theory is to express the properties of the field theory in terms of an underlying symmetry object. For conformal field theories, this means that in particular one would like to understand the operator product algebra in terms of the symmetry algebra of the theory. It has already been mentioned that the operator product coefficients involving secondary fields can be expressed through those involving only primary fields via the Ward identities of the symmetry algebra. On the other hand, it also became clear (see p. 171) that the operator product coefficients of primary fields *cannot* be understood in terms of the representation theory of the symmetry algebra alone.

It is therefore natural to search for structures related to the symmetry algebra which do possess tensor products corresponding to the operator product algebra. As will be seen later (section 5.7), this search leads to the concepts of Hopf algebras and quantum groups. It is the purpose of the present chapter to introduce these mathematical objects.

In section 4.1 we present the definition of Hopf algebras and of quantum groups, by which we essentially mean quasitriangular Hopf algebras (there is as yet no generally accepted use of the term quantum group; accordingly, we will use it not only for quasitriangular Hopf algebras, but also for other related structures; from the context it will always be clear which particular mathematical object we are talking about). Section 4.1 also lists the basic properties of Hopf algebras and quantum groups, and gives a few simple examples; in addition, the Yang–Baxter equation is derived, and its relation with integrable systems is pointed out. A particular class of quasitriangular Hopf algebras is obtained by deformations of the enveloping algebras of semisimple Lie algebras; these algebras which will turn out to be relevant to WZW theories are the subject of sections 4.2 to 4.5. Section 4.2 contains the basic definitions and a preliminary discussion of the R-matrix of these algebras. Section 4.3 describes the representation theory, concentrating on the special features that arise when the deformation parameter is a root of unity. In the latter case it is possible to introduce a modified Kronecker product, which is presented in section 4.5; the definition of this tensor product is motivated in section 4.4 by investigating the so-called quantum dimensions of the modules.

In section 4.7 we discuss the quantum double construction that provides an algorithm to calculate universal R-matrices and apply it to the simplest cases of deformed enveloping algebras; also, the R-matrix in the defining representation is presented for a few more complicated cases. Quantum groups can alternatively be introduced by a formal quantization procedure, either in an abstract algebraic setting, or more concretely by analysing groups of matrices, which leads to the concept of matrix quantum groups. The quantization procedure and the interrelations between the various approaches to quantum groups are also described in section 4.7. Finally, section 4.8 is devoted to the problem of finding a direct expression for the generators of deformed enveloping algebras in terms of the generators of affine Lie algebras. A solution to this problem has not yet been found, and it is not even clear whether such a relation can exist at all. Therefore this section is mainly for entertainment; however, it also prepares the stage for the description of the known relations between deformed enveloping algebras and affine algebras that will be the subject of chapter 5.

4.1 Hopf algebras

The fundamental new concept to be introduced is the one of a Hopf algebra; this is an associative algebra endowed with a lot of additional structure:

Definition : A **Hopf algebra** a is a vector space endowed with five operations

$$\begin{aligned} &\mathcal{M} &&: \mathsf{a}\times\mathsf{a}\to\mathsf{a} &&\text{(multiplication)},\\ &\eta &&: F\to\mathsf{a} &&\text{(unit map)},\\ &\boldsymbol{\Delta} &&: \mathsf{a}\to\mathsf{a}\times\mathsf{a} &&\text{(co-multiplication)},\\ &\varepsilon &&: \mathsf{a}\to F &&\text{(co-unit map)},\\ &\gamma &&: \mathsf{a}\to\mathsf{a} &&\text{(antipode)},\end{aligned} \tag{4.1.1}$$

which possess the following properties:

$$\begin{array}{lll} (4.1.1a) & \mathcal{M}\circ(id\times\mathcal{M})=\mathcal{M}\circ(\mathcal{M}\times id) & \text{(associativity)},\\ (4.1.1b) & \mathcal{M}\circ(id\times\eta)=id=\mathcal{M}\circ(\eta\times id) & \text{(existence of unit)},\\ (4.1.1c) & (id\times\mathbf{\Delta})\circ\mathbf{\Delta}=(\mathbf{\Delta}\times id)\circ\mathbf{\Delta} & \text{(co-associativity)},\\ (4.1.1d) & (\varepsilon\times id)\circ\mathbf{\Delta}=id=(id\times\varepsilon)\circ\mathbf{\Delta} & \text{(existence of co-unit)},\\ (4.1.1e) & \mathcal{M}\circ(id\times\gamma)\circ\mathbf{\Delta}=\eta\circ\varepsilon=\mathcal{M}\circ(\gamma\times id)\circ\mathbf{\Delta}, & \\ (4.1.1f) & \mathbf{\Delta}\circ\mathcal{M}=(\mathcal{M}\times\mathcal{M})\circ(\mathbf{\Delta}\times\mathbf{\Delta}) & \text{(connecting axiom)}. \end{array}$$

Here F is the base field of the vector space; in the following it will be assumed that $F = \mathbb{C}$ or $F = \mathbb{R}$.

At first sight, this structure, involving five different operations on a vector space, might seem quite complicated. This is not at all so, however. First of all, associative algebras, i.e. vector spaces endowed with a multiplication which obeys (4.1.1a), are very common structures as they just formalize the notion of multiplying objects, and many of these algebras (called *unital algebras*) possess a unit. Analogously, by formalizing the notion of "unmultiplying" objects, one gets the operations of co-multiplication and co-unit; vector spaces endowed with a co-multiplication are called *co-algebras.* The theory of co-algebras is essentially dual to the theory of algebras, and hence it is not necessary to consider them in their own right; however, the concepts of co-multiplication and co-unit also arise naturally when one tries to define tensor products of representations of an algebra. Recall from the end of section 1.2 that in general such tensor products cannot be defined for associative algebras which do not possess any further structure. To avoid this problem, one introduces an additional operation $\mathbf{\Delta}$ such that for any two representations $R_i : \mathsf{a} \to gl(V_i)$, $i = 1, 2$, the map

$$R: \begin{array}{rcl} \mathsf{a} & \to & gl(V_1 \times V_2) \\ x & \mapsto & (R_1 \times R_2)\circ\mathbf{\Delta}(x) \end{array} \tag{4.1.2}$$

is a representation too, i.e. is linear and preserves the multiplication. Because of the associativity of the multiplication, it is also natural to require that the formation of tensor products is associative (up to isomorphism). These requirements then restrict the map $\mathbf{\Delta}$ to being co-associative, i.e. to obeying (4.1.1c) (and also to be linear and preserve the multiplication). Having a co-multiplication as a dual operation to the multiplication, it is again natural to require also the existence of a co-unit as the dual operation to the unit. The algebra a then possesses the four operations $\mathcal{M}$, η, $\mathbf{\Delta}$ and ϵ obeying (4.1.1$a-d$); an algebra which is endowed with these operations such that they are compatible with each other is called a *bi-algebra.* Finally, one may look for a sensible way to connect the operations of multiplication and co-multiplication non-trivially. This amounts to constructing a map as essentially the composition of the multiplication and co-multiplication; this then naturally leads to the concept of the antipode, and hence to Hopf algebras.

The defining properties (4.1.1) of Hopf algebras have been written in terms of the operations of (co-) multiplication, (co-) unit and antipode and the identity map *id*. They may also be expressed more explicitly writing out the elements $x \in \mathsf{a}$. For example, introducing the symbol "$\diamond$" to write the multiplication $\mathcal{M}$ as

$$\mathcal{M}: x \otimes y \mapsto \mathcal{M}(x \otimes y) \equiv x \diamond y, \tag{4.1.3}$$

the associativity property reads

$$x \diamond (y \diamond z) = (x \diamond y) \diamond z \quad \text{for all} \quad x, y, z \in \mathsf{a}. \tag{4.1.4}$$

Similarly, the existence of a unit means that there is an element $e \in \mathsf{a}$ (called the unit element) such that

$$e \diamond x = x = x \diamond e \quad \text{for all} \quad x \in \mathsf{a}; \tag{4.1.5}$$

the map η is then given by

$$\eta : \xi \mapsto \xi e \quad \text{for all} \quad \xi \in F. \tag{4.1.6}$$

Another possibility of writing such properties in a more explicit way is to visualize them by commutative diagrams which display the various spaces involved together with the maps among them (a diagram is said to be *commutative* iff the composite maps which are obtained by following the arrows are independent of the path which is used to link any two given spaces in the diagram). The defining properties of a Hopf algebra are then displayed as shown in (4.1.7) to (4.1.11).

associativity:

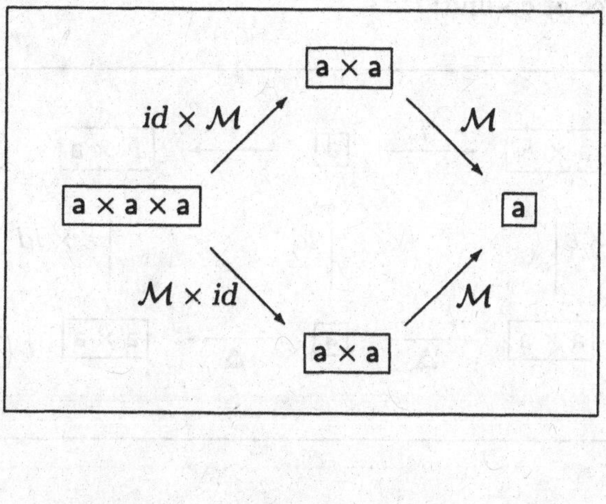

(4.1.7)

existence of unit:

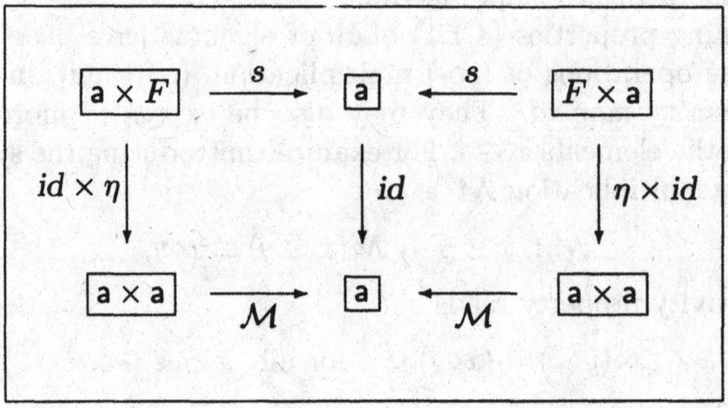

(4.1.8)

co-associativity:

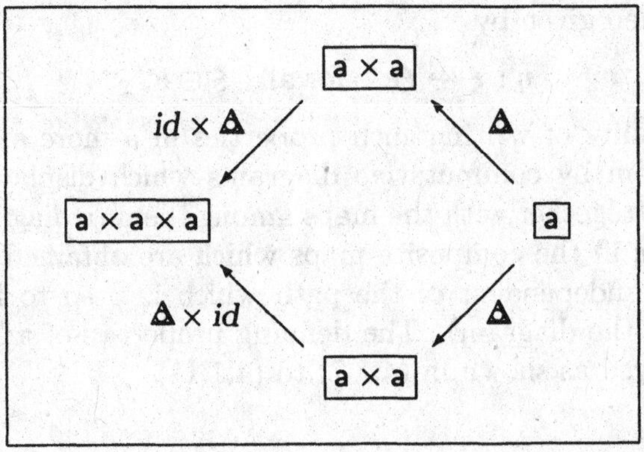

(4.1.9)

existence of co-unit:

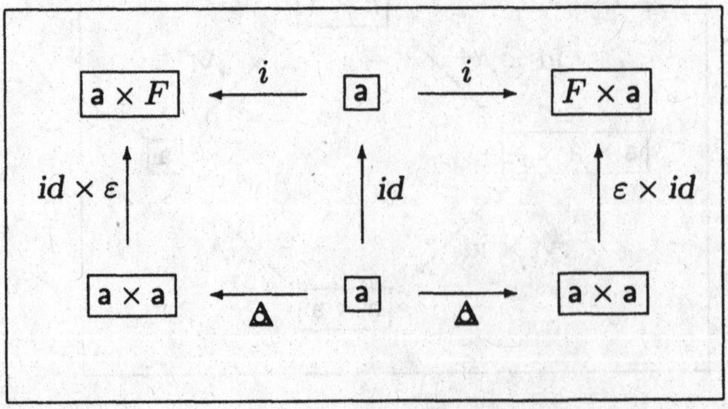

(4.1.10)

existence of antipode:

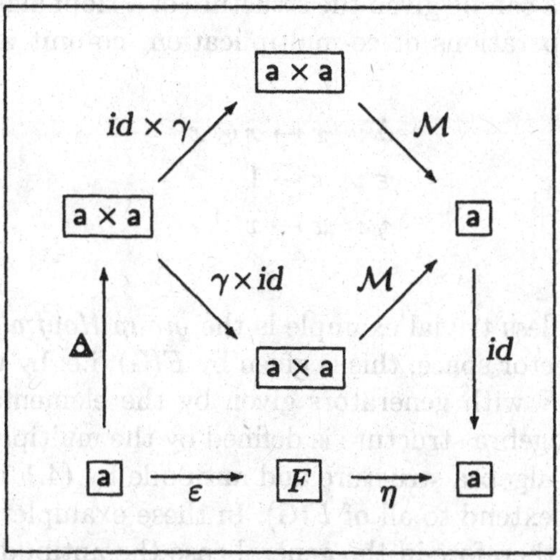

(4.1.11)

In (4.1.8) and (4.1.10) the maps $s : \mathsf{a} \times F \to \mathsf{a}$ and $i : \mathsf{a} \to \mathsf{a} \times F$ denote the scalar multiplication and the inclusion $x \mapsto x \otimes 1$ (with 1 the multiplicative unit of F), respectively. Also, the third and fourth diagrams are displayed in a way which makes it obvious that they are obtained from the first two diagrams by simply "reversing the arrows" (this justifies the terms *co-multiplication* and *co-unit*).

When a basis $\{e_p\}$ of a is fixed, then multiplication, co-multiplication and antipode get expressed in terms of structure constants, just as is the case for the bracket relations of a Lie algebra. Let us denote the respective matrices by μ, ν, τ according to

$$\begin{aligned} e_s \diamond e_t &= \textstyle\sum_p \mu^p{}_{st}\, e_p, \\ \Delta(e_p) &= \textstyle\sum_{s,t} \nu_p{}^{st}\, e_s \otimes e_t, \\ \gamma(e_p) &= \textstyle\sum_s \tau_p{}^s\, e_s. \end{aligned} \tag{4.1.12}$$

The associativity property (4.1.1a) is then expressed as

$$\mu^p{}_{qr}\mu^s{}_{pt} = \mu^p{}_{rt}\mu^s{}_{qp}, \tag{4.1.13}$$

a formula which we already wrote down, in a different notation, in (1.1.46). Similarly, the structural properties (4.1.1c) and (4.1.1e) of a Hopf algebra read

$$\begin{aligned} \nu_p{}^{qr}\nu_q{}^{st} &= \nu_p{}^{sq}\nu_q{}^{tr}, \\ \nu_p{}^{qr}\tau_r{}^s\mu^t{}_{qs} &= \nu_p{}^{sr}\tau_s{}^q\mu^t{}_{qr}. \end{aligned} \tag{4.1.14}$$

A trivial example of a Hopf algebra is obtained by considering any algebra which is also a multiplicative group (i.e. any algebra such that its bilinear operation $\mathcal{M}$ is associative and possesses a unit and an inverse). Each such algebra can be given the structure of a Hopf algebra by defining the additional operations of co-multiplication, co-unit and antipode as follows:

$$\begin{aligned} \Delta:&\quad x \mapsto x \otimes x, \\ \varepsilon:&\quad x \mapsto 1, \\ \gamma:&\quad x \mapsto x^{-1} \end{aligned} \tag{4.1.15}$$

for all $x \in \mathsf{a}$.

A similar, but less trivial example is the *group Hopf algebra* of a finite group G. As a vector space, this is given by $F(\mathsf{G})$, i.e. by the vector space over some field F with generators given by the elements of G. On the generators, the algebra structure is defined by the multiplication and unit of G, and the co-algebra structure and antipode by (4.1.15); by linearity, these definitions extend to all of $F(\mathsf{G})$. In these examples the antipode is just the inverse; therefore in the general case the antipode is often called the "analogue of an inverse" or also a "co-inverse." However, defining a product "$\bullet$" of maps on a by $\varphi \bullet \psi := \mathcal{M} \circ (\varphi \times \psi) \circ \Delta$, one has $id \bullet \gamma = \gamma \bullet id = \eta \circ \varepsilon$, so that one also might look at the antipode as a kind of inverse of the identity map; also, for a generic Hopf algebra it is *not* required that $\gamma \circ \gamma = id$ nor even that γ as a linear map possesses an inverse.

Another interesting simple example is provided by the universal enveloping algebras $\mathsf{U}(\mathfrak{g})$ of complex Lie algebras $\mathfrak{g}$. $\mathsf{U}(\mathfrak{g})$ becomes a Hopf algebra by taking $\mathcal{M}$ as the usual formal multiplication on $\mathsf{U}(\mathfrak{g})$, and defining the unit by $\eta(\xi) = \xi 1$ for all $\xi \in \mathbb{C}$, and the co-multiplication, co-unit and antipode by

$$\left.\begin{aligned} \Delta(x) &= x \otimes 1 + 1 \otimes x \\ \varepsilon(x) &= 0 \\ \gamma(x) &= -x \end{aligned}\right\} \quad \text{for all } x \in \mathfrak{g}, \tag{4.1.16}$$

and

$$\begin{aligned} \Delta(1) &= 1 \otimes 1, \\ \varepsilon(1) &= 1, \\ \gamma(1) &= 1. \end{aligned} \tag{4.1.17}$$

These formulae define the operations on the generators of $\mathsf{U}(\mathfrak{g})$ and hence, as a consequence of the homomorphism property of Δ (see below), on all of $\mathsf{U}(\mathfrak{g})$; also note that in this case $\gamma \circ \gamma = id$.

In particular, for any $x \in \mathsf{g}$ and $\xi \in \mathbb{C}$, the formal element $X = \mathrm{e}^{\xi x} \in$ $\mathsf{U(g)}$ obeys

$$\begin{aligned} \Delta(\mathrm{e}^{\xi x}) &= \mathrm{e}^{\xi x} \otimes \mathrm{e}^{\xi x}, \\ \gamma(\mathrm{e}^{\xi x}) &= \mathrm{e}^{-\xi x}. \end{aligned} \tag{4.1.18}$$

As a consequence, this element satisfies $X \diamond \gamma(X) = 1 = \gamma(X) \diamond X$, which justifies again the description of the antipode as the analogue of an inverse. The element $X \in \mathsf{U(g)}$ can also be interpreted as an element of a Lie group G whose Lie algebra is g; more generally, one calls any element w of a Hopf algebra a a *group-like element* iff it satisfies an equation analogous to (4.1.18), namely

$$\begin{aligned} \Delta(w) &= w \otimes w, \\ w \diamond \gamma(w) &= e = \gamma(w) \diamond w. \end{aligned} \tag{4.1.19}$$

(Thus in the previous trivial example, all elements of a are group-like.)

Some general properties of Hopf algebras are the following.

- For a given multiplication and co-multiplication, the co-unit is unique.
- If $(\mathsf{a}, \mathcal{M}, \eta, \Delta, \varepsilon, \gamma)$ is a finite-dimensional Hopf algebra, then the dual vector space $\mathsf{a}^\star$ (i.e. the space of linear maps φ on a) inherits a Hopf algebra structure from a by "interchanging" $\mathcal{M}$, η with Δ, ε. More precisely, the dual Hopf algebra is $(\mathsf{a}^\star, \mathcal{M}^\star, \eta^\star, \Delta^\star, \varepsilon^\star, \gamma^\star)$ with the various operations defined by

$$\begin{aligned} (\mathcal{M}^\star(\varphi \otimes \psi))\,(x) &:= (\varphi \times \psi)\,(\Delta(x)), \\ (\eta^\star(\xi))\,(x) &:= \xi\, \varepsilon(x), \\ (\Delta^\star(\varphi))\,(x \otimes y) &:= \varphi(x \diamond y), \\ \varepsilon^\star(\varphi) &:= \varphi(\eta), \\ (\gamma^\star(\varphi))\,(x) &:= \varphi(\gamma(x)) \end{aligned} \tag{4.1.20}$$

for $x, y \in \mathsf{a}$, $\varphi, \psi \in \mathsf{a}^\star$ and $\xi \in F$. In terms of the basis $\{e_s\}$ of a introduced in (4.1.12), and denoting by $\{e^s\}$ the dual basis of $\mathsf{a}^\star$ (i.e. the basis obeying $e^s(e_t) = \delta^s{}_t$), this means that

$$\begin{aligned} \mathcal{M}^\star(e^s \otimes e^t) &= \nu_p{}^{st}\, e^p, \\ \Delta^\star(e^p) &= \mu^p{}_{st}\, e^s \otimes e^t, \\ \gamma^\star(e^s) &= \tau_t{}^s\, e^t. \end{aligned} \tag{4.1.21}$$

In the infinite-dimensional case, one has in general $(\mathsf{a}^\star)^\star \neq \mathsf{a}$ so that the definition of a dual Hopf algebra becomes more involved; one possibility is to define the dual Hopf algebra again via the relations (4.1.20), but to restrict the vector space $\mathsf{a}^\star$ to an appropriate subspace on which the pairing effected by these relations is non-degenerate. Thus in short, the dual of a Hopf algebra is the maximal Hopf algebra contained in the dual vector space.

- The co-multiplication and co-unit are homomorphisms of a, i.e. preserve the multiplication.

For the co-unit, this simply means that

$$\varepsilon \circ \mathcal{M} = m \circ (\varepsilon \times \varepsilon), \tag{4.1.22}$$

where $m\colon F \to F$ is the multiplication in the base field, or in other words

$$\varepsilon(x \diamond y) = \varepsilon(x)\varepsilon(y) \tag{4.1.23}$$

for all $x, y \in \mathsf{a}$. For the co-multiplication, the corresponding formulae require a careful definition of what is meant by the multiplication in the four-fold direct product of the algebra: writing for brevity

$$\Delta : \begin{array}{c} \mathsf{a} \to \mathsf{a} \times \mathsf{a} \\ x \mapsto x_1 \otimes x_2, \end{array} \tag{4.1.24}$$

the property that Δ is a homomorphism is meant in the sense that $(x \diamond y)_1 = x_1 \diamond y_1$, $(x \diamond y)_2 = x_2 \diamond y_2$ for all $x, y \in \mathsf{a}$. Of course, the object $x_1 \otimes x_2$ in (4.1.24) is just a shorthand notation for $\sum_\ell x_1^{(\ell)} \otimes x_2^{(\ell)}$; in more detail the homomorphism property means that $\Delta(x \diamond y) = \sum_{\ell,m} (x_1^{(\ell)} \diamond y_1^{(m)}) \otimes (x_2^{(\ell)} \diamond y_2^{(m)})$. Thus the homomorphism property may be formalized as

$$\Delta \circ \mathcal{M} = (\mathcal{M}_{13} \times \mathcal{M}_{24}) \circ (\Delta \times \Delta). \tag{4.1.25}$$

(This is the explicit form of the connecting axiom (4.1.1f); $\mathcal{M}_{13}$ acts as

$$\mathcal{M}_{13} : \begin{array}{c} \mathsf{a} \times \mathsf{a} \times \mathsf{a} \times \mathsf{a} \to \mathsf{a} \times \mathsf{a} \times \mathsf{a} \\ w \otimes x \otimes y \otimes z \mapsto (w \diamond y) \otimes x \otimes z, \end{array} \tag{4.1.26}$$

and similarly for $\mathcal{M}_{24}$, so that in particular

$$\mathcal{M}_{13} \times \mathcal{M}_{24} : \begin{array}{c} \mathsf{a} \times \mathsf{a} \times \mathsf{a} \times \mathsf{a} \to \mathsf{a} \times \mathsf{a} \\ w \otimes x \otimes y \otimes z \mapsto (w \diamond y) \otimes (x \diamond z). \end{array} \tag{4.1.27}$$

In the following we will use for this the shorthand notation

$$\mathcal{M} : \begin{array}{c} \mathsf{a} \times \mathsf{a} \times \mathsf{a} \times \mathsf{a} \to \mathsf{a} \times \mathsf{a} \\ w \otimes x \otimes y \otimes z \mapsto (w \diamond y) \otimes (x \diamond z) \end{array} \tag{4.1.28}$$

so that the connecting axiom can be rewritten as

$$\Delta(x \diamond y) = \Delta(x) \diamond \Delta(y) \quad \text{for all} \quad x \otimes y \in \mathsf{a} \times \mathsf{a}. \tag{4.1.29}$$

Similarly, $\mathcal{M} : \mathsf{a}^{\otimes 6} \to \mathsf{a}^{\otimes 3}$ will be used as a shorthand notation for $\mathcal{M}_{14} \times \mathcal{M}_{25} \times \mathcal{M}_{36}$, etc. As an example for (4.1.29), consider the simple case where the co-multiplication acts on generators of $\mathbf{a} \times \mathbf{a}$ as $\Delta(x) = x \otimes 1 + 1 \otimes x$; then $\Delta(x \diamond y) = (x \diamond y) \otimes 1 + x \otimes y + y \otimes x + 1 \otimes (x \diamond y)$.

- The antipode is an *anti-homomorphism*, i.e.

$$\gamma(x \diamond y) = \gamma(y) \diamond \gamma(x) \qquad (4.1.30)$$

for all $x, y \in \mathbf{a}$.

If we introduce the permutation map

$$\pi : \begin{array}{c} \mathbf{a} \times \mathbf{a} \to \mathbf{a} \times \mathbf{a} \\ x \otimes y \mapsto y \otimes x, \end{array} \qquad (4.1.31)$$

this may be written as

$$\gamma \circ \mathcal{M} = \mathcal{M} \circ (\gamma \times \gamma) \circ \pi, \qquad (4.1.32)$$

or, in terms of a commutative diagram,

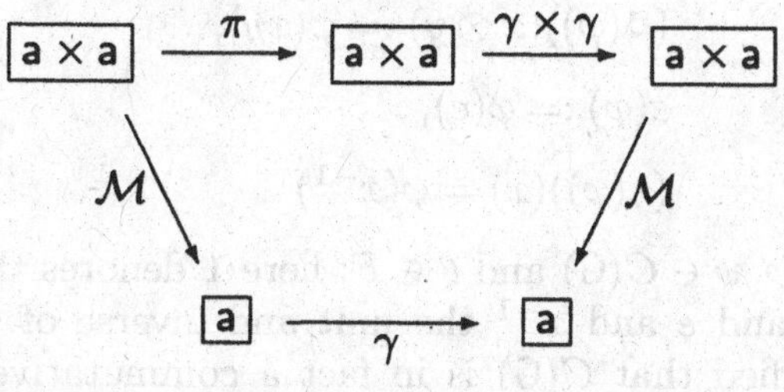

$$
\begin{array}{ccccc}
\boxed{\mathbf{a} \times \mathbf{a}} & \xrightarrow{\ \pi\ } & \boxed{\mathbf{a} \times \mathbf{a}} & \xrightarrow{\ \gamma \times \gamma\ } & \boxed{\mathbf{a} \times \mathbf{a}} \\
& {}_{\mathcal{M}}\searrow & & & \swarrow_{\mathcal{M}} \\
& & \boxed{\mathbf{a}} & \xrightarrow[\ \gamma\ ]{} & \boxed{\mathbf{a}}
\end{array}
\qquad (4.1.33)
$$

- The antipode is an *anti-cohomomorphism*, i.e.

$$(\gamma \times \gamma) \circ \Delta = \pi \circ \Delta \circ \gamma \qquad (4.1.34)$$

and $\varepsilon \circ \gamma = \varepsilon$.

- The map

$$\Delta' := \pi \circ \Delta \qquad (4.1.35)$$

is also a (co-associative) co-multiplication. Together with the "inverse" γ' of the antipode γ (and with the same multiplication, unit, and co-unit) it provides another Hopf algebra structure for $\mathbf{a}$. Here γ' is the inverse of γ in the sense that the matrices which represent the antipodes in a given basis $\{e_s\}$ of $\mathbf{a}$ are inverse to each other, i.e.

$$\gamma(e_s) = \tau_s{}^t e_t \iff \gamma'(e_s) = (\tau^{-1})_s{}^t e_t. \qquad (4.1.36)$$

An algebra $\mathbf{a}$ is by definition a commutative algebra iff the multiplication does not depend on the order of the factors, which may be formalized by saying that $\mathcal{M} \circ \pi = \mathcal{M}$. Analogously, a Hopf algebra $\mathbf{a}$ is called *co-commutative* iff the co-multiplication satisfies $\pi \circ \Delta = \Delta$, or in other words iff Δ' coincides with Δ. Examples for co-commutative Hopf algebras are provided by the group Hopf algebras and the enveloping algebras $\mathsf{U}(\mathfrak{g})$ which were discussed above (in fact, almost any theorem that holds for both of these classes of algebras is also true for all co-commutative Hopf algebras). For any commutative or co-commutative Hopf algebra, the antipode obeys $\gamma \circ \gamma = id$.

The commutative Hopf algebras turn out to be intimately related to compact topological groups (i.e. groups which are compact topological spaces, for example Lie group manifolds). For any such group G, consider the vector space $C(\mathsf{G})$ of continuous complex-valued functions on G (considered as a C^* algebra, i.e. as endowed with the natural notion of complex conjugation). This space is endowed with a Hopf algebra structure via

$$\begin{aligned} &(\mathcal{M}(\varphi,\psi))(x) := \varphi(x)\,\psi(x), \\ &\eta(\xi) := \xi\,\mathbf{1}, \\ &(\Delta(\varphi))(x \otimes y) := \varphi(xy), \\ &\varepsilon(\varphi) := \varphi(e), \\ &(\gamma(\varphi))(x) = \varphi(x^{-1}) \end{aligned} \tag{4.1.37}$$

for all $x, y \in \mathsf{G}$, $\varphi, \psi \in C(\mathsf{G})$ and $\xi \in F$; here $\mathbf{1}$ denotes the map $\mathbf{1} : x \mapsto 1 \in F \; \forall x \in \mathsf{G}$, and e and x^{-1} the unit and inverse of G, respectively. It is readily verified that $C(\mathsf{G})$ is in fact a commutative Hopf algebra. Moreover, it is co-commutative iff the group G is abelian.

In short, it is possible to associate to any compact topological group a commutative Hopf algebra. It can be shown that this correspondence works also the other way round. This is the content of the *Gelfand–Naimark theorem*, which states that every commutative Hopf algebra is isomorphic, as a Hopf algebra, to the Hopf algebra of continuous functions on an appropriate compact topological group. In fact, the theory of compact topological groups and the theory of commutative Hopf algebras turn out to be completely equivalent. Thus it is possible to describe the properties of compact topological groups G solely in terms of the associated Hopf algebra of functions $C(\mathsf{G})$, without making explicit reference to G and its elements at all. In the case where G is a Lie group, one can show that there exists a non-degenerate duality between $C(\mathsf{G})$ and the universal enveloping algebra $\mathsf{U}(\mathfrak{g})$ of the Lie algebra $\mathfrak{g}$ of G, i.e. $C(\mathsf{G})$ is isomorphic to (a subspace of) the dual space $\mathsf{U}(\mathfrak{g})^\star$.

Inspired by the Gelfand–Naimark theorem, one is led to describe also non-commutative Hopf algebras as function spaces on appropriate objects. These objects cannot be topological groups, and not much is known about their explicit structure. In analogy to the commutative case, they are called *pseudogroups*, *non-commutative geometric groups*, or *quantum groups*. A special class of quantum groups G_q which may indeed be interpreted as quantized versions of Lie groups G is obtained if one identifies the non-commutative Hopf algebra $C(\mathsf{G}_q)$ that is defined as the space of functions on G_q with the dual space $\mathsf{U}_q(\mathfrak{g})^\star$ of a quantum analogue $\mathsf{U}_q(\mathfrak{g})$ of the universal enveloping algebra $\mathsf{U}(\mathfrak{g})$. The precise definition of these quantum universal enveloping algebras will be provided in the following section. Since $\mathsf{U}_q(\mathfrak{g})$ is dual to the non-commutative Hopf algebra $C(\mathsf{G}_q)$, it will be a non-cocommutative Hopf algebra. The duality between $\mathsf{U}_q(\mathfrak{g})$ and the space of functions on G_q will be discussed in detail in section 4.7.

It must be noted that the term *quantum group* does not possess a generally accepted meaning. Often this name is used not only for the geometric object introduced above, but generically for any Hopf algebra, or sometimes also only for Hopf algebras possessing a property called quasitriangularity. The latter property is shared by quantum universal enveloping algebras; thus these algebras are commonly referred to as quantum groups, although they are in fact dual to the function space on a quantum group; we will adhere to this practice in the following. (Sometimes one also requires that the Hopf algebra is non-cocommutative if it is to be called a quantum group; we will not impose this requirement.)

The Hopf algebras of our main interest will be the quantum universal enveloping algebras. As just mentioned, these are quasitriangular Hopf algebras; the property of quasitriangularity is defined as follows.

Definition : A **quasitriangular** Hopf algebra is a Hopf algebra for which the co-multiplications $\boldsymbol{\Delta}$ and $\boldsymbol{\Delta}'$ are related by conjugation, i.e.

(4.1.38*a*) $\boldsymbol{\Delta}'(x) = \mathsf{R}\diamond\boldsymbol{\Delta}(x)\diamond\mathsf{R}^{-1}$ for all $x \in \mathsf{a}$

for some element R of $\mathsf{a}\times\mathsf{a}$ which is invertible and satisfies

(4.1.38*b*) $(id\times\boldsymbol{\Delta})(\mathsf{R}) = \mathsf{R}_{13}\diamond\mathsf{R}_{12}$

(4.1.38*c*) $(\boldsymbol{\Delta}\times id)(\mathsf{R}) = \mathsf{R}_{13}\diamond\mathsf{R}_{23}$

(4.1.38*d*) $(\gamma\times id)(\mathsf{R}) = \mathsf{R}^{-1}$.

(4.1.38)

Here the inverse R^{-1} of $\mathsf{R}\in\mathsf{a}\times\mathsf{a}$ is by definition that element of $\mathsf{a}\times\mathsf{a}$ which satisfies

$$\mathsf{R}^{-1}\diamond\mathsf{R} = e\otimes e = \mathsf{R}\diamond\mathsf{R}^{-1}. \tag{4.1.39}$$

Instead of (4.1.38*d*), one can equivalently require that $(id\times\gamma)(\mathsf{R}^{-1}) = \mathsf{R}$; this shows in particular that $(\gamma\times\gamma)(\mathsf{R}) = \mathsf{R}$. Thus, if R takes the simple product form

$$\mathtt{R} = r_1 \otimes r_2, \quad \mathtt{R}^{-1} = s_1 \otimes s_2, \tag{4.1.40}$$

one has $r_1 \diamond s_1 = e = s_1 \diamond r_1$, $r_2 \diamond s_2 = e = s_2 \diamond r_2$, i.e. $s_1 = r_1^{-1}$, $s_2 = r_2^{-1}$ so that e.g.

$$\gamma(r_1) = r_1^{-1}, \quad r_2 = r_2^{-1} \tag{4.1.41}$$

becomes the explicit form of the requirement (4.1.38d). Of course, generically the structure of R is more complicated,

$$\mathtt{R} = \sum_\ell r_1^{(\ell)} \otimes r_2^{(\ell)} \tag{4.1.42}$$

(see e.g. the example (4.2.37) below) and analogously for $\mathtt{R}^{-1}$, so that the implications of (4.1.39) are less trivial. In the following, we will for brevity nevertheless often use the form (4.1.40) of R.

Also, in the definition (4.1.38), $\mathtt{R}_{13}$ is meant as the identity in the second factor of $\mathsf{a} \times \mathsf{a} \times \mathsf{a}$ and as R in the first and third factors; with the short-hand form (4.1.40) of R, this means

$$\mathtt{R}_{13} = r_1 \otimes e \otimes r_2, \tag{4.1.43}$$

and analogously for $\mathtt{R}_{12}$, $\mathtt{R}_{23}$.

A quasitriangular Hopf algebra is called *triangular* iff

$$\mathtt{R}_{12} \diamond \mathtt{R}_{21} = e \otimes e, \tag{4.1.44}$$

or in other words iff $\pi(\mathtt{R}) = \mathtt{R}^{-1}$ (if $\mathtt{R} = r_1 \otimes r_2$, this requirement becomes $r_1 \diamond r_2 = e = r_2 \diamond r_1$).

Note that any co-commutative Hopf algebra obeying $\boldsymbol{\Delta}(e) = e \otimes e$ and $\gamma(e) = e$ is also quasitriangular since the axioms of quasitriangularity are then fulfilled by just taking $\mathtt{R} = e \otimes e$. In particular, any group Hopf algebra is quasitriangular, and so is the universal enveloping algebra $\mathsf{U}(\mathfrak{g})$ of any Lie algebra $\mathfrak{g}$. Generically, a quasitriangular Hopf algebra is neither commutative nor co-commutative; however, the non-cocommutativity is under control, being described by the axiom (4.1.38a). As a consequence, many properties of co-commutative Hopf algebras generalize to generic quasitriangular Hopf algebras. Also, any finite-dimensional Hopf algebra can be embedded in a suitable finite-dimensional quasitriangular Hopf algebra (this is implied by the so-called quantum double construction which will be discussed in section 4.6).

An immediate consequence of the axiom (4.1.38a) is that

$$\mathtt{R}_{12} \diamond (\boldsymbol{\Delta} \times id)(x \otimes y) = (\boldsymbol{\Delta}' \times id)(x \otimes y) \diamond \mathtt{R}_{12} \tag{4.1.45}$$

for all $x \otimes y \in \mathsf{a} \times \mathsf{a}$. Taking $x \otimes y = \mathtt{R}$ and using (4.1.38b, c), this yields

$$\mathtt{R}_{12} \diamond \mathtt{R}_{13} \diamond \mathtt{R}_{23} = \mathtt{R}_{23} \diamond \mathtt{R}_{13} \diamond \mathtt{R}_{12}. \tag{4.1.46}$$

This is the so-called *Yang–Baxter equation*, which plays a fundamental role in the theory of completely integrable systems; in the context of integrable systems, the quantity R is called the *universal R-matrix,* and this terminology has been also adapted for the element R of an arbitrary quantum group. In the case of the simple product form (4.1.40) of the R-matrix, the Yang–Baxter equation simply reads

$$(r_1 \diamond r_1) \otimes (r_2 \diamond r_1) \otimes (r_2 \diamond r_2) = (r_1 \diamond r_1) \otimes (r_1 \diamond r_2) \otimes (r_2 \diamond r_2). \qquad (4.1.47)$$

Thus if R is really of the simple form (4.1.40), the Yang-Baxter equation is equivalent to $r_1 \diamond r_2 = r_2 \diamond r_1$, a is a commutative algebra). Of course, usually the R-matrix is given as an expansion (4.1.42) so that the Yang–Baxter equation is a rather non-trivial equation.

The meaning of the relation $\mathbf{\Delta}' = \mathrm{R}\mathbf{\Delta}\mathrm{R}^{-1}$ becomes a bit more transparent if an N-dimensional matrix representation $R : \mathsf{a} \to gl_{\mathrm{N}}$ is considered. Denoting a basis of the associated module by $\{v_i\}$ and the corresponding dual basis by $\{v^i\}$, one obtains representation matrices for $\mathbf{\Delta}$ and R as

$$\begin{aligned} (R \times R)(\mathbf{\Delta}(x)) &= ((R \times R)(\mathbf{\Delta}(x)))^{ij}_{\ kl}\, v_i \otimes v_j \otimes v^k \otimes v^l, \\ (R \times R)(\mathrm{R}) &= (\mathcal{R})^{ij}_{\ kl}\, v_i \otimes v_j \otimes v^k \otimes v^l, \end{aligned} \qquad (4.1.48)$$

so that in the representation R, the formula $\mathbf{\Delta}'(x) \diamond \mathrm{R} = \mathrm{R} \diamond \mathbf{\Delta}(x)$ reads

$$((R \times R)(\mathbf{\Delta}'(x)))^{ij}_{\ kl}\, \mathcal{R}^{kl}_{\ pq} = \mathcal{R}^{ij}_{\ kl}\, ((R \times R)(\mathbf{\Delta}(x)))^{kl}_{\ pq} \qquad (4.1.49)$$

(more generally, one can consider pairs of distinct matrix representations $R \times R'$ instead of $R \times R$). Now define the linear functions $\varphi^i_{\ j} \in \mathsf{a}^\star$, $i, j = 1, \dots, \mathrm{N}$ by $x \mapsto \varphi^i_{\ j}(x) := (R(x))^i_{\ j}$; then because of the duality between multiplication and co-multiplication one has

$$\begin{aligned} ((R \times R)(\mathbf{\Delta}(x)))^{ij}_{\ kl} &= \left(\varphi^i_{\ k} \times \varphi^j_{\ l}\right)(\mathbf{\Delta}(x)) = \varphi^i_{\ k} \cdot \varphi^j_{\ l}(x), \\ ((R \times R)(\mathbf{\Delta}'(x)))^{ij}_{\ kl} &= \left(\varphi^j_{\ k} \times \varphi^i_{\ l}\right)(\mathbf{\Delta}(x)) = \varphi^j_{\ k} \cdot \varphi^i_{\ l}(x). \end{aligned} \qquad (4.1.50)$$

Thus (4.1.49) can be rewritten as

$$\mathcal{R}^{kl}_{\ pq}\, \varphi^i_{\ k}\, \varphi^j_{\ l} = \mathcal{R}^{ij}_{\ kl}\, \varphi^k_{\ p}\, \varphi^l_{\ q}. \qquad (4.1.51)$$

If one defines matrices T and $\tilde{T}$ by

$$\begin{aligned} T &:= (\varphi^i_{\ j}\, v_i \otimes v^j) \otimes \mathbf{1} = \varphi^i_{\ j}\, v_i \otimes v_k \otimes v^j \otimes v^k, \\ \tilde{T} &:= \mathbf{1} \otimes (\varphi^i_{\ j}\, v_i \otimes v^j) = \varphi^i_{\ j}\, v_k \otimes v_i \otimes v^k \otimes v^j \end{aligned} \qquad (4.1.52)$$

(i.e. in terms of matrix elements, $T^{ij}_{\ kl} = \varphi^i_{\ k}\delta^j_l$ and $\tilde{T}^{ij}_{\ kl} = \varphi^j_{\ l}\delta^i_k$), then this finally becomes the matrix equation

$$\mathcal{R}\, T\, \tilde{T} = \tilde{T}\, T\, \mathcal{R}. \qquad (4.1.53)$$

Thus the relation $\Delta' = \mathtt{R}\Delta\mathtt{R}^{-1}$ is the abstract version of the matrix equation (4.1.53) which is valid in some given representation. For the applications to integrable systems, the relevant informations are the matrix solutions $\mathcal{R}$ to this equation; the theory of quasitriangular Hopf algebras thus provides a way of generating many such solutions from a single abstract solution $\mathtt{R}$. Analogously one also obtains a matrix version of the Yang–Baxter equation, reading

$$\mathcal{R}_{12}\mathcal{R}_{13}\mathcal{R}_{23} = \mathcal{R}_{23}\mathcal{R}_{13}\mathcal{R}_{12} \tag{4.1.54}$$

with $(\mathcal{R}_{12})^{ijk}_{\ \ lmn} = \mathcal{R}^{ij}_{\ lm}\delta^k_{\ n}$ etc. In terms of the entries of the matrix $\mathcal{R}$, this reads

$$\mathcal{R}^{ij}_{\ pq}\mathcal{R}^{pk}_{\ lr}\mathcal{R}^{qr}_{\ mn} = \mathcal{R}^{jk}_{\ pq}\mathcal{R}^{iq}_{\ rn}\mathcal{R}^{rp}_{\ lm}. \tag{4.1.55}$$

As an equation for $\mathrm{N}^3 \times \mathrm{N}^3$-matrices, this gives in fact a system of N^6 equations for the N^4 unknown matrix elements of $\mathcal{R}$, and hence is extremely overdetermined. However, very often there are symmetries that reduce this overdeterminacy considerably. Instead of the matrix $\mathcal{R}$, often also the matrix $\check{\mathcal{R}}$ defined by $\check{\mathcal{R}}^{ij}_{\ pq} = \mathcal{R}^{ij}_{\ qp}$ is used. In terms of the entries of this matrix, the Yang-Baxter equation reads

$$\check{\mathcal{R}}^{ij}_{\ pq}\check{\mathcal{R}}^{qk}_{\ rl}\check{\mathcal{R}}^{pr}_{\ mn} = \check{\mathcal{R}}^{jk}_{\ pq}\check{\mathcal{R}}^{ip}_{\ mr}\check{\mathcal{R}}^{rq}_{\ nl}; \tag{4.1.56}$$

in matrix notation, this means

$$\check{\mathcal{R}}_{12}\check{\mathcal{R}}_{23}\check{\mathcal{R}}_{12} = \check{\mathcal{R}}_{23}\check{\mathcal{R}}_{12}\check{\mathcal{R}}_{23}. \tag{4.1.57}$$

The equation (4.1.53) plays an important role in the the theory of two-dimensional factorizable N-particle scattering matrices and (via the so-called quantum inverse scattering method) in the closely related subject of completely integrable systems in statistical mechanics, where it was obtained a long time before the relation to quantum groups was known. In that context one considers in fact a whole class of isomorphic representations parametrized by some continuous parameter λ (which is called the *spectral parameter* and plays for example the role of a parameter in the basis transformation which identifies the various representations with each other) rather than a single representation. The equation then reads

$$\mathcal{R}(\lambda,\mu)\,T(\lambda)\otimes\tilde{T}(\mu) = \tilde{T}(\mu)\otimes T(\lambda)\,\mathcal{R}(\lambda,\mu). \tag{4.1.58}$$

This implies that

$$[\operatorname{tr} T(\lambda), \operatorname{tr} T(\mu)] = 0 \tag{4.1.59}$$

for all λ, μ, which means that $\operatorname{tr} T(\lambda)$ is the generating function for an infinite number of conserved quantities. Also note that the matrix Yang–Baxter equation is a sufficient condition for the relation (4.1.53), but not a necessary one. While for most of the known solutions to (4.1.53) the

Yang–Baxter equation is indeed fulfilled, there are also a few other solutions which do not share this property.

From the version (4.1.57) of the Yang–Baxter equation, one concludes that any solution to the Yang–Baxter equation provides a representation of the *braid group* $\mathcal{B}_{\rm N}$. By definition, the braid group $\mathcal{B}_{\rm N}$ on N strands is the discrete group generated by a unit e and $\rm N-1$ elements σ_i, $i = 1, 2, \dots, \rm N-1$, subject to the two relations

$$\begin{aligned} &\sigma_i\sigma_j = \sigma_j\sigma_i \qquad \text{for } |i-j| \geq 2, \\ &\sigma_i\sigma_{i+1}\sigma_i = \sigma_{i+1}\sigma_i\sigma_{i+1}. \end{aligned} \tag{4.1.60}$$

Thus the braid group is a generalization of the symmetric group: the symmetric group $\mathcal{S}_{\rm N}$ has all these properties as well, but in addition the square of any of its generators is equal to the unit element; see (4.3.36) below. This additional relation implies that the order (i.e., number of elements) of $\mathcal{S}_{\rm N}$ is finite, namely N!, whereas the order of $\mathcal{B}_{\rm N}$ is infinite.

The generators σ_i can be graphically represented as follows:

σ_i :

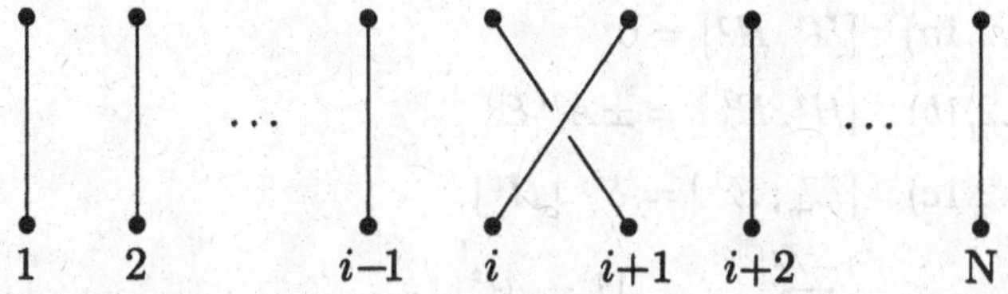

(4.1.61)

The group multiplication, the inverse, and the constraints (4.1.60) then have an obvious geometrical intrepretation: braids are formed when N points on a horizontal line are connected by N strings to N points on another horizontal line directly below the first N points; the identity element of $\mathcal{B}_{\rm N}$ is the trivial configuration where the strings do not intersect; a general element is constructed from the identity element by successive application of the operations σ_i depicted above; and the inverse σ_i^{-1} is given by an analogous picture where the $(i{+}1)$th line lies "in front of" the ith line rather than the other way round.

4.2 Deformations of enveloping algebras

The quantum groups which will turn out to be relevant to WZW theories are obtained as deformations of enveloping algebras of semisimple Lie algebras. Recall that the universal enveloping algebra $\mathsf{U}(\mathfrak{g})$ of a Lie algebra $\mathfrak{g}$ is the algebra of (finite) formal power series in the generators of $\mathfrak{g}$, with the relations of $\mathfrak{g}$ among these generators implicit (with the Lie bracket interpreted as a formal commutator). If $\mathfrak{g}$ is a semisimple (or affine) Lie algebra, the relations may be taken, in a Chevalley basis, to be those of (1.5.1) (respectively (2.1.1)). The *quantum universal enveloping algebra* $\mathsf{U}_q(\mathfrak{g})$ can then be defined as a *q-deformation* or *quantum deformation* of the enveloping algebra. Deforming the relations of $\mathsf{U}(\mathfrak{g})$ means changing them in a manner depending on some formal parameter q such that the original algebra $\mathsf{U}(\mathfrak{g})$ is obtained in the limit $q \to 1$, which is referred to as the *classical limit.* The terms "quantum" and "classical" are used in this context merely because the q-deformation can be understood as a formal "quantization" procedure (see section 4.7). They do *not*, however, refer directly to the connection between classical and quantum field theories; note e.g. that already $\mathsf{U}(\mathfrak{g})$ is a quasitriangular Hopf algebra and hence, according to the terminology adopted in section 4.1, a quantum group.

Definition : The **quantum universal enveloping algebra** $\mathsf{U}_q(\mathfrak{g})$ is the algebra of power series in the $3r+1$ generators $\{E^i_\pm, H^i \mid i = 1, \dots, r\} \cup \{\mathbf{1}\}$ modulo the relations

$$(4.2.1)$$

$$(4.2.1a)\quad [H^i, H^j] = 0,$$

$$(4.2.1b)\quad [H^i, E^j_\pm] = \pm A^{ji} E^j_\pm,$$

$$(4.2.1c)\quad [E^i_+, E^j_-] = \delta^{ij} \lfloor H^i \rfloor,$$

$$(4.2.1d)\quad \sum_{p=0}^{1-A^{ji}} (-1)^p \begin{bmatrix} 1-A^{ji} \\ p \end{bmatrix}_i (E^i_\pm)^p (E^j_\pm)(E^i_\pm)^{1-A^{ji}-p} = 0 \quad \text{for } i \neq j,$$

$$(4.2.1e)\quad \mathbf{1} \diamond x = x = x \diamond \mathbf{1} \quad \text{for all } x \in \mathsf{U}_q(\mathfrak{g}),$$

with A^{ij} the Cartan integers of the Lie algebra $\mathfrak{g}$.

Here the square brackets are to be understood as commutators,

$$[x, y] \equiv x \diamond y - y \diamond x, \tag{4.2.2}$$

with $\mathcal{M}(x \otimes y) = x \diamond y$ the formal product in $\mathsf{U}(\mathfrak{g})$. Also, we have used the *q-number* symbol already introduced in section 1.7,

$$\lfloor x \rfloor \equiv \lfloor x \rfloor_q := \frac{q^{x/2} - q^{-x/2}}{q^{1/2} - q^{-1/2}}, \tag{4.2.3}$$

together with

$$\lfloor x \rfloor_i \equiv \lfloor x \rfloor_{q_i} \quad \text{with} \quad q_i := q^{(\alpha^{(i)}, \alpha^{(i)})/(\theta, \theta)}, \tag{4.2.4}$$

$$\lfloor n \rfloor ! := \prod_{m=1}^{n} \lfloor m \rfloor, \tag{4.2.5}$$

$$\left\lfloor \begin{matrix} n \\ m \end{matrix} \right\rfloor := \frac{\lfloor n \rfloor!}{\lfloor m \rfloor! \lfloor n-m \rfloor!}. \tag{4.2.6}$$

The exponential functions of generators appearing in the expressions $\lfloor H^i \rfloor$ in (4.2.1c) are defined through the corresponding power series, i.e.

$$e^{\xi H} = \sum_{n=0}^{\infty} \frac{\xi^n}{n!} H^n \tag{4.2.7}$$

(with $H^n \equiv H^{\diamond n}$ defined inductively, i.e. $H^{\diamond n} = H \diamond H^{\diamond(n-1)}$) so that in particular

$$e^{\xi H} \diamond e^{-\xi H} = 1. \tag{4.2.8}$$

Note that owing to the appearance of $q^{\pm H^i/2}$ in (4.2.1) one is forced to consider infinite power series in the H^i (contrary to the case of the enveloping algebra $\mathsf{U}(\mathfrak{g})$). In contrast, it is consistent to restrict to only finite power series in the generators $E^i_\pm$. As a consequence, usually this restriction is included in the definition uf $\mathsf{U}_q(\mathfrak{g})$; this convention will be adopted from now on. Instead of allowing for infinite power series in the H^i, one may alternatively replace each generator H^i by a pair of generators $K^i_\pm \equiv q^{\pm H^i/2}$ and consider only finite power series in $K^i_\pm$ as well as in $E^i_\pm$. In terms of these new generators, the relations of $\mathsf{U}_q(\mathfrak{g})$ read

$$\begin{aligned} & K^i_+ K^i_- = K^i_- K^i_+ = \mathbf{1}, \\ & [K^i_\pm, K^j_\pm] = [K^i_+, K^j_-] = 0, \\ & K^i_+ E^j_\pm = q^{\pm A_{ji}/2} E^j_\pm K^i_+, \\ & K^i_- E^j_\pm = q^{\mp A_{ji}/2} E^j_\pm K^i_-, \\ & [E^i_+, E^j_-] = \delta^{ij} (q^{1/2} - q^{-1/2})^{-1} (K^i_+ - K^i_-) . \end{aligned} \tag{4.2.9}$$

This distinction between finite and infinite power series is however not so important because the generators H^i act multiplicatively on the modules of our interest, which will be discussed in the following section.

In the limit $q \to 1$, the relations (4.2.1a, b, c) reduce to the Lie brackets of the algebra $\mathfrak{g}$. Similarly, the relations (4.2.1d) are the q-analogues of the Serre relations. E.g. for $A^{ji} = -1$, they read explicitly

$$E^i_\pm E^i_\pm E^j_\pm - (q^{1/2} + q^{-1/2})\, E^i_\pm E^j_\pm E^i_\pm + E^j_\pm E^i_\pm E^i_\pm = 0; \tag{4.2.10}$$

in the limit $q \to 1$, one has $(q^{1/2} + q^{-1/2}) \equiv \lfloor 2 \rfloor \to 2$ so that (4.2.10) reduces to

$$\begin{aligned} 0 &= E^i_\pm E^i_\pm E^j_\pm - 2\, E^i_\pm E^j_\pm E^i_\pm + E^j_\pm E^i_\pm E^i_\pm \\ &= [E^i_\pm, [E^i_\pm, E^j_\pm]] = (\mathrm{ad}_{E^i_\pm})^2 E^j_\pm . \end{aligned} \tag{4.2.11}$$

The quantum analogue (4.2.1d) of the Serre relations can also be written more suggestively as

$$\left(\mathrm{Ad}_{E^i_\pm}\right)^{1-A_{ji}} (E^j_\pm) = 0, \tag{4.2.12}$$

if one defines the quantum version of the map ad_x as a q-deformed commutator,

$$\mathrm{Ad}_{E^\alpha}(E^\beta) \equiv [E^\alpha, E^\beta]_q := q^{-(\alpha,\beta)/4}\, E^\alpha \diamond E^\beta - q^{(\alpha,\beta)/4}\, E^\beta \diamond E^\alpha. \tag{4.2.13}$$

For example, for $\mathsf{U}_q(A_2)$, $\mathrm{Ad}_{E^1_\pm}(E^2_\pm)$ does not vanish so that one should define

$$E^{\pm(\alpha^{(1)}+\alpha^{(2)})} := \mathrm{Ad}_{E^1_\pm}(E^2_\pm) = q^{1/4}\, E^1_\pm E^2_\pm - q^{-1/4}\, E^2_\pm E^1_\pm, \tag{4.2.14}$$

and the Serre relations then mean that

$$\begin{aligned} 0 &= \mathrm{Ad}_{E^i_\pm}(E^{\pm(\alpha^{(1)}+\alpha^{(2)})}) \\ &= q^{1/4}\, E^i_\pm E^{\pm(\alpha^{(1)}+\alpha^{(2)})} - q^{-1/4}\, E^{\pm(\alpha^{(1)}+\alpha^{(2)})} E^i_\pm \end{aligned} \tag{4.2.15}$$

for $i = 1, 2$.

The formulae above become of course most transparent if one chooses $\mathfrak{g} = A_1$ so that there are no Serre relations. The defining relations of $\mathsf{U}_q(A_1)$ are thus

$$\begin{aligned} [H, E_\pm] &= \pm 2\, E_\pm, \\ [E_+, E_-] &= \lfloor H \rfloor. \end{aligned} \tag{4.2.16}$$

By definition, the enveloping algebra $\mathsf{U}(A_1)$ is associative, and its unit 1 is clearly given by the trivial power series 1. These properties are inherited by $\mathsf{U}_q(A_1)$. Moreover, it turns out that $\mathsf{U}_q(A_1)$ is endowed with the structure of a quasitriangular Hopf algebra. The Hopf algebra structure arises via the following definitions: the co-multiplication acts on the generators H, $E_\pm$, and on 1 as

$$\begin{aligned} \Delta(H) &= H \otimes 1 + 1 \otimes H, \\ \Delta(E_\pm) &= E_\pm \otimes q^{H/4} + q^{-H/4} \otimes E_\pm, \\ \Delta(1) &= 1 \otimes 1, \end{aligned} \tag{4.2.17}$$

and the antipode is

$$\begin{aligned} \gamma(H) &= -H, \\ \gamma(E_\pm) &= -q^{\pm 1/2}\, E_\pm, \\ \gamma(1) &= 1. \end{aligned} \tag{4.2.18}$$

Finally, the co-unit is simply given by

$$\varepsilon(H) = \varepsilon(E_\pm) = 0, \qquad \varepsilon(1) = 1. \tag{4.2.19}$$

When verifying the defining property (4.1.1e) of the antipode, use has to be made of the identity

$$[q^{\xi H}, E_\pm] = (q^{\pm 2\xi} - 1)\, E_\pm q^{\xi H}, \tag{4.2.20}$$

i.e.

$$q^{\xi H}\, E_\pm = E_\pm\, q^{\xi(H\pm 2)}, \tag{4.2.21}$$

which is easily deduced from the second of the relations (4.2.16). Some other useful identities which can be derived by straightforward manipulations are

$$[(E_+)^m, (E_-)^n] = \frac{\lfloor n\rfloor!}{\lfloor n-m\rfloor!}\,(E_-)^{n-m} \prod_{j=1}^{m} \lfloor H+n-j\rfloor \tag{4.2.22}$$

for $n \geq m$ (and similarly for $n < m$), and

$$\Delta^{(n)}((E_\pm)^m) = \sum_{0\leq m_1\leq\ldots\leq m_{n-1}\leq m} \left\lfloor \begin{matrix} m \\ m_1\; m_2{-}m_1 \ldots m{-}m_{n-1} \end{matrix} \right\rfloor \bigotimes_{j=1}^{n} (E_\pm)^{m_j-m_{j-1}}\, q^{-(m-m_j-m_{j-1})H/4}. \tag{4.2.23}$$

Here $\Delta^{(n)}$ denotes the $n-1$-fold composition of co-multiplications,

$$\Delta^{(n)} : \mathsf{a} \rightarrow \mathsf{a}^{\otimes n} \tag{4.2.24}$$

which is defined inductively as $\Delta^{(2)} \equiv \Delta$ and $\Delta^{(n)} = (\Delta \times id) \circ \Delta^{(n-1)}$. Also,

$$\left\lfloor \begin{matrix} n \\ n_1\, n_2 \ldots n_\ell \end{matrix} \right\rfloor := \frac{\lfloor n\rfloor!}{\prod_{j=1}^{\ell} \lfloor n_j\rfloor!}. \tag{4.2.25}$$

The formulae (4.2.17) and (4.2.18) also lead to a simple interpretation of the co-multiplication $\Delta' = \tau \circ \Delta$. Recalling that the antipode γ' associated to Δ' is obtained from γ by $\tau \mapsto \tau^{-1}$, it is clear that the quantum group $\mathsf{U}_q(A_1)'$ defined by Δ' and γ' is related to $\mathsf{U}_q(A_1)$ by

$$\mathsf{U}_q(A_1)' = \mathsf{U}_{q^{-1}}(A_1). \tag{4.2.26}$$

The Hopf algebra structure identified above for $\mathsf{U}_q(A_1)$ generalizes in a natural way to $\mathsf{U}_q(\mathfrak{g})$ with arbitrary simple (or, in fact, affine) $\mathfrak{g}$. The defining formulae for the co-multiplication are

$$\begin{aligned}\Delta(H^i) &= H^i \otimes 1 + 1 \otimes H^i,\\ \Delta(E^i_\pm) &= E^i_\pm \otimes q^{H^i/4} + q^{-H^i/4} \otimes E^i_\pm, \qquad (4.2.27)\\ \Delta(1) &= 1 \otimes 1,\end{aligned}$$

for the co-unit

$$\begin{aligned}\varepsilon(H^i) &= 0 = \varepsilon(E^i_\pm),\\ \varepsilon(1) &= 1,\end{aligned} \qquad (4.2.28)$$

and for the antipode

$$\begin{aligned}\gamma(H^i) &= -H^i,\\ \gamma(E^i_\pm) &= -q^{H_\rho} E^i_\pm \, q^{-H_\rho}, \qquad (4.2.29)\\ \gamma(1) &= 1.\end{aligned}$$

Here H_ρ is defined as

$$\begin{aligned}H_\rho := \frac{1}{(\theta,\theta)} \sum_{\alpha>0} H_{\alpha^\vee} &\equiv \frac{1}{(\theta,\theta)} \sum_{\alpha>0} \sum_{i=1}^{r} \alpha_i \, H^i\\ &= \sum_{i=1}^{r} \tilde\rho_i \, H^i = (\tilde\rho, H)\end{aligned} \qquad (4.2.30)$$

with

$$\tilde\rho := \frac{2}{(\theta,\theta)} \rho \qquad (4.2.31)$$

and ρ the Weyl vector of $\mathfrak{g}$. It is a straightforward exercise to check that the maps defined above satisfy the defining properties (4.1.1) of a Hopf algebra (the least trivial part is the co-associativity of Δ when applied to $E^i_\pm$; this can be checked with the help of $\Delta(H^n) = (\Delta(H))^{\diamond n}$, which implies

$$\Delta(q^{\xi H}) = q^{\xi H} \otimes q^{\xi H}, \qquad (4.2.32)$$

i.e. with the homomorphism property of Δ).

Above, the Hopf algebra structure was only presented for the generators of the Chevalley–Serre basis. To identify it also for the rest of the generators, one has to employ the Serre relations together with the homomorphism property of the co-multiplication etc. As an example consider $\mathsf{U}_q(A_2)$. Then one would like to know the Hopf algebra action on the generators $E^{\pm(\alpha^{(1)}+\alpha^{(2)})}$ that were defined in (4.2.14). From the homomorphism property of Δ and the action of Δ on the Chevalley generators one easily deduces that

$$\begin{aligned}
&\Delta(E^{\pm(\alpha^{(1)}+\alpha^{(2)})}) \\
&= E^{\pm(\alpha^{(1)}+\alpha^{(2)})} \otimes q^{(H^1+H^2)/4} + q^{-(H^1+H^2)/4} \otimes E^{\pm(\alpha^{(1)}+\alpha^{(2)})} \\
&\quad + q^{1/4} \left(q^{-H^1/4} E^2_{\pm} \otimes E^1_{\pm} q^{H^2/4} + E^1_{\pm} q^{-H^2/4} \otimes q^{H^1/4} E^2_{\pm} \right) \\
&\quad + q^{-1/4} \left(q^{-H^2/4} E^1_{\pm} \otimes E^2_{\pm} q^{H^1/4} + E^2_{\pm} q^{-H^1/4} \otimes q^{H^2/4} E^1_{\pm} \right).
\end{aligned} \tag{4.2.33}$$

With the help of the relations

$$\begin{aligned}
[H^1, E^2_{\pm}] &= \mp E^2_{\pm}, \\
[H^2, E^1_{\pm}] &= \mp E^1_{\pm},
\end{aligned} \tag{4.2.34}$$

this can be rewritten as

$$\begin{aligned}
&\Delta(E^{\alpha^{(1)}+\alpha^{(2)}}) \\
&= E^{\alpha^{(1)}+\alpha^{(2)}} \otimes q^{(H^1+H^2)/4} + q^{-(H^1+H^2)/4} \otimes E^{\alpha^{(1)}+\alpha^{(2)}} \\
&\qquad + (q^{1/2} - q^{-1/2})\, q^{-H^1/4} E^2_{+} \otimes q^{H^2/4} E^1_{+}, \\
&\Delta(E^{-(\alpha^{(1)}+\alpha^{(2)})}) \\
&= E^{-(\alpha^{(1)}+\alpha^{(2)})} \otimes q^{(H^1+H^2)/4} + q^{-(H^1+H^2)/4} \otimes E^{-(\alpha^{(1)}+\alpha^{(2)})} \\
&\qquad + (q^{1/2} - q^{-1/2})\, q^{-H^2/4} E^1_{-} \otimes q^{H^1/4} E^2_{-}.
\end{aligned} \tag{4.2.35}$$

Similarly, one finds for the antipode

$$\gamma(E^{\pm(\alpha^{(1)}+\alpha^{(2)})}) = q^{\mp 1} \left(q^{-1/4}\, E^1_{\pm} E^2_{\pm} - q^{1/4}\, E^2_{\pm} E^1_{\pm} \right). \tag{4.2.36}$$

Next we would like to show that for $\mathsf{U}_q(\mathfrak{g})$ the quasitriangularity properties (4.1.38) are valid, in particular that the co-multiplications Δ and $\Delta' = \pi \circ \Delta$ are conjugate to each other. To see this, first notice that the quantum deformations $\mathsf{U}_q(\mathsf{b}_+)$ and $\mathsf{U}_q(\mathsf{b}_-)$ of the enveloping algebras of $\mathsf{b}_\pm = \mathfrak{g}_0 \oplus \mathfrak{g}_\pm$ (the Borel subalgebras of $\mathfrak{g}$) are Hopf subalgebras of $\mathsf{U}_q(\mathfrak{g})$; it also turns out (see section 4.6) that as vector spaces they are dual to each other. If $\{e_p\}$ is a basis of $\mathsf{U}_q(\mathsf{b}_+)$ and $\{e^p\}$ the corresponding dual basis of $\mathsf{U}_q(\mathsf{b}_-)$, then the universal R-matrix is given by

$$\mathsf{R} = \sum_p e_p \otimes e^p. \tag{4.2.37}$$

To see this, it is best to consider the matrices which express the multiplication, co-multiplication and antipode in the bases $\{e_p\}$ and $\{e^p\}$. If in $\mathsf{U}_q(\mathsf{b}_-)$ these matrices are μ, ν, τ as defined in (4.1.12), then due to the duality between $\mathsf{U}_q(\mathsf{b}_-)$ and $\mathsf{U}_q(\mathsf{b}_+)$, the corresponding matrices in

$\mathsf{U}_q(\mathsf{b}_+)$ are ν, μ^t, τ^{-1}:

$$\begin{aligned} e^s \diamond e^t &= \textstyle\sum_p \nu_p^{\;st}\, e^p, \\ \Delta(e^p) &= \textstyle\sum_{s,t} \mu^p_{\;ts}\, e^s \otimes e^t, \\ \gamma(e^p) &= \textstyle\sum_s (\tau^{-1})^p_{\;s}\, e^s. \end{aligned} \tag{4.2.38}$$

It is now immediately seen that (4.1.38*b*, *c*) hold:

$$\begin{aligned} (\Delta \times id)(\mathrm{R}) &= \textstyle\sum_p \Delta(e_p) \otimes e^p = \sum_{p,s,t} \nu_p^{\;st}\, e_s \otimes e_t \otimes e^p \\ &= \textstyle\sum_{s,t} e_s \otimes e_t \otimes (e^s \diamond e^t) = \mathrm{R}_{13} \diamond \mathrm{R}_{23}, \end{aligned} \tag{4.2.39}$$

and

$$\begin{aligned} (id \times \Delta)(\mathrm{R}) &= \textstyle\sum_p (e_p) \otimes \Delta(e^p) = \sum_{p,s,t} \mu^p_{\;st}\, e_p \otimes e^t \otimes e^s \\ &= \textstyle\sum_{s,t} (e_s \diamond e_t) \otimes e^t \otimes e^s = \mathrm{R}_{13} \diamond \mathrm{R}_{12}. \end{aligned} \tag{4.2.40}$$

Next one has, using (4.2.28) and the defining property (4.1.1*e*) of the antipode,

$$\begin{aligned} \mathrm{R} \diamond (\gamma \times id)(\mathrm{R}) &= \textstyle\sum_{p,q} (e_p \diamond \gamma(e_q)) \otimes (e^p \diamond e^q) \\ &= \textstyle\sum_{p,q,r,s,t} \tau_q^{\;r}\, \mu^s_{\;pr}\, \nu_t^{\;pq}\, e_s \otimes e^t \\ &= \textstyle\sum_t (\mathcal{M} \circ (id \times \gamma) \circ \Delta(e_t)) \otimes e^t \\ &= \textstyle\sum_t (\eta \circ \varepsilon(e_t)) \otimes e^t = 1 \otimes 1, \end{aligned} \tag{4.2.41}$$

and analogously

$$\begin{aligned} (\gamma \times id)(\mathrm{R}) \diamond \mathrm{R} &= \textstyle\sum_{p,q,r,s,t} \tau_p^{\;r}\, \mu^s_{\;rq}\, \nu_t^{\;pq}\, e_s \otimes e^t \\ &= \textstyle\sum_t (\mathcal{M} \circ (\gamma \times id) \circ \Delta(e_t)) \otimes e^t = 1 \otimes 1; \end{aligned} \tag{4.2.42}$$

this shows that (4.1.38*d*) is obeyed, and hence completes the proof of quasitriangularity.

It should be noted that the quantity (4.2.37) is in fact not quite an element of $\mathsf{U}_q(\mathfrak{g}) \times \mathsf{U}_q(\mathfrak{g})$ because it involves an infinite summation whereas with the usual definition of $\mathsf{U}_q(\mathfrak{g})$ (see p. 259) the elements of $\mathsf{U}_q(\mathfrak{g}) \times \mathsf{U}_q(\mathfrak{g})$ involve only finite power series in the generators $E^i_\pm$. As a consequence, $\mathsf{U}_q(\mathfrak{g})$ should be called *essentially quasitriangular* rather than quasitriangular; for practical purposes this distinction is however not at all relevant, and hence usually the qualification "essentially" is suppressed.

It is also not difficult to obtain the Yang–Baxter equation (4.1.46). Combining the formulae (4.2.37), (4.1.12) and (4.2.38), one has

$$\Delta'(e_p) \diamond \mathrm{R} = \sum_{q,r,s} \nu_p^{\;rq}\, (e_q \diamond e_s) \otimes (e_r \diamond e^s) = \sum_{q,r,s,t} \nu_p^{\;rq}\, \mu^t_{\;qs}\, e_t \otimes (e_r \diamond e^s), \tag{4.2.43}$$

as well as

$$\mathrm{R}\diamond\Delta(e_p) = \sum_{q,r,s} \nu_p{}^{qr}\,(e_s\diamond e_q)\otimes(e^s\diamond e_r) = \sum_{q,r,s,t} \nu_p{}^{qr}\mu^t{}_{sq}\, e_t\otimes(e^s\diamond e_r). \qquad (4.2.44)$$

Comparing this with (4.1.38a) (applied to $x = e_p$) and expanding with respect to $\{e_t\}$ in the first factor, one obtains

$$\sum_{q,r,s} \nu_p{}^{qr}\mu^t{}_{sq}\, e^s\diamond e_r = \sum_{q,r,s} \nu_p{}^{rq}\mu^t{}_{qs}\, e_r\diamond e^s. \qquad (4.2.45)$$

Now again using (4.2.37), (4.1.12) and (4.2.38), it follows that

$$\begin{aligned}\mathrm{R}_{12}\mathrm{R}_{13}\mathrm{R}_{23} &= \textstyle\sum_{q,r,s}(e_s\diamond e_q)\otimes(e^s\diamond e_r)\otimes(e^q\diamond e^r)\\ &= \textstyle\sum_{p,q,r,s,t}\mu^t{}_{sq}\nu_p{}^{qr}\, e_t\otimes(e^s\diamond e_r)\otimes e^p\end{aligned} \qquad (4.2.46)$$

and

$$\begin{aligned}\mathrm{R}_{23}\mathrm{R}_{13}\mathrm{R}_{12} &= \textstyle\sum_{q,r,s}(e_q\diamond e_s)\otimes(e_r\diamond e^s)\otimes(e^r\diamond e^q)\\ &= \textstyle\sum_{p,q,r,s,t}\mu^t{}_{qs}\nu_p{}^{rq}\, e_t\otimes(e_r\diamond e^s)\otimes e^p.\end{aligned} \qquad (4.2.47)$$

Together with (4.2.45) this shows that R as defined by (4.2.37) indeed obeys the Yang–Baxter equation.

The result (4.2.37) for the R-matrix is still rather formal. To obtain a more explicit formula, one may use the fact that in the limit $q \to 1$, $\mathsf{U}_q(\mathfrak{g})$ reduces to $\mathsf{U}(\mathfrak{g})$ for which the R-matrix is given by $\mathrm{R} = 1\otimes 1$. This leads to the power series ansatz

$$\mathrm{R} = 1\otimes 1 + \sum_{n=1}^{\infty} (\ln q)^n\, \mathrm{R}_{(n)}. \qquad (4.2.48)$$

When this is inserted into the defining relations (4.1.38) of quasitriangularity, one obtains a set of relations which in principle determine the $\mathrm{R}_{(n)}$ recursively. In practice, however, these relations are very complicated. As a consequence, only for $\mathfrak{g} = A_1$ the explicit solution for R has been determined by this method; the result reads

$$\begin{aligned}\mathrm{R} = q^{H\otimes H/4}\sum_{n=0}^{\infty} &\frac{(q^{1/2}-q^{-1/2})^n}{\lfloor n\rfloor!}\, q^{-n(n+1)/4}\\ &\cdot\left(q^{nH/4}(E_+)^n\right)\otimes\left(q^{-nH/4}(E_-)^n\right).\end{aligned} \qquad (4.2.49)$$

There exists, however, a different method for finding the explicit form of the universal R-matrix. This will be described in section 4.6.

It is also of interest to insert the power series (4.2.48) into the Yang–Baxter equation (4.1.46). The zeroth and first order in $\ln q$ are then fulfilled trivially, while to the order $(\ln q)^2$ one obtains

$$[\mathrm{r}_{12},\mathrm{r}_{13}] + [\mathrm{r}_{12},\mathrm{r}_{23}] + [\mathrm{r}_{13},\mathrm{r}_{23}] = 0, \qquad (4.2.50)$$

where $\mathtt{r} \equiv \mathtt{R}_{(1)}$. This equation arises as a consistency condition in the theory of integrable classical dynamical systems; it is called the *classical Yang–Baxter equation*, and its solution $\mathtt{r}$ the *classical R-matrix*. Since the classical Yang-Baxter equation is entirely written in terms of commutators, one may take it to be defined on the enveloping algebra $\mathsf{U}(\mathfrak{g})$ of a Lie algebra $\mathfrak{g}$.

Conversely, for any solution of the classical Yang–Baxter equation corresponding to some enveloping algebra $\mathsf{U}(\mathfrak{g})$, one can obtain a solution to the quantum Yang–Baxter equation by a procedure of "integrating up" or "quantizing." It turns out that this quantization can be applied not only to concrete solutions of the classical Yang–Baxter equation, but also abstractly to the corresponding Hopf algebras. Thus the classical R-matrix $\mathtt{r}$ for $\mathsf{U}(\mathfrak{g})$ can be used as a starting point for the transition from $\mathsf{U}(\mathfrak{g})$ to $\mathsf{U}_q(\mathfrak{g})$ by a formal quantization procedure; the universal R-matrix of $\mathsf{U}_q(\mathfrak{g})$ is then related to $\mathtt{r}$ by

$$\mathtt{R} = 1 \otimes 1 + \ln q \cdot \mathtt{r} + \mathcal{O}((\ln q)^2), \tag{4.2.51}$$

or conversely,

$$\mathtt{r} = \frac{1}{\hbar}\,(\mathtt{R} - 1 \otimes 1) \bmod \hbar \tag{4.2.52}$$

with $\hbar \equiv \ln q$. This will be demonstrated in more detail in section 4.7.

The classical Yang–Baxter equation bears a close similarity to the Jacobi identity; in a sense it is the analogue for $\mathsf{U}(\bar{\mathfrak{g}})$ of the Jacobi identity for $\bar{\mathfrak{g}}$. Therefore the (quantum) Yang–Baxter equation may be looked at as a quantum version of the Jacobi identity.

It is also of interest to determine the center of $\mathsf{U}_q(\mathfrak{g})$. Recall that the quadratic Casimir operator $\mathcal{C}_2$ of a simple Lie algebra $\mathfrak{g}$ acts as a constant on any irreducible $\mathfrak{g}$-module. In terms of the universal enveloping algebra, this means that $\mathcal{C}_2$ belongs to the center of $\mathsf{U}(\mathfrak{g})$, as can also be seen explicitly as follows. First, one has trivially $[\mathcal{C}_2, H^i] = 0$; in addition, the commutator with a step operator is easily seen to be

$$[\mathcal{C}_2, E^\alpha] = \sum_{\beta\neq\alpha} e_{-\beta,\alpha}\,(E^\beta E^{\alpha-\beta} + E^{\alpha-\beta} E^\beta) \tag{4.2.53}$$

(the term with $\beta = \alpha$ cancels the contribution from the Cartan subalgebra part of $\mathcal{C}_2$), and that this is zero follows by shifting $\beta \to \alpha - \beta$ in the second term and using the property (1.4.44) of the structure constants $e_{\alpha,\beta}$. In the case $\mathfrak{g} = A_1$, the quadratic Casimir operator is the only independent Casimir operator, and as a consequence it in fact generates the center of $\mathsf{U}(A_1)$. It turns out that this situation generalizes to $q \neq 1$ in a simple manner: the center of $\mathsf{U}_q(A_1)$ is generated by the element

$$(\tilde{\mathcal{C}}_2)_q := 2E_-E_+ + \lfloor \tfrac{1}{2}\,(H+1) \rfloor^2, \tag{4.2.54}$$

or equivalently by

$$(\mathcal{C}_2)_q = (\tilde{\mathcal{C}}_2)_q - \lfloor \tfrac{1}{2} \rfloor^2, \tag{4.2.55}$$

which is obviously the q-analogue of the quadratic Casimir of A_1,

$$2\mathcal{C}_2 \equiv \tfrac{1}{2} H^2 + E_- E_+ + E_+ E_- = 2\,E_- E_+ + \tfrac{1}{2}\,(H+1)^2 - \tfrac{1}{2}. \tag{4.2.56}$$

The eigenvalue of the quadratic Casimir operator $(\tilde{\mathcal{C}}_2)_q$ in the module R_Λ is as usual most easily computed by applying the Casimir operator to the highest weight Λ, yielding

$$\tilde{C}_\Lambda = \lfloor \tfrac{1}{2}\,(\Lambda + 1) \rfloor^2. \tag{4.2.57}$$

The situation turns out to be completely analogous for $\mathsf{U}_q(\mathfrak{g})$ with $\mathfrak{g} \neq A_1$. For $q = 1$, the center of $\mathsf{U}(\mathfrak{g})$ is generated by the the quadratic Casimir operator together with the independent higher order Casimir operators, and for $q \neq 1$ one again has q-deformations of all these Casimir operators, and for generic q they generate the center of $\mathsf{U}_q(\mathfrak{g})$.

4.3 Representation theory

From the fact that the commutation relations for the elements $H^i \in \mathsf{U}_q(\mathfrak{g})$ are the same as the Lie brackets for the generators of the Cartan subalgebra of $\mathfrak{g}$, it seems plausible that the theory of finite-dimensional representations of $\mathsf{U}_q(\mathfrak{g})$ runs parallel to the representation theory of $\mathfrak{g}$. Indeed, for simple $\mathfrak{g}$ it can be shown that *generically* the representation theories of $\mathfrak{g}$ and $\mathsf{U}_q(\mathfrak{g})$ are isomorphic. Namely, all finite-dimensional modules of $\mathsf{U}_q(\mathfrak{g})$ are fully reducible, and the irreducible finite-dimensional modules of $\mathsf{U}_q(\mathfrak{g})$ are parametrized by the dominant integral highest weights Λ of $\mathfrak{g}$. Also, the finite-dimensional modules can be decomposed into weight spaces in complete analogy to (1.6.5),

$$R = \bigoplus_\lambda R_{(\lambda)}, \tag{4.3.1}$$

such that $R(H^i) \cdot v_\lambda = \lambda^i\, v_\lambda$, and the dimensionalities of these weight spaces are the same as those of the corresponding weight spaces of the $\mathfrak{g}$-module R,

$$\dim R_{(\lambda)}(\mathsf{U}_q(\mathfrak{g})) = \dim R_{(\lambda)}(\mathfrak{g}). \tag{4.3.2}$$

The complete reducibility of finite-dimensional modules implies that any such module may be considered as an irreducible component of the Kronecker product of a suitable number of *basic* modules (namely, the modules mentioned on p. 69).

For the applications we have in mind, the representation theory of the generic case is in fact not relevant; thus it must be explained which special situation is *non-generic* so that the statements above are no longer true. This is the case where the formal parameter q is interpreted as a complex number, and in addition is a root of unity. That this leads to a markedly different representation theory is most easily understood by considering the simplest example of a Kronecker product of basic modules, namely the product $R \times R$ with $R = R_{\Lambda_{(1)}}$ the two-dimensional defining module of $\bar{\mathfrak{g}} = A_1$. (From now on we will again use the notation $\bar{\mathfrak{g}}$ for simple Lie algebras in order to distinguish them from the associated affine algebras $\mathfrak{g} = \bar{\mathfrak{g}}^{(1)}$; however, until further notice we denote the $\bar{\mathfrak{g}}$-weights by λ rather than $\bar{\lambda}$ because the weights of the affine algebras will play no role.) The module R has two basis vectors v_+ and v_- obeying

$$\begin{aligned} R(H) \cdot v_\pm &= \pm v_\pm , \\ R(E_\pm) \cdot v_\pm &= 0, \\ R(E_\pm) \cdot v_\mp &= v_\pm . \end{aligned} \tag{4.3.3}$$

Considering only the underlying vector spaces, the tensor product $R \times R$ is therefore four-dimensional, with basis vectors

$$\begin{aligned} v_{++} &= v_+ \otimes v_+ , \\ v_{+-} &= v_+ \otimes v_- , \\ v_{-+} &= v_- \otimes v_+ , \\ v_{--} &= v_- \otimes v_- . \end{aligned} \tag{4.3.4}$$

From the definition (4.2.27) of $\boldsymbol{\Delta}$ it follows immediately that

$$\begin{aligned} \boldsymbol{\Delta}(H) \cdot v_{++} &= 2\, v_{++} , \\ \boldsymbol{\Delta}(H) \cdot v_{--} &= -2\, v_{--} , \\ \boldsymbol{\Delta}(H) \cdot v_{+-} &= 0 = \boldsymbol{\Delta}(H) \cdot v_{-+} \end{aligned} \tag{4.3.5}$$

and

$$\boldsymbol{\Delta}(E_-) \cdot v_{++} = \boldsymbol{\Delta}(E_+) \cdot v_{--} = q^{1/4} v_{-+} + q^{-1/4} v_{+-} . \tag{4.3.6}$$

Similarly, one has

$$\begin{aligned} \boldsymbol{\Delta}((E_-)^2) \cdot v_{++} &= (q^{1/2} + q^{-1/2}) v_{--} = \lfloor 2 \rfloor\, v_{--} , \\ \boldsymbol{\Delta}((E_+)^2) \cdot v_{--} &= \lfloor 2 \rfloor\, v_{++} . \end{aligned} \tag{4.3.7}$$

Here we simplified the notation by writing $\boldsymbol{\Delta}(x)$ for $\boldsymbol{\Delta}^{\Lambda_{(1)}\Lambda_{(1)}}(x)$, where in general $\boldsymbol{\Delta}^{\Lambda\Lambda'}$ is defined as

$$\boldsymbol{\Delta}^{\Lambda\Lambda'} := (R_\Lambda \times R_{\Lambda'})(\boldsymbol{\Delta}). \tag{4.3.8}$$

As long as $\lfloor 2 \rfloor \neq 0$, the vectors v_{++}, v_{--} and $v_0 := q^{1/4}v_{-+} + q^{-1/4}v_{+-}$ generate an irreducible three-dimensional module of $\mathsf{U}_q(A_1)$, while $q^{1/4}v_{-+} - q^{-1/4}v_{+-}$ (which is annihilated by $\mathbf{\Delta}(E_\pm)$) generates an irreducible one-dimensional module. Thus we have the situation familiar from the corresponding Kronecker product of A_1; graphically, this may be represented as

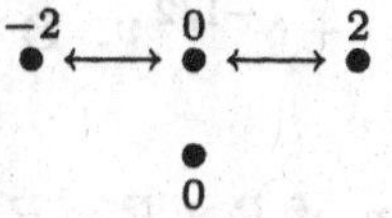

(4.3.9)

where arrows to the right and left denote the action of $\mathbf{\Delta}(E_+)$ and $\mathbf{\Delta}(E_-)$, respectively. In contrast, if $\lfloor 2 \rfloor = 0$ which is equivalent to $q = -1$, then $\mathbf{\Delta}((E_\pm)^2)$ annihilates $v_{\mp\mp}$ so that the corresponding arrowheads in (4.3.9) have to be removed. However, the action of $\mathbf{\Delta}((E_\pm)^2)/\lfloor 2 \rfloor$ (defined by first leaving q arbitrary and taking the limit $q \to -1$ only after this operator has been applied to a state) still maps $v_{\mp\mp}$ to $v_{\pm\pm}$ so that v_{--} still belongs to the module. Graphically, the situation is thus as follows.

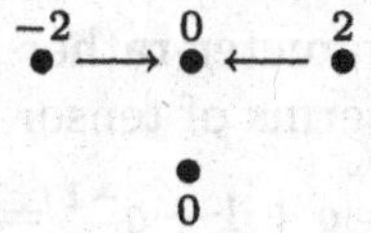

(4.3.10)

In terms of the inner product (1.6.18) on the three-dimensional module, the missing arrowheads just mean that all states in the module have zero norm, and are in fact null states; e.g. one has $(v_0 \,|\, v_0) \simeq (E_+ \cdot v_{--} \,|\, E_+ \cdot v_{--}) = (v_{--} \,|\, E_- E_+ \cdot v_{--}) = -(v_{--} \,|\, \lfloor H \rfloor \cdot v_{--}) = \lfloor 2 \rfloor \, (v_{--} \,|\, v_{--}) = 0$ (for notational simplicity, we write just H instead of $\mathbf{\Delta}^{R,R}(H)$, etc.). Thus, while the three-dimensional module is still irreducible at $q = -1$, it is no longer unitarizable. (In defining the inner product, one first assumes q to be arbitrary and then takes the limit to the special value $q \to -1$; this means that as far as the inner product is concerned, q has to be treated as a formal parameter, i.e. must not be complex conjugated, $(q\,v \mid w) = (v \mid q\,w) = q\,(v \mid w)$.) Actually, the example just presented does not yet display all additional features which arise for q a root of unity. Namely, in this case all modules involved are still fully reducible, while generically also non-fully reducible modules appear. This is seen by considering the triple Kronecker product $R \times R \times R$. For arbitrary q, this decomposes as

$$R \times R \times R = R \oplus R \oplus R' \tag{4.3.11}$$

with $R' \equiv R_{3\Lambda_{(1)}}$ being four-dimensional, with basis vectors

$$\begin{aligned} v_3 &= v_+ \otimes v_+ \otimes v_+, \\ v_1 &= q^{1/2}\, v_- \otimes v_+ \otimes v_+ + v_+ \otimes v_- \otimes v_+ \\ &\qquad + q^{-1/2}\, v_+ \otimes v_+ \otimes v_-, \\ v_{-1} &= q^{1/2}\, v_+ \otimes v_- \otimes v_+ + v_- \otimes v_+ \otimes v_- \\ &\qquad + q^{-1/2}\, v_- \otimes v_- \otimes v_+, \\ v_{-3} &= v_- \otimes v_- \otimes v_-. \end{aligned} \tag{4.3.12}$$

Pictorially, the decomposition of $R \times R \times R$ is given by

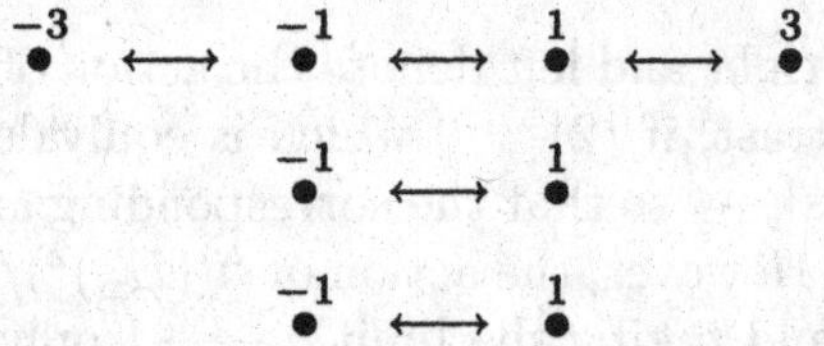

(4.3.13)

The situation changes if q takes the special value $q = e^{2\pi i/3}$. Then v_1 is annihilated by E_+. As a consequence it has zero norm, i.e. is orthogonal to itself; this is also seen by direct calculation (keeping in mind that q has to be treated as a formal parameter rather than as a complex number): from the expression of v_1 in terms of tensor products of $v_\pm$ one deduces

$$(v_1 \,|\, v_1) = q + 1 + q^{-1} = \lfloor 3 \rfloor = 0. \tag{4.3.14}$$

Now the orthogonal complement of v_1 is two-dimensional (this certainly holds for arbitrary q, but by continuity it remains true for $q = e^{2\pi i/3}$), whereas according to the above picture the subspace of $R \times R \times R$ with H-eigenvalue 1 is three-dimensional; thus the latter space now contains a vector w_1 which is *not* orthogonal to v_1 and hence must be included in the module to which v_1 belongs. In particular one has

$$0 \neq (v_1 \,|\, w_1) = (E_- \cdot v_3 \,|\, w_1) = (v_3 \,|\, E_+ \cdot w_1), \tag{4.3.15}$$

(here $E_\pm$ stands for $\Delta^{(3)}(E_\pm)$) so that in the pictorial description one has to include an arrow pointing from w_1 to v_3. Similarly, because one now has $(E_-)^3 \cdot v_3 = 0$, the arrowhead from v_{-1} to v_{-3} has to be removed (nevertheless the state v_{-3} still belongs to the module because it can be reached from v_3 by applying the operator $(E_-)^3/\lfloor 3 \rfloor!$ with the limit $q \to e^{2\pi i/3}$ taken at the end of the calculation), and there is a new arrow pointing from $w_{-1} = E_- \cdot w_1$ to v_{-3}. Putting these informations together, one sees that at $q = e^{2\pi i/3}$ the product $R \times R \times R$ decomposes into an irreducible module R and a six-dimensional module which is reducible, but not fully reducible because it contains the four-dimensional module generated by v_i, $i = -3, -1, 1, 3$, as a (non-irreducible) submodule. Pictorially, this

situation is described by

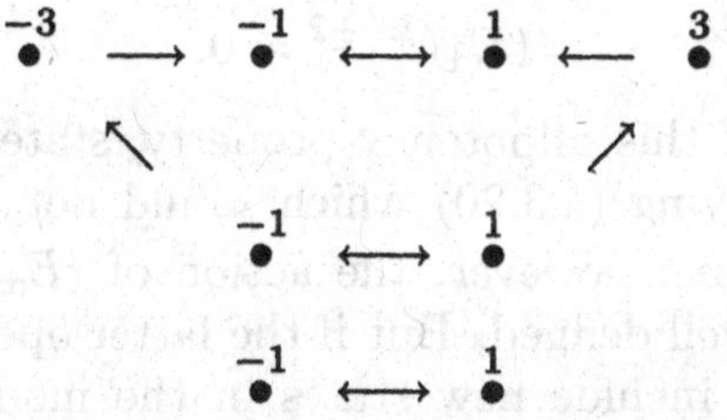

(4.3.16)

Let us also remark that the fact that v_1 is annihilated by E_+ can be derived by considering the action of the quadratic Casimir operator. Namely, one has of course

$$\tilde{C}_2 \cdot v_1 = \tilde{C}_{3\Lambda_{(1)}} v_1, \tag{4.3.17}$$

but also

$$\tilde{C}_2 \cdot v_1 = E_- E_+ \cdot v_1 + \tilde{C}_{\Lambda_{(1)}} v_1. \tag{4.3.18}$$

Now for $q = \mathrm{e}^{2\pi i/3}$ the quadratic Casimir eigenvalues, computed according to (4.2.57), obey

$$\tilde{C}_{\Lambda_{(1)}} \equiv \tilde{C}_1 = \lfloor 1 \rfloor^2 = \lfloor 2 \rfloor^2 = \tilde{C}_3 \equiv \tilde{C}_{3\Lambda_{(1)}}. \tag{4.3.19}$$

Thus one concludes that $E_- E_+ \cdot v_1 = 0$. However, one also knows that $E_+ \cdot v_1 \propto v_3$ and $E_- \cdot v_3 = v_1 \neq 0$. Together, it follows that indeed $E_+ \cdot v_1 = 0$.

Analogous complications arise whenever q is a root of unity. Application of $\boldsymbol{\Delta}^{(n)}(E_\pm)$ to a weight vector v_λ in an n-fold Kronecker product of the basic module yields the weight vector $v_{\lambda\pm 2}$ multiplied with a product of square roots of *q-integers,* i.e. factors of the form $\sqrt{\lfloor m_i \rfloor}$ for some integer m_i. Now if

$$q^{k^\vee + 2} = 1 \tag{4.3.20}$$

for some integer $k^\vee$, then

$$\lfloor k^\vee + 2 \rfloor = 0. \tag{4.3.21}$$

As a consequence, $E_\pm$ become nilpotent, i.e. $\boldsymbol{\Delta}^{(m)}(E_\pm)$ annihilates weight vectors which for generic q would not be annihilated. Analogous considerations apply to the Kronecker product of two arbitrary finite-dimensional modules. Therefore it follows that for any finite-dimensional module R, one has

$$(R(E_\pm)^{k^\vee + 2}) \cdot v = 0 \quad \text{for all } v \in R \tag{4.3.22}$$

(of course, some of the vectors of R are already annihilated by smaller powers of $R(E_\pm)$); this fact will be expressed, a bit sloppily, by writing

$$(E_\pm)^{k^\vee+2} = 0. \tag{4.3.23}$$

As a consequence of this nilpotency property, states get annihilated by $(E_\pm)^{k^\vee+2}$ for q obeying (4.3.20) which would not be annihilated for q taking a generic value; however, the action of $(E_\pm)^{k^\vee+2}/\lfloor k^\vee+2\rfloor!$ on these states is still well defined. But if the latter operators are admitted, then one must also include new states in the module from which one obtains the original states by application of these operators.

The analysis of higher tensorial powers of basic modules can thus be carried through precisely as in the case of the Kronecker product of three basic modules considered above. The result is as follows: for q the primitive $(k^\vee+2)$th root of unity, $q = \exp(\frac{2\pi i}{k^\vee+2})$, irreducible highest weight modules exist only for $\Lambda \le k^\vee$ or (if $k^\vee > 1$) Λ an integer multiple of $k^\vee+1$; modules with maximal weight in the range $k^\vee+1 < \Lambda < 2k^\vee+2$ which appear in the Kronecker product of irreducible highest weight modules pair up with modules of maximal weight $2k^\vee+2-\Lambda$ to form a reducible, but not fully reducible module of dimension $2k^\vee+4$ according to the scheme

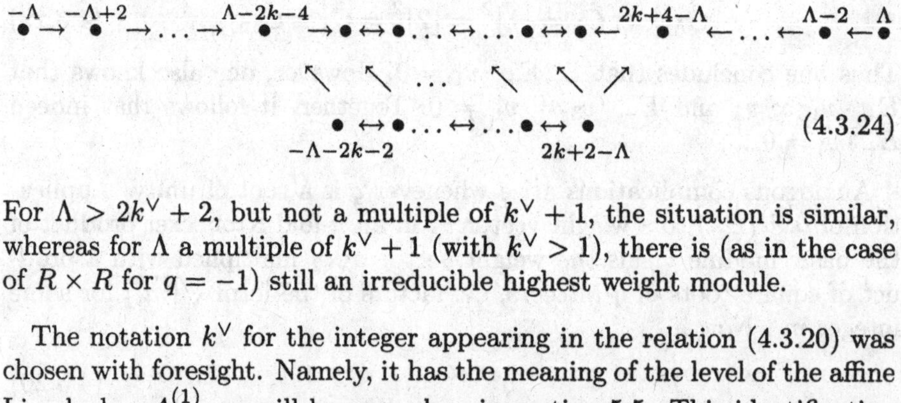

(4.3.24)

For $\Lambda > 2k^\vee+2$, but not a multiple of $k^\vee+1$, the situation is similar, whereas for Λ a multiple of $k^\vee+1$ (with $k^\vee > 1$), there is (as in the case of $R\times R$ for $q=-1$) still an irreducible highest weight module.

The notation $k^\vee$ for the integer appearing in the relation (4.3.20) was chosen with foresight. Namely, it has the meaning of the level of the affine Lie algebra $A_1^{(1)}$, as will become clear in section 5.5. This identification also shows that the case $q=-1$ corresponding to $k^\vee=0$ is not very interesting so that the fact that no non-fully reducible modules exist in this case is rather irrelevant (nevertheless this case was chosen above as an example because notationally it is much more convenient to consider the Kronecker product of two rather than of more than two defining modules). From the knowledge of the representation theory of untwisted affine algebras, it is therefore not difficult to guess how the situation described above generalizes to the case of $\mathsf{U}_q(\mathfrak{g})$ with arbitrary simple $\mathfrak{g}$. Namely, for q the primitive $(k^\vee+g^\vee)$th root of unity, $q=\exp(\frac{2\pi i}{k^\vee+g^\vee})$, so that

$$q^{k^\vee+g^\vee} = 1, \tag{4.3.25}$$

which implies

$$\lfloor k^\vee + g^\vee \rfloor = 0, \tag{4.3.26}$$

one finds the following. Unitary irreducible highest weight modules exist iff the associated modules of the untwisted affine algebra $\mathfrak{g}$ (having $\bar{\mathfrak{g}}$ as its horizontal subalgebra) are integrable, i.e. iff

$$(\Lambda, \theta^\vee) \leq k^\vee. \tag{4.3.27}$$

Modules which are irreducible, but not unitarizable, exist if $(\Lambda, \theta^\vee)$ is a multiple of $k^\vee + 1$; and some of the modules with $(\Lambda, \theta^\vee)$ different from these values are of this type, too. But most modules with highest weight Λ different from these special values are not irreducible, but combine with modules of some maximal weight Λ' to form a reducible, but not fully reducible module. Note that in particular the basic modules of $\bar{\mathfrak{g}}$ can be deformed to irreducible highest weight modules of $\mathsf{U}_q(\bar{\mathfrak{g}})$ for all simple $\bar{\mathfrak{g}}$. As in the case of $\bar{\mathfrak{g}} = A_1$, the weight Λ' can be determined for given Λ by considering the action of the step operators on Kronecker products of irreducible modules (say, the basic modules). By considering the sl_2-subalgebras $\mathfrak{h}_\alpha$ of $\bar{\mathfrak{g}}$, it is clear that all step operators are nilpotent,

$$(E^\alpha)^{p_\alpha} = 0 \tag{4.3.28}$$

for appropriate numbers p_α which fulfill

$$p_\alpha \leq k^\vee + g^\vee, \tag{4.3.29}$$

and

$$p_\theta = k^\vee + g^\vee. \tag{4.3.30}$$

However, the explicit analysis becomes considerably more complicated with higher rank, and it has not yet been worked out in full generality.

A different algorithm to determine Λ' will be discussed at the end of this section; however, it should already be pointed out here that it has not yet been proven rigorously that this algorithm is indeed equivalent to the considerations above, which are based on the nilpotency of the step operators.

If a subspace of a non-fully reducible module can be obtained in the limit where q approaches a root of unity from an irreducible highest weight module R_Λ, then it will be denoted by $\check{R}_\Lambda$. To save space, we will refer to such subspaces sometimes as "would-be irreducible" modules. For ease of notation, let us also assume that the non-fully reducible modules encountered for q a root of unity are always (non-direct) sums of precisely two such subspaces $\check{R}_\Lambda$ and $\check{R}_{\Lambda'}$ and denote them by

$$R_{\Lambda,\Lambda'} \equiv \check{R}_\Lambda \oplus \check{R}_{\Lambda'} \tag{4.3.31}$$

(for the majority of non-fully reducible modules this assumption is fulfilled, and it may even be true for all non-fully reducible modules that can appear in tensor products of irreducible modules with positive quantum dimension). Thus for $\bar{\mathfrak{g}} = A_1$, the modules $R_{\Lambda,\Lambda'}$ decompose as

$$R_{\Lambda,\Lambda'} = \breve{R}_\Lambda \oplus \breve{R}_{2k^\vee+2-\Lambda}. \tag{4.3.32}$$

Note that these reducible modules possess more than one maximal weight vector and hence are in particular not highest weight modules.

Note that a necessary condition for $\breve{R}_1$ and $\breve{R}_2$ to combine into a non-fully reducible module $\breve{R}_1 \oplus \breve{R}_2$ is that the Casimir eigenvalues of $\breve{R}_1$ and $\breve{R}_2$ are identical, since the Casimir operator commutes with the step operators that connect vectors belonging to the different subspaces (for $\bar{\mathfrak{g}} \neq A_1$, the same must be true for the eigenvalues with respect to all independent higher order Casimir operators). Indeed, upon use of the identity $\lfloor\xi\rfloor = \lfloor\xi + 2(k^\vee+2)\rfloor$ (see (4.4.21) below) one has

$$\begin{aligned} \tilde{C}_{2k^\vee+2-\Lambda} &= \lfloor \tfrac{1}{2}\,(2k^\vee - \Lambda + 3)\rfloor^2 \\ &= \lfloor k^\vee + 2 - \tfrac{1}{2}(\Lambda+1)\rfloor^2 = \tilde{C}_\Lambda. \end{aligned} \tag{4.3.33}$$

At a more abstract level, the special features occurring for q a root of unity can be understood in terms of the centralizer algebras of $\mathsf{U}_q(\bar{\mathfrak{g}})$. To understand the role of the centralizer algebra, first consider the case $q = 1$. The decomposition (1.6.57) of the Kronecker product of $\bar{\mathfrak{g}}$-modules into its irreducible components is nothing but the branching rule for the embedding of $\bar{\mathfrak{g}}$ into the algebra $\bar{\mathfrak{h}} = R(\bar{\mathfrak{g}}) \times R(\bar{\mathfrak{g}})$ defined by the Kronecker product. But this embedding can of course be extended to an embedding of $\bar{\mathfrak{g}} \oplus \mathcal{C}(\bar{\mathfrak{g}})$ where $\mathcal{C}(\bar{\mathfrak{g}})$ is the centralizer of $\bar{\mathfrak{g}}$ in $\bar{\mathfrak{h}}$. Thus

$$R_\Lambda \times R_{\Lambda'} = \bigoplus_i S_i \otimes R_{\Lambda_i}, \tag{4.3.34}$$

where S_i are irreducible modules of $\mathcal{C}(\bar{\mathfrak{g}})$, and the dimensionalities of these modules are just the Littlewood-Richardson coefficients,

$$\dim S_i = \mathcal{L}_i. \tag{4.3.35}$$

Moreover, it turns out that $\mathcal{C}(\bar{\mathfrak{g}})$ has only a finite number of irreducible modules, and for the case of tensor products of one of the basic modules with a non-trivial highest weight module each of these modules appears precisely once in the sum (4.3.34). Thus the decomposition of Kronecker products of $\bar{\mathfrak{g}}$ can be completely understood in terms of the representation theory of the centralizer algebra $\mathcal{C}(\bar{\mathfrak{g}})$.

The centralizer algebra of course depends on the particular Kronecker product considered. In general, it is rather complicated. For the m-fold Kronecker product of the defining module of $\bar{\mathfrak{g}} = A_r$, the centralizer is the group algebra $\mathbb{C}(\mathcal{S}_m)$ of the symmetric group $\mathcal{S}_m$ (for $m > r+1$, one has to divide out by some ideal, in order to implement the fact that $(r+2)$-fold fully antisymmetrized combinations of indices vanish). By definition, the group algebra of a group is the set of all linear combinations of group elements with respect to some base field; thus $\mathbb{C}(\mathcal{S}_m)$ is the algebra over $\mathbb{C}$ generated by the unit 1 and generators s_i, $i = 1, \ldots, m-1$, with relations

$$\begin{aligned} &s_i s_i = 1, \\ &s_i s_{i+1} s_i = s_{i+1} s_i s_{i+1}, \\ &s_i s_j = s_j s_i \qquad\qquad \text{for} \quad |i-j| \geq 2. \end{aligned} \tag{4.3.36}$$

The general statements about the centralizer algebra apply equally well to the case of the quantum group $\mathsf{U}_q(\mathfrak{g})$; only the explicit form of the centralizer algebras becomes more complicated. In the case of the m-fold Kronecker product of the defining module of $\mathsf{U}_q(A_r)$, the centralizer is the so-called *Hecke algebra* (of type A_r) $H_m(q)$ (in addition, for $m > r+1$ one must again divide $H_m(q)$ by a certain ideal, see below). This is the associative algebra generated by a unit 1 and further generators t_i, $i = 1, \ldots, m-1$, with relations

$$\begin{aligned} &t_i t_i = (1-q)\, t_i + q, \\ &t_i t_{i+1} t_i = t_{i+1} t_i t_{i+1}, \\ &t_i t_j = t_j t_i \qquad\qquad \text{for} \quad |i-j| \geq 2. \end{aligned} \tag{4.3.37}$$

In short, the relations of both $\mathcal{S}_m$ and $H_m(q)$ are those of the braid group $\mathcal{B}_m$ (see (4.1.60)), supplemented by a quadratic relation; the quadratic relation of $H_m(q)$ is a deformation of the quadratic relation for $\mathcal{S}_m$. Also, $H_m(1)$ is clearly isomorphic to the group algebra of $\mathcal{S}_m$, and for any q the dimension of the Hecke algebra is $\dim H_m(q) = \dim \mathcal{S}_m = m!$ Moreover, with some effort it can be shown that the representation theory of $H_m(q)$ remains isomorphic to that of $\mathcal{S}_m$ if q is deformed away from 1, provided however that q is not a root of unity. In contrast, for q a root of unity, $H_m(q)$ is no longer semisimple so that its representation theory becomes very different, allowing in particular for modules being reducible, but not fully reducible.

More precisely, define the *characteristic* p of $H_m(q)$ as the smallest positive integer such that

$$1 + q + q^2 + \ldots + q^{p-1} = 0 \tag{4.3.38}$$

(and $p = \infty$ if such an integer does not exist); thus $p = k^\vee + g^\vee$ in the case of our interest. Then for $m < p$, $H_m(q)$ is semisimple and its representation theory is isomorphic to that of $\mathbb{C}(\mathcal{S}_m)$, while for $m \geq p$, $H_m(q)$ is not semisimple, and its irreducible modules have to be constructed as quotient modules of the q-deformations of $\mathbb{C}(\mathcal{S}_m)$-modules by the kernel of the standard bilinear form.

As already mentioned, for $m > r+1$ there is a slight complication, because one has to implement that only Young tableaux with at most n rows are allowed for $sl_n \cong A_{n-1}$, and likewise for $\mathsf{U}_q(sl_n)$. This is implemented by the inclusion of one further relation that forces the $(n+1)$ row antisymmetrizer to vanish. The centralizer is then the factor of $H_m(q)$ by the ideal that is generated by the element

$$\hat{t} = \sum_{s \in \mathcal{S}_{n+1}} (\text{sign}\, s)\, t_s, \tag{4.3.39}$$

where $t_s \in H_m(q)$ is given by $t_s = t_{i_1} t_{i_2} ... t_{i_p}$ if $s \in \mathcal{S}_{n+1}$ is written as a (non-reducible) product $s = s_{i_1} s_{i_2} ... s_{i_p}$ of the generators s_i of $\mathcal{S}_{n+1}$. Consider for example the case $n = 2$; then the three row antisymmetrizer vanishes:

$$1 - t_1 - t_2 + t_1 t_2 + t_2 t_1 - t_1 t_2 t_1 = 0. \tag{4.3.40}$$

The algebra generated by the t_i is then isomorphic to the so-called Temperley–Lieb–Jones algebra.

Let us also note that instead of the generators t_i of $H_m(q)$, one sometimes considers the generators

$$u_i := \frac{t_i - 1}{q+1}. \tag{4.3.41}$$

The Hecke algebra relations (4.3.37) then become

$$\begin{aligned} &u_i u_i = u_i, \\ &u_i u_{i+1} u_i - \kappa\, u_i = u_{i+1} u_i u_{i+1} - \kappa\, u_{i+1}, \\ &u_i u_j = u_j u_i \qquad\qquad \text{for} \quad |i-j| \geq 2, \end{aligned} \tag{4.3.42}$$

where $\kappa = q/(q+1)^2$. For the Temperley–Lieb–Jones algebra, the additional relation yields

$$u_i u_{i\pm1} u_i = (q + 2 + q^{-1})^{-1}\, u_i. \tag{4.3.43}$$

We conclude this section by presenting an algorithm for the determination of the weight Λ' which appears for $R_{\Lambda,\Lambda'}$ with given Λ. This is a generalization of the Racah–Speiser algorithm which was described at the end of section 1.7. The original Racah–Speiser algorithm for $\bar{\mathfrak{g}}$ of course still works for $\mathsf{U}_q(\bar{\mathfrak{g}})$, but now some of the weights obtained by this method are not highest weights Λ of irreducible highest weight modules R_Λ, but rather the weights Λ and Λ' of the non-fully reducible modules $R_{\Lambda,\Lambda'}$. Since we know that for $(\Lambda, \theta) \leq k^\vee$, there exist irreducible highest weight modules, while for $(\Lambda, \theta) > k^\vee$, the modules are usually non-fully reducible, it is clear that there is a close connection to the affine Lie algebra $\mathfrak{g} = \bar{\mathfrak{g}}^{(1)}$. Therefore a natural extension of the procedure is to include also the reflection with respect to the horizontal projection of the zeroth simple root of $\mathfrak{g}$. This is the reflection at the hyperplane perpendicular to the highest $\bar{\mathfrak{g}}$-root θ which however does not go through zero, but is shifted such that its defining equation reads

$$\sum_{i=1}^{r} a_i^\vee \lambda^i = k^\vee + g^\vee, \tag{4.3.44}$$

where $a_i^\vee$ are the dual Coxeter labels of $\bar{\mathfrak{g}}$. The weight obtained by acting with this reflection on a weight λ will be denoted as $\sigma_{(0)} \star \lambda$. Explicitly, one has (using (1.7.21))

$$\sigma_{(0)} \star \lambda = \lambda + (k^\vee + 1 - (\lambda, \theta^\vee))\, \theta. \tag{4.3.45}$$

In more detail, the extended algorithm works as follows. Given the strictly dominant weights $\Lambda_i + \rho$ obtained through the Racah–Speiser algorithm for some Kronecker product, one has to check whether $(\Lambda_i, \theta) - (k^\vee + 1)$ is positive, zero or negative. If it is zero, then there exists an irreducible (but non-unitarizable) module R_{Λ_i}. If it is negative, reflect the weight with respect to the hyperplane (4.3.44). If the weight $\sigma_{(0)} \star \Lambda_i + \rho$ obtained this way is strictly dominant, then there is a non-fully reducible module containing $\breve{R}_{\Lambda_i}$ as a subspace; if $\sigma_{(0)} \star \Lambda_i + \rho$ lies on the boundary of the fundamental Weyl chamber (i.e. is dominant, but not strictly dominant), then there is still an irreducible (non-unitarizable) module R_{Λ_i}. Finally, if $\mu = \sigma_{(0)} \star \Lambda_i + \rho$ is not dominant, take the ordinary Weyl reflection σ which maps μ into the dominant Weyl chamber; if $(\sigma \star \sigma_{(0)} \star \Lambda_i, \theta^\vee) \le k^\vee$, then again there is a non-fully reducible module containing $\breve{R}_{\Lambda_i}$ as a subspace; otherwise the procedure just described has to be repeated, now applied to the weight $\sigma \star \sigma_{(0)} \star \Lambda_i$ instead of Λ_i (this process terminates after a finite number of steps, as will be explained in more detail in section 5.5).

The precise form of the non-fully reducible modules can be found after this procedure has been carried out for all Λ_i; if a strictly dominant weight μ is obtained as $\sigma_{(0)} \star \Lambda_i + \rho$ or $\sigma \star \sigma_{(0)} \star \Lambda_i + \rho$ etc. for a single value of i, then the non-fully reducible module is

$$R = R_{\Lambda_i, \sigma_{(0)} \star \Lambda_i}, \tag{4.3.46}$$

respectively

$$R = R_{\Lambda_i, \sigma \star \sigma_{(0)} \star \Lambda_i} \tag{4.3.47}$$

etc. However, it can also happen that a given strictly dominant weight μ is obtained in the way just described for more than one value of the index i. In this case, the algorithm does not fix the non-fully reducible modules completely, and a more detailed analysis is required.

To prove that the algorithm just described gives correct results, one must identify the null vector structure of the Verma modules that can be constructed for any given highest weight in a manner completely analogous to the case of $\bar{\mathfrak{g}}$ or of affine $\mathfrak{g}$. Such an analysis of quantum group modules has been performed in the literature, coming very close to a complete proof of the agorithm; the description of this analysis is however beyond the scope of the present book.

4.4 Quantum dimensions

The special features of the representation theory of $\mathsf{U}_q(\bar{\mathfrak{g}})$ for q a root of unity have been explained in the previous section rather abstractly in terms of the centralizer algebras of $\mathsf{U}_q(\bar{\mathfrak{g}})$ in the Kronecker products of representations. A more suggestive understanding of these results can be gained by introducing the q-analogue of the dimension of a module, the so-called *quantum dimension.* Recall from section 1.6 that the dimension of a $\bar{\mathfrak{g}}$-module R is just the sum of the dimensions of its weight spaces,

$$d_R = \sum_{\lambda:\, v_\lambda \in R} \mathrm{mult}_R(\lambda) = \mathrm{tr}_R\, \mathbf{1} \tag{4.4.1}$$

(here R may also be viewed as a $\mathsf{U}(\bar{\mathfrak{g}})$-module and hence, by continuity, as a $\mathsf{U}_q(\bar{\mathfrak{g}})$-module). The quantum dimension $\mathcal{D}_R$ of R is defined as a q-deformation of d_R,

$$\mathcal{D}_R \equiv \mathcal{D}^q{}_R := \sum_{\lambda:\, v_\lambda \in R} \mathrm{mult}_R(\lambda)\, q^{(\tilde{\rho},\lambda)} = \mathrm{tr}_R\, q^{R(H_\rho)}, \tag{4.4.2}$$

with $\tilde{\rho}$ and the Cartan subalgebra element H_ρ defined as in (4.2.31) and (4.2.30), respectively, so that

$$R(H_\rho) \cdot v_\lambda = \frac{2\,(\rho,\lambda)}{(\theta,\theta)}\, v_\lambda. \tag{4.4.3}$$

While the ordinary dimensions of $\mathsf{U}_q(\bar{\mathfrak{g}})$-modules are of course positive integers, according to the definition just given the quantum dimensions of $\mathsf{U}_q(\bar{\mathfrak{g}})$-modules are a priori generic complex numbers; in fact they are real iff $|q| = 1$. Below we will see that the quantum dimension of a highest weight module is equal to the character of the highest weight module, evaluated at a specific point of the weight space. As will be pointed out in section 4.5, this implies that any Kronecker product of highest weight modules of $\mathsf{U}_q(\bar{\mathfrak{g}})$ obeys a quantum dimension sum rule analogous to the sum rule for ordinary dimensions. Moreover, contrary to the sum rule for ordinary dimensions, the quantum dimension sum rule remains valid upon a certain truncation of the Kronecker products that can be performed for q a root of unity; this is explained in detail in section 4.5.

For $\mathfrak{g} = A_1$, the above definition yields the quantum dimension of a highest weight module R_Λ as

$$\mathcal{D}_\Lambda(A_1) = \sum_{\substack{\lambda=-\Lambda \\ \Lambda-\lambda\in 2\mathbb{Z}}}^{\Lambda} q^{\lambda/2} = \lfloor \Lambda+1 \rfloor. \tag{4.4.4}$$

Thus $\mathcal{D}_\Lambda(A_1)$ is a Laurent polynomial in the variable $q^{1/2}$. Because of $\rho = \sum_i \Lambda_{(i)}$, this is in fact true for the quantum dimensions of highest weight modules of any simply laced $\mathfrak{g}$, while for non-simply laced $\mathfrak{g}$ one gets Laurent polynomials in $q^{S^2/2L^2}$ with S and L the lengths of short and long roots, respectively.

As we will see in section 5.5 (p. 360), the definition of the quantum dimension introduced here coincides with another concept of quantum dimension that is introduced in (5.3.18) and is motivated by the modular properties of the characters in conformal field theory.

In terms of centralizer algebras, the above definition of the quantum dimension arises as follows. For a given irreducible module R of $\mathsf{U}_q(\mathfrak{g})$, denote by $\mathcal{C}_R^n$ the centralizer of $\mathsf{U}_q(\mathfrak{g})$ in the n-fold Kronecker product $R^{\otimes n} = R \otimes ... \otimes R$, and define the formal inductive unification $\mathcal{C}_R^\infty := \bigcup_{n=0}^\infty \mathcal{C}_R^n$. There exists a homomorphism from $\mathcal{B}_n$, the braid group on n strands, to $\mathcal{C}_R^n$, and this extends to a homomorphism from the inductive limit $\mathcal{B}_\infty$ to $\mathcal{C}_R^\infty$. As a consequence of the quasitriangularity property (4.1.38a), one has

$$\mathcal{R}\,(q^{R(H_\rho)} \otimes q^{R(H_\rho)}) = (q^{R(H_\rho)} \otimes q^{R(H_\rho)})\,\mathcal{R} \tag{4.4.5}$$

for $\mathcal{R} \equiv (R \times R)(\mathtt{R})$. It can be shown that the quantity $\mathcal{R}$ represents the non-trivial generator of the braid group $\mathcal{B}_2$; thus one can use $\mathcal{R}$ and q^{H_ρ} to define a trace operation "Tr" on $\mathcal{C}_R^\infty$, namely via

$$\mathsf{Tr}(x) = \frac{1}{[\mathrm{tr}_R\, q^{R(H_\rho)}]^n}\, \mathrm{tr}_{R^{\otimes n}} \Big[\Big(q^{R(H_\rho)} \otimes \ldots \otimes q^{R(H_\rho)} \Big) \cdot x \Big] \tag{4.4.6}$$

for $x \in \mathcal{C}_r^n$, where "tr_R" denotes the ordinary matrix trace in the representation R. In the case of the defining module of $\mathsf{U}_q(A_r)$, this trace is known as the *Ocneanu trace* on the Hecke algebra $H_n(q)$.

The operation Tr is normalized and symmetric,

$$\begin{aligned} \mathsf{Tr}(1) &= 1, \\ \mathsf{Tr}(xy) &= \mathsf{Tr}(yx), \end{aligned} \tag{4.4.7}$$

and it obeys the following relation, which is known as the *Markov property*:

$$\mathsf{Tr}(x_j^{\pm 1} y) = \frac{q^{\pm 2C_R/(\theta,\theta)}}{\mathrm{tr}_R\, q^{R(H_\rho)}}\, \mathsf{Tr}(y) \tag{4.4.8}$$

for $y \in C_r^j$. Here C_R denotes the quadratic Casimir eigenvalue in the representation R of $\mathfrak{g}$, and x_j is the image of the braid group element

$$\underbrace{1\otimes\ldots\otimes 1\otimes}_{j-1 \text{ times}} \underbrace{\mathcal{R}}_{\substack{j\text{th and}\\(j+1)\text{th place}}} \otimes 1\otimes\ldots \tag{4.4.9}$$

under the homomorphism from $\mathcal{B}_\infty$ to $\mathcal{C}_R^\infty$. It is possible to expand $\mathsf{Tr}(x)$ into a linear combination of characters of $\mathcal{C}_R^n$; the coefficients in this expansion turn out to be, up to a normalization factor, the quantum dimensions for the irreducible representations contained in $R^{\otimes n}$. Also note that one can obtain the quantum dimension of R as $q^{2C_R/(\theta,\theta)}/\mathsf{Tr}(x_j)$ by applying the Markov property to $y = 1$.

From the expression (1.7.42) for the characters of A_1 one reads off that the quantum dimension (4.4.4) is nothing but the character of R_Λ, evaluated on the Cartan subalgebra element $h = \ln q \cdot J_3$. More generally, comparing (4.4.2) with the definition (1.7.33) of the $\mathfrak{g}$-characters, it follows that for irreducible highest weight modules, one has

$$\mathcal{D}_\Lambda = \chi_\Lambda(\ln q \cdot H_\rho). \tag{4.4.10}$$

Recalling that the Cartan subalgebra can be identified with the weight space, this may also be written as

$$\mathcal{D}_\Lambda = \chi_\Lambda(\ln q \cdot \tilde{\rho}). \tag{4.4.11}$$

(This formula makes it again clear that the quantum dimension is a deformation of the ordinary dimension because in the limit $q \to 1$, the q-dimension approaches $\chi_\Lambda(0) = d_\Lambda$.) An analogous formula of course holds for direct sums of irreducible highest weight modules. Moreover, its validity can be extended, by analytic continuation, to the case of the non-fully reducible modules which arise if q is a root of unity, by simply setting

$$\mathcal{D}_{\Lambda,\Lambda'} = \mathcal{D}_\Lambda + \mathcal{D}_{\Lambda'} \tag{4.4.12}$$

for $R_{\Lambda,\Lambda'} = \breve{R}_\Lambda \oplus \breve{R}_{\Lambda'}$.

Expressing the q-dimensions as in (4.4.10) through the characters, one opens the possibility to use the Weyl character formula to compute quantum dimensions. In fact, it turns out that one can employ the denominator identity to deduce a q-analogue of the Weyl dimension formula (1.6.27).

Recall that the denominator identity states that the characters of a simple Lie algebra $\mathfrak{g}$ obey

$$\chi_\Lambda(\xi\rho) = \prod_{\alpha>0} \frac{\exp[\frac{1}{2}\xi(\Lambda+\rho,\alpha)] - \exp[-\frac{1}{2}\xi(\Lambda+\rho,\alpha)]}{\exp[\frac{1}{2}\xi(\rho,\alpha)] - \exp[-\frac{1}{2}\xi(\rho,\alpha)]}. \tag{4.4.13}$$

Choosing the parameter ξ as $\xi = (\ln q)^{2/(\theta,\theta)}$, one obtains the following formula for the quantum dimension of an irreducible highest weight module:

$$\mathcal{D}_\Lambda = \prod_{\alpha>0} \frac{\lfloor(\Lambda+\rho,\tilde{\alpha})\rfloor}{\lfloor(\rho,\tilde{\alpha})\rfloor} \tag{4.4.14}$$

with $\tilde{\alpha} := 2\alpha/(\theta,\theta)$. For $\mathfrak{g} = A_1$, this reads simply $\mathcal{D}_\Lambda = \lfloor\Lambda+1\rfloor/\lfloor 1\rfloor = \lfloor\Lambda+1\rfloor$ and hence reproduces (4.4.4).

As a less trivial example, consider $\mathfrak{g} = A_2$. There are then three positive roots, namely $\alpha^{(1)}$, $\alpha^{(2)}$, $\alpha^{(1)}+\alpha^{(2)}$; accordingly, the formula (4.4.14) yields

$$\mathcal{D}_\Lambda(A_2) = \frac{\lfloor\Lambda^1+1\rfloor\lfloor\Lambda^2+1\rfloor\lfloor\Lambda^1+\Lambda^2+2\rfloor}{\lfloor 2\rfloor}. \tag{4.4.15}$$

Similarly, for $\mathfrak{g} = G_2$ there are three long positive roots $\alpha^{(1)}$, $\alpha^{(1)}+3\alpha^{(2)}$, $2\alpha^{(1)}+3\alpha^{(2)}$, and three short positive roots (of length $1/\sqrt{3}$ times the length of the long roots) $\alpha^{(2)}$, $\alpha^{(1)}+\alpha^{(2)}$, $\alpha^{(1)}+2\alpha^{(2)}$, so that (4.4.14) gives

$$\begin{aligned} \mathcal{D}_\Lambda(G_2) = {} & \lfloor\Lambda^1+1\rfloor\lfloor\Lambda^1+\Lambda^2+2\rfloor\lfloor 2\Lambda^1+\Lambda^2+3\rfloor\lfloor\tfrac{1}{3}(\Lambda^2+1)\rfloor \\ & \cdot\lfloor\tfrac{1}{3}(3\Lambda^1+\Lambda^2+4)\rfloor\lfloor\tfrac{1}{3}(3\Lambda^1+2\Lambda^2+5)\rfloor \\ & /\,\lfloor\tfrac{1}{3}\rfloor\lfloor\tfrac{4}{3}\rfloor\lfloor\tfrac{5}{3}\rfloor\lfloor 2\rfloor\lfloor 3\rfloor. \end{aligned} \tag{4.4.16}$$

It is readily checked that in the limit $q \to 1$, thes latter formulae reproduce the results (1.6.28) and (1.6.29) for the ordinary dimensions, as they should.

As further applications of (4.4.14), we list in the table (4.4.18) the quantum dimensions for a number of low-dimensional modules. The factorials involving q-half-integers which appear there are defined in an obvious manner,

$$\begin{aligned} \lfloor n+\tfrac{1}{2}\rfloor! &= \lfloor\tfrac{1}{2}\rfloor\lfloor\tfrac{3}{2}\rfloor\cdots\lfloor n-\tfrac{1}{2}\rfloor\lfloor n+\tfrac{1}{2}\rfloor, \\ \lfloor 2n+\tfrac{1}{2}\rfloor!! &= \lfloor\tfrac{1}{2}\rfloor\lfloor\tfrac{5}{2}\rfloor\cdots\lfloor 2n-\tfrac{3}{2}\rfloor\lfloor 2n+\tfrac{1}{2}\rfloor, \\ \lfloor 2n-\tfrac{1}{2}\rfloor!! &= \lfloor\tfrac{3}{2}\rfloor\lfloor\tfrac{7}{2}\rfloor\cdots\lfloor 2n-\tfrac{5}{2}\rfloor\lfloor 2n-\tfrac{1}{2}\rfloor \end{aligned} \tag{4.4.17}$$

for $n \in \mathbb{Z}_{\geq 0}$.

$\mathfrak{g}$	Λ	$\mathcal{D}_\Lambda$
A_r	$\Lambda_{(i)}$	$\begin{bmatrix} r+1 \\ i \end{bmatrix}$
	$j\,\Lambda_{(1)}$	$\begin{bmatrix} r+j \\ j \end{bmatrix}$
	$\Lambda_{(i)}+\Lambda_{(j)},\ j\geq i$	$\dfrac{\lfloor r+1\rfloor!\lfloor r+2\rfloor!\lfloor j-i+1\rfloor}{\lfloor r-j+1\rfloor!\lfloor r-i+2\rfloor!\lfloor i\rfloor!\lfloor j+1\rfloor!}$
	θ	$\lfloor r\rfloor\lfloor r+2\rfloor$
	$j\,\theta$	$\lfloor r\rfloor\lfloor r+2j\rfloor\left(\dfrac{\lfloor r+j-1\rfloor!}{\lfloor j\rfloor!\lfloor r\rfloor!}\right)^2$
B_r	$\Lambda_{(1)}$	$\dfrac{\lfloor r+\frac{1}{2}\rfloor\lfloor 2r-1\rfloor}{\lfloor r-\frac{1}{2}\rfloor}$
	$\Lambda_{(r)}$	$\dfrac{\lfloor 2r-1\rfloor!!}{\lfloor r-\frac{1}{2}\rfloor!}$
C_r	$\Lambda_{(1)}$	$\dfrac{\lfloor \frac{r}{2}\rfloor\lfloor r+1\rfloor}{\lfloor \frac{1}{2}\rfloor\lfloor \frac{r+1}{2}\rfloor}$
	$\Lambda_{(r)}$	$\dfrac{\lfloor \frac{1}{2}\rfloor\lfloor r+1\rfloor!\lfloor r+\frac{1}{2}\rfloor!}{\left(\lfloor \frac{r+1}{2}\rfloor!\right)^2\lfloor \frac{r}{2}+1\rfloor!\lfloor \frac{r}{2}\rfloor!}$
D_r	$\Lambda_{(1)}$	$\dfrac{\lfloor r\rfloor\lfloor 2r-2\rfloor}{\lfloor r-1\rfloor}$
	$\Lambda_{(r-1)},\ \Lambda_{(r)}$	$\dfrac{\lfloor 2r-2\rfloor!!}{\lfloor r-1\rfloor!}$
E_6	$\Lambda_{(1)},\ \Lambda_{(5)}$	$\dfrac{\lfloor 9\rfloor\lfloor 12\rfloor}{\lfloor 4\rfloor}$
	$\theta=\Lambda_{(6)}$	$\dfrac{\lfloor 8\rfloor\lfloor 9\rfloor\lfloor 13\rfloor}{\lfloor 2\rfloor\lfloor 4\rfloor}$
E_7	$\Lambda_{(6)}$	$\dfrac{\lfloor 10\rfloor\lfloor 14\rfloor\lfloor 18\rfloor}{\lfloor 5\rfloor\lfloor 9\rfloor}$
	$\theta=\Lambda_{(1)}$	$\dfrac{\lfloor 12\rfloor\lfloor 14\rfloor\lfloor 19\rfloor}{\lfloor 4\rfloor\lfloor 6\rfloor}$

(4.4.18)

$\mathfrak{g}$	Λ	$\mathcal{D}_\Lambda$
E_8	$\theta=\Lambda_{(1)}$	$\frac{\lfloor 20\rfloor\lfloor 24\rfloor\lfloor 31\rfloor}{\lfloor 6\rfloor\lfloor 10\rfloor}$
	$\Lambda_{(7)}$	$\frac{\lfloor 14\rfloor\lfloor 20\rfloor\lfloor 25\rfloor\lfloor 30\rfloor\lfloor 31\rfloor}{\lfloor 4\rfloor\lfloor 6\rfloor\lfloor 7\rfloor\lfloor 10\rfloor}$
F_4	$\theta=\Lambda_{(1)}$	$\frac{\lfloor \frac32\rfloor\lfloor \frac{13}{2}\rfloor\lfloor 4\rfloor\lfloor 9\rfloor}{\lfloor \frac12\rfloor\lfloor \frac92\rfloor\lfloor 2\rfloor\lfloor 3\rfloor}$
	$\Lambda_{(4)}$	$\frac{\lfloor \frac{13}{2}\rfloor\lfloor 6\rfloor\lfloor 10\rfloor}{\lfloor \frac52\rfloor\lfloor 3\rfloor}$
G_2	$\Lambda_{(2)}$	$\frac{\lfloor \frac23\rfloor\lfloor \frac73\rfloor\lfloor 4\rfloor}{\lfloor \frac13\rfloor\lfloor \frac43\rfloor\lfloor 2\rfloor}$
	$\theta=\Lambda_{(1)}$	$\frac{\lfloor \frac73\rfloor\lfloor \frac83\rfloor\lfloor 5\rfloor}{\lfloor \frac43\rfloor\lfloor \frac53\rfloor}$

(4.4.18) (continued)

For direct sums of irreducible highest weight modules, the quantum dimension can of course be obtained by adding up the Weyl formulae for the irreducible submodules. By analytic continuation in q, one can then use the Weyl formula also for non-fully reducible modules which are obtained as the limit of the direct sum of irreducible highest weight modules R_Λ and $R_{\Lambda'}$ in the limit when q tends to a root of unity, namely

$$\mathcal{D}_{\Lambda,\Lambda'}=\prod_{\alpha>0}\frac{\lfloor(\Lambda+\rho,\tilde\alpha)\rfloor}{\lfloor(\rho,\tilde\alpha)\rfloor}+\prod_{\alpha>0}\frac{\lfloor(\Lambda'+\rho,\tilde\alpha)\rfloor}{\lfloor(\rho,\tilde\alpha)\rfloor}. \tag{4.4.19}$$

From (4.4.14) it also follows that for q a root of unity, the quantum dimensions possess interesting symmetry properties. Namely, take $q=\exp(2\pi i/(k^\vee+g^\vee))$. Then according to (4.3.26) one has $\lfloor k^\vee+g^\vee\rfloor=0$; in fact, with the help of the definition (4.2.3) of $\lfloor x\rfloor$ it follows immediately that more generally

$$\lfloor\xi+m(k^\vee+g^\vee)\rfloor=(-1)^m\lfloor\xi\rfloor \quad\text{for } m\in\mathbb{Z}. \tag{4.4.20}$$

In particular,

$$\lfloor\xi\rfloor=\lfloor\xi+2(k^\vee+g^\vee)\rfloor, \tag{4.4.21}$$

$$\lfloor\xi\rfloor=\lfloor k^\vee+g^\vee-\xi\rfloor. \tag{4.4.22}$$

Since for any $\mathfrak{g}$-coroot β and any $\mathfrak{g}$-root α, (β,α) is an integer, it follows that

$$\lfloor(\lambda+(k^\vee+g^\vee)\,\tilde\beta,\tilde\alpha)\rfloor=(-1)^{(\tilde\beta,\tilde\alpha)}\lfloor(\lambda,\tilde\alpha)\rfloor \tag{4.4.23}$$

with $\tilde\beta=\frac{(\theta,\theta)}{2}\,\beta$. As a consequence, (4.4.14) implies that

$$\mathcal{D}_{\Lambda+(k^\vee+g^\vee)\tilde\beta}=\mathcal{D}_\Lambda\cdot\prod_{\alpha>0}(-1)^{(\beta,\alpha)}=\mathcal{D}_\Lambda\cdot(-1)^{2(\beta,\rho)}, \tag{4.4.24}$$

and thus (since $(\beta,\rho)\in\mathbb{Z}$ owing to $\rho=\sum_i\Lambda_{(i)}$)

$$\mathcal{D}_{\Lambda+(k^\vee+g^\vee)\tilde\beta}=\mathcal{D}_\Lambda. \tag{4.4.25}$$

Note that this formula holds even if $\Lambda+(k^\vee+g^\vee)\tilde\beta$ is not a dominant weight.

Another symmetry property shows up by considering the action of the Weyl group on the quantum dimensions. Recall that any element σ of the Weyl group $W(\mathfrak{g})$ permutes the $\mathfrak{g}$-roots. In other words, σ permutes the positive roots of $\mathfrak{g}$ up to possibly sign factors. For a fundamental Weyl reflection, the number of minus signs is just equal to one because $\sigma_{(i)}$ permutes the positive roots other than $\alpha^{(i)}$; from the definition of the length of σ (p. 72) it is then clear that the number of minus signs encountered for σ equals $\ell(\sigma)$. Therefore

$$\prod_{\alpha>0}(\sigma(\Lambda+\rho),\tilde\alpha)=\prod_{\alpha>0}(\Lambda+\rho,\sigma(\tilde\alpha))=(-1)^{\ell(\sigma)}\prod_{\alpha>0}(\Lambda+\rho,\tilde\alpha), \tag{4.4.26}$$

so that (4.4.14) implies

$$\mathcal{D}_{\sigma\star\Lambda}=(-1)^{\ell(\sigma)}\,\mathcal{D}_\Lambda, \tag{4.4.27}$$

where "$\star$" denotes the modified Weyl reflection (1.7.19), $\sigma\star\Lambda=\sigma(\Lambda+\rho)-\rho$. This formula holds for arbitrary q (and hence, setting $q=1$ also for the ordinary dimensions), but is in itself not too interesting because typically $\sigma\star\Lambda$ is not a dominant weight. However, for q a root of unity, it can be combined with the result (4.4.25), leading to

$$\mathcal{D}_{w_{\sigma,\beta}(\Lambda)}=(-1)^{\ell(\sigma)}\,\mathcal{D}_\Lambda, \tag{4.4.28}$$

where

$$w_{\sigma,\beta}(\Lambda):=\sigma\star\Lambda+(k^\vee+g^\vee)\,\tilde\beta \tag{4.4.29}$$

with σ an arbitrary Weyl reflection and β an arbitrary element of the coroot lattice of $\mathfrak{g}$. For any given Weyl group element σ there exists a coroot β such that $w_{\sigma,\beta}(\Lambda)$ is a dominant weight so that (4.4.28) is not only a formal identity, but really relates the quantum dimensions of highest weight modules, or more precisely, possibly the quantum dimensions of would-be irreducible subspaces $\check{R}_\Lambda$. Moreover, it is easily seen that for any given $\sigma \in W(\mathfrak{g})$ (and given Λ) there is a unique choice of β such that the weight $w_{\sigma,\beta}(\Lambda)$ corresponds to an integrable irreducible highest weight module of the associated untwisted affine algebra $\mathfrak{g}^{(1)}$ at level $k^\vee$. (To be precise, β is unique only if the affine weight associated to $w_{\sigma,\beta}(\Lambda)$ does not lie on a boundary of the fundamental affine Weyl chamber; see section 5.5.) As a simple example for (4.4.28), consider the case $\mathfrak{g} = A_1$. The elements of the root lattice are then just the even integers, so that $\mathcal{D}_{w_{\sigma,\beta}(\Lambda)} = \mathcal{D}_{w_{\sigma,0}(\Lambda)}$ as a consequence of (4.4.21) and (4.4.4). Also, there is single non-trivial Weyl reflection acting as $\sigma \ast \Lambda = -\Lambda - 2$ so that (4.4.4) implies $\mathcal{D}_{\sigma \ast \Lambda} = -\mathcal{D}_\Lambda$.

As another application of (4.4.28), observe that a highest weight module has quantum dimension zero,

$$\mathcal{D}_\Lambda = 0, \tag{4.4.30}$$

whenever there exists a pair σ, β such that

$$w_{\sigma,\beta}(\Lambda) = \Lambda \quad \text{and} \quad \ell(\sigma) \in 2\mathbb{Z} + 1. \tag{4.4.31}$$

It turns out that in fact all modules with this property are irreducible. Moreover, all irreducible modules of $\mathsf{U}_q(\mathfrak{g})$ are either such that (4.4.30) holds, or else their highest weight is the horizontal part of an integrable $\mathfrak{g}^{(1)}$-module at level $k^\vee$ (in the latter case the module is unitarizable, in the former case it is not).

Even more importantly, the quantum dimension also gives a partial solution to the problem of determining which would-be irreducible modules combine into non-fully reducible modules. Namely, it turns out that any non-fully reducible module has zero quantum dimension, too:

$$\mathcal{D}_{\Lambda,\Lambda'} = 0. \tag{4.4.32}$$

In terms of the subspaces $\check{R}_\Lambda$ and $\check{R}_{\Lambda'}$ of $R_{\Lambda,\Lambda'}$, according to (4.4.12) this means that

$$\mathcal{D}_{\Lambda'} = -\mathcal{D}_\Lambda. \tag{4.4.33}$$

(For non-fully reducible modules consisting of more than two subspaces $\check{R}_\Lambda$, this generalizes as follows. The absolute value of the quantum dimension is the same for all subspaces; there is an even number of subspaces, and half of them have positive, the others have negative quantum dimension.) Generically, this simple numeric equation does not identify Λ' uniquely in terms of Λ (or vice versa). However, inspection shows that for small values of the level $k^\vee$ often unique answers are obtained in the region $(\Lambda, \theta^\vee), (\Lambda', \theta^\vee) \le 2k^\vee$, provided that one also takes into account that the associated $\mathfrak{g}$-modules R_Λ and $R_{\Lambda'}$ must belong to the same conjugacy class of $\bar{\mathfrak{g}}$-weights (this is so because they can both be obtained as submodules of one particular Kronecker product of irreducible modules).

The proof of (4.4.33) is based on the symmetry property (4.4.28) of the quantum dimensions. From the constructive determination of Λ' in terms of Λ that was described in the previous section, it is clear that (4.4.28) implies that $\mathcal{D}_\Lambda$ and $\mathcal{D}_{\Lambda'}$ are equal up to possibly a sign. The non-trivial part of the proof is thus to show that this sign is actually a minus sign; that this is true follows e.g. from the relation to the so-called fusion rules of WZW theories which will be discussed in section 5.5.

To a large extent, the non-uniqueness of the solutions to (4.4.33) is a result of further symmetries of the quantum dimensions, which are related to Dynkin diagram automorphisms. Namely, one finds that any automorphism of the Dynkin diagram of the affine algebra $\mathfrak{g} = \bar{\mathfrak{g}}^{(1)}$ induces a symmetry of the set of quantum dimensions of those $\mathsf{U}_q(\bar{\mathfrak{g}})$-modules which correspond to integrable $\mathfrak{g}$-modules. More explicitly, denoting by $\Gamma \equiv \Gamma(\mathfrak{g})$ the automorphism group of the Dynkin diagram of $\mathfrak{g}$, each $\dot\omega \in \Gamma$ is a permutation of the labels $i \in \{0, 1, \ldots, r\}$ of the nodes of the Dynkin diagram, and hence acts on the set of $\mathfrak{g}$-weights in the obvious manner

$$\Lambda = \sum_{i=0}^{r} \Lambda^i \Lambda_{(i)} \mapsto \omega(\Lambda) := \sum_{i=0}^{r} \Lambda^i \Lambda_{(\dot\omega(i))} = \sum_{i=0}^{r} \Lambda^{\dot\omega^{-1}(i)} \Lambda_{(i)}, \tag{4.4.34}$$

and this map leaves the quantum dimensions invariant,

$$\mathcal{D}_{\omega(\Lambda)} = \mathcal{D}_\Lambda . \tag{4.4.35}$$

If Λ is an integrable highest weight of $\bar{\mathfrak{g}}$, then all Dynkin labels Λ^i are nonnegative, and hence the same is true for $\omega(\Lambda)$ so that the weight $\omega(\Lambda)$ is integrable as well. If $\dot\omega$ obeys $\dot\omega(0) = 0$, then it is very easy to see that (4.4.35) is fulfilled. In this case $\dot\omega$ is already contained in the automorphism group $\bar\Gamma \equiv \Gamma(\bar{\mathfrak{g}})$ of the Dynkin diagram of $\bar{\mathfrak{g}}$. The induced map $\bar\omega$ of the weight lattice of $\bar{\mathfrak{g}}$ clearly permutes the set of positive roots; together with the identity $(\bar\omega(\mu), \alpha) = (\mu, \bar\omega^{-1}(\alpha))$, the equality of quantum dimensions then follows immediately. For $\dot\omega \in \Gamma/\bar\Gamma$, i.e. $\dot\omega$ a genuine automorphism of the *affine* Dynkin diagram, the situation is more complicated. A general proof of (4.4.35) for this case will be presented in section 5.5 (p. 359). Here we consider only the case of $\bar{\mathfrak{g}} = sl_n$ which has the largest automorphism group,

$$\Gamma \cong \mathbb{Z}_n \ltimes \mathbb{Z}_2, \quad \bar\Gamma \cong \mathbb{Z}_2. \tag{4.4.36}$$

In this case a proof is possible by expressing the problem in terms of Young tableaux. Recall from section 1.6 that for sl_n the Weyl dimension formula is equivalent to the hook formula (1.6.71) which allows for a quick calculation of d_Λ in terms of the Young tableau Y_Λ of R_Λ. The same manipulations which lead to the hook formula can be performed to show that the quantum dimension formula implies a q-analogue of the hook formula:

$$\mathcal{D}_\Lambda(sl_n) = \prod_{(i,j)\in Y_\Lambda} \frac{\lfloor n-i+j\rfloor}{\lfloor h_{ij}\rfloor}. \tag{4.4.37}$$

Here, as in section 1.6, (i,j) labels the boxes of Y_Λ, and h_{ij} is the length of the hook with corner (i,j). For example, the hook formula immediately reproduces the quantum dimensions of the fundamental modules and the symmetric tensor modules given in table (4.4.18),

$$\begin{aligned} \mathcal{D}_{\Lambda_{(j)}}(sl_n) &= \begin{bmatrix} n \\ j \end{bmatrix} && \text{for} \quad j = 1,\dots,r, \\ \mathcal{D}_{m\Lambda_{(1)}}(sl_n) &= \begin{bmatrix} m+n-1 \\ m \end{bmatrix} && \text{for} \quad m \in \mathbb{Z}_{\geq 0}. \end{aligned} \tag{4.4.38}$$

Consider now the automorphism (4.4.36). The non-trivial element $\ddot{\omega}$ of $\bar{\Gamma} = \Gamma(sl_n)$ induces a map $\bar{\omega}$ which sends any sl_n-weight to its conjugate weight,

$$\bar{\omega}(\Lambda) = \Lambda^+ \tag{4.4.39}$$

with $(\Lambda^+)^i = \Lambda^{n-i}$ for $i = 1,2,\dots,n-1$. $\ddot{\omega}$ is also the non-trivial element of the $\mathbb{Z}_2$-factor of Γ; the elements of the $\mathbb{Z}_n$-factor of Γ are $(\dot{\omega})^j$ for $j = 0,1,\dots,n-1$ where $\dot{\omega}(i) = i+1 \bmod n$ for $i = 0,1,\dots,n-1$. $\dot{\omega}$ induces the map

$$\Lambda \mapsto \hat{\omega}(\Lambda), \quad (\hat{\omega}(\Lambda))^i = \Lambda^{i-1} \quad \text{for} \quad i = 1,\dots,n-1 \tag{4.4.40}$$

with Λ^0 the same as for the affine weight associated to Λ,

$$\Lambda^0 = k^\vee - \sum_{i=1}^{r} \Lambda^i. \tag{4.4.41}$$

The automorphism group Γ is generated by $\ddot{\omega}$ and $\dot{\omega}$ (to be precise, Γ is the group generated by $\ddot{\omega}$ and $\dot{\omega}$ subject to the constraints $\dot{\omega}^n = \ddot{\omega}^2 = (\dot{\omega}\ddot{\omega})^2 = 1$). Therefore it is sufficient to prove the identity (4.4.35) for the two maps (4.4.39) and (4.4.40). In terms of Young tableaux, these act as follows. It will be assumed that Λ corresponds to an integrable affine module so that Λ^0 is nonnegative. Then the map $\hat{\omega}$ corresponds to adding a row of length $k^\vee$ to the Young tableau Y_Λ (due to $\Lambda^0 \geq 0$ this becomes the first row of the new tableau $Y_{\hat{\omega}(\Lambda)}$) and deleting any columns of length n (such columns arise if Y_Λ has columns of length $n-1$). Similarly, the Young tableau $Y_{\bar{\omega}(\Lambda)}$ is the "complement" of Y_Λ in the sense that Y_Λ and $Y_{\bar{\omega}(\Lambda)}$, with one of them rotated by 180 degrees, add up to a rectangular tableau with $k^\vee - \Lambda^0$ columns of length n. Now in order to prove (4.4.35), one has to show that the hook formulae (4.4.37) for

Y_Λ and for $Y_{\omega(\Lambda)}$ give identical results. This can be done by rearranging the various factors and making use of the identity (4.4.20) (with $g^\vee = n$), which is a rather lengthy but straightforward exercise. In the case of $\hat{\omega}$, the manipulations essentially reduce to showing that the quantum dimension remains unchanged if Y_Λ is reflected along the diagonal defined by the points $(i,i) \in Y_\Lambda$ (i.e. if the meaning of rows and columns of the tableau is interchanged), and if at the same time the roles of n and $k^\vee$ are exchanged as well. Denoting the transformation resulting in the reflection of the Young tableau by η and writing $\mathcal{D}_\Lambda(n,m)$ for $\mathcal{D}_\Lambda(sl_n)$ at level m, this means that

$$\mathcal{D}_\Lambda(n,k^\vee) = \mathcal{D}_{\eta(\Lambda)}(k^\vee,n). \tag{4.4.42}$$

This relation is fulfilled because $h_{ij}(Y_\Lambda) = h_{ji}(Y_{\eta(\Lambda)})$ so that the product of q-hook lengths $\lfloor h_{ij} \rfloor$ is invariant under the reflection, and because in the numerator of the hook formula one can use

$$\lfloor n-i+j \rfloor = \lfloor k^\vee - j + i \rfloor, \tag{4.4.43}$$

which is a special case of (4.4.22). When (4.4.42) is applied to both sides of (4.4.35) (for $\hat{\omega}$), one obtains the assertion that the quantum dimension does not change if a column of length $k^\vee$ is added to a Young tableau of $sl_{k^\vee}$, and this in turn is a rather direct consequence of the hook formula.

Let us now consider some examples. First, j-fold application of $\hat{\omega}$ takes the singlet to the weight $k^\vee\Lambda_{(j)}$; since the singlet has quantum dimension $\lfloor 1 \rfloor = 1$, this means that

$$\mathcal{D}_{k^\vee\Lambda_{(j)}}(n,k^\vee) = 1. \tag{4.4.44}$$

As examples for the symmetry (4.4.28), consider the cases $\alpha = \beta = \theta = \Lambda_{(1)} + \Lambda_{(n-1)}$, and $\alpha = \beta = \alpha^{(i)}$. In the former case one obtains

$$\mathcal{D}_{\Lambda+(\Lambda^0+1)\theta}(n,k^\vee) = -\mathcal{D}_\Lambda(n,k^\vee), \tag{4.4.45}$$

while the latter choice yields

$$\mathcal{D}_{\Lambda'}(n,k^\vee) = \mathcal{D}_\Lambda(n,k^\vee) \tag{4.4.46}$$

with

$$(\Lambda')^j = \begin{cases} 2\,(k^\vee+n-1) - \Lambda^j & \text{for } \; j=i \\ \Lambda^i + \Lambda^j - k^\vee - n + 1 & \text{for } \; |j-i| = 1 \\ \Lambda^j & \text{for } \; |j-i| \geq 2. \end{cases} \tag{4.4.47}$$

A special case of (4.4.45) is

$$\mathcal{D}_{k^\vee\theta}(n,k^\vee) = -\mathcal{D}_\theta(n,k^\vee). \tag{4.4.48}$$

It also follows from (4.4.45) that

$$\mathcal{D}_\Lambda(n, k^\vee) = 0 \quad \text{if} \quad \Lambda^0 = -1, \tag{4.4.49}$$

in agreement with the general result (4.4.30).

Finally, consider the "$(k^\vee - n)$-duality" or *level-rank duality* (4.4.42) for $k^\vee = 2$. At level two, the highest weights of sl_n having $\Lambda^0 \geq 0$ are of the form $\Lambda = \Lambda_{(i)} + \Lambda_{(j)}$ with $0 \leq i \leq j \leq n-1$ (and $\Lambda_{(0)} \equiv 0$). The transformation η maps such a Λ to the sl_2-weight $\tilde{\Lambda} = j - i$. Together with (4.4.4), it follows that

$$\mathcal{D}_{\Lambda_{(i)}+\Lambda_{(j)}}(n, 2) = \lfloor j - i + 1 \rfloor. \tag{4.4.50}$$

4.5 The truncated Kronecker product

According to the previous sections, for q a root of unity there are both irreducible and non-fully reducible modules of $\mathsf{U}_q(\mathfrak{g})$. The modules may also be distinguished by the value of their q-dimension which can either be positive (for fully reducible modules) or zero (for non-fully reducible modules, but also for some which are fully reducible). In the present section we will see that the latter classification is the more fundamental one. Consequently we introduce the following notations: fully reducible modules (i.e. irreducible ones or direct sums thereof) which do not contain irreducible submodules of zero quantum dimension are generically denoted by $R_{[+]}$, while direct sums of non-fully reducible modules and of irreducible modules of zero quantum dimension are denoted by $R_{[0]}$. Thus for a generic finite-dimensional module S which is the direct sum of fully reducible modules and non-fully reducible ones, we write $S = S_{[+]} \oplus S_{[0]}$ for the decomposition into its parts of type $R_{[+]}$ and $R_{[0]}$. That this decomposition makes sense is already apparent from the observation that $S_{[+]}$ is unitarizable whereas $S_{[0]}$ (which according to section 4.2 contains null states) is not.

Since the dimension of the non-fully reducible module $R_{\Lambda,\Lambda'}$ is just the sum of the would-be irreducible modules R_Λ and $R_{\Lambda'}$, the dimension sum rule (1.6.59) of $\mathfrak{g}$ is also valid for $\mathsf{U}_q(\mathfrak{g})$:

$$d_R \cdot d_{R'} = \sum_i \mathcal{L}_i \, d_{R_i} \tag{4.5.1}$$

for

$$R \times R' = \bigoplus_i \mathcal{L}_i \, R_i, \tag{4.5.2}$$

where (without loss of generality) R, R' and R_i are either irreducible highest weight modules R_Λ or non-fully reducible modules $R_{\Lambda,\Lambda'}$. An important observation is that an analogous sum rule also holds for the q-dimensions:

$$\mathcal{D}_R \cdot \mathcal{D}_{R'} = \sum_i \mathcal{L}_i \, \mathcal{D}_{R_i} . \tag{4.5.3}$$

This simply follows from the character sum rule (1.7.46) by evaluating the characters on the weight $\ln q \cdot \tilde{\rho}$ (just as the dimension sum rule is obtained by evaluating the characters on the zero weight). From (4.5.3) several conclusions can be drawn immediately. Namely, if R (or R') is of type $R_{[0]}$, then the left hand side of the equation is zero; thus the right hand side must also vanish, and hence all modules on the right hand side are of type $R_{[0]}$ as well. Shortly,

$$R_{[0]} \times R' = \bigoplus \mathcal{L}_i \, (R_{[0]})_i . \tag{4.5.4}$$

Conversely, if both R and R' are of type $R_{[+]}$, then their Kronecker product contains at least one module of this type,

$$R_{[+]} \times R'_{[+]} = \bigoplus_{i \in I} \mathcal{L}_i \, (R_{[+]})_i \oplus \bigoplus_{j \in J} \mathcal{L}_j \, (R_{[0]})_j \quad \text{with } I \neq \emptyset . \tag{4.5.5}$$

These properties of the Kronecker product suggest the definition of a new product "$\star$" of $\mathsf{U}_q(\mathfrak{g})$-modules as

$$R \star R' := (R \times R')_{[+]} . \tag{4.5.6}$$

For R or R' of type $R_{[0]}$, the right hand side of this equation vanishes, so that when discussing $\star$-products, one may restrict the set of modules under consideration to those of type $R_{[+]}$. Using the associativity of the ordinary Kronecker product together with the property (4.5.4), one can see that the product (4.5.6) is associative as well,

$$(R \star R') \star R'' = R \star (R' \star R'') . \tag{4.5.7}$$

This means for example that the terms which are to be discarded obey

$$\Big((R \times R')_{[0]} \times R'' \Big)_{[0]} = \Big(R \times (R' \times R'')_{[0]} \Big)_{[0]} . \tag{4.5.8}$$

Because of the associativity, the $\star$-product can be used as a tensor product in the sense that it defines a new representation theory for $\mathsf{U}_q(\mathfrak{g})$ for which all allowed modules are obtained by taking appropriate tensorial powers of a small set of basic modules (namely, exactly the ones mentioned for $\mathfrak{g}$ on p. 69, just as in the case of the ordinary Kronecker product). Thus the $\star$-product has the same relevance to representation theory as the ordinary Kronecker product; it will therefore be called the *truncated Kronecker product* of $\mathsf{U}_q(\mathfrak{g})$ from now on. If the Kronecker product of two modules is decomposed as in (4.5.2), the corresponding truncated product will be written as

$$R \star R' = \bigoplus_i \mathcal{L}_i^\star \, R_i, \tag{4.5.9}$$

which means that

$$\mathcal{L}_i^\star = \begin{cases} \mathcal{L}_i & \text{if } \mathcal{D}_{R_i} > 0 \\ 0 & \text{if } \mathcal{D}_{R_i} = 0 \end{cases} \tag{4.5.10}$$

Since the truncation only projects out modules of zero quantum dimension, the quantum dimension sum rule still holds for the truncated Kronecker product:

$$\mathcal{D}_R \cdot \mathcal{D}_{R'} = \sum_i \mathcal{L}_i^\star \, \mathcal{D}_{R_i}. \tag{4.5.11}$$

It must be emphasized that the prescription (4.5.6) means that the truncated Kronecker product is obtained from the ordinary Kronecker product by not only throwing away the non-fully reducible part, but also some of the irreducible submodules, namely those whose quantum dimension vanishes. If the latter modules were kept, the quantum dimension sum rule would still be valid. However, the product obtained that way would no longer be associative. Consider e.g. the case $\mathfrak{g} = A_1$. From the form (4.3.32) of the non-fully reducible modules and the ordinary Kronecker product decomposition (see (4.5.17) below), it is easily deduced that keeping the irreducible module $R_{k^\vee+1}$ which has quantum dimension zero would give a modified product "$\dot\star$" such that e.g.

$$\begin{aligned} & R_{k^\vee} \dot\star R_{k^\vee} = R_0, \\ & R_{k^\vee} \dot\star R_{k^\vee+1} = \begin{cases} \emptyset & \text{for } k^\vee \in 2\mathbb{Z} \\ R_{k^\vee+1} & \text{for } k^\vee \in 2\mathbb{Z}+1, \end{cases} \\ & R_{k^\vee+1} \dot\star R_{k^\vee+1} = \begin{cases} \emptyset & \text{for } k^\vee \in 2\mathbb{Z}+1 \\ R_{k^\vee+1} & \text{for } k^\vee \in 2\mathbb{Z}, \end{cases} \\ & R_1 \dot\star R_{k^\vee} = R_{k^\vee+1} \oplus R_{k^\vee-1}. \end{aligned} \tag{4.5.12}$$

Thus for odd level, one would have $R_1 \dot\star (R_{k^\vee} \dot\star R_{k^\vee+1}) = \emptyset \neq (R_1 \dot\star R_{k^\vee}) \dot\star R_{k^\vee+1}$, and a similar contradiction to associativity arises for even level.

By definition, the truncated Kronecker product of any two irreducible highest weight modules is fully reducible. From this property it is rather clear how to interprete the truncated Kronecker product in terms of the centralizer algebras of $\mathsf{U}_q(\mathfrak{g})$. Namely, the appearence of non-fully reducible modules in ordinary Kronecker products is a consequence of the fact that the centralizer algebra is not semisimple. To remove such modules from the representation theory, one has to take the semisimple quotient of the centralizer algebra, i.e. the quotient with respect to its radical (the centralizer algebras are never solvable, so that this construction gives a non-trivial algebra). Thus the truncated Kronecker product is the Kronecker product for the algebra obtained by taking the semisimple quotient.

To understand the meaning of the truncated Kronecker product from a slightly different point of view, recall that the presence of non-fully reducible modules could be traced back to the fact that $(E^\alpha)^{p_\alpha} = 0$ for some p_α. Thus the characterization of highest weight vectors v via

$$R(E^\alpha) \cdot v = 0 \quad \text{for all } \alpha > 0 \tag{4.5.13}$$

is less stringent than in the case of $\bar{\mathfrak{g}}$, because it could be just a consequence of $(R(E^\alpha))^p \cdot w \equiv 0$. To exclude this possibility, one may restrict the set of "allowed" highest weights by requiring that

$$R(E^\alpha) \cdot v = 0 \;\; \text{and} \;\; v \notin \mathrm{Im}\,(R(E^\alpha))^{p_\alpha - 1} \;\; \text{for all } \alpha > 0, \tag{4.5.14}$$

or in other words

$$\{v_\Lambda \mid \Lambda \text{ allowed}\} = \bigcap_{\alpha>0} \mathrm{Ker}\, E^\alpha / \mathrm{Im}\,(E^\alpha)^{p_\alpha - 1}. \tag{4.5.15}$$

Here "Im" and "Ker" stand for the image and the kernel of a map, respectively. The set of allowed modules obtained this way is just the set of modules of type $R_{[+]}$, and imposing the restriction (4.5.14) also to the Kronecker products of allowed modules just leads to the truncated Kronecker product as defined in (4.5.6). For $\bar{\mathfrak{g}} = A_1$, the condition (4.5.14) is easily made explicit because there is only a single positive root, with $p_\alpha = k^\vee + 2$, whereas for generic $\bar{\mathfrak{g}}$ it has not yet been worked out in full detail.

Let us now present some examples of truncated Kronecker products. For $\bar{\mathfrak{g}} = A_1$, a general formula can be obtained. One starts from the ordinary Kronecker product rule which is just the well-known rule for A_1 except for the fact that at q a root of unity the sum on the right hand side is in general not a direct sum:

$$\Lambda \times \Lambda' = \bigoplus_{j=0}^{\Lambda'} (\Lambda - \Lambda' + 2j), \tag{4.5.16}$$

where (without loss of generality) $\Lambda \geq \Lambda'$ is assumed, and Λ is used as a short hand notation both for irreducible highest weight modules R_Λ and for would-be irreducible modules $\check{R}_\Lambda$. The truncated version of (4.5.16) is immediately obtained with the help of the equation (4.3.32) which describes the explicit form of the non-fully reducible modules; it reads

$$\Lambda \star \Lambda' = \bigoplus_{j=0}^{\min\{\Lambda', k^\vee - \Lambda\}} (\Lambda - \Lambda' + 2j). \tag{4.5.17}$$

For $\mathfrak{g} \neq A_1$, one can use the generalized Racah–Speiser algorithm of section 4.3 to calculate the truncated Kronecker products. In this context it is important that the ambiguity concerning the precise form of the non-fully reducible products is easily seen to drop out from the truncated Kronecker products (compare section 5.5). As a simple example, consider the two modules with highest weights $\Lambda_{(1)}$ and $\Lambda_{(7)}$ of $\bar{\mathfrak{g}} = E_8$, i.e. the 248-dimensional (adjoint) and 3875-dimensional module, respectively. All highest weight modules which appear on the right hand side of the E_8-Kronecker products involving $R_{\Lambda_{(1)}}$ and $R_{\Lambda_{(7)}}$ have the property that the corresponding affine weights are integrable at level four and higher. Therefore for $k^\vee \geq 4$ the truncated Kronecker products are isomorphic to the E_8-Kronecker products which according to (3.5.14) read

$$\begin{aligned} \Lambda_{(1)} \times \Lambda_{(1)} &= \Lambda_{(0)} \oplus \Lambda_{(1)} \oplus \Lambda_{(2)} \oplus \Lambda_{(7)} \oplus (2\Lambda_{(1)}), \\ \Lambda_{(1)} \times \Lambda_{(7)} &= \Lambda_{(1)} \oplus \Lambda_{(2)} \oplus \Lambda_{(7)} \oplus \Lambda_{(8)} \oplus (\Lambda_{(1)} + \Lambda_{(7)}), \\ \Lambda_{(7)} \times \Lambda_{(7)} &= \Lambda_{(0)} \oplus \Lambda_{(1)} \oplus \Lambda_{(2)} \oplus \Lambda_{(3)} \oplus \Lambda_{(6)} \oplus \Lambda_{(7)} \oplus \Lambda_{(8)} \\ &\qquad \oplus (2\Lambda_{(1)}) \oplus (\Lambda_{(1)} + \Lambda_{(7)}) \oplus (2\Lambda_{(7)}). \end{aligned} \tag{4.5.18}$$

The dimensionalities of the modules appearing here and the inner products with $\theta^\vee$ are given in table (4.5.19).

Λ	d_Λ	$(\Lambda, \theta^\vee)$
$\Lambda_{(1)}$	248	2
$\Lambda_{(2)}$	30 380	3
$\Lambda_{(3)}$	2 450 240	4
$\Lambda_{(6)}$	6 696 000	4
$\Lambda_{(7)}$	3 875	2
$\Lambda_{(8)}$	147 250	3
$2\Lambda_{(1)}$	27 000	4
$\Lambda_{(1)} + \Lambda_{(7)}$	779 247	4
$2\Lambda_{(7)}$	4 881 384	4

(4.5.19)

It is also quite trivial to write down the truncated Kronecker products for $k^\vee = 3$. Namely, one just has to discard all modules with $(\Lambda, \theta^\vee) = 4$ because these highest weights are mapped onto themselves by $\sigma_{(0)}$. Thus for $k^\vee = 3$ one has

$$\begin{aligned} \Lambda_{(1)} \star \Lambda_{(1)} &= \Lambda_{(0)} \oplus \Lambda_{(1)} \oplus \Lambda_{(2)} \oplus \Lambda_{(7)}, \\ \Lambda_{(1)} \star \Lambda_{(7)} &= \Lambda_{(1)} \oplus \Lambda_{(2)} \oplus \Lambda_{(7)} \oplus \Lambda_{(8)}, \\ \Lambda_{(7)} \star \Lambda_{(7)} &= \Lambda_{(0)} \oplus \Lambda_{(1)} \oplus \Lambda_{(2)} \oplus \Lambda_{(7)} \oplus \Lambda_{(8)}. \end{aligned} \tag{4.5.20}$$

Finally at level two, one has to discard the modules with maximal weights $\Lambda_{(2)}$ and $\Lambda_{(8)}$ that are mapped onto themselves by $\sigma_{(0)}$, but also the fields with highest weights $\Lambda = \Lambda_{(3)}, \Lambda_{(6)}, 2\Lambda_{(7)}$ because for these weights $\sigma_{(0)} \star \Lambda$ is mapped onto itself by $\sigma_{(1)} \equiv w_{\alpha^{(1)},0}$. Moreover, one finds

$$2\,\Lambda_{(1)} \overset{\sigma_{(0)}}{\mapsto} \Lambda_{(1)}, \qquad \Lambda_{(1)} + \Lambda_{(7)} \overset{\sigma_{(0)}}{\mapsto} \Lambda_{(7)}, \tag{4.5.21}$$

implying the existence of non-fully reducible modules

$$R_{2\,\Lambda_{(1)}} \oplus R_{\Lambda_{(1)}}, \qquad R_{\Lambda_{(1)}+\Lambda_{(7)}} \oplus R_{\Lambda_{(7)}}. \tag{4.5.22}$$

Thus for $k^\vee = 2$ one arrives at the truncated Kronecker products

$$\begin{aligned} \Lambda_{(1)} \star \Lambda_{(1)} &= \Lambda_{(0)} \oplus \Lambda_{(7)}, \\ \Lambda_{(1)} \star \Lambda_{(7)} &= \Lambda_{(1)}, \\ \Lambda_{(7)} \star \Lambda_{(7)} &= \Lambda_{(0)}. \end{aligned} \tag{4.5.23}$$

For $k^\vee = 2$, $R_{\Lambda_{(1)}}$ and $R_{\Lambda_{(7)}}$ are in fact the only irreducible non-trivial modules of positive quantum dimension; therefore the latter formulæ already exhaust the full set of truncated Kronecker products. (For $k^\vee = 1$, only the trivial one-dimensional module has positive quantum dimension so that no non-trivial truncated Kronecker products exist.)

With slightly more effort, one can also determine the full set of truncated Kronecker products at level three or higher. For $k^\vee = 3$ one finds

$$\begin{aligned} \Lambda_{(1)} \star \Lambda_{(2)} &= \Lambda_{(1)} \oplus \Lambda_{(7)} \oplus \Lambda_{(8)}, \\ \Lambda_{(1)} \star \Lambda_{(8)} &= \Lambda_{(2)} \oplus \Lambda_{(7)}, \\ \Lambda_{(7)} \star \Lambda_{(2)} &= \Lambda_{(1)} \oplus \Lambda_{(2)} \oplus \Lambda_{(7)}, \\ \Lambda_{(7)} \star \Lambda_{(8)} &= \Lambda_{(1)} \oplus \Lambda_{(7)}, \\ \Lambda_{(2)} \star \Lambda_{(2)} &= \Lambda_{(0)} \oplus \Lambda_{(2)} \oplus \Lambda_{(7)}, \\ \Lambda_{(2)} \star \Lambda_{(8)} &= \Lambda_{(1)} \oplus \Lambda_{(8)}, \\ \Lambda_{(8)} \star \Lambda_{(8)} &= \Lambda_{(0)} \oplus \Lambda_{(2)}, \end{aligned} \tag{4.5.24}$$

in addition to the ones already given.

As a further example, we derive the truncated Kronecker products for $\mathfrak{g} = F_4$ up to level three. At level one, there is only a single non-trivial module with positive quantum dimension, namely the one with highest weight $\Lambda_{(4)}$, and one easily finds

$$\Lambda_{(4)} \star \Lambda_{(4)} = \Lambda_{(0)} \oplus \Lambda_{(4)}. \tag{4.5.25}$$

For any level $k^\vee \geq 2$, the truncated Kronecker product of $R_{\Lambda_{(4)}}$ with itself is given by

$$\Lambda_{(4)} \star \Lambda_{(4)} = \Lambda_{(0)} \oplus \Lambda_{(4)} \oplus \Lambda_{(1)} \oplus \Lambda_{(3)} \oplus (2\,\Lambda_{(4)}), \tag{4.5.26}$$

and is isomorphic to the corresponding F_4 Kronecker product (for $k^\vee = 1$, the last three highest weights are mapped onto themselves by $\sigma_{(0)}$ and hence have to be discarded).

Next consider $k^\vee = 2$. There are then four non-trivial modules with positive quantum dimension, just as for E_8 at level three, and one finds that in fact the truncated Kronecker products become isomorphic to those of $\mathcal{U}_q(E_8)$ at $k^\vee = 3$ if one identifies the irreducible highest weight modules of the respective quantum groups according to

$$\begin{array}{ccccc} \mathcal{U}_q(E_8),\ k^\vee = 3 & & \mathcal{U}_q(F_4),\ k^\vee = 2 & \\ \Lambda_{(7)} & \longleftrightarrow & \Lambda_{(4)} & (26) \\ \Lambda_{(8)} & \longleftrightarrow & \Lambda_{(1)} & (52) \\ \Lambda_{(1)} & \longleftrightarrow & \Lambda_{(3)} & (273) \\ \Lambda_{(2)} & \longleftrightarrow & 2\,\Lambda_{(4)} & (324) \end{array} \tag{4.5.27}$$

These truncated Kronecker products are derived by noticing that at level two the transformation $\sigma_{(0)}$ maps the weights

$$\begin{array}{lll} \Lambda_{(2)}, & 2\,\Lambda_{(3)}, & \text{and} \\ (\Lambda + \Lambda_{(4)}), & \Lambda \in \{\Lambda_{(1)}, \Lambda_{(2)}, \Lambda_{(3)}, 2\,\Lambda_{(4)}, 3\,\Lambda_{(4)}\} & \end{array} \tag{4.5.28}$$

onto themselves, while non-fully reducible modules arise according to

$$(\Lambda + \Lambda_{(1)}) \overset{\sigma_{(0)}}{\mapsto} \Lambda \quad \text{for } \Lambda \in \{\Lambda_{(1)}, \Lambda_{(3)}, 2\,\Lambda_{(4)}\}. \tag{4.5.29}$$

Finally we come to the level $k^\vee = 3$. There are then eight non-trivial highest weight modules of positive quantum dimension. For brevity, we list just the truncated Kronecker products involving the module $R_{\Lambda_{(4)}}$:

$$\begin{aligned}
\Lambda_{(4)} \star \Lambda_{(1)} &= \Lambda_{(3)} \oplus \Lambda_{(4)} \oplus (\Lambda_{(1)} + \Lambda_{(4)}), \\
\Lambda_{(4)} \star \Lambda_{(2)} &= \Lambda_{(3)} \oplus (\Lambda_{(1)} + \Lambda_{(4)}) \oplus (\Lambda_{(3)} + \Lambda_{(4)}), \\
\Lambda_{(4)} \star \Lambda_{(3)} &= \Lambda_{(1)} \oplus \Lambda_{(2)} \oplus \Lambda_{(3)} \oplus \Lambda_{(4)} \oplus (\Lambda_{(1)} + \Lambda_{(4)}) \\
&\qquad \oplus (\Lambda_{(3)} + \Lambda_{(4)}) \oplus (2\Lambda_{(4)}), \\
\Lambda_{(4)} \star \Lambda_{(4)} &= \Lambda_{(0)} \oplus \Lambda_{(1)} \oplus \Lambda_{(3)} \oplus \Lambda_{(4)} \oplus (2\Lambda_{(4)}), \\
\Lambda_{(4)} \star (\Lambda_{(1)}+\Lambda_{(4)}) &= \Lambda_{(1)} \oplus \Lambda_{(2)} \oplus \Lambda_{(3)} \oplus (\Lambda_{(1)} + \Lambda_{(4)}) \\
&\qquad \oplus (\Lambda_{(3)} + \Lambda_{(4)}) \oplus (2\Lambda_{(4)}), \\
\Lambda_{(4)} \star (\Lambda_{(3)}+\Lambda_{(4)}) &= \Lambda_{(2)} \oplus \Lambda_{(3)} \oplus (\Lambda_{(1)} + \Lambda_{(4)}) \\
&\qquad \oplus 2\cdot(\Lambda_{(3)} + \Lambda_{(4)}) \oplus (2\Lambda_{(4)}) \oplus (3\Lambda_{(4)}), \\
\Lambda_{(4)} \star (2\Lambda_{(4)}) &= \Lambda_{(3)} \oplus \Lambda_{(4)} \oplus (\Lambda_{(1)} + \Lambda_{(4)}) \\
&\qquad \oplus (\Lambda_{(3)} + \Lambda_{(4)}) \oplus (2\Lambda_{(4)}) \oplus (3\Lambda_{(4)}), \\
\Lambda_{(4)} \star (3\Lambda_{(4)}) &= (\Lambda_{(3)} + \Lambda_{(4)}) \oplus (2\Lambda_{(4)}) \oplus (3\Lambda_{(4)}).
\end{aligned} \tag{4.5.30}$$

To come to examples which are a bit more complicated, consider $\mathsf{U}_q(D_4)$ for $k^\vee = 6$, corresponding to $q = \exp(\frac{\pi i}{6})$. In these cases the number of highest weight modules with positive quantum dimension is already quite large (namely, 130) so that it would be too lengthy to write down even only the truncated Kronecker products involving some specific module. The reason why we mention this example is that there are now several cases where the extended Racah–Speiser algorithm does not give unique predictions for the non-fully reducible modules. For example, the weights

$$\begin{aligned}
&\Lambda_1 = (4\,2\,0\,0), \quad \Lambda_2 = (6\,0\,2\,2), \quad \Lambda_3 = (7\,0\,3\,1), \\
&\Lambda_4 = (7\,0\,1\,3), \quad \Lambda_5 = (8\,0\,2\,2)
\end{aligned} \tag{4.5.31}$$

are all mapped to $\Lambda = (4\,1\,0\,0)$ by some transformation $w_{\sigma,\beta}$ (namely, by $\sigma_{(0)}$, $w_{(2)}\circ\sigma_{(0)}$, $w_{(4)}\circ w_{(2)}\circ\sigma_{(0)}$, $w_{(3)}\circ w_{(2)}\circ\sigma_{(0)}$ and $w_{(4)}\circ w_{(3)}\circ w_{(2)}\circ\sigma_{(0)}$ (with $w_{(i)} := w_{\sigma^{(i)},0}$), respectively). Therefore, non-fully reducible modules could a priori be realized as $\breve R_\Lambda \oplus \breve R_{\Lambda_i}$ for $i = 1, 3, 4$, and as $\breve R_{\Lambda_i} \oplus \breve R_{\Lambda_j}$ for $i = 2, 5$ and $j = 1, 3, 4$, and in principle also as more complicated modules containing four or even all six subspaces. The extended Racah–Speiser algorithm cannot decide which of these combinations are actually $\mathsf{U}_q(D_4)$-modules, but nevertheless the results for the truncated Kronecker products are unique because the contribution of any such module not containing $\breve R_\Lambda$ to any coefficient $\mathcal{L}_i^\star$ vanishes anyway.

It should also be noted that in many cases for which the number of highest weight modules possessing positive quantum dimension is small, the quantum dimension sum rule (4.5.3) already fixes the truncated Kronecker products uniquely. For example, for $\mathsf{U}_q(E_8)$ with $k^\vee = 2$, one finds

$$\mathcal{D}_{\Lambda_{(1)}} = \sqrt{2}, \qquad \mathcal{D}_{\Lambda_{(7)}} = 1, \tag{4.5.32}$$

which immediately leads to the conclusion that these modules must obey the decomposition (4.5.23). For $\mathsf{U}_q(E_8)$ at $k^\vee = 3$, inspection of the quantum dimensions is sufficient to determine uniquely the truncated Kronecker products too, but it fails e.g. for $\mathsf{U}_q(F_4)$ at $k^\vee = 3$.

4.6 R-matrices

At the end of section 4.2, it was pointed out that while the universal R-matrix of $\mathsf{U}_q(\mathfrak{g})$ can in principle be obtained by the power series ansatz (4.2.48), in practice this method is only manageable for $\mathfrak{g} = A_1$. However, the general result (4.2.37) for R together with the explicit solution (4.2.49) for $\mathfrak{g} = A_1$ suggest a different procedure which can be more easily applied in the general case. This method, called the *quantum double construction*, will now be described.

For a an arbitrary Hopf algebra, define the matrices μ, ν, τ in terms of the multiplication, co-multiplication and antipode as in (4.1.12). Then the multiplication, opposite co-multiplication and antipode of the dual space $\mathsf{a}^\star$ are given by (4.2.38); denote by a° the algebra $\mathsf{a}^\star$ endowed with the opposite co-multiplication. Now consider a vector space $\mathsf{d}(\mathsf{a})$ which contains a and a° as Hopf subalgebras, and which in addition is isomorphic as a vector space to $\mathsf{a}\times\mathsf{a}^\circ$ (which just means that its dimension is the product of the (infinite) dimensions of a and $\mathsf{a}^\star$). Denoting, as in (4.1.21), the basis of a by $\{e_s\}$ and the dual basis of $\mathsf{a}^\star$ by $\{e^s\}$, the co-multiplication of a° is

$$\mathbf{\Delta}'(e^p) \equiv \pi \circ \mathbf{\Delta}(e^p) = \sum_{s,t} \pi_{st}\mu^p{}_{st}\, e^s \otimes e^t = \sum_{s,t} \mu^p{}_{ts}\, e^s \otimes e^t. \tag{4.6.1}$$

The particular element

$$\mathsf{R} = \sum_p e_p \otimes e^p \tag{4.6.2}$$

of $\mathsf{d(a)} \times \mathsf{d(a)}$ then obeys

$$\begin{aligned}(\Delta \times id)(\mathsf{R}) &= \textstyle\sum_{p,s,t} \nu_p{}^{st} e_s \otimes e_t \otimes e^p \\ &= \textstyle\sum_{s,t} e_s \otimes e_t \otimes (e^s \diamond e^t) = \mathsf{R}_{13} \diamond \mathsf{R}_{23}, \\ (id \times \Delta)(\mathsf{R}) &= \textstyle\sum_{p,s,t} \mu^p{}_{ts} e_p \otimes e^s \otimes e^t \\ &= \textstyle\sum_{s,t} (e_t \diamond e_s) \otimes e^s \otimes e^t = \mathsf{R}_{13} \diamond \mathsf{R}_{12},\end{aligned} \tag{4.6.3}$$

precisely as in section 4.2. Thus $\mathsf{d(a)}$ fulfills the axioms (4.1.38b,c) of quasitriangularity, with (4.6.2) as the universal R-matrix. Similarly one can verify, just as in section 4.2, that (4.1.38d) is satisfied as well. To check also (4.1.38a), one must first realize that so far only the vector space structure of $\mathsf{d(a)}$ has been defined completely, whereas the part of the algebra structure describing the multiplication of elements of the subspace a with those of the subspace a° is not yet defined. This turns out to be fortunate, because it allows us to go backwards and *define* this multiplication by choosing, say, $e_p \diamond e^q$ arbitrarily and then fixing $e^p \diamond e_q$ by imposing (4.1.38a), i.e.

$$\sum_p \Delta'(e_q) \diamond (e_p \otimes e^p) = \sum_p (e_p \otimes e^p) \diamond \Delta(e_q), \tag{4.6.4}$$

or in terms of the matrices μ, ν

$$\sum_{q,r,s,t} \nu_p{}^{st} \mu^q{}_{tr}\, e_q \otimes (e_s \diamond e^r) = \sum_{q,r,s,t} \nu_p{}^{ts} \mu^q{}_{rt}\, e_q \otimes (e^r \diamond e_s). \tag{4.6.5}$$

(At this point one *assumes* that $\mathsf{d(a)}$ is not just a vector space, but in fact an algebra, i.e. that a product of e_s with e^t does indeed exist.)

Thus $\mathsf{d(a)}$ becomes a quasitriangular Hopf algebra with R-matrix (4.6.2) if the multiplication is arranged to satisfy

$$\sum_{r,s,t} \nu_p{}^{st} \mu^q{}_{tr}\, (e_s \diamond e^r) = \sum_{r,s,t} \nu_p{}^{ts} \mu^q{}_{rt}\, (e^r \diamond e_s). \tag{4.6.6}$$

One can also check that this restriction is solved by

$$e^p \diamond e_q = \sum_{i,j,k,l,r,s,t} \mu^p{}_{ij} \mu^i{}_{kl} \nu_q{}^{rk} \nu_r{}^{st} \tau_s{}^j\, e_t \diamond e^l, \tag{4.6.7}$$

and that the multiplication defined this way is associative as required. The algebra $\mathsf{d(a)}$ with the multiplication obeying (4.6.7) is called the *double* of a. Thus, shortly, to any Hopf algebra a one can associate a quasitriangular Hopf algebra $\mathsf{d(a)}$ which contains a and a° as Hopf subalgebras; the universal R-matrix is given by (4.6.2), and the multiplication on $\mathsf{d(a)}$ is defined by the multiplications on a and a° together with (4.6.7).

Let us now employ these general results to reproduce the formula (4.2.49) for the R-matrix of $\mathsf{U}_q(sl_2)$. It turns out to be useful to consider the basis of $U_q(sl_2)$ which is generated by the elements H and

$$\breve{E}_\pm := q^{\pm H/4}\, E_\pm \tag{4.6.8}$$

of sl_2, in addition to the one generated by H and $E_\pm$. Thus one has

$$\begin{aligned} U_q(sl_2) &= \mathrm{span}\,\{H^\ell (E_+)^m (E_-)^n \mid \ell,m,n \in \mathbb{Z}_{\geq 0}\} \\ &= \mathrm{span}\,\{H^\ell (\breve{E}_+)^m (\breve{E}_-)^n \mid \ell,m,n \in \mathbb{Z}_{\geq 0}\}, \end{aligned} \tag{4.6.9}$$

and for the deformed enveloping algebras of the Borel subalgebras $b_\pm$ of sl_2

$$\begin{aligned} U_q(b_\pm) &= \mathrm{span}\,\{H^\ell (E_\pm)^m \mid \ell,m \in \mathbb{Z}_{\geq 0}\} \\ &= \mathrm{span}\,\{H^\ell (\breve{E}_\pm)^m \mid \ell,m \in \mathbb{Z}_{\geq 0}\}. \end{aligned} \tag{4.6.10}$$

In terms of the generators (4.6.8), the relations (4.2.16) of $U_q(sl_2)$ become (using the identity (4.2.20))

$$\begin{aligned} [H, \breve{E}_\pm] &= \pm 2\, \breve{E}_\pm, \\ [\breve{E}_+, \breve{E}_-] &= q^{1/2}\, \lfloor H \rfloor, \end{aligned} \tag{4.6.11}$$

and the co-multiplication reads

$$\begin{aligned} \Delta(\breve{E}_+) &= 1 \otimes \breve{E}_+ + \breve{E}_+ \otimes q^{H/2}, \\ \Delta(\breve{E}_-) &= \breve{E}_- \otimes 1 + q^{-H/2} \otimes \breve{E}_-. \end{aligned} \tag{4.6.12}$$

The basis of the dual space $(U_q(b_+))^\star$ which is dual to $\{H^\ell (E_+)^m\}$ is given by $\{\mathcal{H}^\ell (\mathcal{E}_+)^m \mid \ell,m \in \mathbb{Z}_{\geq 0}\}$ with the linear functions $\mathcal{H}$, $\mathcal{E}_+ \in (U_q(b_+))^\star$ defined by

$$\begin{aligned} \mathcal{H}(H) = 1, \quad & \mathcal{E}_+(H) = 0, \\ \mathcal{H}(E_+) = 0, \quad & \mathcal{E}_+(E_+) = 1. \end{aligned} \tag{4.6.13}$$

The commutation relations and co-multiplication on $(U_q(b_+))^\star$ follow from the duality between multiplication and co-multiplication. According to (4.1.20) one has $[\varphi,\psi](x) = (\varphi \otimes \psi)(\Delta(x) - \Delta'(x))$ and $(\Delta(\varphi))(x \otimes y) = \varphi(x \diamond y)$; in the present case this gives

$$[\mathcal{H}, \mathcal{E}_+] = -\tfrac{1}{2} \ln q\, \mathcal{E}_+ \tag{4.6.14}$$

and

$$\begin{aligned} \Delta(\mathcal{H}) &= \mathcal{H} \otimes 1 + 1 \otimes \mathcal{H}, \\ \Delta(\mathcal{E}_+) &= 1 \otimes \mathcal{E}_+ + \mathcal{E}_+ \otimes \exp(-2\mathcal{H}). \end{aligned} \tag{4.6.15}$$

Defining $\breve{\mathcal{H}} := (4/\ln q)\, \mathcal{H}$, these relations become

$$[\breve{\mathcal{H}}, \mathcal{E}_+] = -2\, \mathcal{E}_+ \tag{4.6.16}$$

and, for the opposite co-multiplication,

$$\begin{aligned} \Delta'(\breve{\mathcal{H}}) &= \breve{\mathcal{H}} \otimes 1 + 1 \otimes \breve{\mathcal{H}}, \\ \Delta'(\mathcal{E}_+) &= \mathcal{E}_+ \otimes 1 + q^{-\breve{\mathcal{H}}/2} \otimes \mathcal{E}_+. \end{aligned} \tag{4.6.17}$$

But these relations are just isomorphic to the ones of $\mathsf{U}_q(\mathsf{b}_-)$, namely via the substitutions $\mathcal{E}_+ \hat{=} \breve{E}_-$ and $\breve{\mathcal{H}} \hat{=} H$, i.e. one has

$$(\mathsf{U}_q(\mathsf{b}_+))^\circ \cong \mathsf{U}_q(\mathsf{b}_-). \tag{4.6.18}$$

Thus the quantum double $\mathsf{d}(\mathsf{U}_q(\mathsf{b}_+))$ of $\mathsf{U}_q(\mathsf{b}_+)$ which is generated by $\{H,\, E_+,\, \mathcal{H},\, \mathcal{E}_+\}$ is isomorphic as a vector space to $\mathsf{U}_q(\mathsf{b}_+) \times \mathsf{U}_q(\mathsf{b}_-)$. In other words, $\mathsf{d}(\mathsf{U}_q(\mathsf{b}_+))$ can be viewed as a lift of the original Hopf algebra $\mathsf{U}_q(\mathfrak{g})$ which is performed in such a way that the two Borel subalgebras get separated, $(\mathsf{U}_q(\mathsf{b}_+))' \cap (\mathsf{U}_q(\mathsf{b}_-))' = \emptyset$. Moreover, using (4.6.7) and defining $\breve{\mathcal{E}}_+ := 2\,\frac{1-q^{-1}}{\ln q}\,\mathcal{E}_+$, the commutation relations can be written as

$$\begin{aligned} [H, \breve{E}_+] &= [\breve{\mathcal{H}}, \breve{E}_+] &&= 2\,\breve{E}_+, \\ [H, \breve{\mathcal{E}}_+] &= [\breve{\mathcal{H}}, \breve{\mathcal{E}}_+] &&= -2\,\breve{\mathcal{E}}_+, \\ [\breve{E}_+, \breve{\mathcal{E}}_+] &= \frac{q^{H/2} - q^{-\breve{\mathcal{H}}/2}}{1-q^{-1}}, \\ [H, \breve{\mathcal{H}}] &= 0. \end{aligned} \tag{4.6.19}$$

Obviously, these coincide with the $\mathsf{U}_q(A_1)$-relations (4.6.11) if one identifies

$$H \stackrel{!}{=} \breve{\mathcal{H}} \tag{4.6.20}$$

in addition to the substitution $\breve{\mathcal{E}}_+ \hat{=} \breve{E}_-$. One can verify that $\breve{\mathcal{H}} - H$ commutes with all elements of $\mathsf{d}(\mathsf{U}_q(\mathsf{b}_+))$, and that in fact it generates both an ideal and a co-ideal of $\mathsf{d}(\mathsf{U}_q(\mathsf{b}_+))$ so that one may indeed impose $\breve{\mathcal{H}} = H$ without spoiling the Hopf algebra properties. Thus the quantum double of $\mathsf{U}_q(\mathsf{b}_+)$ becomes isomorphic to $\mathsf{U}_q(sl_2)$ if the equivalence relation $\breve{\mathcal{H}} \cong H$ is divided out,

$$\frac{\mathsf{d}(\mathsf{U}_q(\mathsf{b}_+))}{\langle \breve{\mathcal{H}} - H \rangle} \cong \mathsf{U}_q(sl_2). \tag{4.6.21}$$

With this result we are finally in a position to compute the R-matrix. The R-matrix of $\mathsf{d}(\mathsf{U}_q(\mathsf{b}_+))$ is given by (4.6.2) with $\{e_p\}$ and $\{e^q\}$ dual bases of $\mathsf{U}_q(\mathsf{b}_+)$ and $\mathsf{U}_q(\mathsf{b}_+)^\circ$. If a different basis $\{f^q = \sum_r M^q_{\ r}\, e^r\}$ of $\mathsf{U}_q(\mathsf{b}_+)^\circ$ is chosen, then the general expression for the R-matrix reads

$$\mathsf{R} = \sum_{p,q} (M^{-1})^p_{\ q}\, e_p \otimes f^q. \tag{4.6.22}$$

The matrix M can be recovered from the basis elements e_p and f^q as $M^p_q = f^p(e_q)$. In the present case it is convenient to choose $\{H^\ell(\breve{E}_+)^m\}$ as the basis of $\mathsf{U}_q(\mathsf{b}_+)$ and $\{\mathcal{H}^\ell(\mathcal{E}_+)^m\}$ as the basis of $\mathsf{U}_q(\mathsf{b}_+)^\circ$. Thus in order to find the R-matrix, one has to determine

$$M^{k\ell}_{mn} = [\mathcal{H}^m(\mathcal{E}_+)^n]\,(H^k(\breve{E}_+)^\ell). \tag{4.6.23}$$

From (4.6.13) it is rather clear that this is diagonal, $M^{k\ell}_{mn} \propto \delta^k_{\ m}\delta^\ell_{\ n}$. This is fortunate, because then it is trivial to invert the matrix (if M were non-diagonal, the inversion would be problematic because M is infinite-dimensional). With some combinatorics, the constant of proportionality can be determined, leading to

$$M^{k\ell}_{mn} = \delta^k_{\ m}\delta^\ell_{\ n}\,\frac{m!\,\lfloor\!\lfloor n\rfloor\!\rfloor!}{(\ln q)^n}, \tag{4.6.24}$$

where

$$\lfloor\!\lfloor x\rfloor\!\rfloor := \frac{1-q^{-x}}{1-q^{-1}} \tag{4.6.25}$$

and $\lfloor\!\lfloor n\rfloor\!\rfloor! := \prod_{j=1}^n \lfloor\!\lfloor j\rfloor\!\rfloor$. The R-matrix of $\mathsf{d}(\mathsf{U}_q(\mathfrak{g}_+))$ then becomes

$$\begin{aligned} \mathsf{R} &= \sum_{m,n} \frac{(\ln q)^n}{m!\,\lfloor\!\lfloor n\rfloor\!\rfloor!}\, H^m(\breve{E}_+)^n \otimes \mathcal{H}^m(\mathcal{E}_+)^n \\ &= \mathrm{e}^{H\otimes\mathcal{H}} \sum_n \frac{(\ln q)^n}{\lfloor\!\lfloor n\rfloor\!\rfloor!}\,(\breve{E}_+)^n \otimes (\mathcal{E}_+)^n. \end{aligned} \tag{4.6.26}$$

Identifying the generators as described above, one then obtains the universal R-matrix of $\mathsf{U}_q(sl_2)$ as

$$\mathsf{R}(\mathsf{U}_q(sl_2)) = q^{H\otimes H/4} \sum_{m=0}^{\infty} \frac{(1-q^{-1})^m}{\lfloor\!\lfloor m\rfloor\!\rfloor!}\,(\breve{E}_+)^m \otimes (\breve{E}_-)^m. \tag{4.6.27}$$

Recalling the relation (4.6.8) between $\breve{E}_\pm$ and $E_\pm$, one sees that this indeed reproduces the formula given previously in (4.2.49). Also note that this result can be written in a more suggestive manner by introducing the *q-deformed exponential*

$$\mathsf{E}^x \equiv \mathsf{E}(x;q) := \sum_{n=0}^{\infty} \frac{x^n}{\lfloor\!\lfloor n\rfloor\!\rfloor!}, \tag{4.6.28}$$

namely as

$$\mathsf{R}(\mathsf{U}_q(sl_2)) = q^{H\otimes H/4}\,\mathsf{E}^{(1-q^{-1})\breve{E}_+\otimes\breve{E}_-}. \tag{4.6.29}$$

With the quantum double construction, the calculation of R-matrices for $\mathsf{U}_q(\mathfrak{g})$ with $\mathfrak{g} \neq sl_2$ becomes manageable as well. For example, for

$\mathfrak{g} = sl_n$ one finds

$$\mathtt{R}(\mathsf{U}_q(sl_n)) = q^{\sum_{i=1}^{n-1} G_{ij}\, H^i \otimes H^j/4} \prod_{\alpha>0} \mathtt{E}^{(1-q^{-1})\check{E}^{\alpha} \otimes \check{E}^{-\alpha}} . \tag{4.6.30}$$

To be precise, this formula holds as it stands only if a particularly convenient partial ordering of the roots α is chosen, namely according to the magnitude of the number l_α defined as $l_\alpha = i + j - 2$ if the simple root expansion of α is $\alpha = \alpha^{(i)} + \alpha^{(i+1)} + ... + \alpha^{(j)}$ (this number is equal to the length of that Weyl group element σ for which $\alpha = \sigma(\alpha^{(1)})$). Other choices of ordering the roots typically lead to much more complicated expressions, and also make the intermediate steps of the computation (e.g. the identification of the dual basis) harder.

In order to investigate the centralizer algebras of $\mathsf{U}_q(\mathfrak{g})$, one is also interested in the form the R-matrix takes in a given pair of representations of $\mathsf{U}_q(\mathfrak{g})$. If $\mathtt{R}$ is the universal R-matrix of the quantum group $\mathsf{U}_q(\mathfrak{g})$, then for any pair R_Λ, $R_{\Lambda'}$ of irreducible highest weight modules of $\mathfrak{g}$ one can define the map

$$\mathcal{R}^{\Lambda\Lambda'} := \pi \circ (R_\Lambda \times R_{\Lambda'})(\mathtt{R}) : \; R_\Lambda \times R_{\Lambda'} \to R_{\Lambda'} \times R_\Lambda . \tag{4.6.31}$$

Here $\pi \equiv \pi^{\Lambda\Lambda'}$ denotes the permutation

$$\pi : \begin{array}{c} R_\Lambda \times R_{\Lambda'} \to R_{\Lambda'} \times R_\Lambda \\ v \otimes w \mapsto w \otimes v . \end{array} \tag{4.6.32}$$

The quantities (4.6.31) are called the R-matrices in the representation $R_\Lambda \times R_{\Lambda'}$, or shortly (for $\mathsf{U}_q(\mathfrak{g})$ with $\mathfrak{g}$ simple) *finite-dimensional R-matrices.* As a consequence of the Yang–Baxter equation (4.1.46), they satisfy

$$\begin{aligned} &(\mathcal{R}^{\Lambda\Lambda'} \otimes \mathbf{1})(\mathbf{1} \otimes \mathcal{R}^{\Lambda''\Lambda'})(\mathcal{R}^{\Lambda''\Lambda} \otimes \mathbf{1}) \\ &\qquad = (\mathbf{1} \otimes \mathcal{R}^{\Lambda''\Lambda})(\mathcal{R}^{\Lambda''\Lambda'} \otimes \mathbf{1})(\mathbf{1} \otimes \mathcal{R}^{\Lambda\Lambda'}). \end{aligned} \tag{4.6.33}$$

Here the matrices on either side of the equation act on the triple product $R_\Lambda \times R_{\Lambda'} \times R_{\Lambda''}$.

Define also $\mathbf{\Delta}^{\Lambda\Lambda'} := (R_\Lambda \times R_{\Lambda'})(\mathbf{\Delta})$ as in (4.3.8) so that

$$\mathbf{\Delta}^{\Lambda\Lambda'}(x) : \; R_\Lambda \times R_{\Lambda'} \to R_\Lambda \times R_{\Lambda'} \tag{4.6.34}$$

for all $x \in \mathsf{U}_q(\mathfrak{g})$. Then the quasitriangularity of $\mathsf{U}_q(\mathfrak{g})$ implies

$$\mathcal{R}^{\Lambda\Lambda'} \mathbf{\Delta}^{\Lambda\Lambda'}(x) = \mathbf{\Delta}^{\Lambda'\Lambda}(x)\, \mathcal{R}^{\Lambda\Lambda'} \quad \text{for all } \; x \in \mathsf{U}_q(\mathfrak{g}). \tag{4.6.35}$$

As a consequence, the elements

$$\underbrace{1 \otimes 1 \otimes \ldots \otimes 1}_{i-1 \text{ times}} \otimes \mathcal{R}^{\Lambda_i \Lambda_{i+1}} \otimes 1 \otimes \ldots \otimes 1 \tag{4.6.36}$$

belong to the centralizer of $\mathsf{U}_q(\mathfrak{g})$ in the Kronecker product $R_{\Lambda_1} \times \ldots \times R_{\Lambda_i} \times R_{\Lambda_{i+1}} \times \ldots$. More importantly, it can be shown that, conversely, for the case of the m-fold Kronecker product of the defining module of $\mathfrak{g}$ with itself, these elements actually generate the centralizer algebra.

Note that, taking the centralizer algebra as given, one could in principle also derive the definition of the co-multiplication of $\mathsf{U}_q(\mathfrak{g})$ and the relation (4.6.35) by making use of the reciprocity relation (1.1.40) between centralizer algebras.

It is not too difficult to determine $\mathcal{R}^{\Lambda\Lambda'}$ for R_Λ and $R_{\Lambda'}$ being the defining module, using the knowledge of the co-multiplication, the formula (4.6.35) and the fact that the defining modules of $\mathfrak{g}$ and $\mathsf{U}_q(\mathfrak{g})$ are isomorphic. By considering Kronecker products of the defining module, one can then determine $\mathcal{R}^{\Lambda\Lambda'}$ also for arbitrary Λ, Λ'. As an example, let us present the R-matrix for the defining module of $\mathsf{U}_q(sl_n)$. The representation matrices of $\mathsf{U}_q(sl_n)$ for this module are q-independent and hence the same as those of sl_n, namely

$$\begin{aligned} R_{\Lambda_{(1)}}(H^i) &= \mathcal{E}_{ii} - \mathcal{E}_{i+1,i+1}, \\ R_{\Lambda_{(1)}}(E_+^i) &= \mathcal{E}_{i,i+1}, \\ R_{\Lambda_{(1)}}(E_-^i) &= \mathcal{E}_{i+1,i} \end{aligned} \tag{4.6.37}$$

for $i = 1, \ldots, n$, where $\mathcal{E}_{ij}$, $i, j = 1, \ldots, n$, are the $n \times n$-matrices with entries

$$(\mathcal{E}_{ij})_{pq} = \delta_{ip}\delta_{jq}. \tag{4.6.38}$$

The R-matrix in this representation turns out to be

$$\begin{aligned} \mathcal{R}^{\Lambda_{(1)}\Lambda_{(1)}} &= q^{1/2}\sum_{i=1}^{n} \mathcal{E}_{ii} \otimes \mathcal{E}_{ii} \\ &+ \sum_{\substack{i,j=1 \\ i\neq j}}^{n} \mathcal{E}_{ii} \otimes \mathcal{E}_{jj} + (q^{1/2} - q^{-1/2})\sum_{\substack{i,j \\ i<j}}^{n} \mathcal{E}_{ij} \otimes \mathcal{E}_{ji}. \end{aligned} \tag{4.6.39}$$

In particular, for $\mathfrak{g} = sl_2$ one has

$$\mathcal{R}^{\Lambda_{(1)}\Lambda_{(1)}} = \begin{pmatrix} q^{1/2} & 0 & 0 & 0 \\ 0 & 1 & q^{1/2} - q^{-1/2} & 0 \\ 0 & 0 & 1 & 0 \\ 0 & 0 & 0 & q^{1/2} \end{pmatrix}. \tag{4.6.40}$$

As a more complicated example, let us consider the defining module of

$\mathcal{U}_q(so_{2n+1})$. The representation matrices are

$$
\begin{aligned}
&R_{\Lambda_{(1)}}(E_-^i) = \left(R_{\Lambda_{(1)}}(E_+^i)\right)^t && \text{for } i=1,\dots,n\,,\\
&\left.\begin{aligned}
R_{\Lambda_{(1)}}(H^i) &= \mathcal{E}_{ii} - \mathcal{E}_{i+1,i+1} + \mathcal{E}_{2n+1-i,2n+1-i}\\
&\qquad\qquad - \mathcal{E}_{2n+2-i,2n+2-i}\\
R_{\Lambda_{(1)}}(E_+^i) &= \mathcal{E}_{i,i+1} - \mathcal{E}_{2n+1-i,2n+2-i}
\end{aligned}\right\} && \text{for } i=1,\dots,n-1\,,\\
&R_{\Lambda_{(1)}}(H^n) = \mathcal{E}_{nn} - \mathcal{E}_{n+2,n+2},\\
&R_{\Lambda_{(1)}}(E_+^n) = \tfrac{1}{\sqrt{2}}(\mathcal{E}_{n,n+1} + \mathcal{E}_{n+1,n+2})\,.
\end{aligned}
\tag{4.6.41}
$$

Here the matrices $\mathcal{E}_{ij}$ are again defined by (4.6.38), but now as $(2n+1)\times(2n+1)$-matrices. The R-matrix in this representation is found to be

$$
\begin{aligned}
\mathcal{R}^{\Lambda_{(1)}\Lambda_{(1)}} = {}& q^{1/2}\sum_{\substack{i=1\\ i\neq n}}^{2n+1} \mathcal{E}_{ii}\otimes\mathcal{E}_{ii} + \mathcal{E}_{n+1,n+1}\otimes\mathcal{E}_{n+1,n+1} + \sum_{\substack{i,j=1\\ i\neq j,\, i\neq 2n+2-j}}^{2n+1} \mathcal{E}_{ij}\otimes\mathcal{E}_{ji}\\
&+ q^{-1/2}\sum_{\substack{i=1\\ i\neq n+1}}^{2n+1} \mathcal{E}_{i,2n+2-i}\otimes\mathcal{E}_{2n+2-i,i} + (q^{1/2}-q^{-1/2})\sum_{\substack{i,j\\ i<j}}^{2n+1} \mathcal{E}_{ii}\otimes\mathcal{E}_{jj}\\
&- (q^{1/2}-q^{-1/2})\sum_{\substack{i,j\\ i<j}}^{2n+1} q^{(i-j+\zeta_{ij})/2}\mathcal{E}_{2n+2-j,i}\otimes\mathcal{E}_{j,2n+2-i},
\end{aligned}
\tag{4.6.42}
$$

where

$$
\zeta_{ij} = \begin{cases} 1 & \text{for } i\leq n,\ j>n\\ 0 & \text{otherwise.}\end{cases}
\tag{4.6.43}
$$

From the formula (4.6.39) an important property of $\mathcal{R}^{\Lambda_{(1)}\Lambda_{(1)}}$ can be deduced: namely,

$$
(\mathcal{R}^{\Lambda_{(1)}\Lambda_{(1)}})^2 = (q^{1/2}-q^{-1/2})\,\mathcal{R}^{\Lambda_{(1)}\Lambda_{(1)}} + 1.
\tag{4.6.44}
$$

Therefore, if one defines

$$
t_i := -q^{-1/2}\ \underbrace{1\otimes 1\otimes\ldots\otimes 1}_{i-1\text{ times}} \otimes(\mathcal{R}^{\Lambda_{(1)}\Lambda_{(1)}})_{i,i+1}\otimes \underbrace{1\otimes\ldots\otimes 1}_{m-i-1\text{ times}}
\tag{4.6.45}
$$

(acting on the m-fold Kronecker product $(R_{\Lambda_{(1)}})^{\otimes m}$), one has

$$
t_i t_i = (1-q)\,t_i + q,
\tag{4.6.46}
$$

i.e. one obtains part of the defining relations (4.3.37) of the Hecke algebra. But the rest of those relations is also fulfilled, since when expressed in terms of $\mathcal{R}^{\Lambda_{(1)}\Lambda_{(1)}}$, they are equivalent to the Yang–Baxter equation (4.6.33). Thus one concludes that the centralizer of $\mathsf{U}_q(sl_n)$ in $(R_{\Lambda_{(1)}})^{\otimes m}$ is the Hecke algebra $H_m(q)$. (Actually, as described in section 4.3, for $m > n$ the centralizer is the factor of $H_m(q)$ by the ideal that is generated by $\hat{t} = \sum_{s \in S_{n+1}} (\mathrm{sign}\, s)\, t_s$.)

In an analogous manner, one can derive the centralizer algebras for less simple cases, but the calculations then tend to become much more complicated.

4.7 Quantized groups

In sections 4.1 and 4.2 we have mentioned that quantum groups (respectively their function spaces) play a role for quantum-mechanical systems that is analogous to the role of ordinary groups in classical mechanical systems. Also, the Hopf algebras $\mathsf{U}_q(\mathfrak{g})$ which are the quantum groups which we are mainly interested in are obtained as q-deformations of more familiar objects. Now the mathematical procedure of deformation is precisely what is employed in physics when the quantization of some theory is performed. It is therefore tempting to try to obtain quantum groups from more traditional mathematical objects by some appropriate (formal) *quantization procedure.* To investigate this possibility, as a first step it is necessary to define in abstract terms what is meant by the quantization of a classical system.

A *classical dynamical system* is defined by the requirement that the dynamics is subject to the requirement

$$\partial f = \{H, f\}, \tag{4.7.1}$$

valid for any function f on the manifold on which the dynamics takes place. Denote the space of these functions by $\mathfrak{a}_0$. This space and the symbols appearing in (4.7.1) are then to be interpreted as follows. $\mathfrak{a}_0$ is a commutative associative unital algebra, and "∂" a derivation on $\mathfrak{a}_0$, i.e. a map $\mathfrak{a}_0 \to \mathfrak{a}_0$ acting on products according to the Leibniz rule,

$$\partial(x \diamond y) = x \diamond \partial y + (\partial x) \diamond y. \tag{4.7.2}$$

The action of the derivation defines the dynamical evolution of the system. The curly brackets in (4.7.1) denote a *Poisson bracket* which may be abstractly defined as a map

$$\{\ ,\ \} : \ \mathfrak{a}_0 \times \mathfrak{a}_0 \to \mathfrak{a}_0 \tag{4.7.3}$$

with respect to which $\mathfrak{a}_0$ is a Lie algebra and which acts on products as

$$\{x \circ y, z\} = x\{y,z\} + \{x,z\}\, y. \tag{4.7.4}$$

An algebra endowed with this additional structure is called a *Poisson algebra.* The particular element $H \in \mathfrak{a}_0$ that governs the dynamics according to (4.7.1) is called the *Hamilton function.* In terms of coordinates q_i, p_i on the underlying manifold, the elements of the Poisson algebra are functions of the coordinates, and the Poisson bracket is defined by $\{q_i, p_j\} = \delta_{ij}$ so that on arbitrary elements of $\mathfrak{a}_0$ it acts as

$$\{x,y\} = \sum_i \left(\frac{\partial x}{\partial q_i}\frac{\partial y}{\partial p_i} - \frac{\partial x}{\partial p_i}\frac{\partial y}{\partial q_i} \right). \tag{4.7.5}$$

For a *quantum dynamical system*, the equation describing the dynamics is given by

$$\partial f = [H, f], \tag{4.7.6}$$

where ∂ again is a derivation, but f now takes values in a associative unital, but non-commutative, algebra $\mathfrak{a}$, and the square brackets denote the commutator. $H \in \mathfrak{a}$ is in this case called the *Hamilton operator.* A quantum dynamical system is obtained as a *quantization* of a classical dynamical system iff the classical system can be obtained from the quantum system in the limit (called the classical limit) where some formal parameter $\hbar$ tends to zero (in ordinary quantum mechanics, $\hbar$ is Planck's constant). More formally, one defines: For $\mathfrak{a}_0$ a Poisson algebra over $\mathbb{C}$, a quantization is a non-commutative associative algebra $\mathfrak{a}$ over $\mathbb{C}[\![\hbar]\!]$ such that

$$\mathfrak{a}_0 \cong \frac{\mathfrak{a}}{\hbar\mathfrak{a}} \quad \text{and} \quad \{x_0, y_0\} = \left(\hbar^{-1}[x,y]\right)_0 \quad \text{for all } x,y \in \mathfrak{a}. \tag{4.7.7}$$

Here $\mathbb{C}[\![\hbar]\!]$ is the ring of formal polynomials (i.e., including infinite series) in the variable $\hbar$ with coefficients in $\mathbb{C}$, so that as a vector space $\mathfrak{a}$ is isomorphic to the space of formal polynomials in $\hbar$ with coefficients in $\mathfrak{a}_0$. Furthermore, $x_0 \in \mathfrak{a}_0$ denotes the image of $x \in \mathfrak{a}$ under the quotient map $\mathsf{p}\colon \mathfrak{a} \to \mathfrak{a}/\hbar\mathfrak{a}$; the map p consists in putting $\hbar = 0$ in the polynomials of $\mathbb{C}[\![\hbar]\!]$. Finally, note that in order for the second condition to make sense, one needs $\hbar^{-1}[x,y] \in \mathfrak{a}$ for all x, $y \in \mathfrak{a}$. This is indeed the case, as follows from the commutativity of $\mathfrak{a}_0$; namely, one has $0 = [x_0, y_0] = [x + \hbar\mathfrak{a}, y + \hbar\mathfrak{a}] = [x,y] + \hbar\mathfrak{a}$ so that $[x,y]_0 = 0$ and hence $[x,y] \in \hbar\mathfrak{a}$.

Now consider the special situation where the Poisson algebra $\mathfrak{a}_0$ possesses in addition also the structure of a Hopf algebra. Since $\mathfrak{a}_0$ is commutative, according to the Gelfand–Naimark theorem (see p. 252) this is possible iff the underlying manifold is a compact topological group G, with the *Poisson–Hopf algebra* a_0 being the space of functions on the group, $\mathsf{a}_0 \cong C(\mathsf{G})$. If there exists a quantization a of a_0, it is then natural to interpret a as the space of functions on some topological object $\mathsf{G}_\hbar$, and to call $\mathsf{G}_\hbar$ a quantization of the group G or shortly a quantum group.

The relation of this type of quantum group with the quantized universal enveloping algebras (which we also called quantum groups; see section 4.1) arises as follows. Take the group G to be a compact Lie group with Lie algebra $\mathfrak{g}$. Then there exists a pairing between the functions $C(\mathsf{G})$ on G and the dual $\mathsf{U}(\mathfrak{g})^\star$ of the universal enveloping algebra $\mathsf{U}(\mathfrak{g})$, which is nondegenerate on a subspace of $\mathsf{U}(\mathfrak{g})$ which for the sake of brevity we will also denote by $\mathsf{U}(\mathfrak{g})$ (the restriction to a subspace is necessary as a consequence of the infinite-dimensionality of $\mathsf{U}(\mathfrak{g})$; compare the situation for a general Hopf algebra, p. 249). Thus one has $C(\mathsf{G}) \cong \mathsf{U}(\mathfrak{g})^\star$. Moreover, from $\mathsf{U}(\mathfrak{g})^\star$ one can determine $\mathsf{U}(\mathfrak{g})$, so that one arrives at the following scheme of correspondences:

G	$\rightleftarrows$	$C(\mathsf{G}) \cong \mathsf{U}(\mathfrak{g})^\star$	$\rightleftarrows$	$\mathsf{U}(\mathfrak{g})$
group		commutative Hopf algebra		co-commutative Hopf algebra

(4.7.8)

The formal quantization procedure discussed above relates $C(\mathsf{G})$ to $C(\mathsf{G}_\hbar)$. As already mentioned in section 4.1, it is not possible to give a precise meaning to the quantum group $\mathsf{G}_\hbar$ directly as a quantization of G (but, as also mentioned there, this is not really a problem because any issue concerning $\mathsf{G}_\hbar$ may be reformulated in terms of $C(\mathsf{G}_\hbar)$). In contrast, it *is* possible to translate the quantization procedure into the language of universal enveloping algebras. The Poisson structure on $C(\mathsf{G})$ transcribes into a dual operation

$$\delta:\ \mathsf{U}(\mathfrak{g}) \to \mathsf{U}(\mathfrak{g}) \times \mathsf{U}(\mathfrak{g}), \tag{4.7.9}$$

called a co-Poisson map, via

$$(f\otimes g)(\delta(x)) := (\{f,g\})(x) \tag{4.7.10}$$

for all $f,g \in C(\mathsf{G})$ and all $x \in \mathsf{U}(\mathfrak{g})$. The Hopf algebra $\mathsf{U}(\mathfrak{g})$ endowed with the operation δ is called a *co-Poisson–Hopf algebra.* Given $\mathsf{U}(\mathfrak{g})$ and δ, a quantization procedure can be defined by dualizing the quantization of $C(\mathsf{G})$: A quantization $\mathsf{U}(\mathfrak{g})_\hbar$ of $\mathsf{U}(\mathfrak{g})$ is by definition a non-cocommutative Hopf algebra over $\mathbb{C}[\![\hbar]\!]$ such that

$$\begin{aligned} &\mathsf{U}(\mathfrak{g}) \cong \frac{\mathsf{U}(\mathfrak{g})_\hbar}{\hbar\mathsf{U}(\mathfrak{g})_\hbar} \qquad \text{and} \\ &\delta(x_0) = \left(\hbar^{-1}\left[\Delta(x) - \Delta'(x)\right]\right)_0 \quad \text{for all } x \in \mathsf{U}(\mathfrak{g})_\hbar. \end{aligned} \tag{4.7.11}$$

Here again $x_0 \in \mathsf{U}(\mathfrak{g})$ denotes the restriction of $x \in \mathsf{U}(\mathfrak{g})_\hbar$ under the projection $\mathsf{U}(\mathfrak{g})_\hbar \to \mathsf{U}(\mathfrak{g})_\hbar/\hbar\mathsf{U}(\mathfrak{g})_\hbar$. This way the above picture of correspondences can be extended as shown in figure (4.7.12). In the figure, downwards and upwards arrows denote the quantization procedure and taking the classical limit $\hbar \to 0$, respectively. Note that in this diagram there are no arrows directly connecting G and $\mathsf{G}_\hbar$. Although this does not prevent the study of the object $\mathsf{G}_\hbar$ (the properties of which can be entirely described in terms of the functions $C(\mathsf{G}_\hbar)$), it is still tempting to find a more direct connection. A little further on (p. 311), we will indeed encounter such a connection, namely via the so-called matrix quantum groups which, in short, are matrix representations of suitable quantum groups $G_\hbar$.

G group	$\rightleftarrows$	$C(\mathsf{G}) \cong \mathsf{U}(\mathfrak{g})^\star$ commutative Hopf algebra	$\rightleftarrows$	$\mathsf{U}(\mathfrak{g})$ co-commutative Hopf algebra
		$\downarrow\ \uparrow$ (4.7.7)		$\downarrow\ \uparrow$ (4.7.11)
$\mathsf{G}_\hbar$ quantum group	$\rightleftarrows$	$C(\mathsf{G}_\hbar) \cong (\mathsf{U}(\mathfrak{g})_\hbar)^\star$ non-commutative Hopf algebra	$\rightleftarrows$	$\mathsf{U}(\mathfrak{g})_\hbar$ non-cocommutative Hopf algebra

(4.7.12)

The prescription (4.7.11) only lays down what type of algebra is to be called a quantization $\mathsf{U}(\mathfrak{g})_\hbar$ of $\mathsf{U}(\mathfrak{g})$, but leaves open the question whether such algebras exist, and if so, whether for given $\mathsf{U}(\mathfrak{g})$ they are unique. The existence question can be answered in the affirmative by considering the classical Yang–Baxter equation (4.2.50). Let

$$\mathbf{r} = \sum_{\mu,\nu} r^{\mu\nu}\, e_\mu \otimes e_\nu \in \mathsf{U}(\mathfrak{g}) \times \mathsf{U}(\mathfrak{g}) \qquad (4.7.13)$$

be a solution of the classical Yang–Baxter equation; then a Poisson bracket on $\mathsf{U}(\mathfrak{g})$ can be defined as

$$\{f,g\} = \sum_{\mu,\nu} r^{\mu\nu}\, \partial_\mu f\, \partial_\nu g \qquad (4.7.14)$$

(the Jacobi identity for this bracket just reduces to the Yang–Baxter equation equation for $r^{\mu\nu}$, and the other defining property of the bracket follows from the Leibniz rule for the partial derivative ∂_μ). With the help of this Poisson bracket, one can construct the map δ and thus a quantization $\mathsf{U}(\mathfrak{g})_\hbar$. It turns out that this quantization is isomorphic to the

q-deformation $\mathsf{U}_q(\mathfrak{g})$ of $\mathsf{U}(\mathfrak{g})$, with the deformation parameters q and $\hbar$ reflecting the exponential relation between the group G and its Lie algebra $\mathfrak{g}$,

$$q = \mathrm{e}^{\hbar}. \tag{4.7.15}$$

In short, as already stated in section 4.1, the "quantum group" $\mathsf{U}_q(\mathfrak{g})$ is actually the space dual to the space of functions on the quantum group $\mathsf{G}_\hbar$.

For any complex simple Lie algebra $\mathfrak{g}$, one knows only a single solution (up to equivalence) to the classical Yang–Baxter equation, namely the so-called *Drinfeld–Jimbo solution*

$$\mathbf{r} = \sum_{i,j=1}^{r} G_{ij}\, H^i \otimes H^j + 2 \sum_{\alpha>0} (\alpha, \alpha)^{-1}\, E^\alpha \otimes E^{-\alpha}. \tag{4.7.16}$$

This may be taken as an indication that the quantization $\mathsf{U}_q(\mathfrak{g})$ of $\mathsf{U}(\mathfrak{g})$ is unique. Indeed, the uniqueness has been claimed in the literature, although a rigorous proof has not yet been published.

Let us now discuss the construction of $\mathsf{U}_q(\mathfrak{g})$ along these lines for the simplest case, $\mathfrak{g} = A_1$. The Drinfeld–Jimbo solution then reads

$$\mathbf{r} = H \otimes H + E_+ \otimes E_-. \tag{4.7.17}$$

The map $\delta : \mathsf{U}(A_1) \to \mathsf{U}(A_1) \times \mathsf{U}(A_1)$ determined by this classical R-matrix is

$$\begin{aligned} \delta(H) &= 0, \\ \delta(E_\pm) &= \tfrac{1}{2}\,(E_\pm \otimes H - H \otimes E_\pm). \end{aligned} \tag{4.7.18}$$

The quantization now amounts to the construction of a co-associative co-multiplication $\boldsymbol{\Delta}$ such that the second of the relations (4.7.11) holds. That such a construction is indeed possible is seen by inserting into that equation the ansatz

$$\boldsymbol{\Delta} = \sum_{n=0}^{\infty} \frac{\hbar^n}{n!}\, \boldsymbol{\Delta}_{(n)}. \tag{4.7.19}$$

This yields the condition

$$\delta(x_0) = \boldsymbol{\Delta}_{(1)}(x) - \pi \circ \boldsymbol{\Delta}_{(1)}(x), \tag{4.7.20}$$

on $\boldsymbol{\Delta}_{(1)}$, while the quantities $\boldsymbol{\Delta}_{(n)}$, $n \geq 2$, remain undetermined (the term involving $\boldsymbol{\Delta}_{(0)}$ must vanish identically, $\boldsymbol{\Delta}_{(0)} - \pi \circ \boldsymbol{\Delta}_{(0)} = 0$, as it is proportional to $\hbar^{-1}$; this condition is indeed fulfilled, since taking the limit $\hbar \to 0$ shows that $\boldsymbol{\Delta}_{(0)}$ coincides with the co-multiplication of $\mathsf{U}(A_1)$). The latter quantities can, however, be determined from $\boldsymbol{\Delta}_{(1)}$ by exploiting the co-associativity requirement. Namely, inserting the ansatz for $\boldsymbol{\Delta}$

into the co-associativity equation, one gets the recursion relations

$$\sum_{j=0}^{n} \binom{n}{j} \left(\mathbf{\Delta}_{(j)} \otimes 1 - 1 \otimes \mathbf{\Delta}_{(j)} \right) \mathbf{\Delta}_{(n-j)} = 0. \qquad (4.7.21)$$

An obvious solution to the constraint on $\mathbf{\Delta}_{(1)}$ is $\mathbf{\Delta}_{(1)} = \frac{1}{2}\,\delta$, and this is the only known solution apart from the trivial possibility of adding terms that are symmetric under the permutation π. Taking this solution, inspection shows that the recursion relations are all solved simultaneously by taking

$$\begin{aligned} &\mathbf{\Delta}_{(n)}(H) = 0, \\ &\mathbf{\Delta}_{(n)}(E_\pm) = 2^{-2n} \left(E_\pm \otimes H^n + (-1)^n H^n \otimes E_\pm \right). \end{aligned} \qquad (4.7.22)$$

Inserting this into the ansatz, one recovers the co-product rules (4.2.17) of $\mathsf{U}_q(A_1)$, with $q = \exp(\hbar)$.

Finally, the commutation relations of $\mathsf{U}(A_1)_\hbar$ can be found by imposing the homomorphism property of $\mathbf{\Delta}$, which yields

$$\begin{aligned} &\mathbf{\Delta}([H, E_\pm]) = [H, E_\pm] \otimes \mathrm{e}^{\hbar H/4} + e^{-\hbar H/4} \otimes [H, E_\pm], \\ &\mathbf{\Delta}([E_+, E_-]) = [E_+, E_-] \otimes \mathrm{e}^{\hbar H/2} + e^{-\hbar H/2} \otimes [E_+, E_-]. \end{aligned} \qquad (4.7.23)$$

The first of these equations is satisfied by taking $[H, E_\pm] = \pm 2E_\pm$, and the second one by $[E_+, E_-] = \mathrm{e}^{\hbar H/2} - \mathrm{e}^{-\hbar H/2}$, but of course in both cases the commutators may also be multiplied by any "constant", i.e. by any power series $\mathcal{P}$ over $\mathbb{C}$ in the parameter $\hbar$. Only the leading terms in these power series can be fixed, namely by insisting on the correct behavior in the classical limit $\hbar \to 0$. For the specific choice $\mathcal{P} = 1$ for $[H, E_\pm]$ and $\mathcal{P} = \frac{1}{2} \sinh^{-1}(\frac{\hbar}{2})$ for $[E_+, E_-]$, one recovers the commutation relations (4.2.16) of $\mathsf{U}_q(A_1)$, $q = \exp(\hbar)$, while any other choice consistent with the classical limit leads to an isomorphic algebra.

There exists also a different deformation procedure which can be regarded as describing the quantization of Lie groups. One starts with a Lie group G of matrices, and the deformation consists in interpreting the matrix elements no longer as elements of $\mathbb{C}$ (or of some other field F), but rather as elements of an associative algebra a. Also, for the deformed system it should still be possible to define a meaningful matrix multiplication, i.e. if the elements of two matrices M and M' belong to a, then the same property should hold for the combinations

$$(M \cdot M')^i_{\ j} = \sum_k M^i_{\ k} M'^k_{\ j}. \qquad (4.7.24)$$

The object obtained by such a deformation is then called a *quantum matrix group*. Now one may take the generators of the algebra a as elements of a formal matrix $\varphi = (\varphi^i_{\ j})$ and form the matrices $T = \varphi \otimes \mathbf{1}$ and

$\tilde{T} = \mathbf{1} \otimes \varphi$. In terms of these quantities, the requirement that the matrix multiplication still makes sense means that there must exist some matrix $\mathcal{R}$ (acting on $V \times V$ if the matrices M act on some vector space V) such that

$$\mathcal{R} T \tilde{T} = \tilde{T} T \mathcal{R}; \tag{4.7.25}$$

this is nothing but the relation (4.1.53) for the R-matrix of a quasitriangular Hopf algebra. To be precise, it is in general not possible to put all generators of a into a single matrix φ; rather, it may be necessary to define several such matrices each of which must obey a relation of the form (4.7.25).

By requiring also that multiple matrix products possess the right properties, further relations must be satisfied, such as

$$\mathcal{R}_{23}\mathcal{R}_{13}\mathcal{R}_{12}\,\varphi_1\varphi_2\varphi_3 = \varphi_3\varphi_2\varphi_1\mathcal{R}_{23}\mathcal{R}_{13}\mathcal{R}_{12}, \tag{4.7.26}$$

where the indices are used as in section 4.1, i.e. $\mathcal{R}_{23}$ acts on $V \times V \times V$ as the identity on the first factor and as $\mathcal{R}$ on the second and third factor, etc. These higher order equations generically provide additional constraints on the matrix φ. However, it can be shown (e.g. by category theoretic arguments) that all of them are fulfilled identically if the matrix $\mathcal{R}$ obeys

$$\mathcal{R}_{12}\mathcal{R}_{13}\mathcal{R}_{23} = \mathcal{R}_{23}\mathcal{R}_{13}\mathcal{R}_{12}, \tag{4.7.27}$$

i.e. the (matrix version of the) Yang–Baxter equation.

If the Yang–Baxter equation holds, it is in fact possible to endow the algebra a with the structure of a quasitriangular Hopf algebra. Moreover, if G is a classical simple Lie group with Lie algebra $\mathfrak{g}$, one can identify the deformed enveloping algebra $\mathsf{U}_q(\mathfrak{g})$ as a Hopf subalgebra of the dual $\mathsf{a}^\star$ of a. Thus it follows that the algebra a plays the role of the algebra of functions on the quantum group G_q, and the matrices consisting of elements of a form a matrix representation of G_q.

It is also possible to give a geometric interpretation to the "deformed vector spaces" V_q on which G_q is acting, namely as the objects appearing in the mathematical theory of *non-commutative geometry.* Thus one can arrive at quantum matrix groups also by studying these objects V_q in their own right and defining G_q as the object that effects the linear transformations which implement the symmetries of V_q.

Let us consider the simplest non-trivial example, given by 2×2-matrices

$$M = \begin{pmatrix} a & b \\ c & d \end{pmatrix}. \tag{4.7.28}$$

For a, b, c, $d \in \mathbb{C}$ these matrices form the (non-semisimple) Lie group $GL_2(\mathbb{C})$. The requirement that for its deformation a sensible matrix multiplication can be defined then leads to 16 equations. Six of them are independent; they read

$$\begin{aligned} ab &= q^{1/2}\, ba, \qquad ac = q^{1/2}\, ca, \\ bd &= q^{1/2}\, db, \qquad cd = q^{1/2}\, dc, \\ bc &= cb, \\ ad - da &= (q^{1/2} - q^{-1/2})\, bc, \end{aligned} \tag{4.7.29}$$

where q is an arbitrary complex number. It is easily checked that the corresponding matrix $\mathcal{R}$ is precisely the R-matrix of $\mathsf{U}_q(sl_2)$ which was given in (4.6.40). Thus the matrices obeying (4.7.29) form the quantum matrix group $(GL_2)_q(\mathbb{C})$. From this one obtains a deformation $(SL_2)_q(\mathbb{C})$ of the simple Lie group $SL_2(\mathbb{C})$ (whose Lie algebra is $sl_2(\mathbb{C})$) by imposing the constraint

$$\det{}_q M = 1 \tag{4.7.30}$$

on the q-deformed determinant, or quantum determinant,

$$\det{}_q M \equiv ad - q^{1/2}\, bc. \tag{4.7.31}$$

This restriction on the quantum determinant can be imposed consistently because, as one easily checks, $\det_q M$ commutes with a, b, c and d; in fact, $\det_q M$ generates the center of the algebra a. The antipode on a is given by the matrix

$$\gamma(M) = \begin{pmatrix} d & -q^{-1/2}b \\ -q^{1/2}c & a \end{pmatrix}. \tag{4.7.32}$$

This is just the ordinary matrix inverse of M:

$$M\,\gamma(M) = \gamma(M)\,M = (\det{}_q M)\,\mathbf{1} = \mathbf{1}. \tag{4.7.33}$$

The generalization of this example to the deformation of $GL_n(\mathbb{C})$ with $n > 2$ is straightforward. The q-deformed determinant is now defined as

$$\det{}_q M = \sum_{s \in \mathcal{S}_n} (-q)^{\ell(s)/2}\, M_{1,s(1)} M_{2,s(2)} \cdots M_{n,s(n)}, \tag{4.7.34}$$

where the sum extends over the symmetric group $\mathcal{S}_n$, and the length $\ell(s)$ is defined to be the minimum number of generators $s_i \in \mathcal{S}_n$ that is needed to write s in the form $s = s_{i_1} s_{i_2} \ldots s_{i_\ell}$. The restriction from $GL_n(\mathbb{C})$ to $SL_n(\mathbb{C})$ is again implemented by constraining the quantum determinant to unity.

It is also possible to obtain the relations of $\mathsf{U}_q(\mathfrak{g})$ quite directly. We will not describe this in full generality, but only present the example $\mathfrak{g} = sl_2$.

The generators of $U_q(sl_2)$ can be put into two matrices as

$$\varphi_+ = \begin{pmatrix} q^{H/4} & (q^{1/2} - q^{-1/2})\, E_+ \\ 0 & q^{-H/4} \end{pmatrix},$$
$$\varphi_- = \begin{pmatrix} q^{-H/4} & 0 \\ (q^{1/2} - q^{-1/2})\, E_- & q^{H/4} \end{pmatrix}. \tag{4.7.35}$$

Requiring that (4.7.25) holds for $\varphi = \varphi_\pm$, one finds $[H, E_\pm] = \pm 2E_\pm$. If one imposes in addition that

$$\mathcal{R} T_+ \tilde{T}_- = \tilde{T}_- T_+ \mathcal{R} \tag{4.7.36}$$

for $T_+ = \varphi_+ \otimes \mathbf{1}$ and $T_- = \mathbf{1} \otimes \varphi_-$, one obtains also $[E_+, E_-] = \lfloor H \rfloor$, and thus the defining relations of $U_q(sl_2)$.

Since the co-multiplication of $a^* \supset U_q(sl_2)$ is the multiplication of a, it reads in terms of the matrices $\varphi_\pm$

$$\Delta((\varphi_\pm)^i_{\ j}) = \sum_k (\varphi_\pm)^i_{\ k} \otimes (\varphi_\pm)^k_{\ j}. \tag{4.7.37}$$

In the general case, one derives in a similar way the relations of $U_q(\mathfrak{g})$ for any classical Lie algebra $\mathfrak{g}$. However, they are obtained in a Cartan–Weyl basis rather than in the Chevalley–Serre basis that was used in the definition (4.2.1) of $U_q(\mathfrak{g})$.

4.8 Affine Lie algebras and quantum groups

From chapter 3 we know that for WZW theories an underlying Lie algebraic structure can be identified in terms of untwisted affine Lie algebras g. In chapter 5, it will be seen that one can also associate to any WZW theory a Hopf algebra structure, namely the quantum group $U_q(\mathfrak{g})$. While the Lie algebra g generates the chiral symmetry algebra of the theory, the quantum group $U_q(\mathfrak{g})$ of a WZW theory may be identified as the centralizer of the braid group in the space of chiral block functions, leading to a quantum group interpretation of the fusing and braiding matrices (section 5.7). Also, the fields of the WZW theory are organized into irreducible highest weight modules of the Lie algebra g, while certain modules of $U_q(\mathfrak{g})$ can be constructed (section 5.8) from the so-called screened vertex operators. Finally, the truncated Kronecker products of $U_q(\mathfrak{g})$ are isomorphic to the fusion rules of the WZW theory (section 5.5). All these connections between g and $U_q(\mathfrak{g})$ through their role in WZW theories are somewhat indirect, but together they certainly suggest that the two algebraic structures might be related also at a more fundamental level. Ideally one would like to have an expression of the generators of $\mathsf{U}_q(\mathfrak{g})$ in terms of those of $\mathfrak{g}$. The present section describes one approach to this issue which seems to point into the right direction although so far it has not yielded decisive results. It is however by no means clear that a solution to this problem exists at all; anyhow, this section also serves as providing some non-trivial exercises for calculating operator products in conformal field theory.

From the fact (see section 5.6) that the duality properties encoded in the quantum group structure are rather non-local operations in terms of the chiral block functions, it is apparent that the expression of the $\mathsf{U}_q(\mathfrak{g})$-generators in terms of the $\mathfrak{g}$-generators $J^a(z)$, if it exists, is of a non-local functional type. Hence it is natural to try to express the quantum group generators as contour integrals of (appropriately normal ordered products of) the currents $J^a(z)$. It turns out to be convenient to work in a Chevalley–Serre basis of $\mathfrak{g}$ so that the relations of $\mathsf{U}_q(\mathfrak{g})$ are as in (4.2.1), while the currents are $\{J^a(z)\} = \{e^i_\pm(z),\, h^i(z)\}$ with operator products

$$\begin{aligned} \underbrace{h^i(z)\, h^j(w)} &= k\, G^{ij}\,(z-w)^{-2}, \\ \underbrace{h^i(z)\, e^j_\pm(w)} &= \pm A^{ji}\,(z-w)^{-1}\, e^j_\pm(w), \\ \underbrace{e^i_+(z)\, e^j_-(w)} &= \delta^{ij}\,\left[k\,(z-w)^{-2} + (z-w)^{-1}\, h^i(w)\right], \end{aligned} \tag{4.8.1}$$

supplemented with rules involving multiple operator products of the $e^i_\pm$. The latter rules correspond to the Serre relations; they implement the operator products between current fields for arbitrary roots which read

$$\underbrace{e^\alpha(z)\, e^\beta(w)} = \begin{cases} \pm\,(z-w)^{-1}\, e^{\alpha+\beta}(w) & \text{for } \alpha+\beta \text{ a } \mathfrak{g}\text{-root} \\ k\,(z-w)^{-2} + (z-w)^{-1}\, \alpha_j h^j(w) & \text{for } \alpha+\beta = 0 \\ 0 & \text{otherwise.} \end{cases} \tag{4.8.2}$$

The relations (4.2.1) of $\mathsf{U}_q(\mathfrak{g})$ differ from those of $\mathfrak{g}$ only as far as the commutators of the step operators among themselves are concerned. A natural ansatz is therefore to identify the Cartan subalgebra generators H^i with the Cartan subalgebra generators of $\mathfrak{g}$, and hence with the zero modes h^i_0 of $h^i(z)$; in terms of contour integrals this means

$$H^i = \tfrac{1}{2\pi i} \oint_0 \mathrm{d}z\, h^i(z). \tag{4.8.3}$$

Just as this ansatz reproduces immediately the commutativity (4.2.1a) of the H^i, the analogous ansatz

$$E^i_\pm \overset{?}{=} \tfrac{1}{2\pi i} \oint_0 \mathrm{d}z\, e^i_\pm(z) \tag{4.8.4}$$

would reproduce the rest of the relations of $\mathfrak{g}$ as well, rather than the deformed relations (4.2.1c,d), and hence is too crude. However, one may seek for a deformation of this ansatz which reduces to it in the classical limit $q \to 1$, and which in addition involves only the Cartan subalgebra fields so that it preserves the natural root space grading (1.4.63) of the integrand. Moreover, in order to reproduce (4.2.1*b*), the additional terms must not change the commutator with H^i which is q-independent. An educated guess is then

$$E^i_\pm = \frac{1}{2\pi i} \oint_0 \mathrm{d}z \; : Q^i_\pm(z)\, e^i_\pm(z) : \tag{4.8.5}$$

with

$$\begin{aligned} Q^i_+(z) &:= q^{[\gamma_+ \int_z^{z_0} \mathrm{d}w\, h^i(w)]}, \\ Q^i_-(z) &:= q^{[\gamma_- \int_{z_0}^{z} \mathrm{d}w\, h^i(w)]}, \end{aligned} \tag{4.8.6}$$

where $\gamma_\pm$ are appropriate constants and where the integration contours are to be interpreted as parts of the closed contour appearing in (4.8.5), with z_0 an arbitrarily chosen point on this contour.

To calculate commutators involving (4.8.5), one deforms contours as usual (see p. 152), taking into account, however, that the integrand contains effectively two distinct singular points z and z_0. Thus one finds

$$\begin{aligned} [H^i, E^j_\pm] = \frac{1}{2\pi i} \oint_0 \mathrm{d}z \,\big(\tfrac{1}{2\pi i} \oint_z \mathrm{d}w + \tfrac{1}{2\pi i} \oint_{z_0} \mathrm{d}w\big) \\ \cdot\, \underbrace{h^i(w) \; : Q^j_\pm(z) e^j_\pm(z) :}\; . \end{aligned} \tag{4.8.7}$$

Using the Wick rule (3.1.35), the right hand side of this equation splits into two pieces; the piece containing the contraction of h^i with $e^j_\pm$ can be written as

$$\begin{aligned} &\frac{1}{2\pi i} \oint_0 \mathrm{d}z \,\big(\tfrac{1}{2\pi i} \oint_z \mathrm{d}w + \tfrac{1}{2\pi i} \oint_{z_0} \mathrm{d}w\big) \; : Q^j_\pm(z)\, \underbrace{h^i(w)\, e^j_\pm(z)} : \\ &\qquad = \pm A^{ji} \frac{1}{2\pi i} \oint_0 \mathrm{d}z \, \frac{1}{2\pi i} \oint_z \mathrm{d}w\, (w-z)^{-1} : Q^j_\pm(z)\, e^j_\pm(z) : \\ &\qquad = \pm A^{ji}\, E^j_\pm . \end{aligned} \tag{4.8.8}$$

The second piece involves the contraction between h^i and $Q^j_\pm$; it can be evaluated with the help of the contraction formula (3.1.37), yielding

$$\begin{aligned} \pm\, k\, \gamma_\pm\, G^{ij} \ln q \cdot \frac{1}{2\pi i} \oint_0 \mathrm{d}z \,\big(\tfrac{1}{2\pi i} \oint_z \mathrm{d}w + \tfrac{1}{2\pi i} \oint_{z_0} \mathrm{d}w\big) \\ \cdot\; : Q^j_\pm(z) e^j_\pm(z) : \int_{z_0}^{z} \mathrm{d}v\, (w-v)^{-2}; \end{aligned} \tag{4.8.9}$$

this vanishes owing to

$$\big(\tfrac{1}{2\pi i} \oint_z \mathrm{d}w + \tfrac{1}{2\pi i} \oint_{z_0} \mathrm{d}w\big) \left(\frac{1}{w-z} - \frac{1}{w-z_0} \right) = 0. \tag{4.8.10}$$

Thus the commutation relation

$$[H^i, E^j_\pm] = \pm\, A^{ji}\, E^j_\pm \tag{4.8.11}$$

is reproduced correctly.

Proceeding now to the calculation of $[E^i_+, E^j_-]$, one has to investigate the contraction

$$\underbrace{:Q^i_+(z)e^i_+(z)::Q^j_-(w)e^j_-(w):} \qquad (4.8.12)$$

This quantity poses various kinds of problems. For example, owing to the field dependence of the term with $h(w)$ in the contraction of $e^i_+(z)$ and $e^j_-(w)$, the application of the Wick rule becomes much more involved than in the previous cases. More severely, the terms coming from the contraction of Q^i_+ and Q^j_- lead to an integrand possessing cuts in the w- and z-planes (except for $q = 1$) so that the simple contour deformation procedure can no longer be used. To get rid of the problem, let us for the moment assume that the latter terms vanish identically; this forces us to take $k^\vee = 0$. The contributions involving contractions between $Q_\pm$ and $e_\mp$ are then still very complicated, but it can be argued that they cancel against each other provided that

$$\gamma_- = -\gamma_+ =: \gamma. \qquad (4.8.13)$$

(This is most easily seen by translating the calculation into the corresponding one for the fields on the real line (compare (3.1.13)) rather than on the circle.) The remaining contribution involving the contraction of e^i_+ and e^j_- then yields

$$[E^i_+, E^j_-] = \delta^{ij} \tfrac{1}{2\pi i} \oint_0 \mathrm{d}z \, \tfrac{1}{2\pi i} \oint_0 \mathrm{d}w \, (z-w)^{-1} Q^i_+(z) \, Q^j_-(w) \, h(w). \quad (4.8.14)$$

One may now deform the integration contours as usual and perform one of the integrations to obtain

$$\begin{aligned} [E^i_+, E^j_-] &= \delta^{ij} \tfrac{1}{2\pi i} \oint_0 \mathrm{d}z \, : Q^i_+(z) \, Q^j_-(z) \, h(z) : \\ &= \delta^{ij} \, (2\gamma \ln q)^{-1} \tfrac{1}{2\pi i} \oint_0 \mathrm{d}z \, \tfrac{\mathrm{d}}{\mathrm{d}z} \Big(: Q^i_+(z) \, Q^j_-(z) : \Big) . \end{aligned} \qquad (4.8.15)$$

Despite appearences, this closed integral over a total derivative does not vanish, owing to the particular contours involved in the definition of $Q^i_\pm$. To make this explicit, let us take the z-integration contour more carefully as an almost closed contour between points z_+ and z_- that lie infinitesimally below and above z_0, respectively. We then also have to write $z_\mp$ for z_0 in $Q^i_\pm$, respectively. Thus the result of the integration reads

$$[E^i_+, E^j_-] \propto \left[q^{[-\gamma \int_z^{z_-} \mathrm{d}w \, h^i(w)]} \, q^{[\gamma \int_{z_+}^{z} \mathrm{d}w \, h^i(z)]} \right]_{z_+}^{z_-}, \qquad (4.8.16)$$

or, what is the same,

$$[E^i_+, E^j_-] \propto q^{[\gamma \oint \mathrm{d}w \, h^i(w)]} - q^{[-\gamma \oint \mathrm{d}w \, h^i(w)]}. \qquad (4.8.17)$$

Recalling the definition of H^i in terms of $h^i(z)$, we see that for $\gamma = 1/4\pi i$ this is the desired result (4.2.1c), up to an overall constant which is finite for $q \to 1$ and which can be absorbed by a suitable change in the normalization of $E^i_\pm$. In a similar manner one can show that the q-deformed Serre relations (4.2.1d) are fulfilled. Thus for $k^\vee = 0$, we have indeed managed to express the generators of $\mathsf{U}_q(\mathfrak{g})$ through those of $\mathfrak{g}$.

Of course, this result is of no help for understanding the relation between the quantum group and the affine algebra that underlie a WZW theory, since in the WZW case the level has to be a positive integer (so that, contrary to the case $k^\vee = 0$, non-trivial irreducible highest weight modules exist). Moreover, in the derivation above the value of q was completely arbitrary, whereas for the WZW case it has to be related to the level via (see (4.3.25), and section 5.5)

$$q = \exp\left(\frac{2\pi i}{k^\vee + g^\vee}\right). \tag{4.8.18}$$

It is not unreasonable to expect that such a quantization will arise automatically when a nonzero value of k is introduced, because only for a definite relation between k and q will the cancellations that are necessary to obtain sensible results be possible. Unfortunately, further inspection shows that the ansatz proposed above cannot correctly reproduce the desired results for $k^\vee \neq 0$. Nevertheless we think that the positive result found for $k^\vee = 0$ indicates that this ansatz may be quite close to the correct answer.

To explain the problems arising for $k^\vee \neq 0$, it is convenient to consider the special case of $k^\vee = 1$ with $\mathfrak{g}$ simply laced. In this case the vertex operator construction can be employed which expresses the affine currents through r free bosons $\phi^i(z)$. More precisely, the fields in the Cartan subalgebra are just the derivatives of the free bosons, while the fields associated to the roots are given up to sign factors by vertex operators, i.e. by exponentials of the free bosons:

$$\begin{aligned} h^j(z) &= i\,\partial\phi^j(z), \\ e^\alpha(z) &= c_\alpha \; : \mathrm{e}^{i(\alpha,\phi(z))} : . \end{aligned} \tag{4.8.19}$$

Here $c_\alpha \in \{\pm 1\}$ are the so-called cocycle factors which have to be chosen to ensure the correct signs in the operator products of the currents (also, the normalization of the inner product has been fixed such that $(\theta, \theta) = 2$). The free bosons $\phi(z)$ are not proper conformal fields, as is demonstrated by their propagator

$$\underbrace{\phi^i(z)\,\phi^j(w)} = -A^{ij}\,\ln(z-w). \tag{4.8.20}$$

Nevertheless the currents defined above behave as ordinary conformal fields, as follows by computing the contractions among them with the help of the Wick theorem (compare the formula (3.1.38) for the contraction of exponentials). Moreover, the same calculation also shows that the fields h^i and e^α indeed satisfy the current algebra relations with level $k^\vee = 1$, provided the cocycle factors are chosen properly. (In fact, it is a rather lengthy exercise to show that a consistent choice of cocycle factors is possible.)

In this free field realization, our ansatz for the quantum group generators becomes

$$\begin{aligned} H^j &= \tfrac{1}{2\pi}\,[\phi^j(z_+) - \phi^j(z_-)], \\ E^j_\pm &= \tfrac{1}{2\pi i}\oint_0 \mathrm{d}z\ :\exp\left(i\,[(\pm 1 + \kappa)\phi^j(z) - \kappa\phi^j(z_\mp)]\right): \end{aligned} \tag{4.8.21}$$

with $\kappa = \ln q/4\pi i$. It is readily checked that this reproduces the correct commutation relations $[H^i, H^j]$ and $[H^i, E^j_\pm]$. For the commutator of E^i_+ with E^j_-, one obtains

$$\begin{aligned} &\Big(\tfrac{1}{2\pi i}\oint_{0\atop |z|>|w|} \mathrm{d}z - \tfrac{1}{2\pi i}\oint_{0\atop |z|<|w|} \mathrm{d}z\Big)\,\tfrac{1}{2\pi i}\oint_0 \mathrm{d}w\ [f(z,w)]^{A^{ij}} \\ &\quad :\exp\left(i\,[(1+\kappa)\phi^i(z) - \kappa\phi^i(z_-) + (-1+\kappa)\phi^j(w) - \kappa\phi^j(w_+)]\right): \end{aligned} \tag{4.8.22}$$

with

$$f(z,w) = (z-w)^{-1}\left(\frac{z_- - w}{z - w_+}\right)^{\kappa}\left(\frac{(z-w)(z_- - w_+)}{(z-w_+)(z_- - w)}\right)^{\kappa^2}. \tag{4.8.23}$$

For the "physical" value of κ, i.e. $\kappa = \ln q/4\pi i = 1/2(k^\vee + g^\vee)$ with $k^\vee = 1$, this is certainly not what we need because the branch cuts in $f(z,w)$ prevent us from performing the usual contour deformation. Nevertheless this result is not without interest: if we were to replace $f(z,w)$ in the integrand by $(z-w)^{-1}$, then the contour deformation would become possible and for arbitrary κ the result would be (using $A^{ii} = 2$, $A^{ij} \le 0$ for $i \ne j$)

$$\begin{aligned} &-\tfrac{1}{2\pi i}\oint_0 \mathrm{d}z\,\tfrac{1}{2\pi i}\oint_z \mathrm{d}w\,(z-w)^{A^{ij}} \\ &\qquad \cdot\ :\exp\left(i\,[(\kappa+1)\phi^i(z) - \kappa\phi^i(z_-) + (\kappa-1)\phi^j(w) - \kappa\phi^j(w_+)]\right): \\ &\quad = i\,(1-\kappa)\,\delta^{ij}\,\tfrac{1}{2\pi i}\oint_0 \mathrm{d}z\,\partial\phi^j(z)\ :\exp\left(i\kappa[2\phi^j(z) - \phi^j(z_+) - \phi^j(z_-)]\right): \\ &\quad = \tfrac{1}{2}\,(\kappa^{-1} - 1)\,\tfrac{1}{2\pi i}\oint_0 \mathrm{d}z\,\tfrac{\mathrm{d}}{\mathrm{d}z}\ :\exp\left(i\kappa[2\phi^j(z) - \phi^j(z_+) - \phi^j(z_-)]\right):\,, \end{aligned} \tag{4.8.24}$$

which (up to an irrelevant multiplicative constant) gives the desired result.

Of course we are not allowed to replace $f(z,w)$ by $(z-w)^{-1}$, and we also do not want to obtain the correct result for unrestricted q. However, these considerations are certainly suggestive. It seems that the only thing that is required is to change the ansatz in a way such that the contour deformation is allowed precisely for the physical values of q. Unfortunately, as already pointed out at the beginning of this section, this puzzle has not yet been resolved. It is of course also conceivable that for $k^\vee \neq 0$ a direct expression of the $\mathsf{U}_q(\bar{\mathfrak{g}})$-generators through the generators of $\mathfrak{g}$ does not exist (for example, one may be forced to change the prefactors in the exponentials in (4.8.21), which formally would correspond to considering fractional powers of $e_\pm(z)$ rather than $e_\pm(z)$ itself). This would in turn suggest that the objects on which the affine Lie algebra $\mathfrak{g}$ (or more generally, the chiral symmetry algebra of a conformal field theory) and the Hopf algebra $\mathsf{U}_q(\bar{\mathfrak{g}})$ (respectively, the quantum group underlying the conformal field theory) are acting must be quite different. Indeed, except for the highest weight vectors, the elements of the quantum group modules that will be encountered in section 5.8 do not correspond to conformal fields, in contrast to the elements of the irreducible highest weight modules of $\mathfrak{g}$. Whether this indicates that a relation of the type discussed above does not exist, or rather that so far the role of the quantum group in conformal field theories is not well enough understood, is a question that still has to be answered.

4.9 Literature

There are various textbooks on Hopf algebras, e.g. [B32, B2], but much of the material used in chapter 4 is also contained in the conference talk by Drinfeld [173], and in [15]. Recent reviews are [R1] and [R21]; the latter paper contains an extensive list of references to the original literature. For a short introduction, see [163]. The concepts of quasitriangular Hopf algebras and quantum groups are due to Drinfeld [172, 173], but many aspects of them are already implicit in the work of Sklyanin [522, 523, 524] and of Faddeev and Takhtajan [186] which was motivated by the quantum inverse scattering method [525]. The quantum deformations $\mathsf{U}_q(\mathfrak{g})$ were introduced for $\mathfrak{g} = sl_2$ in [361], and for arbitrary $\mathfrak{g}$ by Jimbo [297, 298]; they are discussed in much detail in [489, 490, 342]. Various related deformations, and identification maps between them, are described in [141, 187, 475, 140, 594]. For the description of q-deformation as a quantization procedure, see [173, 206].

For the connection between quantum groups and the Yang–Baxter equation, see e.g. [523, 524, 172, 171, 489, R11, B3, R21, 585, 380]. The space of functions on a quantum group was introduced independently by

Woronowicz [587]. Noncommutative and non-cocommutative Hopf algebras were also found in a different context in [404, 586].

The general representation theory of $\mathsf{U}_q(\mathfrak{g})$ for q not a root of unity was developed in [298, 586, 342, 119, 120, 496, 394, 551, 414, 396, 397, 527, 539, 448, 547, 420, 415, 18]. Representations in terms of q-deformed harmonic oscillators were discussed in [400, 60, 360, 286, 116, 115, 113, 188] for $\mathfrak{g} = A_1$, and for other $\mathfrak{g}$ in [536, 285, 329, 423]. A realization of $\mathsf{U}_q(A_1)$ in terms of finite difference operators was analyzed in [551, 414, 243, 500]. For $\mathfrak{g} = A_1$, the case where q is a root of unity was treated in [395, 493, 491, 461, 569, 333]. For $\mathfrak{g}$ different from A_1, this special case has so far only been dealt with to some extent in [160, 461, 537] for A_2, and in [15, 148] for arbitrary simple $\mathfrak{g}$. The related results for representations of Hecke algebras at roots of unity can be found in [570, 334, 271]; for the Temperley-Lieb–Jones algebra, see [540, 303, 305, 383]. Finally, so-called "cyclic" non-highest weight modules at q a root of unity are discussed in [524, 493, 145, 22]. A review of the current status of quantum group representation theory can be found in [499, 161].

The q-analogues of Casimir operators and their role in generating the center of $\mathsf{U}_q(\mathfrak{g})$ have been investigated in [297, 298, 424, 539, 174, 494]. A q-analogue of the Weyl group is discussed in [528, 381, 343], and tensor operators for $\mathsf{U}_q(sl_2)$ are described in [62]. The center of $\mathsf{U}_q(\mathfrak{g})$ for q a root of unity was first analyzed in [461], and in much more detail in [148, 22].

The truncated Kronecker product was discussed in [213, 15, 238, 241, 239]. The properties of centralizer algebras underlying the special features which appear for q a root of unity were pointed out in [15], following the mathematical papers [305, 570, 67, 571, 438, 440, 572]. The extended Racah–Speiser algorithm was introduced in the context of WZW theories in [314, B20, 567, 532]. It was applied to $\mathsf{U}_q(\mathfrak{g})$ in [241, 239] (for further references, see section 5.9); the examples listed in section 4.5 were worked out in [239].

Quantum dimensions were discussed in [15, 461, 238]. The Weyl formula for quantum dimensions was first noted in [461], and the hook formula in [238] (following [489]). The symmetry properties (4.4.28) were found in [461], and the properties (4.4.35) in [238]. The list (4.4.18) of quantum dimensions can be obtained with a simple computer program implementing the Weyl formula. Quantum dimensions also play a role, as so-called statistical dimensions, in the algebraic approach to quantum field theory [209, 390, 402, 486], and [147, R1, 389, 390, 234] as square roots of von Neumann indices [441, 130, 303, 356] in the theory of von Neumann algebras.

The property (4.4.42) of quantum dimensions was observed in [238]. This level-rank duality is a more general feature of the relevant quantum groups; it manifests itself e.g. in the Littlewood–Richardson coefficients of Hecke algebras [271] and consequently in the structure of the truncated Kronecker products [506]. For the associated WZW theories, this results in analogous symmetries of the fusion rules [366, 139], the modular matrix S [559], and the braid matrices [444]. The consequences for link polynomials are discussed in [425].

The quantum double construction is due to Drinfeld [173]. The example of $\mathfrak{g} = A_r$ mentioned in section 4.6 was worked out in [95, 497], and the case of arbitrary simple $\mathfrak{g}$ is analyzed in [343, 380]. For A_1, the universal R-matrix can also be found by other methods, see e.g. [361, 297, 362]. For finite-dimensional R-matrices, the main reference is [489, 490], from which also the examples shown in section 4.6 are taken; some of those results habe been obtained earlier in [43, 299].

The Drinfeld–Jimbo solution of the classical Yang–Baxter equation is described in [300]. A classification of the classical R-matrices has been obtained by Belavin and Drinfeld [44, 46, 45]; further results on solutions of the classical Yang–Baxter equation can be found in [520, 29, 26]. The quantum Yang–Baxter equation has its origin in the work of Yang [589, 590], Baxter [40, 41] and others [244] on integrable systems in statistical mechanics. It is also encountered in the context of the quantum inverse scattering method [522, 186, 525], as a factorization condition for many-body scattering matrices [328, 599] and in several other areas of physics and mathematics; for an overview, see [357]. The quantum version of the Drinfeld–Jimbo solution has been found in [298]; for the connection between classical and quantum R-matrices see also [362]. Other solutions of the quantum Yang–Baxter equation were obtained in [51, 598, 114, 363, 364, 43, 296, 362, 299, 302, 301].

Quantum matrix groups and the duality between $\mathsf{U}_q(\mathfrak{g})$ and G_q are discussed in [495, 184, 185, 538, 405, 543, R21, 498, 96]. The connection between non-commutative geometry [131, 469, R7, R8] and quantum groups is described in [586, 587, 409, 410, 132, 588, 470, 513, 574]. Quantum matrix groups are also analyzed in their own right in [175, 562, 207, 563, 104].

The discussion of the connection between quantum groups and affine algebras in section 4.8 is taken from unpublished work by the author and P. van Driel, which is based on an earlier paper by Babelon [27]. Formulæ similar to those conjectured in this discussion have been obtained in [290, 291, 292, 505].

5
Duality, fusion rules, and modular invariance

The main purpose of this final chapter is to describe how a quantum group structure can be identified in rational conformal field theories, with particular emphasis on WZW theories. This is done in sections 5.5, 5.7 and 5.8. In section 5.5 it is shown that the truncated Kronecker products of the quantum group $\mathsf{U}_q(\bar{\mathfrak{g}})$ are isomorphic to the fusion rules of the WZW theory based on the affine algebra $\mathfrak{g}$; in section 5.7 a quantum group structure is identified for any rational conformal field theory as the centralizer of the braid group in the space of chiral blocks; and in section 5.8 vector spaces generated by so-called screened vertex operators are identified as modules of a quantum group, with the quantum group operators acting essentially as creation and annihilation of the screenings.

The description of these matters requires more information about conformal field theory than what is contained in chapter 3; rather, one has also to understand further concepts of conformal field theory, notably the fusion rules, modular invariance, and the duality properties of chiral block functions. These issues are discussed in the remaining sections of this chapter. In section 5.1, the concept of fusion rule algebras is introduced, a few simple examples are presented, and the special primary fields known as simple currents are analysed (simple currents may also be characterized as the fields with unit quantum dimension, as is described at the end of section 5.5). Section 5.2 deals with the issue of modular invariance; it describes how this concept arises in string theory, and a few elementary properties of the modular transformation matrices are presented. There exists a surprising connection between fusion rule algebras and modular transformations: the modular transformation matrix S diagonalizes the fusion rules, and as a consequence the fusion rules can be expressed entirely in terms of S; these matters are described in section 5.3 (the proof of one of the underlying formulae is completed in section 5.7). Section 5.4 summarizes what is currently known about the modular invariant partition functions of WZW theories; in this context a prominent role is played by the simple currents, and by automorphisms of fusion rule algebras. In section 5.6, the duality properties of chiral blocks are introduced, and their relevance to the classification of rational conformal field theories is discussed. The duality properties are investigated further in section 5.7 where the fusing and braiding matrices are defined and the polynomial equations satisfied by them are displayed. In section 5.8 the screened vertex operators are introduced; these can only be defined in the context of the so-called Coulomb gas picture of conformal field theories, and hence this section also provides some basic knowledge about the free field realizations of the Virasoro and of affine Lie algebras that underly the Coulomb gas construction.

For quantities pertaining to simple Lie algebras we return to the conventions of chapter 2, i.e. use overbars to indicate quantities referring to a simple Lie algebra $\bar{\mathfrak{g}}$, whereas unbarred quantities refer to the associated untwisted affine algebra $\mathfrak{g} = \bar{\mathfrak{g}}^{(1)}$.

5.1 Fusion rule algebras

A substantial amount of information about a conformal field theory is provided by the so-called *fusion rules* of the theory. Recall from section 3.3 that in the operator product of any two primary fields, one finds a certain number of primary fields together with their descendants. We have already learned in section 3.4 that the operator product coefficients involving the secondary fields are completely determined, via the symmetry algebra $\mathcal{W}$, through those involving only primaries, so that it is often unnecessary to know the explicit values of the operator product coefficients involving secondaries. Moreover, for many purposes it is not even relevant to know the precise values of the operator product coefficients for primary fields; rather, one needs only the information which primary fields appear on the right hand side of an operator product. Precisely this information is the content of the fusion rules of the conformal field theory. Thus the fusion rules read generically

$$\phi_i * \phi_j = \sum_k \mathcal{N}_{ij}{}^k \, \phi_k, \tag{5.1.1}$$

where $\{\phi_i\}$ is the set of primary fields of the conformal field theory, and $\mathcal{N}_{ij}{}^k$ are non-negative integers. The situation is thus analogous to the one encountered when calculating Kronecker products for simple Lie algebras. The operator product coefficients are the analogue of the Clebsch–Gordan coefficients (1.6.61), while the fusion rule coefficients $\mathcal{N}_{ij}{}^k$ are the analogue of the Littlewood–Richardson coefficients $\mathcal{L}_{ij}{}^k$ which contain the information on the decomposition of the Kronecker product into (irreducible) submodules according to (see (1.6.58))

$$\bar{\Lambda}_i \times \bar{\Lambda}_j = \sum_k \mathcal{L}_{ij}{}^k \bar{\Lambda}_k . \tag{5.1.2}$$

From the way the operator product algebra was written down in (3.3.9), one might get the impression that the fusion rule coefficients can only be zero or one, thus describing whether or not the primary field ϕ_k appears in the operator product of ϕ_i and ϕ_j. However, it turns out that integers larger than one can also appear, corresponding to the fact that two primary fields can couple in several distinct ways to a third primary field, again in full analogy to the Kronecker products of simple Lie algebras. In terms of the operator product algebra (3.3.9), this means that the fields φ_k on the right hand side need not all be distinct, i.e. it may happen that $\varphi_{k_1} = \varphi_{k_2}$ for $k_1 \neq k_2$. Note that one is not allowed just to add up the corresponding operator product coefficients owing to the fact that the relative values of operator product coefficients involving different descendants of a given primary field are fixed by the symmetry algebra (cf. (3.4.3)). Thus the presence of a fusion rule coefficient $\mathcal{N}_{ij}{}^k$ larger than one can (in principle) be inferred from the operator product algebra as follows: subtract the leading contribution to the operator product $\phi_i \phi_j$ which corresponds to the descendant φ of ϕ_k with the lowest allowed grade, and also subtract all contributions involving descendants at higher grades which must be present according to (3.4.3); then $\mathcal{N}_{ij}{}^k > 1$ iff after this subtraction there is still some contribution to $\phi_i \phi_j$ left over which involves a descendant of ϕ_k. From this analysis it follows in particular that for $\mathcal{N}_{ij}{}^k > 1$ the minimal grades of the fields that actually can contribute to one of the $\mathcal{N}_{ij}{}^k$ different couplings of the families headed by the primaries ϕ_i, ϕ_j, ϕ_k must be all distinct.

It is clear from the definition of $\mathcal{N}_{ij}{}^k$ that $\mathcal{N}_{ij}{}^k = 0$ implies that also $C_{ij}{}^k = 0$. It can be shown that the converse is true as well. In words, an operator product coefficient vanishes if and only if the corresponding fusion rule coefficient vanishes; this property of conformal field theories is called *naturality*.

The fusion rule coefficients inherit several simple properties from the operator product coefficients. For example, from the properties of the identity primary field $\mathbf{1} = \phi_0$ it follows immediately that

$$\mathcal{N}_{i0}{}^j = \mathcal{N}_{0i}{}^j = \delta_i^{\,j} \tag{5.1.3}$$

and

$$\mathcal{N}_{ij}{}^0 = \delta_{i\,j^+} \tag{5.1.4}$$

(recall that by definition ϕ_{j^+} is the field conjugate to ϕ_j, and also compare the analogous formulae for the operator product coefficients, (3.4.5) and (3.4.8)). In the following we will therefore usually suppress the fusion rules involving the identity primary field when we explicitly display the fusion rule algebras for specific theories. With the help of (5.1.4) one can also define fusion rule coefficients with lower indices only (compare (3.4.7)):

$$\mathcal{N}_{ijk} := \mathcal{N}_{ij}{}^{k^+} . \tag{5.1.5}$$

Furthermore, from the definition of the operator products as radially ordered products it follows immediately that the fusion rule coefficients are symmetric in their lower indices:

$$\mathcal{N}_{ij}{}^{k} = \mathcal{N}_{ji}{}^{k}; \tag{5.1.6}$$

or in other words that the fusion rules of a conformal field theory form a commutative algebra. Using this fact together with (5.1.3) and (5.1.4), it is easily seen that the $\mathcal{N}_{ijk}$ are totally symmetric in i, j, k:

$$\mathcal{N}_{ijk} = \mathcal{N}_{ikj} = \mathcal{N}_{kji}. \tag{5.1.7}$$

Moreover, from the associativity of the operator product algebra it follows that that the fusion rule algebra is associative as well,

$$\sum_k \mathcal{N}_{ij}{}^{k} \mathcal{N}_{kl}{}^{m} = \sum_k \mathcal{N}_{jl}{}^{k} \mathcal{N}_{ik}{}^{m}, \tag{5.1.8}$$

or in terms of the matrices $\mathcal{N}_i$ with entries $(\mathcal{N}_i)_j{}^k = \mathcal{N}_{ij}{}^k$,

$$[\mathcal{N}_i, \mathcal{N}_j] = 0. \tag{5.1.9}$$

Finally, the fusion rules are clearly invariant under the conjugation $\phi_i \mapsto \phi_{i^+}$:

$$\mathcal{N}_{i^+j^+}{}^{k^+} = \mathcal{N}_{ij}{}^{k}. \tag{5.1.10}$$

The determination of the fusion rules is a major step in the process of solving a conformal field theory. As will be explained in section 4.6, it is in principle even possible to compute the full operator algebra (up to some free parameters) starting only from the knowledge of the symmetry algebra and the fusion rules. To be precise, this is possible provided that the primary fields and their fusion rules are understood as referring to the *maximally extended* symmetry algebra of the theory. There are further reasons to consider always the fusion rules with respect to the maximal symmetry algebra. Clearly, by taking into account the maximal symmetry algebra, the spectrum of primary fields and relations among operator product coefficients will look less mysterious. The most important reason, however, consists in the relation between the fusion rules and modular transformations, which will be the subject of section 5.3; a more practical one is that the number of primary fields of the maximal symmetry algebra is always smaller than the number of primaries of any of its subalgebras. In this context, it should be remarked that in the case of the Virasoro algebra, the fusion rules are *simply reducible*, i.e. the fusion rule coefficients never exceed the value one,

$$\mathcal{N}_{ij}{}^{k} \in \{0,1\} \quad \text{for} \quad \mathcal{W} = \mathcal{V}, \tag{5.1.11}$$

so that it might seem preferable to consider only $\mathcal{V}$ even if there is an extended algebra present. However, the number of $\mathcal{V}$-primaries contained in a primary field of an extended algebra is always infinite, so that in the presence of an extended symmetry algebra insistence on using $\mathcal{V}$ would lead not only to simple reducibility of the fusion rules but also to an infinite increase in the number of primary fields. This would be a particularly bad deal in those theories for which there is only a finite number of primary fields of the extended algebra; conformal field theories with this property are called *rational conformal field theories.* Note that all WZW theories are rational conformal field theories, since the primary fields carry integrable highest weight modules of the underlying affine algebra, and these modules are finite in number.

There is no general procedure to determine the fusion rules of a given conformal field theory. However, there are classes of theories for which one does know how to compute the fusion rules, the most important class being the WZW theories. For a WZW theory, the fusion rule coefficients can be completely determined with the help of the depth rule described in section 3.5 (p. 194). The analysis of the depth rule is particularly simple for the case of $\mathfrak{g} = A_1$ because then the Clebsch–Gordan coefficients are known explicitly and the criterion (3.5.12*b*) simply reads

$$\bar{\Lambda}_1 + \bar{\Lambda}_2 + \bar{\lambda}_1 + \bar{\lambda}_2 \leq 2\,(k^\vee - \bar{\Lambda}). \tag{5.1.12}$$

The depth rule then leads immediately to the following fusion rules for the $A_1^{(1)}$ WZW theory:

$$\phi_\Lambda * \phi_{\Lambda'} = \sum_{j=0}^{\min\{\bar{\Lambda}',k^\vee-2\bar{\Lambda}\}} \phi_{\Lambda_j}; \tag{5.1.13}$$

here Λ_j stands for the affine weight with horizontal part

$$\bar{\Lambda}_j = \bar{\Lambda} - \bar{\Lambda}' + 2j, \tag{5.1.14}$$

and without loss of generality we have assumed that $\bar{\Lambda} \geq \bar{\Lambda}'$ to simplify notation. Thus the fusion rule coefficients are

$$\mathcal{N}_{\phi_\Lambda \phi_{\Lambda'}}{}^{\phi_{\Lambda''}} = \begin{cases} 1 & \text{for} \quad (\bar\Lambda + \bar\Lambda' + \bar\Lambda'') \in 2\mathbb{Z} \\ & \text{and} \quad |\bar\Lambda - \bar\Lambda'| \le \bar\Lambda'' \le k^\vee - |k^\vee - \bar\Lambda - \bar\Lambda'| \\ 0 & \text{otherwise.} \end{cases} \tag{5.1.15}$$

For $\bar{\mathfrak{g}} \neq A_1$, application of the depth rule is usually very tedious (even with the help of a computer), because one needs the weight space decomposition of both $R_{\bar\Lambda}$ and $R_{\bar\Lambda'}$, as well as a certain amount of information about the Clebsch–Gordan coefficients of the Kronecker product, which is particularly problematic if the Kronecker product is not simply reducible. Compare e.g. the discussion of the level two $E_8^{(1)}$ WZW theory in section 3.5 which may be summarized in the fusion rules

$$\begin{aligned} \psi * \psi &= \mathbf{1}, \\ \sigma * \psi &= \sigma, \\ \sigma * \sigma &= \mathbf{1} + \psi, \end{aligned} \tag{5.1.16}$$

where

$$\psi \equiv \phi_{3875}, \qquad \sigma \equiv \phi_{248}. \tag{5.1.17}$$

The fusion rules (5.1.16) are called the *Ising fusion rules* because they also arise for the two-dimensional critical Ising model.

An algorithm determining the WZW fusion rules which is much more easily applied than the depth rule will be discussed in section 5.5. For now we only remark on one general feature of the fusion rules which follows from the fact that the zero mode subalgebra of the WZW symmetry algebra $\mathfrak{g} \oplus \mathcal{V}$ is just the simple Lie algebra $\bar{\mathfrak{g}}$ (more precisely, its direct sum with the u_1-algebra generated by L_0). Namely, this implies that the WZW operator products, and thus also the fusion rule algebra, respect the $\bar{\mathfrak{g}}$-symmetry, i.e. the fusion rules have to be compatible with the Kronecker products (5.1.2) of $\bar{\mathfrak{g}}$. In terms of the fusion rule and Littlewood–Richardson coefficients, this means that

$$\mathcal{N}_{\phi_i \phi_j}{}^{\phi_k} \le \mathcal{L}_{\bar\Lambda_i \bar\Lambda_j}{}^{\bar\Lambda_k} \tag{5.1.18}$$

with $\phi_i \equiv \phi_{\Lambda_i}$, or shortly

$$\mathcal{N}_{ij}{}^k \le \mathcal{L}_{ij}{}^k . \tag{5.1.19}$$

For example, the fact the the Kronecker products of A_1 and of level two E_8 are simply reducible implies that the corresponding fusion rules must be simply reducible as well, in agreement with the results (5.1.15) and (5.1.16).

Among the set of primary fields of a conformal field theory, the *simple currents* are of particular interest. These are by definition those primary fields Φ which have unique fusion rules with all primaries of the theory, i.e.

$$\Phi * \phi_i = \phi_{i'}, \tag{5.1.20}$$

with a single primary field $\phi_{i'}$ appearing on the right hand side for any choice of the primary field ϕ_i. A primary field $\Phi = \phi_j$ is a simple current iff the fusion rule of Φ with its conjugate field Φ^+ contains only the identity, i.e. iff

$$\mathcal{N}_{j\,j^+}^{\;\;i} = \delta^{i0}. \tag{5.1.21}$$

This requirement is certainly a necessary condition; that it is also a sufficient one follows e.g. by using the associativity of the triple product $\Phi^+ * \Phi * \phi_i$ (for some details, see section 5.5, p. 365). The spectrum of any conformal field theory contains at least one simple current, namely the identity field **1**.

The fusion rule of a simple current Φ with itself gives either the identity primary field **1** (iff Φ is selfconjugate) or else a unique other primary field which is conveniently denoted by $\Phi^{(2)}$. This field is different from Φ (otherwise associativity of $\Phi^+ * \Phi * \Phi$ would imply $\Phi = \mathbf{1}$ and hence also $\Phi^{(2)} = \mathbf{1}$), and it is also a simple current (as a consequence of the associativity of $\Phi * \Phi * \phi_i$). Repeating the process, one can define $\Phi^{(3)} := \Phi * \Phi^{(2)}$ which again is either the identity or else a new simple current, and so on,

$$\Phi * \Phi^{(n)} =: \Phi^{(n+1)}, \tag{5.1.22}$$

thus defining an *orbit* of simple currents. For a rational conformal field theory this orbit must be finite because there is only a finite number of primaries available as $\Phi^{(n)}$; hence in this case for any simple current Φ there exists in particular an integer N which is the smallest positive integer such that

$$\Phi^{(N)} = \mathbf{1}; \tag{5.1.23}$$

this number is called the *order* of the simple current Φ. It then follows immediately that the fusion rules of the simple currents among themselves read

$$\Phi^{(n)} * \Phi^{(m)} = \Phi^{(n+m \bmod N)} \tag{5.1.24}$$

with

$$\Phi^{(0)} \equiv \mathbf{1}, \quad \Phi^{(1)} \equiv \Phi, \tag{5.1.25}$$

and hence are isomorphic to the group $\mathbb{Z}_N$, with the fusion rule corresponding to the group multiplication, and conjugation (i.e. $\Phi \mapsto \Phi^+$) corresponding to the inverse. More generally, the set of simple currents of any conformal field theory possesses fusion rules which are isomorphic to the abelian group $\mathbb{Z}_{N_1} \times \mathbb{Z}_{N_2} \ldots \times \mathbb{Z}_{N_p}$ with p the number of distinct orbits of simple currents (excluding, of course, the trivial orbit formed by **1** alone).

The conformal dimensions of simple currents are strongly restricted. To see this, consider the operator product of a simple current Φ with itself; this must read

$$\underbrace{\Phi(z)\,\Phi(w)} \simeq (z-w)^{-\gamma}\,\Phi^{(2)}(w). \tag{5.1.26}$$

This implies that when $\Phi(z)$ is transported along a closed path surrounding w, it picks up a phase factor $\mathrm{e}^{-2\pi i\gamma}$ if there is a field Φ at the point w. Similarly, when the closed path surrounds several points $w_1, w_2, \ldots, w_m$ at each of which there is placed a field Φ, the resulting phase factor is $\mathrm{e}^{-2\pi i m\gamma}$. As a consequence, the operator product of Φ with $\Phi^{(m)}$ must read

$$\underbrace{\Phi(z)\,\Phi^{(m)}(w)} \simeq (z-w)^{M-m\gamma}\,\Phi^{(m+1)}(w) \tag{5.1.27}$$

with some integer M. On the other hand, the operator product of Φ with $\Phi^{(N)} = \mathbf{1}$ just gives back Φ; thus comparison with (5.1.27) yields $\gamma = \frac{1}{N} \bmod \mathbb{Z}$. Now recalling that the exponent γ in (5.1.26) is a combination of conformal dimensions (see e.g. (3.4.1)), one concludes that $\Delta_m \equiv \Delta(\Phi^{(m)})$ obeys

$$\Delta_m + \Delta_1 - \Delta_{m+1} = \frac{m}{N} \bmod \mathbb{Z}. \tag{5.1.28}$$

Also, the field conjugate to $\Phi^{(m)}$ is $(\Phi^{(m)})^+ = \Phi^{(N-m)}$; this implies that $\Delta_m = \Delta_{N-m}$. When we combine the latter two properties of Δ_m, it follows that

$$\Delta_m = \zeta\,\frac{m(N-m)}{2N} \bmod \mathbb{Z}, \tag{5.1.29}$$

with some integer ζ which is defined modulo N if the order N of Φ is odd, respectively modulo $2N$ if N is even.

Simple currents can also be characterized through the fact that any four-point function containing at least one simple current must be power-like. Thus from the results of section 3.5 it follows that for WZW theories a large class of simple currents is provided by the cominimal fields. In fact, as will be shown in section 5.5 (p. 366) there exists only one exceptional simple current which does not belong to this class; this is the field ϕ_{3875} of the level two $E_8^{(1)}$ WZW theory which possesses the unique fusion rules (5.1.16). The conformal dimension of this field is $\Delta_{3875} = \frac{3}{2}$; thus (5.1.29) is satisfied with $N = \zeta = 2$. Let us also list (table (5.1.30)) the values of N and ζ of the level one cominimal fields $\Phi = \phi_\Lambda$ which can be read off the conformal dimensions in table (3.5.53). For the cominimal fields at higher levels, N remains unchanged while, according to (3.5.43), ζ is multiplied by the level $k^\vee$.

$\bar{\mathfrak{g}}$	cominimal $\bar\Lambda$	Δ	N	ζ
A_r	$\bar\Lambda_{(i)},\ i=1,\dots,r$	$\frac{i(r+1-i)}{2(r+1)}$	$r+1$	1
B_r	$\bar\Lambda_{(1)}$	$\frac{1}{2}$	2	2
C_r	$\bar\Lambda_{(r)}$	$\frac{r}{4}$	2	$r \bmod 4$
D_r	$\bar\Lambda_{(1)}$	$\frac{1}{2}$	2	2
	$\bar\Lambda_{(r-1)},\ \bar\Lambda_{(r)}$	$\frac{r}{8}$	$\begin{cases} r\in 2\mathbb{Z}: & 2 \\ r\in 2\mathbb{Z}+1: & 4 \end{cases}$	$\begin{cases} r \bmod 4 \\ 3r \bmod 8 \end{cases}$
E_6	$\bar\Lambda_{(1)},\ \bar\Lambda_{(5)}$	$\frac{2}{3}$	3	2
E_7	$\bar\Lambda_{(6)}$	$\frac{3}{4}$	2	3

(5.1.30)

In section 3.8 we have seen that free fermions with an energy-momentum tensor of the Sugawara form are in one to one correspondence with conformal embeddings of untwisted affine algebras in ${so_n}^{(1)}$ or ${u_n}^{(1)}$; as a consequence, in all these cases the conformal field theory possesses an enlarged symmetry algebra. Similarly, it turns out that often the existence of a simple current leads to the presence of an extended algebra. To see this consider, for a fixed value of the index i, the primary fields $\phi_{i(n)}$ defined by the fusion rules of $\phi_i \equiv \phi_{i(0)}$ with an orbit of order N of simple currents,

$$\Phi^{(n)} * \phi_i = \phi_{i(n)} \tag{5.1.31}$$

for $n = 1, \dots, N$. These fields need not all be distinct, but certainly the number of distinct fields obtained this way must be a divisor of N. In other words, the simple currents combine all primary fields of the theory into orbits whose length is N or a divisor of N. Now if $\Delta(\Phi) \in \mathbb{Z}$, then it is consistent with the locality constraints of the conformal field theory that the conformal dimension with respect to the antiholomorphic Virasoro algebra vanishes, $\bar{\Delta}(\Phi) = 0$; the corresponding simple currents can then be included in the symmetry algebra, and the orbits of primary fields correspond to the modules of an extended symmetry algebra which possesses the modes of the simple current Φ as additional generators. Of course, often the extended algebra will be a more complicated extension of the original symmetry algebra than in the case of conformal embeddings.

Repeating the discussion after (5.1.26) for the case of the operator product $\phi_i * \Phi^{(n)}$, one obtains a formula relating the conformal dimensions of primary fields that belong to the same orbit:

$$\Delta(\phi_{i^{(n)}}) = \Delta(\phi_i) + \zeta\,\frac{n(N-n)}{2N} - \eta\,\frac{n}{N} \bmod \mathbb{Z} \tag{5.1.32}$$

for some integer η.

5.2 Modular invariance

So far, we have considered conformal field theories, and WZW theories in particular, always as defined on a two-dimensional world sheet that is just the compactified complex plane, and hence has the topology of the two-sphere S^2. Restrictions on the spectrum of a WZW theory are then obtained by requiring that there exists a closed associative operator product algebra. As already mentioned at the end of section 3.3, a spectrum satisfying this condition for any choice of the affine algebra g is the *diagonal spectrum* (3.3.30),

$$\{\phi_{\Lambda,\Lambda} \mid \Lambda \in P_k^+(\mathsf{g})\}. \tag{5.2.1}$$

Here P_k^+ is the set of dominant integrable highest weights of g at level $k^\vee$. Also recall that $\phi_{\Lambda,\tilde{\Lambda}}$ denotes the primary field carrying the highest weight module R_Λ with respect to the holomorphic symmetry algebra g, and the highest weight module $R_{\tilde{\Lambda}}$ with respect to the antiholomorphic symmetry algebra which is given by g as well. Thus, shortly, the diagonal spectrum contains one copy of each of the "left–right symmetric" primary WZW fields.

There exist, however, many more spectra compatible with the presence of an associative operator algebra. Left–right asymmetric primary fields can be present, some left–right symmetric fields may not appear at all, and some may even appear more than once. A complete list of these spectra is known only for $\mathsf{g} = A_1^{(1)}$ where it is given by tables (5.2.2) and (5.2.3).

name	level	spectrum
A_r	$r-1$	$\{(\bar\Lambda,\bar\Lambda)\mid\bar\Lambda\in\mathbb{Z}\}$
D_{2r}	$4r-4$	$\{(\bar\Lambda,\bar\Lambda)\mid\bar\Lambda\in 2\mathbb{Z}\}\cup\{(\bar\Lambda,4r-4-\bar\Lambda)\mid\bar\Lambda\in 2\mathbb{Z}\}$
D_{2r+1}	$4r-2$	$\{(\bar\Lambda,\bar\Lambda)\mid\bar\Lambda\in 2\mathbb{Z}\}\cup\{(\bar\Lambda,4r-2-\bar\Lambda)\mid\bar\Lambda\in 2\mathbb{Z}+1\}$
E_6	10	$\{\,(0\oplus 6,0\oplus 6),\ (3\oplus 7,3\oplus 7),\ (4\oplus 10,4\oplus 10)\,\}$
E_7	16	$\{\,(0\oplus 16,0\oplus 16),\ (4\oplus 12,4\oplus 12),\ (6\oplus 10,6\oplus 10),$ $(8,8),\ (2\oplus 14,8),\ (8,2\oplus 14)\,\}$
E_8	28	$\{\,(0\oplus 10\oplus 18\oplus 28,0\oplus 10\oplus 18\oplus 28),$ $(6\oplus 12\oplus 16\oplus 22,6\oplus 12\oplus 16\oplus 22)\,\}$

(5.2.2)

name	level	spectrum
$Z_\ell^{(0)}$	ℓ	$\{\,(0,0)\,\}$
$Z_\ell^{(0,+)}$	ℓ	$\{\,(0\oplus\ell,0)\,\}$
$Z_\ell^{(0,-)}$	ℓ	$\{\,(0,0\oplus\ell)\,\}$
$Z_\ell^{(1)}$	ℓ	$\{\,(0,0),\ (\ell,\ell)\,\}$
$Z_\ell^{(1,\pm)}$	ℓ	$\{\,(0\oplus\ell,0\oplus\ell)\,\}$
$Z_\ell^{(2)}$	ℓ	$\{\,(\bar\Lambda,\bar\Lambda)\mid\bar\Lambda\in 2\mathbb{Z}\}$

(5.2.3)

In the tables (5.2.2) and (5.2.3) $(\Lambda,\tilde\Lambda)$ stands for $\phi_{\Lambda,\tilde\Lambda}$, and for notational simplicity only the horizontal part of the affine weights is displayed. Also, it is always implicit that the weights are integrable, i.e. $0\le\bar\Lambda\le k^\vee$, and the terms involving the symbol "$\oplus$" are meant in the sense that

$$\{\,(\bar\Lambda,\bar\Lambda'\oplus\bar\Lambda'')\,\}\equiv\{\,(\bar\Lambda,\bar\Lambda')\,,\,(\bar\Lambda,\bar\Lambda'')\,\} \tag{5.2.4}$$

etc. Note that in the case of D_{2r}, the combination $(\frac12 k^\vee,\frac12 k^\vee)=(r-2,r-2)$ appears with multiplicity two. Also, one has the obvious inclusions among the spectra $Z_\ell^{(0)}$, $Z_\ell^{(0,\pm)}$, $Z_\ell^{(1)}$ and $Z_\ell^{(1,\pm)}$, as well as $Z_\ell^{(2)}\subset D_{\ell/2+2}$ etc.

Not all these invariants are relevant to the application in the theory of closed strings. This is because in string theory, as a consequence of the string interactions, the world sheet can be an arbitrary Riemann surface (without boundary), not just the two-sphere. Thus a conformal field theory can describe the ground state of a closed string theory only if it can be formulated consistently on an arbitrary Riemann surface.

The topology of a Riemann surface is described by its *genus,* i.e. the number of "holes" that its embedding in $\mathbb{R}^3$ contains; the surface without holes is the two-sphere S^2, the surface with one hole is the torus $T^2 = S^1 \times S^1$, etc. For all surfaces with at least one hole, there exist coordinate transformations which cannot be connected continuously to the identity, the so-called *modular transformations.* (This corresponds to the fact that these surfaces can be endowed with different inequivalent complex structures.) The reparametrization invariance of the conformal field theory thus requires in particular that any correlation function has a definite transformation property with respect to modular transformations. In fact, the only known way to make sense of the theory is to have all correlation functions invariant under these transformations; this is summarized under the name *modular invariance.* In terms of the path integral formulation of the string theory, the modular transformations must be "divided out" of the path integral to make sure that one integrates only over inequivalent world sheets; if the correlation functions were not modular invariant, it would not be possible to define the integration region of the path integral such that each inequivalent world sheet is covered precisely once. Also, modular invariance should not be imposed on each genus separately, because there are certain limits of the parameters characterizing a Riemann surface (corresponding to the limit where the circumference of a handle shrinks to zero) in which the surface degenerates to a lower genus surface or to the disconnected sum of two lower genus surfaces. In such a limit, the correlation functions must factorize into the corresponding ones for the lower genus surface(s).

The modular invariance conditions are fortunately not all independent. For example, it is very probable that modular invariance at genus two is already sufficient to guarantee modular invariance at any higher genus. Also, the zero point "correlation function" (i.e. the vacuum to vacuum transition amplitude corresponding to a path integral without insertion of any field) plays a very special role; this quantity is called the *partition function* of the theory. Modular invariance of the partition function together with factorization is probably enough to have modular invariance for all correlation functions. Finally, it has been shown that a sufficient set of consistency conditions at genus one is crossing symmetry of the genus zero four-point functions (compare p. 192) together with modular invariance of the one-point functions at genus one.

In the following we will consider only the simplest non-trivial constraint imposed by modular invariance. This arises for the genus one partition function which corresponds to a euclidean path integral with the path being a closed loop swept out by a closed string, i.e. a torus. This may be formally written as

$$\int \mathrm{D}\gamma\, \mathrm{D}\varphi\, \exp(-S(\varphi)), \tag{5.2.5}$$

where φ stands generically for all two-dimensional fields and S is their euclidean action (for many conformal field theories, the formulation in terms of an action is not known; this is why the given expression for the partition function only holds in a formal sense). $\mathrm{D}\gamma$ denotes the path integral over all two-dimensional metrics, corresponding to the different possible shapes of the torus. Via the usual Faddeev–Popov gauge fixing procedure, one can reduce this integration to be only over those metrics which cannot be transformed into each other by coordinate transformations connected continuously to the identity. The latter metrics of the torus are parametrized by a parameter τ which takes values in the complex upper half plane. Namely, a torus may be represented by a parallelogram in the complex plane, with opposite sides to be identified; choosing $z = 0$ and $z = 1$ as two edges of the parallelogram, the parameter τ corresponds to choosing $z = \tau$ and $z = \tau + 1$ as the other two edges. After this gauge fixing, the φ-integration can be rewritten with the help of a formula that relates the path integral to a trace over the exponential of the evolution operator, i.e. the operator that generates translations in euclidean time. For a one-dimensional quantum mechanical system, this operator is just the Hamiltonian H. In the present case, there is a second evolution operator, the "momentum" P which describes translations along the closed string. It is necessary to take into account both of these operators because in terms of the torus swept out by the closed string there is no preferred direction of time. Thus for a closed string of length η which merges back to itself after a euclidean time lapse β, the path integral can be expressed as

$$\int \mathrm{D}\varphi\, \mathrm{e}^{-S(\varphi)} = \mathrm{Tr}\, \mathrm{e}^{-\beta H}\, \mathrm{e}^{i\eta P}. \tag{5.2.6}$$

It turns out that (in the so-called light cone gauge where the Fock space of states coincides with the Hilbert space of physical states, so that the trace operation "Tr" has the usual quantum mechanical sense of summing over all physical states) the Hamiltonian and momentum are related to the zero modes of the holomorphic and antiholomorphic Virasoro algebras of the theory by

$$H = L_0 + \bar{L}_0, \qquad P = L_0 - \bar{L}_0. \tag{5.2.7}$$

Therefore one has, writing $2\pi i\tau := -\beta + i\eta$,

$$Z(\tau) := \int_\tau \mathrm{D}\varphi \, \mathrm{e}^{-S(\varphi)} = \mathrm{Tr}\, \mathrm{e}^{2\pi i \tau L_0} \, \mathrm{e}^{-2\pi i \bar{\tau} \bar{L}_0}, \tag{5.2.8}$$

where $\bar{\tau}$ is the complex conjugate of τ. The variable τ defined this way is nothing else than the modular parameter of the torus encountered before. The full path integral then takes the form

$$\int \mathrm{D}\gamma \, \mathrm{D}\varphi \, \exp(-S(\varphi)) = \int \frac{\mathrm{d}^2\tau}{(\mathrm{Im}\,\tau)^2} \, Z(\tau). \tag{5.2.9}$$

By a slight abuse of nomenclature, the integrand $Z(\tau)$ in (5.2.9) is also called the partition function of the theory. The condition of modular invariance now arises because the Faddeev–Popov procedure only can implement "local" reparametrizations which are continuously connected to the identity. If the additional "global" reparametrizations were not taken into account, one would still get an infinite overcounting of the set of inequivalent tori. In terms of the τ integration region in (5.2.9), this means that the complex upper half plane falls into an infinite number of regions each of which covers all really inequivalent tori exactly once. Thus one should restrict the τ integration to a single region of this type. But this makes sense only if the result does not depend on the specific choice of this region, so that the partition function $Z(\tau)$ must be invariant under the modular transformations which map the different regions onto each other.

In the case of the torus, the group of modular transformations is just the modular group $PSL_2(\mathbb{Z})$ discussed in section 2.7, with the action on $Z(\tau)$ defined as

$$Z(\tau) \;\mapsto\; Z\left(\frac{a\tau + b}{c\tau + d}\right) \tag{5.2.10}$$

with $a,\, b,\, c,\, d \in \mathbb{Z}$, $ad - bc = 1$. Since this group is generated by the transformations $\mathcal{S}$ and $\mathcal{T}$ of (2.7.4), it is sufficient to have the integrand invariant under these two generators, i.e. under

$$\begin{aligned} \mathcal{S}:&\quad \tau \mapsto -\frac{1}{\tau}, \\ \mathcal{T}:&\quad \tau \mapsto \tau + 1. \end{aligned} \tag{5.2.11}$$

Note that the measure $\mathrm{d}^2\tau/(\mathrm{Im}\,\tau)^2$ in (5.2.9) is invariant under $\mathcal{S}$ and $\mathcal{T}$ so that $Z(\tau)$ must be modular invariant by itself.

So far, the discussion of modular invariance has been rather formal. To bring the discussion to a more practical level, one has to implement the presence of the symmetry algebra $\mathcal{W} \oplus \bar{\mathcal{W}}$ in the conformal field theory. Then the trace "Tr" defining $Z(\tau)$ naturally splits into a sum over the primary fields of the theory and, for each term in this sum, summations over the various states in the $\mathcal{W} \oplus \bar{\mathcal{W}}$-modules carried by the primary fields. Thus, using the fact that the conformal fields are eigenstates of L_0 and $\bar{L}_0$ with eigenvalues given by their conformal dimensions, and that the conformal dimensions of secondary fields are obtained from those of their primaries by adding their grade, one has

$$Z = \sum_{\phi_i} e^{2\pi i\tau\Delta_i} e^{-2\pi i\bar{\tau}\bar{\Delta}_i} \sum_{m_{(i)}} e^{2\pi i\tau m_{(i)}} \sum_{\bar{m}_{(i)}} e^{-2\pi i\bar{\tau}\bar{m}_{(i)}} \tag{5.2.12}$$

(from now on for brevity we drop the argument τ of $Z(\tau)$). In (5.2.12), the first sum is over all primary fields in the theory, and the second and third sums are over the secondary fields at grade m in the $\mathcal{W}$-module carried by ϕ_i (including multiplicities) and the corresponding sum for the $\bar{\mathcal{W}}$-module carried by ϕ_i, respectively. Thus the latter sums are nothing but (modified) characters $\tilde{\chi}_j$ of the chiral symmetry algebras, in a specialization corresponding to their Virasoro subalgebras,

$$\tilde{\chi}_j = e^{-\pi i c\tau/12}\, \chi_j = \mathrm{tr}_{\phi_j}\, e^{2\pi i\tau\,(L_0 - c/24)}. \tag{5.2.13}$$

Therefore one can finally write

$$Z = \sum_{s,t} \mathcal{Z}_{s,t}\, \chi_s(\tau)\, \bar{\chi}_t(\bar{\tau}), \tag{5.2.14}$$

with the summation going over the set of (allowed) modules of $\mathcal{W}\oplus\bar{\mathcal{W}}$, and $\mathcal{Z}_{s,t}$ being the number of primary fields carrying the respective modules. (In the context of string theory, the phases $\exp(-\pi i\tau \cdot c/12)$ correspond to the zero-point energies, and they become irrelevant if the conformal central charge takes its critical value $c = 26$ (cf. (3.3.8)) because in the light-cone gauge this gets reduced to $c_{l.c.} = 24$, reflecting the Minkowskian nature of the flat space–time part of the string theory. Also, the contributions of the space–time part to the partition function just give powers of $\eta/\sqrt{\tau}$ with η the Dedekind eta function (2.6.44) for both the holomorphic and the antiholomorphic part, and hence according to (2.7.6) the space–time part is separately modular invariant.)

The transformation of the characters under the modular transformation $\mathcal{T}$ is rather simple, as follows directly from the definition of the characters:

$$\tilde{\chi}_s(\tau + 1) = e^{2\pi i(\Delta_s - c/24)}\, \tilde{\chi}_s(\tau). \tag{5.2.15}$$

Invariance of the partition function under the $\mathcal{T}$-transformation is therefore equivalent to

$$\mathcal{Z}_{s,t} \neq 0 \implies \Delta_s - \bar{\Delta}_t \in \mathbb{Z}, \tag{5.2.16}$$

i.e. the holomorphic and antiholomorphic conformal dimensions of any primary field must at most differ by an integer. This shows in particular that the partition function is real, as it should be, and therefore that the matrix $\mathcal{Z}$ is symmetric. In matrix notation, (5.2.15) means that the modular transformation $\mathcal{T}$ is implemented on the characters by a diagonal matrix T,

$$T_{st} = \exp(2\pi i(\Delta_s - c/24))\,\delta_{st}. \tag{5.2.17}$$

In contrast, the transformation of the characters under $\mathcal{S}$ is more complicated; it is described by some nondiagonal matrix S,

$$\chi_s(-1/\tau) = \sum_t S_{st}\,\chi_t(\tau). \tag{5.2.18}$$

Of course, this matrix is subject to several constraints. First, S_{st} and T_{st} must be the representation matrices of the abstract group elements $\mathcal{S}$ and $\mathcal{T}$ for a representation of the modular group, or possibly of a twofold covering of the modular group. The latter possibility arises because although applying $\tau \mapsto -1/\tau$ twice gives back τ, one need not have $S^2 = \mathbf{1}$; this is because the homology cycles of the torus transform under $\mathcal{S}$ as $a \mapsto -b,\ b \mapsto a$, and hence under $\mathcal{S}^2$ as $a \mapsto -a,\ b \mapsto -b$ so that in terms of the world sheet, twofold application of $\mathcal{S}$ inverts the direction of proper time. Thus instead of the defining relations (2.7.3) of the modular group, the matrices S_{st} and T_{st} need only satisfy

$$S^2 = C = (ST)^3 \tag{5.2.19}$$

for some matrix C_{st} obeying

$$C^2 = \mathbf{1}. \tag{5.2.20}$$

In fact, one has $C = \mathbf{1}$ iff all primary fields of the theory are selfconjugate, while in general

$$\sum_t C_{st}\phi_t = \phi_s^+ \tag{5.2.21}$$

so that the rows and columns of C_{st} are just permutations of those of the unit matrix. Accordingly, C_{st} is called the *conjugation matrix* or the *charge conjugation matrix* of the conformal field theory. In addition to (5.2.19), it can also be shown that S_{st} is a unitary and symmetric matrix,

$$S = S^t, \quad S^{-1} = S^\dagger = S^*. \tag{5.2.22}$$

For the case where the symmetry algebra is an untwisted affine Lie algebra, the modular transformation matrix S is known explicitly: it has already been presented in (2.7.27). The above-mentioned properties of the S-matrix are just the analogues of the properties of (2.7.27) which have been discussed at length in section 2.7.

One further property of S is found by realizing that the transformation $\tau \mapsto -1/\tau$ possesses a fixed point $\tau = \sqrt{-1}$. This implies that

$$\sum_t S_{st}\,\tilde{\chi}_t(\sqrt{-1}) = \tilde{\chi}_s(\sqrt{-1}). \tag{5.2.23}$$

Moreover, from the definition (5.2.13) of $\tilde{\chi}$ it is clear that $\tilde{\chi}_t(\sqrt{-1})$ is real and positive for all t. Thus the matrix S must have an eigenvector with eigenvalue 1 with positive real components (and this eigenvector is unique owing to the unitarity of S). The corresponding primary field is the identity **1**, i.e. one has

$$S_{i0} > 0 \quad \text{for all } i. \tag{5.2.24}$$

In fact it can be shown that the stronger result $S_{i0} \geq S_{00} > 0$ holds for all i. Note that the previous results can fix S only up to an overall sign; the property just presented also fixes this sign.

Moreover, if the holomorphic and antiholomorphic symmetry algebras are identical so that $\bar{\chi}_s(\bar{\tau}) = \chi_s(\bar{\tau})$, then the transformation matrix S is the same for χ and $\bar{\chi}$. Since from the definition of the characters it is clear that S_{st} is analytic in τ, it then follows that $\bar{S}(\bar{\tau}) = S^*(\tau)$. Hence invariance of the partition function Z under the S-transformation yields

$$S\,\mathcal{Z}\,S^{-1} = \mathcal{Z}. \tag{5.2.25}$$

An obvious solution to the conditions (5.2.25) and (5.2.16) is the *diagonal modular invariant*

$$\mathcal{Z}_{s,t} = \delta_{st}. \tag{5.2.26}$$

Thus for $\mathcal{W} = \bar{\mathcal{W}}$, the modular invariance constraints always have at least one solution. In contrast, if $\mathcal{W} \neq \bar{\mathcal{W}}$ then one usually does not know any modular invariant partition function at all.

Non-diagonal modular invariants are to be obtained as non-trivial solutions to the constraints (5.2.25) and (5.2.16). This is only feasible if the matrices under consideration are finite-dimensional, corresponding to a conformal field theory with a finite number of primary fields, i.e. to a rational conformal field theory. The full set of modular invariant genus one partition functions is so far known for two classes of theories, namely the unitary conformal field theories of conformal central charge $c < 1$ (corresponding to $\mathcal{W} = \bar{\mathcal{W}} = \mathcal{V}$, with highest weight modules out of the list (3.2.30)), and the $A_1^{(1)}$ WZW theories. For the latter, the modular invariants are exactly those with one of the spectra (5.2.2). In contrast, the spectra listed in (5.2.3) are not modular invariant although they satisfy all constraints that are needed for the theory to make sense at genus zero. It is an open question whether the requirement of genus one modular invariance which amounts to a completeness requirement on the set of characters (they have to provide a representation of the modular group) can also be reformulated in terms of some completeness property at genus zero, which then could be used to reject the latter invariants already at an earlier stage.

Also, the modular invariants of the $c < 1$ conformal field theories can be obtained from the $A_1^{(1)}$-invariants via the coset construction (3.2.43). Partial results on non-diagonal modular invariants are also known for WZW theories based on other affine algebras $\mathfrak{g}$. They are typically related to automorphisms of the Dynkin diagram of $\mathfrak{g}$, to automorphisms of the fusion rule algebra of the WZW theory, or to conformal embeddings; this will be explained in more detail in section 5.4.

The names of the modular invariants (5.2.2) are those of the simply laced simple Lie algebras. This labelling has been chosen because the level of the modular invariant equals $g^\vee - 2$, with $g^\vee$ the dual Coxeter number of the labelling algebra, and the highest weights of the left–right symmetric primary fields are equal to the exponents minus one of the labelling algebra (including the multiplicity two for the exponent $r-1$ of D_{2r}).

5.3 Fusion rules and modular transformations

We may interpret the fusion rules of a conformal field theory as an abstract algebra, with the primary fields as generators. Such fusion rule algebras possess a unit (the identity primary field) and are both associative and commutative; furthermore their structure constants are (in the *canonical* basis provided by the primary fields) non-negative integers, and finally they are endowed with a conjugation (i.e. a map $i \mapsto i^+$ satisfying (5.1.3) and (5.1.10)). As a consequence of these properties, fusion rule algebras are semisimple. This implies that the adjoint representation, generated by the matrices $\mathcal{N}_i$ with entries $(\mathcal{N}_i)_j^{\ k} = \mathcal{N}_{ij}^{\ \ k}$, is fully reducible, and that the irreducible representations are one-dimensional and given by the eigenvalues of the matrices $\mathcal{N}_i$. Also recall from p. 325 that the associativity and commutativity of the fusion rule algebra imply that the matrices $\mathcal{N}_i$ commute, $[\mathcal{N}_i, \mathcal{N}_j] = 0$. Taken together, this means the following: if there are N distinct primary fields obeying the fusion rules

$$\phi_i * \phi_j = \sum_{k=1}^{N} \mathcal{N}_{ij}^{\ \ k}\, \phi_k, \tag{5.3.1}$$

then there exist N^2 (not necessarily distinct) complex numbers $\ell_i^{(j)}$, $i, j = 1, \ldots, N$, such that

$$\ell_i^{(m)}\,\ell_j^{(m)} = \sum_{k=1}^{N} \mathcal{N}_{ij}{}^{k}\,\ell_k^{(m)} \tag{5.3.2}$$

for $i, j, m = 1, \ldots, N$. The eigenvalues satisfy the orthogonality relations

$$\eta_i \sum_k \ell_k^{(i)} (\ell_k^{(j)})^* = \delta_{ij} = \sum_k \eta_k\, \ell_i^{(k)} (\ell_j^{(k)})^* \tag{5.3.3}$$

with $\eta_i^{-1} = \sum_j |\ell_i^{(j)}|^2$.

The structure of fusion rule algebras bears some similarity with the representation theory of finite groups. Indeed, the tensor product rules for representations of a finite group constitute a commutative associative algebra with unit and conjugation C, and with non-negative integer structure constants. The eigenvalues of this algebra are just the characters of the irreducible representations. It turns out that such an algebra is isomorphic to the fusion rule algebra of a conformal field theory iff the modular transformation matrix S on these characters satisfies $S^2 = C$. All abelian groups possess the latter property (as a consequence, those fusion rule algebras for which on the right hand side only a single primary field ever appears, are in one to one correspondence with finite abelian groups); in contrast, for many non-abelian groups the condition is not satisfied.

In the example provided by finite groups, the primary fields correspond to the irreducible characters of finite groups so that there arises an intimate connection between the fusion rules and the action of the modular matrix S. Remarkably, also in case of a generic conformal field theory there exists a relation between the fusion rule algebra and the modular transformation matrix S that acts on the characters $\tilde{\chi}_i$ associated to the primary fields ϕ_i. To derive this relation, one has to place the conformal field theory on the torus. The basic idea is to employ the primary fields to manipulate the characters, and then compare the situation before and after a modular transformation. To do so one has to choose a basis a and b of homology cycles on the torus (if the torus with modular parameter τ is represented as a parallelogram with corners at 0, 1, τ and $\tau + 1$, the a cycle is homologous to the straight line from 0 to 1, and the b cycle to the straight line from 0 to τ). Then one introduces a set of operators $\Psi_i(a)$ and $\Psi_i(b)$, $i = 1, \ldots, N$, that act on the characters,

$$\Psi_i(c): \quad \tilde{\chi}_j \mapsto \Psi_i(c) \cdot \tilde{\chi}_j, \tag{5.3.4}$$

in a way described as follows. First, the identity field $\mathbf{1}$ is inserted at some point on the torus into the trace (5.2.13) defining $\tilde{\chi}_j$. The latter is expressed through the inner product defined on the module, i.e. as

$$\sum_v (v \mid q^{L_0 - \frac{c}{24}} \cdot v) \tag{5.3.5}$$

with the summation going over a complete basis of vectors v in the $\mathcal{W}$-module carried by ϕ_j. Next the identity is rewritten as the operator product $\phi_i * \phi_{i^+}$ (the additional terms in this operator product other than the identity do not contribute to the trace). Afterwards ϕ_i is moved around the cycle c; when it comes back to its original position, the operator product with ϕ_{i^+} is formed again, thus reobtaining the identity operator. This operation defines some linear operation on the space of characters. For the character $\tilde{\chi}_0$ of $\mathbf{1}$, one simply has

$$\Psi_i(b) \cdot \tilde{\chi}_0 = \tilde{\chi}_i; \tag{5.3.6}$$

in particular the coefficients in this linear transformation are integers. It can be shown that the latter statement is true also for the action of $\Psi_i(b)$ on all other characters, and that in fact

$$\Psi_i(b) \cdot \tilde{\chi}_j = \sum_k \mathcal{N}_{ij}{}^k \, \tilde{\chi}_k. \tag{5.3.7}$$

The proof of this relation will be provided in section 5.7. Pictorially, the result (5.3.7) can be understood by mapping the torus on an annulus $\mathrm{e}^{-\pi \operatorname{Im}\tau} \le z < \mathrm{e}^{\pi \operatorname{Im}\tau}$ in the complex plane so that in the limit $\tau \to i\infty$ in which the a-cycle gets pinched to zero, the field ϕ_j is moved from ∞ to 0.

Similarly, if a field ϕ_i is moved around an a-cycle, the module over which the trace is taken remains unchanged because the a-cycle consists of points with equal time. In other words, the characters $\tilde{\chi}_j$ are eigenvectors of the linear operation $\Psi_i(a)$,

$$\Psi_i(a) \cdot \tilde{\chi}_j = \ell_i^{(j)} \, \tilde{\chi}_j. \tag{5.3.8}$$

Now the modular transformation $\mathcal{S}$ interchanges the cycles a and b, and hence

$$\Psi_j(b) = \mathcal{S}\,\Psi_j(a)\,\mathcal{S}^{-1}. \tag{5.3.9}$$

Combining the last three equations, it follows that

$$\mathcal{N}_{ij}{}^k = \sum_n S_{in} \, \ell_j^{(n)} \, S^{-1}_{nk}, \tag{5.3.10}$$

or in words: the modular matrix S diagonalizes the fusion rules. (Due to $[\mathcal{N}_i, \mathcal{N}_j] = 0$, it is indeed possible that a single matrix S diagonalizes all matrices $\mathcal{N}_i$ simultaneously.) (5.3.10) may also be rewritten as

$$\sum_k \mathcal{N}_{ij}{}^k \, S_{kn} = S_{in} \, \ell_j^{(n)} \tag{5.3.11}$$

By combining this result with the property (5.1.3) of the fusion rule coefficients involving the identity primary field, one obtains

$$\ell_0^{(i)} = 1, \tag{5.3.12}$$

where as usual the index 0 refers to the identity field, and one also can express the fusion rule eigenvalues in terms of the matrix elements S_{ij},

$$\ell_i^{(j)} = S_{ij}/S_{0j}. \tag{5.3.13}$$

One can then also rewrite (5.3.10) such that the fusion rule coefficients are directly expressed through the matrix S :

$$\mathcal{N}_{ij}{}^k = \sum_n \frac{S_{in} S_{jn} S^{-1}_{nk}}{S_{0n}}. \tag{5.3.14}$$

This result is known as the *Verlinde formula*; it can be employed to prove that the matrix S is unitary and symmetric (irrespective of whether the conformal field theory itself is unitary or not). Using the invariance of the fusion rule algebra under conjugation, $C\colon \phi_i \mapsto \phi_{i^+}$, one gets

$$\ell_{i^+}^{(j)} = \left(\ell_i^{(j)}\right)^* . \tag{5.3.15}$$

Together with $\ell_i^{(j)} = S_{ij}/S_{0j}$, it follows that

$$S^* = C\,S = S\,C, \tag{5.3.16}$$

and this implies unitarity of S because of $S^2 = C$. For the matrix elements, the relation (5.3.16) reads $S^*_{ij} = S_{i^+j} = S_{ij^+}$; thus for the fusion rule coefficients with three lower indices, (5.3.14) together with the symmetry of S implies

$$\mathcal{N}_{ijk} \equiv \mathcal{N}_{ij}{}^{k^+} = \sum_n \frac{S_{in} S_{jn} S_{kn}}{S_{0n}}. \tag{5.3.17}$$

This formula nicely exhibits the total symmetry of the fusion coefficients $\mathcal{N}_{ijk}$ with lower indices only (conversely, without using the symmetry of S one has the same formula with S_{kn} replaced by S_{nk}, and hence the complete symmetry of the $\mathcal{N}_{ijk}$ can be employed to prove that S is symmetric), and together with $SC = S^*$ it also shows immediately that $\mathcal{N}_{i^+j^+k^+} = (\mathcal{N}_{ijk})^* = \mathcal{N}_{ijk}$. Note, however, that from the result (5.3.17) it is not at all obvious that the fusion rule coefficients are non-negative integers. Indeed there exist a few modular invariant combinations of characters of untwisted affine Lie algebras which give rise to a well-behaved S matrix corresponding to an extended symmetry algebra but which nevertheless fail to give integer values for the combination (5.3.17); thus these partition functions do not correspond to any conformal field theory.

The quantities $\ell_i^{(0)}$ possess a special interpretation. Namely, one has

$$\ell_i^{(0)} = \frac{S_{i0}}{S_{00}} = \lim_{\tau\to i\infty} \frac{\sum_j S_{ij}\tilde{\chi}_j(\tau)}{\sum_k S_{0k}\tilde{\chi}_k(\tau)} = \lim_{\tau\to 0} \frac{\tilde{\chi}_i(\tau)}{\tilde{\chi}_0(\tau)} = \frac{\mathrm{tr}_{\phi_i}(1)}{\mathrm{tr}_{\mathbf{1}}(1)}. \tag{5.3.18}$$

Here the last equality expresses the fact that in the limit $\tau \to 0$, the character χ_i formally counts the number of states in the $\mathcal{W}$-module carried by ϕ_i. Thus the eigenvalues $\ell_i^{(0)}$ indicate the relative size of the modules carried by the primaries ϕ_i as compared with that of the identity field (for the special case of affine algebras, this was already discussed in section 2.8). In the first equality in (5.3.18) it is assumed that the primary field with the lowest conformal dimension is unique and is the identity field; this is certainly true for all unitary conformal field theories, but holds for many non-unitary theories as well. For unitary theories in addition all contributions in tr_ϕ are positive, which implies that $\ell_i^{(0)} > 0$. In fact, using (5.3.2) for $m = 0$, one even has

$$\ell_i^{(0)} \geq 1, \tag{5.3.19}$$

with equality iff (cf. p. 365) ϕ_i is a simple current.

It is important to note that via (5.3.14), the fusion rule coefficients are in fact defined in terms of only one chiral (i.e. holomorphic or antiholomorphic) half of the conformal field theory. This is puzzling because the fusion rules were introduced in section 5.1 as properties of the primary fields, which of course depend generically both on the holomorphic and on the antiholomorphic halves of the theory. But it is not difficult to resolve this puzzle, and this leads indeed to some new insight in the classification of modular invariant partition functions, provided that one insists on considering the maximal symmetry algebra. The latter requirement means that besides the identity primary field there is no other primary field which is purely (anti-) holomorphic. In terms of the matrix $\mathcal{Z}$ defined by (5.2.14), this implies

$$\mathcal{Z}_{i,0} = \mathcal{Z}_{0,i} = \delta_{i0}. \tag{5.3.20}$$

From now on it will be assumed that the left and right symmetry algebras are isomorphic. Then the invariance of the partition function under the modular transformation S means that $S\mathcal{Z}S^\dagger = \mathcal{Z}$; because of the unitarity of S, this is equivalent to

$$S\,\mathcal{Z} = \mathcal{Z}\,S. \tag{5.3.21}$$

Combining this with the identity $\ell_0^{(i)} = 1$ leads immediately to

$$\ell_i^{(0)} = \textstyle\sum_i \mathcal{Z}_{i,j}\,\ell_j^{(0)}. \tag{5.3.22}$$

If one now uses the fact that the eigenvalues $\ell_i^{(0)}$ are larger than or equal to one, and that the matrix $\mathcal{Z}$ is symmetric and has non-negative integer entries, one sees that $\mathcal{Z}$ is in fact a permutation,

$$\mathcal{Z}_{i,j} = \delta_{i,\pi(j)}. \tag{5.3.23}$$

Moreover, owing to (5.3.20), this permutation π obeys

$$\pi(0) = 0. \tag{5.3.24}$$

Thus we have learned that, provided the maximal symmetry algebra $\mathcal{W} \oplus \bar{\mathcal{W}}$ is considered, all non-diagonal modular invariants have the property that any module of $\mathcal{W}$ gets combined with a unique module of $\bar{\mathcal{W}}$, and vice versa, and in particular the "singlet" of $\mathcal{W}$ (i.e. the $\mathcal{W}$-module carried by the identity primary field) is combined with the singlet of $\bar{\mathcal{W}}$. In fact, the above derivation of this result works only for unitary conformal field theories (because use was made of the property $\ell_i^{(0)} \geq 1$), but it is probable that the result holds for non-unitary theories as well.

Combining the result (5.3.23) with (5.3.21), one sees that the modular matrix S must obey

$$S_{\pi(i),j} = S_{i,\pi(j)}. \tag{5.3.25}$$

Expressing the fusion rule eigenvalues through S and using $\ell_0^{(i)} = 1$, one thus concludes that

$$\ell_{\pi(i)}^{(j)} = \ell_i^{(\pi(j))}. \tag{5.3.26}$$

Analogously, (5.3.14) shows that

$$\mathcal{N}_{\pi(i),\pi(j)}{}^{\pi(k)} = \mathcal{N}_{ij}{}^{k}. \tag{5.3.27}$$

In other words, the map

$$\mathsf{Z}: \ \phi_i \mapsto \phi_{\bar{\imath}} := \sum_j \mathcal{Z}_{i,j}\, \phi_j = \phi_{\pi(i)} \tag{5.3.28}$$

defines an automorphism of the fusion rule algebra. As a consequence, if the maximal symmetry algebra is considered, then for any possible modular invariant combination of the modules of the holomorphic and the antiholomorphic symmetry algebras, the holomorphic and antiholomorphic fusion rules are isomorphic to each other, and hence also isomorphic to the fusion rules defined in terms of primary fields. In contrast, if a non-maximal symmetry algebra is chosen (as is, for example, convenient when solving null vector equations such as the Knizhnik–Zamolodchikov equation), then strictly speaking the concept of fusion rules only makes sense for the chiral halves of the theory.

A good illustration of these results is provided by the modular invariants of the $A_1^{(1)}$ WZW theories which were listed in (5.2.2). Among the non-diagonal invariants, only those of the D_{2n+1} series have the property that $\mathcal{Z}$ is a permutation matrix, namely

$$\phi_{\bar{\Lambda}} = \sum_{\Lambda \in P_k^+} \mathcal{Z}_{\bar{\Lambda},\Lambda}\, \phi_\Lambda = \begin{cases} \phi_\Lambda & \text{for } \bar{\Lambda} \in 2\mathbb{Z} \\ \phi_{k^\vee \rho - \Lambda} & \text{for } \bar{\Lambda} \in 2\mathbb{Z}+1. \end{cases} \tag{5.3.29}$$

(Recall that for $g = A_1^{(1)}$ the fundamental affine Weyl chamber is $P_k^+ = \{\Lambda^0 \Lambda_{(0)} + \bar{\Lambda}\Lambda_{(1)} \mid \Lambda^0, \bar{\Lambda} \in \mathbb{Z}_{\geq 0},\ \Lambda^0 + \bar{\Lambda} = k^\vee\}$; thus $k^\vee \rho - \Lambda \equiv (k^\vee - \Lambda^0)\Lambda_{(0)} + (k^\vee - \bar{\Lambda})\Lambda_{(1)} = \bar{\Lambda}\Lambda_{(0)} + \Lambda^0\Lambda_{(1)}$.) Also, the D_{2n}, E_6 and E_8 modular invariants are diagonal invariants of an extended symmetry algebra. Only in the case of the E_7 invariant is the situation a bit more complicated: this is a non-diagonal invariant of an extended symmetry algebra (for a few more details, see p. 352).

The connection between fusion rules and modular transformations has so far been used mainly to compute the fusion rules for cases where the modular matrix S is known. However, one can also go in the other direction and determine S from known fusion rules when other methods for the calculation of S are too complicated (e.g. for WZW theories for which the Weyl group of $\mathfrak{g}$ is very large, notably for $\mathfrak{g} = E_8$.) To do so, one looks for a matrix M that diagonalizes a given fusion rule matrix $\mathcal{N}_i$. This matrix is of course not unique, but one can also impose the following properties that the correct S-matrix must possess:

$$\begin{aligned} &MC = CM = M^* = M^{-1}, \\ &M = M^t, \\ &M_{i0} \geq M_{00} > 0 \quad \text{for all } i. \end{aligned} \tag{5.3.30}$$

This still does not fix M uniquely. However, one can show that any other matrix $\tilde{M}$ that possesses all these properties as well is related to M as $\tilde{M} = \Pi M$ where Π is an automorphism of the fusion rule algebra (and hence in particular a permutation) and commutes with $\tilde{M}$ (i.e. obeys $\Pi\tilde{M}\Pi^{-1} = \tilde{M}$, or in terms of matrix elements, $\tilde{M}_{i,\pi(j)} = \tilde{M}_{\pi(i),j}$). In particular the true modular matrix S is among all possible matrices $\tilde{M}$ that are obtained this way.

The relation between fusion rules and modular transformations can also be used to obtain restrictions on the allowed values for the central charge c and conformal dimensions Δ of a conformal field theory. Namely, considering the fusion rule algebra as given, the number of equations provided by (5.3.14) is so large that the eigenvalues $\exp(2\pi i(\Delta - c/24))$ of T are actually overdetermined. This means that solutions to the constraint $(ST)^3 = C$ can only be obtained for a restricted set of c- and Δ-values. If the total number of primary fields is very small, then typically there is even a unique solution for $c \bmod 8$ and for $\Delta \bmod 1$. If many primaries are present, the latter numbers are usually no longer determined uniquely. One reason for this is that, as just mentioned, (5.3.14) together with $S^2 = C$ etc. does *not* completely determine the matrix S. In any case, however, the conformal dimensions and c-values of a rational conformal field theory must be rational numbers.

The simplest example for the determination of c and Δ is provided by conformal field theories possessing a single primary field. This must then be the identity field so that we already know that $\Delta = 0$. Moreover, one clearly has $S = 1$ so that the requirement $(ST)^3 = \mathbf{1}$ yields

$$c = 0 \mod 8. \tag{5.3.31}$$

A realization of this rather trivial conformal field theory is given by the $E_8^{(1)}$ WZW theory at level one.

If there are precisely two primaries, $\mathbf{1}$ and ϕ, then there is a single non-trivial fusion rule

$$\phi * \phi = \mathbf{1} + m\,\phi, \tag{5.3.32}$$

and the matrix S takes the form

$$S = \begin{pmatrix} \cos\vartheta & \sin\vartheta \\ \sin\vartheta & -\cos\vartheta \end{pmatrix} \tag{5.3.33}$$

with $\sin\vartheta > 0$. The eigenvalues $\ell_\pm = \ell_\phi^{(\cdot)}$ are the solutions of the quadratic equation $\ell^2 = 1 + m\ell$, and the requirement that S diagonalizes the fusion rules implies that $\ell_\pm = \tan\vartheta$. The constraint $(ST)^3 = \mathbf{1}$ then yields

$$\cos(2\pi\Delta) = -\tfrac{1}{2}\,m\,\ell_\pm, \tag{5.3.34}$$

$$c = 12\,\Delta + 6 \mod 8. \tag{5.3.35}$$

Realizations of these fusion rules for $m = 0$ are given by the $A_1^{(1)}$ and $E_7^{(1)}$ WZW theories at level one; these have $c = 1$, $\Delta = \frac{1}{4}$ and $c = 7$, $\Delta = \frac{3}{4}$, respectively. For $m = 1$, the allowed values for Δ are given by $5\Delta \in \mathbb{Z}$ but $\Delta \notin \mathbb{Z}$; in this case the fusion rules are called the *Lee–Yang* fusion rules because they are realized by the non-unitary conformal field theory which describes the so-called Lee–Yang edge singularity (the maximal symmetry algebra is then the Virasoro algebra, and $c = -\frac{22}{5}$, $\Delta = -\frac{1}{5}$); realizations with an extended algebra are provided by the $G_2^{(1)}$ and $F_4^{(1)}$ WZW theories at level one which have $c = \frac{14}{5}$, $\Delta = \frac{2}{5}$ and $c = \frac{26}{5}$, $\Delta = \frac{3}{5}$, respectively. It also turns out that, at least for unitary theories, fusion rule coefficients $m \geq 2$ in (5.3.32) are never allowed. Namely, (5.3.34) implies $\ell \leq 2/m$, whereas from $\ell^2 = 1 + m\ell$ and $\ell > 0$ it follows that $\ell > m$; thus compatibility requires $\ell \in \{0, 1\}$.

5.4 Modular invariants

One of the prominent problems in conformal field theory is the classification of all modular invariant partition functions. This problem is still largely unsolved, even in the case of WZW theories where the modular matrix S is already known. Below, a summary of the known results about WZW modular invariants will be given, together with a few remarks on modular invariants of other conformal field theories.

For WZW theories, the modular matrix S is given by the formula (2.7.24). All ingredients of this formula are known, and hence one may find modular invariants by simply scanning for all matrices with non-negative integer entries that obey the constraints (5.2.25) of invariance under the $\mathcal{S}$- and $\mathcal{T}$-transformation together with the constraint (5.2.16). However, except for some very simple cases this brute force method requires the use of a computer, and in fact the computer time needed is quite large (the time needed to find a new modular invariant rises asymptotically as $(k^\vee)^{2r-1}$; for ranks and levels of the order of five, the time needed is already of the order of a few hours for a typical FORTRAN program run on a Gould NP1). Consequently, this method has so far led only to a complete classification of WZW modular invariants for small values of the rank and level of g. The only exception to this is the case $g = A_1^{(1)}$ where a (rather complicated) proof based on number-theoretic arguments has been obtained that the list (5.2.2) exhausts the possible modular invariants at arbitrary levels.

Fortunately, modular invariants can also be found by other considerations. The most important observation in this context is the fact that for any (unitary) conformal field theory, $\mathcal{Z}$ must be a permutation (related to an automorphism of the fusion rule algebra) if the maximally extended symmetry algebra is considered. In other words, any modular invariant must be interpretable either as the diagonal invariant with respect to some extended algebra, or as obtained from the diagonal invariant by a permutation corresponding to a fusion rule automorphism. Therefore modular invariants can be found by searching for embeddings of the given symmetry algebra in a larger algebra, or for automorphisms of the fusion rules.

If a modular invariant possesses the interpretation as a diagonal invariant of a larger symmetry algebra, then it is a sum of squares,

$$Z \equiv \sum_{s,t} \tilde{\chi}_s \mathcal{Z}_{s,t} \tilde{\chi}_t = \sum_i \Big| \sum_{j=0}^{m_i-1} \tilde{\chi}_{(i,j)} \Big|^2 , \qquad (5.4.1)$$

where the labels s of the primaries are rewritten as pairs (i,j). Invariants of this type are called *integer spin invariants.* Clearly, the sums $\psi_i = \sum_j \tilde{\chi}_{(i,j)}$ of characters of the original chiral symmetry algebra are to be interpreted as characters of the extended algebra that is present for an integer spin invariant; this implies in particular that the extended algebra is generated by those fields which appear in the sum ψ_0 containing the character $\tilde{\chi}_0 \equiv \tilde{\chi}_{(0,0)}$ of the identity primary field. The lengths of the sums inside the squares in (5.4.1) cannot take arbitrary values, but must always divide the length m_0 of the sum containing $\tilde{\chi}_0$. To see this, first observe that if $\mathcal{Z}$ defines a modular invariant, then any polynomial in $\mathcal{Z}$ defines a modular invariant, too (because $S\mathcal{Z}^n S^{-1} = (S\mathcal{Z}S^{-1})^n$). If the lengths were different from the values just mentioned, then one could always construct a polynomial $\mathcal{P}$ that gives a modular invariant with $(\mathcal{P}(\mathcal{Z}))_{0,0} = 0$. But this is not allowed because of the identity

$$\mathcal{Z}_{0,0} = \sum_{s,t} S_{0s}\, \mathcal{Z}_{s,t}\, S^*_{t0}, \qquad (5.4.2)$$

where all terms on the right hand side are real and positive owing to (5.3.19) and $S_{00} > 0$. As a consequence, for integer spin invariants the partition function can also be written as

$$Z = \sum_i \frac{m_0}{m_i} \Big| \sum_{j=0}^{m_i-1} \tilde{\chi}_{(i,j)} \Big|^2 \qquad (5.4.3)$$

with m_i a divisor of m_0. The prefactors m_0/m_i are needed to ensure that $\mathcal{Z}^n \propto \mathbf{1}$ for some large enough n (if $\mathcal{Z}$ had not this property, then by forming polynomials one could construct an infinite number of modular invariants for a given set of characters, which is certainly not possible).

If some integer m_0/m_i in (5.4.3) is different from 1, then the modular invariant does not look manifestly like a diagonal invariant of an extended algebra. This problem can be resolved by assuming that there are m_0/m_i distinct primary fields which possess the same character. For example, in the D_{2n}-type modular invariant of $A_1^{(1)}$, one has (see table (5.2.2)) $m_i = 2$ for all i except for the character χ_Λ corresponding to $\bar{\Lambda} = k^\vee/2 = 2n-2$, while in the latter case $m_i = 1$ and there are two distinct primary fields having this character. If $m_0/m_i \neq 1$ is a square number, then there is another possibility of resolving the problem, namely by absorbing the integer $\sqrt{m_0/m_i}$ into the characters of the extended algebra.

As already mentioned, any polynomial (over the integers) of modular invariants is again a modular invariant. This often leads to invariants which do not correspond to a sensible conformal field theory (e.g. which have $\mathcal{Z}_{s,t} < 0$ for some s and t, or $\mathcal{Z}_{0,0} = 0$). But sometimes one does get new invariants, and these may then look quite different from what one would expect for invariants describing at most a permutation of the diagonal invariant of an extended algebra.

Another approach to find modular invariants is related to the existence of simple currents, i.e. primary fields Φ obeying the "unique fusion rules" (5.1.20). In (5.1.29), a relation was obtained for the conformal dimensions (modulo integers) of the primaries lying on a given orbit with respect to the $\mathbb{Z}_N$-symmetry generated by Φ. This relation can also be viewed as a relation between the entries of the modular matrix T on the orbit. One can also find an analogous relation for the matrix S. Using the notations of section 5.1, it reads

$$S_{i(m),\,j(n)} = \mathrm{e}^{2\pi i\, n\eta_i/N}\, \mathrm{e}^{2\pi i\, m\eta_j/N}\, \mathrm{e}^{2\pi i\, mn\zeta/N}\, S_{ij}. \tag{5.4.4}$$

Using this formula it follows that, except for N even and ζ odd, the matrix

$$\mathcal{Z}_{i(m),\,j(n)} = \sum_{l=1}^{N} \delta_{ij}\, \delta_{m+l,n}\, \breve{\delta}\Big(\frac{\eta_i}{N} + \frac{2m+l}{2N}\,\zeta\Big) \tag{5.4.5}$$

is invariant under both S and T, where

$$\breve{\delta}(x) = \begin{cases} 1 & \text{for } x \in \mathbb{Z} \\ 0 & \text{otherwise.} \end{cases} \tag{5.4.6}$$

If the conformal dimension of the simple current Φ is integral, then the modular invariant obtained this way is an integer spin invariant. The presence of lengths $m_i \neq m_0$ in such an invariant is then always related to a fixed point of the $\mathbb{Z}_N$-automorphism generated by Φ, i.e. to the existence of a primary field ϕ obeying

$$\Phi^n * \phi = \phi \tag{5.4.7}$$

for some power $\Phi^n \neq 1$ of Φ. If the conformal dimension of Φ is non-integral, one obtains the combination of an automorphism invariant and an integer spin invariant, or a pure automorphism invariant, the latter possibility being realized if no power Φ^n, $n < N$, of Φ has integral conformal dimension.

Having discussed the general methods for finding modular invariants, we now list what has been obtained so far with these methods for the particular case of WZW theories based on affine algebras $\mathfrak{g}$ for simple $\bar{\mathfrak{g}}$. The following modular invariants are known.

- The *diagonal* invariant
$$\mathcal{Z}_{\Lambda,\Lambda'} = \delta_{\Lambda\Lambda'} . \tag{5.4.8}$$
As discussed in section 5.1, this invariant is always present since the set of characters $\{\tilde{\chi}_\Lambda \mid \Lambda \in P_k^+\}$ forms a module of the modular group.

- The *conjugation* invariant
$$\mathcal{Z}_{\Lambda,\Lambda'} = C_{\Lambda\Lambda'} = \delta_{\Lambda^+,\Lambda'} . \tag{5.4.9}$$
This is invariant under $\mathcal{S}$ due to $SCS^{-1} = S\,S^2\,S^{-1} = S^2 = C$, and invariant under $\mathcal{T}$ because of $\Delta_{\Lambda^+} = \Delta_\Lambda$. Since $\Lambda \mapsto \Lambda^+$ is obviously an automorphism of the fusion rules ($\bar{\Lambda} \mapsto \bar{\Lambda}^+$ is already an automorphism of the Kronecker products of $\bar{\mathfrak{g}}$), the conjugation invariant is an automorphism invariant. Also, of course, this invariant is a new one only if $C \not\equiv \mathbf{1}$, i.e. if $\bar{\mathfrak{g}} = A_r\,(r \geq 2)$, D_{2n+1} or E_6.
It has been shown, by number-theoretic arguments, that for $A_2^{(1)}$ at any level $k^\vee$ such that $k^\vee + 3$ is prime, no other invariants besides the diagonal and the conjugation invariant exist.
Another closely related invariant arises for $\bar{\mathfrak{g}} = D_4$, namely the automorphism invariant that results from the generator of the $\mathcal{S}_3$ symmetry of the D_4 Dynkin diagram.

- Invariants induced by *conformal embeddings.* For any conformal embedding $\mathfrak{g} \subset \mathfrak{h}$, the modular invariants of the WZW theory based on $\mathfrak{h}$ can be written in terms of $\mathfrak{g}$. In the case of the diagonal invariant of $\mathfrak{h}$, one obtains an integer spin invariant of $\mathfrak{g}$. All such invariants are known because all conformal embeddings have been classified (see sections 2.8 and 3.8).

- Invariants induced by embeddings $\mathfrak{g} \oplus \mathcal{V} \subset \mathcal{W}$ in extended algebras $\mathcal{W} \not\cong \mathfrak{h} \oplus \mathcal{V}$. Not much is known about such invariants, except for some cases where they also possess a different interpretation. Among the generators of these extended algebras, there are primary fields of conformal spin and dimension larger than 2, and as a consequence they are not even Lie algebras.

- Invariants induced by *simple currents.* As stated above, these can be integer spin invariants, automorphism invariants, or combinations thereof, depending on the integrality properties of the conformal dimension of the simple current. It is expected that these invariants possess a lagrangian interpretation as sigma models on non-simply connected group manifolds, i.e. group manifolds which are obtained from the simply connected covering group G by dividing by (a subgroup of) the center of G. Note that the center of G is isomorphic to the group of automorphisms of the Dynkin diagram of the affine algebra $\mathfrak{g}$. Thus in the case of the simple current ϕ_{3875} of the level two $E_8^{(1)}$ WZW theory (see p. 294 and p. 327), the associated invariant could not have the interpretation just mentioned; however, in this case it turns out that this invariant is not a new one, but just the diagonal invariant.

Exceptional automorphism invariants. There are also examples of automorphisms of the fusion rule algebra which cannot be obtained from simple currents, but rather from a fusion rule automorphism applied to the invariant for an integer spin simple current. There are only three known examples for which the order of the simple current is larger than two, namely equal to three for $A_2^{(1)}$ at level 9 and $A_8^{(1)}$ at level 3, and equal to five for $A_4^{(1)}$ at level 5. For order two there is an infinite series for $so_n^{(1)}$ at level 8 as well as a few isolated cases which however all fall into short series; these are summarized in the following diagram, where for brevity we write $(\mathfrak{g})_{k^\vee}$ for $\mathfrak{g}$ at level $k^\vee$:

$$\begin{array}{ccccccccccc}
(B_1)_8 & - & (B_2)_8 & - & (B_3)_8 & - & (B_4)_8 & - & (B_5)_8 & - & \ldots. \\
\| & & & & & & & & & & \\
(C_1)_{16} & - & (C_2)_8 & - & (C_4)_4 & - & (C_8)_2 & - & (C_{16})_1 & & \\
\| & & & & & & & & & & \\
(su_2)_{16} & - & (su_4)_8 & - & (su_8)_4 & - & (su_{16})_2 & & & & \\
& & \| & & & & & & & & \\
(D_2)_8 & - & (D_3)_8 & - & (D_4)_8 & - & (D_5)_8 & - & (D_6)_8 & - & \ldots. \\
& & & & \| & & & & & & \\
& & (so_4)_{16} & - & (so_8)_8 & - & (so_{16})_4 & - & (so_{32})_2 & &
\end{array} \tag{5.4.10}$$

Except for $(C_4)_4$ where $\mathfrak{g}_1 = \mathfrak{g}_2$, all invariants in this list can be obtained by considering a conformal embedding $\mathfrak{g}_1 \oplus \mathfrak{g}_2 \subset \mathfrak{f}$ and contracting a known nondiagonal invariant for $\mathfrak{f}$ with a known nondiagonal invariant for $\mathfrak{g}_2$ to obtain an invariant for $\mathfrak{g}_1$. Also note that the E_7-invariant for $A_1^{(1)}$ at level 16 belongs to an infinite series for $B_r^{(1)}$ at level 8, i.e. is not that exceptional after all.

Exceptional integer spin invariants. A few examples are known of integer spin invariants which cannot be obtained directly by a simple current or a conformal embedding. These arise for $\mathfrak{g} = A_8$ at level 3, for A_9 and A_{27} at level 2, C_5 at level 3, C_7 at level 4, C_{10} and C_{28} at level 1, and for F_4 at level 6. Except for the last example, all these invariants can be obtained by considering conformal embeddings $\mathfrak{g}_1 \oplus \mathfrak{g}_2 \subset \mathfrak{f}$ (namely, $(A_{n-1})_k \oplus (A_{k-1})_n \subset (A_{nk-1})_1$ and $(C_r)_k \oplus (C_k)_r \subset (C_{rk})_1$, respectively) in the way described in the previous paragraph. In contrast, for the last-mentioned example there is no relation to a conformal embedding at all. In this $(F_4)_6$ case one has an extended algebra with two fields of $\Delta = 4$ and one field of $\Delta = 6$ (namely, $\phi_{(0004)}$ and $\phi_{(1100)}$, and $\phi_{(0030)}$, respectively, where the subscripts denote the F_4-Dynkin labels of the fields); the invariant reads

$$\begin{aligned} Z = & |\tilde{\chi}_{(0000)} + \tilde{\chi}_{(0004)} + \tilde{\chi}_{(0030)} + \tilde{\chi}_{(1100)}|^2 \\ & + |\tilde{\chi}_{(0003)} + \tilde{\chi}_{(0006)} + \tilde{\chi}_{(0021)} + \tilde{\chi}_{(2010)}|^2 \\ & + 2\,|\tilde{\chi}_{(0101)} + \tilde{\chi}_{(1012)}|^2 + 2\,|\tilde{\chi}_{(0102)} + \tilde{\chi}_{(2000)}|^2. \end{aligned} \tag{5.4.11}$$

Often a given invariant can be obtained by several distinct methods. For example, all known invariants coming from embeddings in non-affine algebras can also be obtained with the help of simple currents. As already stated, only for $\mathfrak{g} = A_1^{(1)}$ have all modular invariant partition functions been classified, namely by the $A - D - E$ scheme of (5.2.2). These invariants can be interpreted as follows. The A_r-invariants are diagonal. The D_r-invariants can be obtained via the simple current $\Phi = \phi_{\Lambda,\Lambda}$, $\Lambda = k^\vee \Lambda_{(1)}$ (i.e. $\bar{\Lambda} = k^\vee$), and the D_{2n}-invariants also by embeddings in a symmetry algebra which contains as additional generators the modes of the primary field $\phi_{k^\vee \Lambda_{(1)}, k^\vee \Lambda_{(0)}}$ which has conformal spin and dimension $\Delta = n - 1$; for $n = 2$ the extended algebra is $A_2^{(1)}$ at level one (so that one has a conformal embedding), while for $n > 2$ the extended algebra is no longer a Lie algebra and is not known explicitly. The E_6- and E_8-invariants are obtained via the conformal embeddings

$$(A_1^{(1)})_{k^\vee=10} \subset (C_2^{(1)})_{k^\vee=1} \tag{5.4.12}$$

and

$$(A_1^{(1)})_{k^\vee=28} \subset (G_2^{(1)})_{k^\vee=1}, \tag{5.4.13}$$

respectively. Finally, the E_7-invariant is one of the exceptional automorphism invariants; it arises from the automorphism of the fusion rule algebra corresponding to the D_{10}-invariant that exchanges one of the fields ϕ_{8_1} and ϕ_{8_2} which both have the same character χ_8 with one of the other fields, namely, in a symbolical notation,

$$\chi_{8_1} \longleftrightarrow \chi_2 + \chi_{14} \tag{5.4.14}$$

(here the characters of the extended algebra of the D_{10}-invariant are expressed through the $A_1^{(1)}$-characters, with the subscripts denoting $\bar{\Lambda}$).

A simple way to obtain new conformal field theories from a set of given ones is by forming tensor products. If the given theories are WZW theories based on affine Lie algebras g with $\bar{\mathsf{g}}$ simple, then the tensor product theories are just the WZW theories based on g with $\bar{\mathsf{g}}$ the direct sum of simple Lie algebras, i.e. semisimple. Clearly, modular invariants of such "semisimple WZW theories" can be obtained by tensoring modular invariants of the individual subtheories. However, there are also many further modular invariants that are not of this product type. An example for a conformal embedding which gives rise to a non-product modular invariant is given by

$$(G_2^{(1)})_{k^\vee=1} \oplus (F_4^{(1)})_{k^\vee=1} \subset (E_8^{(1)})_{k^\vee=1}, \tag{5.4.15}$$

which produces a non-diagonal invariant for $G_2^{(1)} \oplus F_4^{(1)}$ at level (1,1) although the individual subtheories possess only the diagonal modular invariant. With simple currents, one can get many more non-product invariants. Namely, if Φ and Φ' are simple currents of two distinct factors of the tensor product, then $\Phi * \Phi'$ is also a simple current, but the modular invariant to which it gives rise is usually not the tensor product of the invariants obtained via Φ and Φ'. Moreover, if $\mathcal{Z}(\Phi * \Phi')$ is indeed a new invariant, one can also multiply it with $\mathcal{Z}(\Phi)$ or $\mathcal{Z}(\Phi')$, leading typically again to new invariants, and so on.

One might hope that all such additional invariants can be explained either by simple currents or by conformal embeddings so that also these additional invariants could be found in a systematic way. Unfortunately, this is not the case. An example for a non-product invariant that is neither related to simple currents nor to conformal embeddings is the $(A_1^{(1)})_{k^\vee=2} \oplus (A_1^{(1)})_{k^\vee=10}$-invariant

$$\begin{aligned} Z = {} & |\chi_0\psi_0 + \chi_0\psi_6 + \chi_2\psi_4 + \chi_2\psi_{10}|^2 + |\chi_1\psi_3 + \chi_1\psi_7|^2 \\ & + [(\chi_0\psi_4 + \chi_0\psi_{10} + \chi_2\psi_0 + \chi_2\psi_6)(\bar\chi_1\bar\psi_3 + \bar\chi_1\bar\psi_7) + \text{c.c.}]\,, \end{aligned} \tag{5.4.16}$$

where $\chi_{\bar\Lambda}$ and $\psi_{\bar\Lambda}$ stand for the characters of the level 2 and level 10 $A_1^{(1)}$ algebra, respectively.

The classification of modular invariants for semisimple WZW theories is not only interesting in itself, but also turns out to be closely related to the classification of modular invariants for coset conformal field theories. (For the definition of coset theories, see p. 167; actually it is rather probable that *all* rational conformal field theories can be obtained via the coset construction, possibly supplemented by dividing out the action of a discrete symmetry). The reason for this is that the characters of a coset theory g/h are essentially given by the so-called *branching functions*, i.e. the coefficients $b_{\Lambda\lambda}$ of the branching rules for the characters $\tilde\chi_\Lambda$ of g into the characters $\tilde\chi_\lambda$ of h,

$$\tilde\chi_\Lambda(\tau) = \sum_\lambda b_{\Lambda\lambda}\,\tilde\chi_\lambda(\tau). \tag{5.4.17}$$

(If $\mathsf{h} \subset \mathsf{g}$ is a conformal embedding, then the branching functions are just constants so that there is no non-trivial coset theory.) The modular transformation properties of branching functions are easily expressed through those of $\tilde\chi_\Lambda$ and $\tilde\chi_\lambda$:

$$\begin{aligned} b_{\Lambda\lambda}(-\tfrac{1}{\tau}) &= \textstyle\sum_{\Lambda',\lambda'} S_{\Lambda\Lambda'}\, b_{\Lambda'\lambda'}(\tau)\, S^*_{\lambda\lambda'}, \\ b_{\Lambda\lambda}(\tau+1) &= \mathrm{e}^{2\pi i(\Delta_\Lambda-\Delta_\lambda)}\, b_{\Lambda\lambda}(\tau), \end{aligned} \tag{5.4.18}$$

where $S_{\Lambda\Lambda'}$ is the matrix implementing $\tau \mapsto -\frac{1}{\tau}$ for the WZW theory based on g, and $S_{\lambda\lambda'}$ the corresponding matrix for the theory based on h. Formally, the branching functions behave therefore just as the characters of some conformal field theory which is the tensor product of the WZW theory based on g with a theory whose modular transformation matrices are the complex conjugates of the corresponding matrices of the WZW theory based on h. It turns out that one can indeed get all modular invariant partition functions of coset theories by formally treating them as tensor product theories. There are, however, some subtleties related to the fact that some of the branching functions may vanish identically (as a consequence of selection rules due e.g. to constraints by the conjugacy class structure of the horizontal subalgebras), and others may be identical to each other (this is again related to the presence of simple currents). As a consequence, not all characters of the coset theory are really equal to the branching functions. Fortunately, these subtleties do not inhibit the classification of modular invariants.

5.5 WZW fusion rules and truncated tensor products

An algorithm for the determination of WZW fusion rule algebras has already been presented in section 3.5, namely the depth rule (3.5.12). As mentioned there (and as explicated with the example of $E_8^{(1)}$ at level two), application of this rule is quite tedious even for low values of the level. Particularly unpleasant is the fact that for the computation of a fusion rule $\phi_\Lambda * \phi_{\Lambda'}$ one must know explicitly the weight system of both modules $R_{\bar\Lambda}$ and $R_{\bar\Lambda'}$ of the horizontal subalgebra $\bar{\mathsf{g}}$ which are carried by the primary fields ϕ_Λ and $\phi_{\Lambda'}$; in addition, rather detailed information about the Clebsch–Gordan coefficients of the Kronecker product $R_{\bar\Lambda} \times R_{\bar\Lambda'}$ is needed. Therefore it is desirable to find a more efficient algorithm for the calculation of WZW fusion rules.

In the following it will be shown that an algorithm possessing the desired properties is provided by the extended Racah–Speiser algorithm, which was already discussed in the context of quantum groups in section 4.3. To apply this algorithm, nothing else is needed than the weight system of one of the $\bar{\mathfrak{g}}$-modules involved, say $R_{\bar\Lambda}$ (which can conveniently been chosen such that e.g. $d_{\bar\Lambda} \leq d_{\bar\Lambda'}$), and the level $k^\vee$ of $\mathfrak{g}$. Moreover, one may of course assume that the $\bar{\mathfrak{g}}$-Kronecker product

$$R_{\bar\Lambda} \times R_{\bar\Lambda'} = \bigoplus \mathcal{L}_{\bar\Lambda\bar\Lambda'}^{\ \ \bar\Lambda''} R_{\bar\Lambda''} \tag{5.5.1}$$

is already known, e.g. by application of the ordinary Racah–Speiser algorithm. The extension of the algorithm will then give a simple expression for the fusion rule coeffients $\mathcal{N}_{\Lambda\Lambda'}^{\ \ \Lambda''}$ in terms of the Littlewood–Richardson coefficients $\mathcal{L}_{\bar\Lambda\bar\Lambda'}^{\ \ \bar\Lambda''}$. Note that the sum in (5.5.1) extends over the set of dominant integral highest weights of $\bar{\mathfrak{g}}$; in the following we will simplify the notation by letting the sum run over the whole weight lattice of $\bar{\mathfrak{g}}$, which is achieved by simply setting

$$\mathcal{L}_{\bar\Lambda\bar\Lambda'}^{\ \ \bar\Lambda''} \equiv 0 \quad \text{for} \quad \{(\bar\Lambda'')^i \mid i = 1, \dots, r\} \not\subset \mathbb{Z}_{\geq 0}. \tag{5.5.2}$$

Still another approach to the calculation of $\mathcal{N}_{\Lambda\Lambda'}^{\ \ \Lambda''}$ is given by the relation (5.3.17) between these coefficients and the modular transformation matrix S. Unfortunately, although the matrix S is in principle known for all WZW theories (see (2.7.24)), more explicit expressions for S are not generally computed, and even when they are, they tend to become very complicated at large rank and/or level. As a consequence, the extended Racah–Speiser algorithm is more convenient for practical calculations than the formula (5.3.17). However, it turns out that the proof of this algorithm is most straightforward if the connection between S and the fusion rules is employed.

To enter the proof of the algorithm, we note that, as a consequence of the identities (2.7.34) and (2.7.32) for the modular matrix S, the fusion rule eigenvalues (5.3.13) of WZW theories can be written as

$$\ell_\Lambda^{(\Lambda')} = \mathrm{tr}_{\bar\Lambda}\, p^{\bar\Lambda' + \bar\rho} \tag{5.5.3}$$

with

$$p = \exp\left(-\frac{2\pi i}{k + I_{\mathrm{ad}}}\right). \tag{5.5.4}$$

In other words, the fusion rule eigenvalues are given by the characters $\chi_{\bar\Lambda}$ of $\bar{\mathfrak{g}}$, evaluated on special elements of the weight space,

$$\ell_\Lambda^{(\Lambda')} = \chi_{\bar\Lambda}\left(\ln p \cdot (\bar\Lambda' + \bar\rho)\right), \tag{5.5.5}$$

which with the help of the Weyl character formula can be written as

$$\ell_\Lambda^{(\Lambda')} = \frac{\sum_{\bar\sigma\in\bar W} (\mathrm{sign}\,\bar\sigma)\, p^{(\bar\sigma(\bar\Lambda+\bar\rho),\bar\Lambda'+\bar\rho)}}{\sum_{\bar\sigma\in\bar W} (\mathrm{sign}\,\bar\sigma)\, p^{(\bar\sigma(\bar\rho),\bar\Lambda'+\bar\rho)}}. \tag{5.5.6}$$

It is important to realize that via (5.5.6), $\ell_\Lambda^{(\Lambda')}$ becomes defined for any integral weight (and in fact for arbitrary elements of the weight space of $\mathfrak{g}$), not only for the integrable ones; of course, for non-integrable weights $\ell_\Lambda^{(\Lambda')}$ has no longer the meaning of an eigenvalue of the fusion rule algebra. Taking (5.5.6) for arbitrary integral weights, one can use the group property of the modified Weyl reflections $\bar\sigma \star \bar\Lambda = \bar\sigma(\bar\Lambda+\bar\rho) - \bar\rho$, the invariance of the inner product of $\bar{\mathfrak{g}}$ under $\bar W$ and the fact that the inner product of a weight and a coroot of $\bar{\mathfrak{g}}$ is always integral, to conclude that the fusion rule eigenvalues possess the symmetry properties

$$\ell_\Lambda^{(\Lambda')} = (\mathrm{sign}\,\bar\sigma)\, \ell_{w_{\bar\sigma,\bar\beta}(\Lambda)}^{(\Lambda')} \tag{5.5.7}$$

and

$$\ell_\Lambda^{(\Lambda')} = \ell_\Lambda^{(w_{\bar\sigma,\bar\beta}(\Lambda'))}. \tag{5.5.8}$$

Here $w_{\bar\sigma,\bar\beta}$ denotes the transformation

$$w_{\bar\sigma,\bar\beta}(\Lambda) := w_{\bar\sigma,\bar\beta}(\bar\Lambda) + \Big(k - (w_{\bar\sigma,\bar\beta}(\bar\Lambda),\bar\theta)\Big)\, \Lambda_{(0)} \tag{5.5.9}$$

with

$$w_{\bar\sigma,\bar\beta}(\bar\Lambda) := \bar\sigma \star \bar\Lambda + (k + I_{\mathrm{ad}})\,\bar\beta, \tag{5.5.10}$$

where $\bar\sigma$ is an arbitrary Weyl reflection and $\bar\beta$ an arbitrary vector in the coroot lattice of $\bar{\mathfrak{g}}$,

$$\bar\sigma \in \bar W, \quad \bar\beta \in L^\vee(\bar{\mathfrak{g}}). \tag{5.5.11}$$

Now recall from section 2.6 that the affine Weyl group W is the semidirect product of $\bar W$ and of a translation group which is isomorphic to the coroot lattice of $\bar{\mathfrak{g}}$, and that it is generated by the fundamental reflections $\sigma_{(i)}$, $i = 0, 1, \ldots, r$. Moreover, using (1.7.22), one has

$$w_{\bar\sigma_{\bar\theta},\bar\theta} \star \bar\Lambda = \bar\sigma_{\bar\theta}(\bar\Lambda) + (k+1)\,\bar\theta^\vee, \tag{5.5.12}$$

which according to (2.6.17) essentially coincides with the horizontal projection $\bar\sigma_{(0)}$ of the zeroth fundamental affine reflection,

$$w_{\bar\sigma_{\bar\theta},\bar\theta}(\bar\Lambda) = \bar\sigma_{(0)}(\bar\Lambda) + \bar\theta. \tag{5.5.13}$$

As a consequence, the transformations $w_{\bar\sigma,\bar\beta}$ form a group which is isomorphic to the horizontal projection of the affine Weyl group, and they are generated by the fundamental transformations $w_{(i)}$, $i = 0, 1, \dots, r$ defined by

$$w_{(i)}(\bar{\Lambda}) = \begin{cases} w_{\bar{\sigma}_{(i)},0}(\bar{\Lambda}) = \bar{\sigma}_{(i)} \star \bar{\Lambda} & \text{for } \; i = 1, \dots, r \\ w_{\bar{\sigma}_{\bar{\theta}},\bar{\theta}}(\bar{\Lambda}) = \bar{\sigma}_{(0)} \star \bar{\Lambda} & \text{for } \; i = 0. \end{cases} \tag{5.5.14}$$

Using the fact that the affine Weyl group permutes transitively and freely the affine Weyl chambers, it therefore follows that the closure P_k^+ of the dominant affine chamber is a fundamental domain of the group of transformations $w_{\bar{\sigma},\bar{\beta}}$, i.e. to any affine weight λ there exists a weight λ' and a transformation

$$w_{(\lambda')} \equiv w_{\bar{\sigma}_{(\lambda')},\bar{\beta}_{(\lambda')}} \tag{5.5.15}$$

such that both $\lambda' + \rho \in P^+_{k+I_{\mathrm{ad}}}$ and $\lambda = w_{(\lambda')}(\lambda')$, and this transformation is unique unless $\lambda' + \rho$ lies on the boundary of $P^+_{k+I_{\mathrm{ad}}}$. (In particular, for $\lambda + \rho$ lying in the interior of $P^+_{k+I_{\mathrm{ad}}}$, the transformation $w_{(\lambda')}$ is the identity map, i.e. $\lambda' = \lambda$.)

It is also easily seen that $\ell_\Lambda{}^{(\mu)}$ vanishes, for any integral weight μ, if $\lambda' + \rho = w_{(\lambda')}{}^{-1}(\lambda) + \rho$ lies on the boundary of $P^+_{k+I_{\mathrm{ad}}}$. For $\lambda' + \rho$ on the "affine boundary," i.e. for $(\bar{\lambda}' + \bar{\rho}, \bar{\theta}) = k + I_{\mathrm{ad}}$ or in other words

$$(\bar{\lambda}', \bar{\theta}^\vee) = k^\vee + 1, \tag{5.5.16}$$

this is seen as follows. (5.5.12) can be written as

$$w_{(0)}(\bar{\lambda}) = \bar{\lambda} + \left(k + \tfrac{1}{2}(\bar{\theta}, \bar{\theta}) - (\bar{\lambda}, \bar{\theta})\right) \bar{\theta}^\vee; \tag{5.5.17}$$

in the present case, this implies $\lambda' = w_{(0)}(\lambda) = \lambda$ so that the symmetry property (5.5.7) shows that

$$\ell_\lambda{}^{(\mu)} = (\operatorname{sign} \bar{\sigma}_{\bar{\theta}})\, \ell_\lambda{}^{(\mu)} = -\ell_\lambda{}^{(\mu)} = 0. \tag{5.5.18}$$

Similarly, if $\lambda' + \rho$ lies on a different boundary of $P^+_{k+I_{\mathrm{ad}}}$, then there exists a simple root $\alpha^{(i)}$, $i \in \{1, \dots, r\}$, of $\bar{\mathfrak{g}}$ such that the associated reflection $w_{(i)}$ gives $w_{(i)}(\lambda') = \lambda'$, implying that $\ell_{\lambda'}{}^{(\mu)} = 0$ and hence also $\ell_\lambda{}^{(\mu)} = 0$.

Next we combine the result (5.5.7) with the character sum rule (1.7.46) for the Kronecker products of $\bar{\mathfrak{g}}$ which in the present notation reads

$$\chi_{\bar{\Lambda}}\, \chi_{\bar{\Lambda}'} = \sum_{\bar{\Lambda}''} \mathcal{L}_{\bar{\Lambda}\bar{\Lambda}'}^{\bar{\Lambda}''}\, \chi_{\bar{\Lambda}''}, \tag{5.5.19}$$

with the summation ranging over the weight lattice of $\bar{\mathfrak{g}}$. Evaluating this sum rule on the weight $\ln p \cdot (\bar{\mu} + \bar{\rho})$, one obtains a sum rule for the quantities $\ell_\Lambda{}^{(\mu)}$,

$$\ell_\Lambda{}^{(\mu)}\, \ell_{\Lambda'}{}^{(\mu)} = \sum_{\bar{\Lambda}''} \mathcal{L}_{\bar{\Lambda}\bar{\Lambda}'}^{\bar{\Lambda}''}\, \ell_{\Lambda''}{}^{(\mu)}. \tag{5.5.20}$$

With the help of (5.5.7), this can be rewritten as

$$\ell_\Lambda^{(\mu)}\,\ell_{\Lambda'}^{(\mu)} = \sum_{\bar\Lambda''} \sum_{\substack{\lambda\in P_k^+ \\ \lambda=w_{(\lambda)}^{-1}(\Lambda'')}} (\mathrm{sign}\,\bar\sigma_{(\lambda)})\, \mathcal{L}_{\bar\Lambda\bar\Lambda'}^{\;\bar\Lambda''}\, \ell_\lambda^{(\mu)} . \tag{5.5.21}$$

Finally, we combine (5.5.21) with the relation between the fusion rule coefficients and the modular matrix S which can be stated in the form

$$\mathcal{N}_{\Lambda\Lambda'}^{\;\;\Lambda''} = \sum_{\mu\in P_k^+} \ell_\Lambda^{(\mu)}\,\ell_{\Lambda'}^{(\mu)}\, S_{k^\vee\Lambda_{(0)},\mu}\, S^*_{\mu\Lambda''} . \tag{5.5.22}$$

This yields

$$\begin{aligned}\mathcal{N}_{\Lambda\Lambda'}^{\;\;\Lambda''} &= \sum_{\mu\in P_k^+} \sum_{\bar\nu} \sum_{\substack{\lambda\in P_k^+ \\ \lambda=w_{(\lambda)}^{-1}(\nu)}} (\mathrm{sign}\,\bar\sigma_{(\lambda)})\, \mathcal{L}_{\bar\Lambda\bar\Lambda'}^{\;\bar\nu}\, \ell_\lambda^{(\mu)}\, S_{k^\vee\Lambda_{(0)},\mu}\, S^*_{\mu\Lambda''} \\ &= \sum_{\bar\nu} \sum_{\substack{\lambda\in P_k^+ \\ \lambda=w_{(\lambda)}(\nu)}} (\mathrm{sign}\,\bar\sigma_{(\lambda)})\, \mathcal{L}_{\bar\Lambda\bar\Lambda'}^{\;\bar\nu} \sum_{\mu\in P_k^+} S_{\lambda\mu}\, S^*_{\mu\Lambda''} ,\end{aligned} \tag{5.5.23}$$

and hence, due to the unitarity and symmetry of the matrix S,

$$\mathcal{N}_{\Lambda\Lambda'}^{\;\;\Lambda''} = \sum_{w_{\bar\sigma,\bar\beta}} (\mathrm{sign}\,\bar\sigma)\, \mathcal{L}_{\bar\Lambda\bar\Lambda'}^{\;\;w_{\bar\sigma,\bar\beta}(\bar\Lambda'')} . \tag{5.5.24}$$

This is the promised relation between the fusion rule coefficients and Littlewood–Richardson coefficients.

Of course, (5.5.24) would not be of practical use if one really had to perform the sum over all possible $w_{\bar\sigma,\bar\beta}$. Fortunately, what is really needed are just the fundamental reflections (5.5.14), since they finitely generate the whole group of transformations. Moreover, one can easily define a canonical procedure to express the relevant reflections $w_{(\lambda')}$ through the generators $w_{(i)}$; a simple prescription is e.g. the lexicographical one defined as follows: if $(\bar\lambda,\bar\theta) \geq k+1$, first apply the transformation $w_{(0)}$; otherwise apply $w_{(i)}$ with i the smallest integer such that $\lambda^i < 0$; if after this step the reflected weight lies in the fundamental affine chamber, one is finished, otherwise the procedure is repeated for the reflected weight; since the group of transformations $w_{\bar\sigma,\bar\beta}$ is finitely generated, the process terminates after a finite number of steps. (The lexicographical prescription does not necessarily lead to an expression for $w_{(\lambda')}$ which contains a minimal number of factors $w_{(i)}$, but it is most convenient for explicit calculations.)

Using this information about the maps $w_{\bar\sigma,\bar\beta}$, one obtains the following recipe for evaluation (5.5.24). For given Λ and Λ', use the ordinary Racah–Speiser algorithm to find the set of weights $\bar\nu$ which obey

$\mathcal{L}_{\bar{\Lambda}\bar{\Lambda}'}^{\bar{\nu}} \neq 0$. Among these, consider the subset of weights satisfying $(\bar{\nu},\bar{\theta}) > k$ (in the cases relevant to the fusion rules, one has $(\bar{\Lambda},\bar{\theta}) \leq k$ and $(\bar{\Lambda}',\bar{\theta}) \leq k$ so that, in addition, $(\bar{\nu},\bar{\theta}) \leq 2k$). Then apply $w_{(0)}$ to any of these weights $\bar{\nu}$. If $w_{(0)}(\bar{\nu})$ is a dominant weight of $\bar{\mathfrak{g}}$, then one has, as is easily concluded from (5.5.17), $w_{(0)}(\nu) + \rho \in P^{+}_{k+I_{\rm ad}}$, and hence $w_{\bar{\sigma},\bar{\beta}} = w_{(0)}^{-1} = w_{(0)}$ in the corresponding contribution to the sum (5.5.24). Otherwise there is some Weyl reflection $\bar{\sigma}' \in \bar{W}$ such that $(w_{\bar{\sigma}',0} \circ w_{(0)})(\bar{\nu}) = ((\prod_{n,\, i_n \geq 1} w_{(i_n)}) \circ w_{(0)})(\bar{\nu})$ is a dominant $\bar{\mathfrak{g}}$-weight; if the associated affine weight fulfills $(w_{\bar{\sigma}',0} \circ w_{(0)})(\nu) + \rho \in P^{+}_{k+I_{\rm ad}}$, then ${w_{\bar{\sigma},\bar{\beta}}}^{-1} = w_{\bar{\sigma}',0} \circ w_{(0)}$ in the corresponding contribution to (5.5.24); if, on the other hand, the latter condition is not fulfilled, then just repeat the previous manipulations, now applied to $(w_{\bar{\sigma}',0} \circ w_{(0)})(\bar{\nu})$ rather than to $\bar{\nu}$, and so on until the process terminates.

Finally, recalling the remarks on the weights which lie on the boundary of $P^{+}_{k+I_{\rm ad}}$, one learns that the term involving $\mathcal{L}_{\bar{\Lambda}\bar{\Lambda}'}^{\bar{\nu}}$ has to be discarded from the sum (5.5.24) iff $\nu \equiv w_{\bar{\sigma},\bar{\beta}}(\Lambda'')$ lies on the boundary of $P^{+}_{k+I_{\rm ad}}$; otherwise this term provides a contribution $(\mathrm{sign}\, \bar{\sigma}) \cdot \mathcal{L}_{\bar{\Lambda}\bar{\Lambda}'}^{\bar{\nu}}$ to $\mathcal{N}_{\Lambda\Lambda'}^{\ \ \Lambda''}$.

The recipe just given is nothing but the extended Racah–Speiser algorithm. It is both simple and very efficient. For low-dimensional modules everything can be easily worked out by hand, and in any case it is no problem to put the algorithm on a computer.

A further simplification arises from the observation that the fusion rule coefficients transform in a simple way under the maps ω which are induced (compare p. 286) by automorphisms $\dot{\omega}$ of the Dynkin diagram of $\mathfrak{g}$. To see this, consider first the case where $\dot{\omega} \in \Gamma/\bar{\Gamma}$ (for the notation, see p. 286). It is not difficult to show that the modular matrix S transforms under ω as

$$S_{\Lambda,\omega(\Lambda')} = S_{\Lambda\Lambda'} \exp\left(-2\pi i\,(\omega(k^\vee \Lambda_{(0)}), \Lambda)\right). \tag{5.5.25}$$

Together with the symmetry of S, this implies that

$$\ell_{\omega(\Lambda)}^{(k^\vee \Lambda_{(0)})} = \ell_{\Lambda}^{(k^\vee \Lambda_{(0)})}, \tag{5.5.26}$$

which proves the symmetry property of quantum dimensions that was postulated on p. 286. More generally, since the factor arising in the transformation of S is merely a phase, it follows immediately from (5.3.14) that the fusion rule coefficients obey

$$\mathcal{N}_{\Lambda,\omega(\Lambda')}^{\ \ \ \omega(\Lambda'')} = \mathcal{N}_{\Lambda\Lambda'}^{\ \ \Lambda''} \tag{5.5.27}$$

for all $\dot{\omega} \in \Gamma/\bar{\Gamma}$. Thus these automorphisms can be applied to the fusion rules in order to reduce the number of independent coefficients. This is particularly helpful for $\bar{\mathfrak{g}} = A_r$ where $\Gamma/\bar{\Gamma}$ is isomorphic to $\mathbb{Z}_{r+1}$, while it does not provide any information for $\bar{\mathfrak{g}} = E_8$, G_2 or F_4 because then $\Gamma/\bar{\Gamma} \cong 1$.

We now recall that the extended Racah–Speiser algorithm was used in section 4.3 to identify the (non-irreducible) subspaces of non-fully reducible modules of $\mathsf{U}_q(\bar{\mathfrak{g}})$. In view of the results of section 4.5, this also means that the truncated tensor products of $\mathsf{U}_q(\bar{\mathfrak{g}})$ can be computed via this algorithm. As a consequence, the truncated tensor products of $\mathsf{U}_q(\bar{\mathfrak{g}})$ are isomorphic to the fusion rules of the WZW theory based on $\mathfrak{g}$, i.e. one has

$$(\mathcal{L}^\star)_{\bar{\Lambda}\bar{\Lambda}'}^{\;\;\bar{\Lambda}''} = \mathcal{N}_{\Lambda\Lambda'}^{\;\;\Lambda''} . \tag{5.5.28}$$

Also, the parameter q of $\mathsf{U}_q(\bar{\mathfrak{g}})$ and the level $k^\vee$ of $\mathfrak{g}$ are related by (4.3.25),

$$q = \exp\left(\frac{2\pi i}{k^\vee + g^\vee}\right), \tag{5.5.29}$$

or in terms of the number p introduced in (5.5.4),

$$q = p^{-(\bar{\theta},\bar{\theta})/2} . \tag{5.5.30}$$

(Note that p depends on the normalization of the bilinear form of $\bar{\mathfrak{g}}$, which is necessary in order to have $\mathrm{tr}_{\bar{\Lambda}}\, p^{\bar{\Lambda}'+\bar{\rho}}$ normalization independent, and hence the parameter q as determined here is normalization independent as it should be.)

There exist convincing arguments that a proof of the extended Racah–Speiser algorithm can be given purely in terms of the truncated Kronecker products of the quantum group, namely via the identification of the null vector structure of the quantum group Verma modules, although for arbitrary $\bar{\mathfrak{g}}$ a rigorous proof has not yet been established. On the other hand, as we have just seen, the same algorithm applies to the fusion rules of WZW theories. Together this then proves in particular the isomorphism between the truncated tensor products and WZW fusion rules. There exist also several independent (i.e., not based on the extended Racah–Speiser algorithm) arguments for this isomorphism; since except for $\bar{\mathfrak{g}} = A_1$ the null vector structure of quantum group modules has not yet been worked out in full detail, we present also these independent arguments:

- By comparing (4.4.11) and (5.5.5), one sees that the quantum dimensions are just the fusion rule eigenvalues for the identity primary field,

$$\mathcal{D}_\Lambda = S_{\Lambda,k^\vee\Lambda_0}/S_{k^\vee\Lambda_0,k^\vee\Lambda_0} \equiv \ell_\Lambda^{(k^\vee\Lambda_{(0)})} \tag{5.5.31}$$

(or in the notation of section 2.8, $\mathcal{D}_\Lambda = a_\Lambda / a_{k^\vee \Lambda_{(0)}}$). Thus the quantum dimension sum rule (4.5.11) for truncated tensor products is nothing but the special case $\bar{\Lambda}' = 0$ of the formula (5.3.2) which expresses the fact that the fusion rule eigenvalues are representations of the fusion rule algebra. For low values of the level $k^\vee$, the quantum dimension sum rule alone often already fixes the corresponding products uniquely, and so in these cases this simple identification suffices to prove the isomorphism.
One can show that for any other conformal field theory that possesses an underlying quantum group structure, the quantum dimensions relevant to the correctly truncated Kronecker products again coincide with the relative size of the corresponding modules of the chiral symmetry algebra that was defined in (5.3.18), $\mathcal{D}_i = S_{i0}/S_{00} = \ell_i^{(0)}$.

- More generally, recall that the non-fully reducible modules are built from subspaces whose quantum dimensions add up to zero. The extended Racah–Speiser algorithm always provides a solution to this requirement. Thus the isomorphism follows immediately for all those cases where there is a unique possibility for satisfying this requirement (also, one can often use the fact that the truncated tensor products must respect the conjugacy classes to reduce the number of a priori solutions to a unique one).
- For the particular case $\mathfrak{g} = A_1$, the isomorphism between truncated Kronecker products and WZW fusion rules simply follows by inspection. Compare, for example, the relevant formulae (4.5.17) and (5.1.13).
- The symmetry properties (4.4.35) of quantum dimensions (which, however, were proven independently of the connection with the corresponding WZW fusion rules only for $\mathfrak{g} = A_r$) provide strong constraints on the possible truncated Kronecker products. This immediately suggests that they are isomorphic to the WZW fusion rules, since according to (5.5.27) these possess the same symmetry properties.
- We also note that it has been claimed that one can derive the depth rule (3.5.12) for WZW fusion rules directly from the description of the truncation of Kronecker products in terms of forming semisimple quotients. However, so far the proof of this assertion has not been made explicit in the literature.

Finally, there does exist a complete proof of the isomorphism for the case of $\mathfrak{g} = A_r$ which, however, uses several ingredients that go beyond the scope of this book. Briefly, the argument goes as follows. Both the fusion rules of the $A_r^{(1)}$ WZW theory at level $k^\vee$ and the truncated Kronecker products of $\mathsf{U}_q(A_r)$ for $q = \exp(2\pi i/(k^\vee + r + 1))$ possess the property that the product between any of the "basic" fields (respectively modules) characterized by $\bar\Lambda = \bar\Lambda_{(i)}$, $i \in \{0, 1, \ldots, r\}$, and any other field is obtained from the decomposition of the corresponding Kronecker product of A_r by simply deleting all modules that would give rise to non-integrable $A_r^{(1)}$-modules. For the fusion rules, this is an immediate consequence of the extended Racah–Speiser algorithm, while for the truncated Kronecker products it can be proven by investigating the properties of the associated centralizer algebras which for products involving the basic fields are just Hecke algebras. Now by purely algebraic methods, one can prove that there is a unique commutative associative ring possessing these properties. As a consequence, both the fusion rules and the truncated Kronecker products must be isomorphic to this unique ring, and hence isomorphic to each other.

Let us now list a few examples of fusion rule algebras of WZW theories. It turns out that at low values of the level, many isomorphisms of fusion rule algebras occur; in addition sometimes the fusion rules become isomorphic in terms of equivalence classes of primary fields with respect to the automorphisms of the extended Dynkin diagram. A well known example for this phenomenon are the Ising fusion rules (5.1.16); these are shared not only by the $E_8^{(1)}$ theory at level two, but by all $B_r^{(1)}$ theories, $r \geq 1$, at level one (here B_1 and B_2 stand for the Lie algebras A_1 and C_2, respectively; also, the genuine critical Ising model can be understood as a member of this series of theories by formally setting $r = 0$).

Consider next $A_1^{(1)}$ at level $k^\vee = 3$. The primary fields are given by $\varphi_{\bar\Lambda}$, $\bar\Lambda \in \{0, 1, 2, 3\}$ (we consider only one chiral half of the theory; also, we will label all fields by the highest weights with respect to the relevant horizontal subalgebra, and use φ as a generic name for $A_1^{(1)}$-primaries and ϕ for the primaries with respect to other affine algebras). These primaries fall into equivalence classes $[\varphi]$ with respect to the $\mathbb{Z}_2$-automorphism of the Dynkin diagram of $A_1^{(1)}$ as $[\varphi_0] = \{\varphi_0, \varphi_3\}$ and $[\varphi_1] = \{\varphi_1, \varphi_2\}$. In terms of these equivalence classes, the fusion rules (4.5.17) become

$$\begin{aligned} [\varphi_0] * [\varphi_0] &= [\varphi_0], \\ [\varphi_0] * [\varphi_1] &= [\varphi_1], \\ [\varphi_1] * [\varphi_1] &= [\varphi_0] + [\varphi_1]. \end{aligned} \tag{5.5.32}$$

These are the Lee–Yang fusion rules which were already encountered at the end of section 5.3. They are also obtained for $G_2^{(1)}$ and $F_4^{(1)}$, both at level $k^\vee = 1$; the correspondence between (equivalence classes of) primary fields is as follows:

$$\begin{array}{ccccc} A_1^{(1)},\ k^\vee = 3 & & G_2^{(1)},\ k^\vee = 1 & & F_4^{(1)},\ k^\vee = 1 \\ [\varphi_0] & \longleftrightarrow & \mathbf{1} & \longleftrightarrow & \mathbf{1} \\ [\varphi_1] & \longleftrightarrow & \phi_{(0\,1)} & \longleftrightarrow & \phi_{(0\,0\,0\,1)} \end{array} \tag{5.5.33}$$

Another example of this type is provided by the fusion rules of $A_1^{(1)}$ at level 9. When expressed in terms of the equivalence classes with respect to the $\mathbb{Z}_2$-automorphism, they become isomorphic to the fusion rules of $F_4^{(1)}$ at level 2, which in turn are isomorphic to those of $E_8^{(1)}$ at level 3 (cf. (4.5.27)). The necessary identifications of (equivalence classes of) primary fields are the following:

$$\begin{array}{ccccc} A_1^{(1)},\ k^\vee = 9 & & F_4^{(1)},\ k^\vee = 2 & & E_8^{(1)},\ k^\vee = 3 \\ [\varphi_0] & \longleftrightarrow & \mathbf{1} & \longleftrightarrow & \mathbf{1} \\ [\varphi_1] & \longleftrightarrow & \phi_{(1\,0\,0\,0)} & \longleftrightarrow & \phi_{(0\,0\,0\,0\,0\,0\,0\,1)} \\ [\varphi_2] & \longleftrightarrow & \phi_{(0\,0\,0\,2)} & \longleftrightarrow & \phi_{(0\,1\,0\,0\,0\,0\,0\,0)} \\ [\varphi_3] & \longleftrightarrow & \phi_{(0\,0\,1\,0)} & \longleftrightarrow & \phi_{(1\,0\,0\,0\,0\,0\,0\,0)} \\ [\varphi_4] & \longleftrightarrow & \phi_{(0\,0\,0\,1)} & \longleftrightarrow & \phi_{(0\,0\,0\,0\,0\,0\,1\,0)} \end{array} \tag{5.5.34}$$

As an example of fusion rules which involve multiplicities larger than one (and hence cannot be isomorphic to those of $A_1^{(1)}$ at any level) we mention those of $F_4^{(1)}$ at level 3. Some of them have been listed, as truncated Kronecker products, in (4.5.30). It turns out that isomorphic fusion rules are obtained for $G_2^{(1)}$ at level 4, via the identifications

$$\begin{array}{ccccc} F_4,\ k^\vee = 3 & & & G_2,\ k^\vee = 4 & \\ \phi_{(0\,0\,0\,1)} & (26) & \longleftrightarrow & \phi_{(1\,0)} & (14) \\ \phi_{(1\,0\,0\,0)} & (52) & \longleftrightarrow & \phi_{(1\,1)} & (77') \\ \phi_{(0\,0\,1\,0)} & (273) & \longleftrightarrow & \phi_{(0\,3)} & (77) \\ \phi_{(0\,0\,0\,2)} & (324) & \longleftrightarrow & \phi_{(0\,2)} & (27) \\ \phi_{(1\,0\,0\,1)} & (1053) & \longleftrightarrow & \phi_{(1\,2)} & (189) \\ \phi_{(0\,1\,0\,0)} & (1274) & \longleftrightarrow & \phi_{(0\,4)} & (182) \\ \phi_{(0\,0\,0\,3)} & (2652) & \longleftrightarrow & \phi_{(0\,1)} & (7) \\ \phi_{(0\,0\,1\,1)} & (4096) & \longleftrightarrow & \phi_{(1\,1)} & (64) \end{array} \tag{5.5.35}$$

(here the numbers in brackets denote the dimensions of the relevant F_4- respectively G_2-modules).

At levels five and seven, the same kind of phenomenon occurs. One finds the following isomorphisms of (quotiented) fusion rule algebras:

$$\begin{aligned} (A_1^{(1)})_{k^\vee=5}/\mathbb{Z}_2 &\longleftrightarrow (E_6^{(1)})_{k^\vee=2}/\mathbb{Z}_3, \\ (A_1^{(1)})_{k^\vee=7}/\mathbb{Z}_2 &\longleftrightarrow (G_2^{(1)})_{k^\vee=2}. \end{aligned} \tag{5.5.36}$$

There even exist infinite series of such correspondences, namely

$$\begin{aligned} (A_{n-1}^{(1)})_{k^\vee}/\mathbb{Z}_n &\longleftrightarrow (A_{k^\vee-1}^{(1)})_n/\mathbb{Z}_{k^\vee}, \\ (C_r^{(1)})_{k^\vee} &\longleftrightarrow (C_{k^\vee})_r, \end{aligned} \tag{5.5.37}$$

and by a somewhat more involved prescription one can relate $(so_n^{(1)})_{k^\vee}$ to $(so_{k^\vee}^{(1)})_n$.

The symmetry groups $\mathbb{Z}_n$ appearing in these relations are always given by $\Gamma/\bar\Gamma$, with Γ and $\bar\Gamma$ the symmetry groups of the Dynkin diagrams of $\mathfrak{g}$ and $\bar{\mathfrak{g}}$, respectively. Recall from the property (5.5.27) of the fusion rule coefficients that $\Gamma/\bar\Gamma$ acts as a symmetry on the fusion rule algebra of the WZW theory based on $\mathfrak{g}$. It is not difficult to see that when this group is divided out from the fusion rules, one obtains again an algebra possessing all the properties required for the fusion rule algebra of a conformal field theory, provided that the symmetry acts without fixed points. (In the case of $A_1^{(1)}$, the action of $\Gamma/\bar\Gamma \cong \mathbb{Z}_2$ is $\bar\lambda \mapsto k^\vee - \bar\lambda$, and hence the latter requirement means that the level $k^\vee$ must be an odd integer, as is the case in the examples desribed above.) Thus it is not surprising that after dividing by the symmetry one arrives at the fusion rule algebra of some other conformal field theory. What *is* surprising is that in so many cases this other conformal field theory is again a WZW theory.

Another interesting subclass of WZW fusion rules is provided by the fusion rules involving simple currents. If Φ is a simple current, then

$$\Phi * \Phi^+ = \mathbf{1}. \tag{5.5.38}$$

As a result, by combining the corresponding quantum dimension sum rule $\mathcal{D}_\Phi \mathcal{D}_{\Phi^+} = \mathcal{D}_{\mathbf{1}} = 1$ with the identity

$$\mathcal{D}_{\phi^+} = \mathcal{D}_\phi \tag{5.5.39}$$

which follows for any field ϕ from (5.3.15), one finds

$$\mathcal{D}_\Phi = 1. \tag{5.5.40}$$

Note that this also implies that on any orbit of primary fields with respect to the simple current, the quantum dimension is constant. (For those simple currents which correspond to automorphisms of the Dynkin diagram of $\mathfrak{g}$, this statement is just a special case of the invariance (5.5.26)).

It is not difficult to see that (5.5.40) is not only a necessary, but also a sufficient condition for Φ to be a simple current. Namely, assume that the fusion rules of Φ with the set $\{\phi_i\}$ of primary fields of the theory are given by

$$\Phi * \phi_i = \sum_j \phi_j^{(i)} . \tag{5.5.41}$$

Summing this over all primary fields ϕ_i, one obtains the quantum dimension sum rule

$$\sum_i \mathcal{D}_i = \mathcal{D}_\Phi \sum_i \mathcal{D}_i = \sum_{i,j} \mathcal{D}_j^{(i)} . \tag{5.5.42}$$

Since any of the fields ϕ_i must appear on the right hand side of at least one of the fusion rules (5.5.41) (otherwise one would be in conflict with the fact that at least one field must appear on the right hand side of any fusion rule), this formula together with the positivity of the quantum dimensions implies that each ϕ_i appears precisely once on the right hand side of (5.5.41), and hence for each of these fusion rules there is a single term on the right hand side, i.e. Φ is a simple current.

Another way to derive this result is to multiply (5.5.41) with Φ^+. If $\mathcal{D}_\Phi = 1$, then (5.5.39) must hold; inserting this and $\mathbf{1} * \phi = \phi$ into (5.5.41), it follows that the left hand side contains only a single term, and requiring the same for the right hand side shows that Φ must be a simple current. Thus, as already stated in section 5.1, (5.5.39) is also both a necessary and a sufficient condition for Φ to be a simple current. Conversely, for any primary field ϕ which is not a simple current, $\phi * \phi^+$ contains at least two primary fields so that

$$\mathcal{D}_\phi \geq \sqrt{2}, \tag{5.5.43}$$

with the lower bound obtained iff $\phi * \phi^+ = \mathbf{1} + \Phi$ with Φ a simple current. With more effort one can show that in addition $\mathcal{D} < 2$ implies that

$$\mathcal{D} = 2 \cos\left(\frac{\pi}{n}\right) \tag{5.5.44}$$

for some $n \in \mathbb{Z}$, $n \geq 3$.

Since the quantum dimensions are easily computed with the help of the q-deformed Weyl formula (4.4.14), the above result provides a very efficient method of searching for simple currents in WZW theories. Indeed one can prove that there are no further simple currents in WZW theories in addition to the ones already mentioned in section 5.1 (i.e. the cominimal fields and the exceptional case of ϕ_{3875} for level two E_8). The proof proceeds in two rather distinct steps: first one shows that the quantum dimension, considered as a continuous function on the weight space of $\mathfrak{g}$, does not possess any local minima inside the horizontal projection

$\bar{P}$ of the fundamental affine Weyl chamber, and iterating the argument shows that any possible global minima of $\mathcal{D}^q$ on $\bar{P}$ must lie on its corners; this conclusion can be reached using only very general properties of the root system of $\mathfrak{g}$. The second step is then to show that among the fields corresponding to these corners no other simple currents except the known ones are present; this can indeed be done, but it requires a lengthy case by case analysis based on the detailed structure of the root systems of the various simple Lie algebras. Note that for the level one cominimal fields, one can read off the property $\mathcal{D} = 1$ immediately from table (4.4.18) with the help of $\lfloor k^\vee + g^\vee \rfloor = 0$. Similarly, for the cominimal fields at arbitrary level one can derive the formulae given in table (5.5.45) for the quantum dimension.

(5.5.45)

$\mathfrak{g}$	i	$\mathcal{D}_{k^\vee \Lambda_{(i)}}$
A_r	$1,\ldots,r$	$\prod_{j=1}^{k}\{j, i+j-1\}$
B_r	1	$\{1, 2r-2\}\,\{r-\frac{1}{2}, r-\frac{1}{2}\}$
C_r	r	$\prod_{j=1}^{[\frac{r+1}{2}]}\{\{j, r-j+1\}\}$
D_r	1	$\{1, 2r-3\}\,\{r-1, r-1\}$
	$r-1,\, r$	$\prod_{j=1}^{[r/2]}\{2j-1, 2r-2j-1\}$
E_6	$1, 5$	$\{1, 11\}\,\{4, 8\}$
E_7	6	$\{1, 17\}\,\{5, 13\}\,\{9, 9\}$

In table (5.5.45) the abbreviations

$$\{a, b\} = \prod_{\substack{j=a \\ \bmod \mathbb{Z}}}^{b} \frac{\lfloor k^\vee + g^\vee - j \rfloor}{\lfloor j \rfloor}, \qquad \{\{a, b\}\} = \prod_{\substack{j=a \\ \bmod \mathbb{Z}/s}}^{b} \frac{\lfloor k^\vee + g^\vee - j \rfloor}{\lfloor j \rfloor} \tag{5.5.46}$$

are used, where s denotes the relative length squared of the long and short roots of $\mathfrak{g}$. This result holds for arbitrary deformation parameter q. At the value relevant to the WZW theory, all the fractions in these products are equal to unity, and so indeed $\mathcal{D} = 1$ for all these fields.

Let us also note that for the primary fields of the level two E_8 WZW theory, table (4.4.18) gives (with $\lfloor 32 \rfloor = 0$)

$$\begin{aligned} \mathcal{D}_{248} &= \lfloor 8 \rfloor \lfloor 12 \rfloor / \lfloor 6 \rfloor \lfloor 10 \rfloor, \\ \mathcal{D}_{3875} &= \lfloor 2 \rfloor \lfloor 12 \rfloor \lfloor 14 \rfloor / \lfloor 4 \rfloor \lfloor 6 \rfloor \lfloor 10 \rfloor, \end{aligned} \tag{5.5.47}$$

and to see that these numbers are equal to $\sqrt{2}$ and 1, respectively (cf. (4.5.32)), one must use some identities for the sine function.

5.6 Chiral blocks

According to chapter 3, among the basic ingredients of conformal field theory are the associativity of the operator product algebra and the fact that the symmetry algebra is a direct sum of a holomorphic and an antiholomorphic part. As a direct consequence, any four-point correlation function of primary fields,

$$F_{ijkl}(z,\bar{z}) \equiv \langle \phi_i(z,\bar{z})\phi_j(0,0)\phi_k(1,1)\phi_l(\infty,\infty)\rangle, \tag{5.6.1}$$

can be expressed through the operator product coefficients of the primary fields as

$$F_{ijkl}(z,\bar{z}) = \sum_m C_{ij}{}^m C_{mkl}\, \mathcal{F}^{(m)}_{ijkl}(z)\, \overline{\mathcal{F}}^{(m)}_{ijkl}(\bar{z}). \tag{5.6.2}$$

The functions $\mathcal{F}^{(m)}(z)$ and $\overline{\mathcal{F}}^{(m)}(\bar{z})$ appearing in this decomposition are known as the *chiral blocks* (or, if the chiral symmetry algebra is just the Virasoro algebra, as *conformal blocks*) for the four-point functions. They describe the contribution to the four-point function that is due to the representation R_m (respectively $\bar{R}_m$) of the holomorphic (respectively antiholomorphic) symmetry algebra that corresponds to some definite primary field ϕ_m and its descendants. If the fusion rule coefficients $\mathcal{N}_{ijm}$ and $\mathcal{N}_{mkl}$ are larger than one, then the sum over m must be supplemented by a sum over a degeneracy index α that labels the distinct ways of coupling the external fields to ϕ_m; for brevity this summation will be assumed to be implicit in the m-summation. Graphically the decomposition of F_{ijkl} can be summarized by associating to each term in the sum (5.6.1) a graph that is analogous to the Feynman diagrams for a φ^3-type quantum field theory: external legs describe the fields ϕ_i to ϕ_l, an internal line the exchanged field ϕ_m and its descendants, and nodes stand for the operator product coefficients:

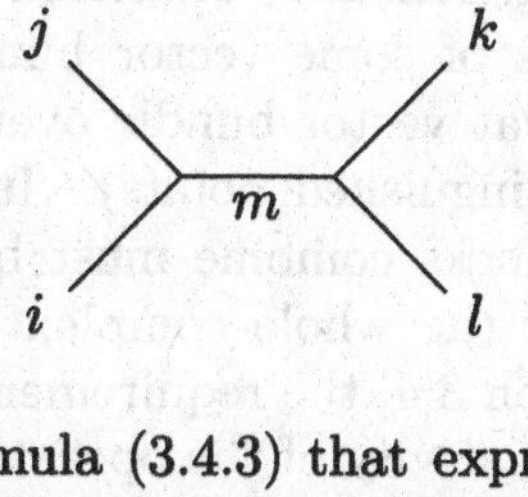

(5.6.3)

With the help of the formula (3.4.3) that expresses the operator product coefficients involving descendants through those involving only primaries, one can in principle compute the chiral blocks completely from the knowledge of the symmetry algebra and of the conformal dimensions of the primary fields. In practice, however, this is far too complicated.

Fortunately, one can also often obtain the blocks by solving certain differential equations which they obey as a consequence of the existence of null vectors in the modules of the symmetry algebra. This is e.g. the case for WZW theories for which the null vector equation is the Knizhnik-Zamolodchikov equation that was discussed in section 3.4. For several other classes of models, such as the discrete series of $c < 1$ conformal field theories, the blocks have been determined with the help of a so-called Coulomb gas formulation (see p. 396) that describes the primary fields in terms of free bosons in the presence of a background charge. For rational conformal field theories, the sum over m in the above decomposition of the four-point function is finite; we will soon see that this leads to other possibilities for the calculation of the blocks.

If the primary field ϕ_m is among those fields which contribute as intermediary fields to $\mathcal{F}^{(m)}$, then in the limit $z \to 0$ the chiral block behaves as

$$\lim_{z\to 0} \mathcal{F}^{(m)}_{ijkl}(z) = z^{\Delta_m - \Delta_i - \Delta_j}\,(1 + \mathcal{O}(z))\,. \tag{5.6.4}$$

However, it is also possible that only descendants of ϕ_m contribute to $\mathcal{F}^{(m)}$. In fact, this necessarily happens if the relevant fusion rule coefficients are larger than one. More precisely, the minimal values of the grades at which descendants of ϕ_m contribute to the distinct chiral blocks $\mathcal{F}^{(m,\alpha)}$, $\alpha = 1, 2, \ldots, \mathcal{N}_{ij}{}^{m}$, must all be different; this is so because otherwise it would not be possible to define unambigously the different couplings $C_{ij}{}^{m,\alpha}$. If $n > 0$ is the lowest possible grade of a field contributing to the chiral block $\mathcal{F}^{(m)}$, then in the limit $z \to 0$ one has

$$\lim_{z\to 0} \mathcal{F}^{(m)}_{ijkl}(z) \propto z^{\Delta_m + n - \Delta_i - \Delta_j}\,(1 + \mathcal{O}(z))\,, \tag{5.6.5}$$

where the precise constant of proportionality is given by the relevant coefficient $c_{ij}^{(m,\{n_p\})}$ that was defined in (3.4.3).

The chiral blocks are (generically) multivalued, i.e. not ordinary functions, but rather sections of some vector bundle (in more detail, this bundle is a projectively flat vector bundle over the moduli space of the two-sphere with four distinguished points). In contrast, the four-point functions to which the blocks combine must be ordinary functions and hence be single-valued in the whole complex z-plane. As already discussed at the end of section 3.4, the requirement of single-valuedness is a manifestation of the associativity of the operator product algebra, and it implies the crossing symmetry relations (also called *duality relations* or *bootstrap equations*) (3.4.61). In the terminology of relativistic kinematics, these identities relate the s-channel exchange to the t- and u-channel exchange, respectively. Thus in the graphical notation introduced above, they may be represented as in figure (5.6.6).

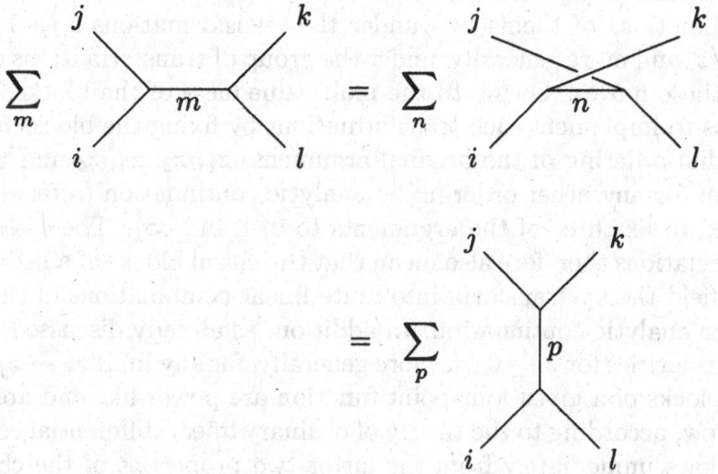

(5.6.6)

In terms of the chiral blocks, the duality relations read

$$\begin{aligned}\sum_m C_{ij}{}^m C_{mkl}\, \mathcal{F}^{(m)}_{ijkl}(z)\, \overline{\mathcal{F}}^{(m)}_{ijkl}(\bar z) &= \sum_n C_{ik}{}^n C_{njl}\, \mathcal{F}^{(n)}_{ikjl}(1-z)\, \overline{\mathcal{F}}^{(n)}_{ikjl}(1-\bar z) \\ &= z^{-2\Delta_i}\bar z^{-2\bar\Delta_i} \sum_p C_{il}{}^p C_{pkj}\, \mathcal{F}^{(p)}_{ilkj}(\tfrac{1}{z})\, \overline{\mathcal{F}}^{(p)}_{ilkj}(\tfrac{1}{\bar z}).\end{aligned} \tag{5.6.7}$$

Note that once the chiral blocks are known, this set of equations can be used to compute the operator product coefficients $C_{ij}{}^k$ of the theory (part of this computation is already implicitly described in section 3.4: the coefficients a_ν obtained in (3.4.60) are just the squares of operator product coefficients). In fact, this way the operator product coefficients are highly overdetermined. That nevertheless there does exist a solution to the system of equations is in a sense a proof of the conformal bootstrap hypothesis. Technically, it can be traced to the fact that the chiral blocks $\mathcal{F}^{(m)}_{ijkl}$ for fixed external legs are not just arbitrary multivalued functions, but rather form a system of independent solutions to a suitable differential equation. In principle, to solve a conformal field theory completely, one just needs to determine the chiral blocks. However, all methods that are known so far to perform this calculation are quite involved; as a consequence, the operator product coefficients for primary fields are completely known only for the minimal $c < 1$ conformal field theories and for the $A_1^{(1)}$ WZW theories.

For a *rational* conformal field theory, the sum over m in the decomposition (5.6.2) contains a finite number N_{ijkl} of terms. The crossing symmetry relations then imply that the chiral blocks $\mathcal{F}^{(m)}_{ijkl}$ transform into finite linear combinations of themselves under the transformations $z \mapsto 1-z$ and $z \mapsto 1/z$, and more generally, under the group of transformations generated by these moves. Owing to the multivaluedness of the blocks, one actually has to implement such transformations by fixing the blocks for a definite radial ordering of the original arguments z_1, z_2, z_3, z_4 and then obtain them for any other ordering by analytic continuation (afterwards it is allowed to fix three of the arguments to 0, 1 and ∞). The crossing symmetry relations therefore also mean that the chiral blocks of a rational conformal field theory transform into finite linear combinations of themselves under analytic continuation. In addition, as already discussed, the leading singularities for $z \to 0$ (or more generally, for any limit $z_i \to z_j$) of the chiral blocks of a given four-point function are power-like and are all distinct. Now, according to the theory of ordinary linear differential equations it follows immediately from the latter two properties of the chiral blocks that they form a system of independent solutions to a differential equation of order N_{ijkl} in the variable z.

More precisely, the differential equation is of the so-called Fuchsian type, with regular singular points at $z = 0, 1$ and ∞ (and also possibly at a few other values of z which, however, must represent so-called *apparent* singularities, i.e. points for which only the differential equation, but none of its solutions, behaves singularly). The transformations acting on the solution space of a differential equation that implement analytic continuation around closed loops form a group; this group is called the *monodromy* group of the differential equation. Now, applying a duality transformation twice has the same effect as analytic continuation around a closed loop; therefore the term half-monodromy is sometimes used as a synonym for duality. The monodromy representations at the singular points 0, 1, ∞ are irreducible, and at any other point in the complex z-plane (including the apparent singularities) the monodromy is trivial. Moreover, the power law behavior at the singular points 0, 1, ∞ is determined uniquely by the fusion rules and the conformal dimensions of the primary fields, and from the knowledge of this power law behavior one can determine the explicit form of the differential equation up to a few free parameters which are called accessory parameters in the theory of differential equations. (Again it must be stressed that in practice this is feasible only if N_{ijkl} is small; in fact only for the rather trivial case $N_{ijkl} = 1$ and for $N_{ijkl} = 2$ have calculations of this type been performed so far.)

As a consequence, at least in principle it is possible to obtain a classification of all rational conformal field theories via the following recipe: first enumerate the possible fusion rule algebras together with the conformal dimensions of the primary fields; then for each such algebra determine the differential equations for the chiral blocks; then solve these equations to find first the blocks, and from them finally the operator product coefficients via the crossing symmetry relations. Actually, the choice of conformal dimensions for the primary fields is not a completely independent step in this program. Namely, from the analytic properties of ordinary differential equations with regular singular points one easily derives the sum rule

$$N_{ijkl}\,(\Delta_i+\Delta_j+\Delta_k+\Delta_l+\tfrac{1}{2}(N_{ijkl}-1))-\sum_{a=s,t,u}\ \sum_{m(a)=1}^{N_{ijkl}}\Delta_{m(a)}=M. \qquad (5.6.8)$$

Here the sum over a corresponds to the three possible ways of fusing the four external legs that are displayed in (5.6.6) and that correspond to the limits $z\rightarrow 0,\,1,\,\infty$, respectively; also, M is a non-negative integer, namely the sum of the grades of those fields in the modules $R_{m(a)}$ exchanged in the various channels ($m(a)=m,\,p,\,n$ for the $a=s,\,t,\,u$-channel, respectively, in the notation of (5.6.6)) that give the leading contribution to the singularity, plus the number of apparent singularities of the differential equation. By combining the formulae (5.6.8) for all possible choices of external legs, one can determine the conformal dimensions Δ_i to a large extent; for a sufficiently low number of primary fields they can actually be fixed uniquely up to a few integer parameters.

So far only four-point correlators have been considered. However, the above considerations easily generalize to N-point correlation functions

$$\begin{aligned} &F_{i_1i_2\ldots i_N}(z_4,\bar z_4,\ldots,z_N,\bar z_N)\\ &\qquad\equiv\langle\phi_{i_1}(0,0)\phi_{i_2}(1,1)\phi_{i_3}(\infty,\infty)\textstyle\prod_{\nu=4}^{N}\phi_{i_\nu}(z_\nu,\bar z_\nu)\rangle \end{aligned} \qquad (5.6.9)$$

(recall that the arguments z_ν are to be taken as the $N-3$ independent sl_2-covariant anharmonic ratios, and that radial ordering is implicit). These can be decomposed into holomorphic and antiholomorphic blocks according to

$$\begin{aligned} &F_{i_1i_2\ldots i_N}(z_4,\bar z_4,\ldots,z_N,\bar z_N)\\ &\qquad=\sum_{\vec m}a_{\vec m}\,\mathcal{F}^{(\vec m)}_{i_1i_2\ldots i_N}(z_4,\ldots,z_N)\,\overline{\mathcal{F}}^{(\vec m)}_{i_1i_2\ldots i_N}(\bar z_4,\ldots,\bar z_N), \end{aligned} \qquad (5.6.10)$$

where $a_{\vec m}$ with $\vec m\equiv(m_4,m_5,\ldots,m_N)$ is the combination of operator product coefficients that corresponds to a definite choice of intermediate channels, namely

$$a_{\vec m}=C_{i_1i_2}{}^{m_4}C_{m_4i_3}{}^{m_5}\cdots C_{m_{N-1}i_{N-2}}{}^{m_N}C_{m_Ni_{N-1}i_N}. \qquad (5.6.11)$$

Graphically, the associated contribution to the correlator is represented by

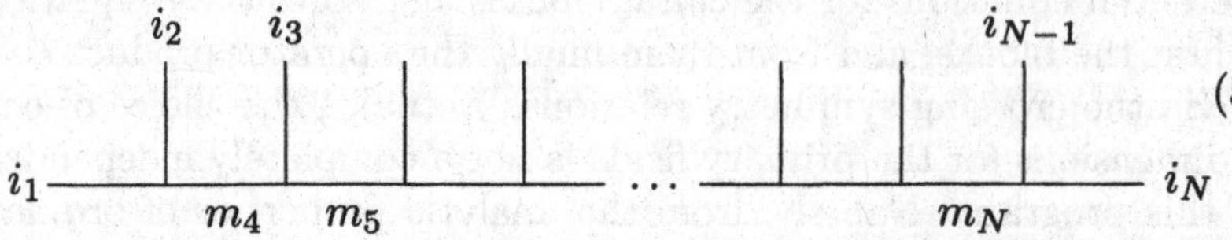

(5.6.12)

From the locality and associativity of the operator product algebra it follows again that the N-point correlation functions have to be single-valued combinations of the (generically) multivalued chiral blocks. This leads again to a series of crossing symmetry relations which, once the chiral blocks are known, provide constraints on the coefficients $a_{\vec{m}}$ and hence on the operator product coefficients of the theory; by the conformal bootstrap, these are again guaranteed to be consistent with the constraints arising from the four-point functions. Also, one can again deduce restrictions on the conformal dimensions, generalizing the sum rule (5.6.8), but these turn out to be all fulfilled identically if (5.6.8) is valid.

The total number of chiral blocks contributing to a given correlator, i.e. the number of distinct possible intermediate channels, is equal to the number of combinations of the labels m_i, counting also multiplicities. Thus it is expressible through the fusion rule coefficients. For the four-point function, the number is

$$N_{ijkl} = \sum_m \mathcal{N}_{ij}{}^m \mathcal{N}_{mkl}, \tag{5.6.13}$$

or what is the same,

$$N_{ijkl} = (\mathcal{N}_j \mathcal{N}_k)_i{}^l, \tag{5.6.14}$$

where, as in section 5.3, $\mathcal{N}_i$ denotes the matrix with entries $(\mathcal{N}_i)_j{}^k = \mathcal{N}_{ij}{}^k$. Analogously, the number of chiral blocks contributing to an N point function is

$$N_{i_1 i_2 \ldots i_N} = (\mathcal{N}_{i_2} \mathcal{N}_{i_3} \ldots \mathcal{N}_{i_{N-1}})_{i_1}{}^{i_N}. \tag{5.6.15}$$

A duality transformation corresponds to a change of the ordering of the external legs and thus to a change of the ordering of the matrices $\mathcal{N}_{i_\nu}$ in this equation. The requirement of crossing symmetry means in particular that such a change amounts to a basis transformation in the space of conformal blocks, and hence the number $N_{i_1 i_2 \ldots i_N}$ must be totally symmetric in its indices; that this is indeed fulfilled follows from the commutativity (5.1.9) of the fusion rule matrices $\mathcal{N}_i$.

The concept of chiral blocks allows us to describe the duality properties of a conformal field theory in terms of only one chiral half of the theory.

A closely related concept, which also allows for such a description, is the one of *chiral vertex operators* Φ. These operators are introduced by a formal splitting of the physical primary fields into a holomorphic and antiholomorphic part as

$$\phi(z,\bar{z}) = \sum_s C_{(s)}\, \Phi_s(z)\, \bar{\Phi}_s(\bar{z}), \tag{5.6.16}$$

where the numbers $C_{(s)}$ are operator product coefficients. They have to be regarded as maps between the modules of the holomorphic respectively antiholomorphic symmetry algebra that are carried by those primary fields to which ϕ couples according to the fusion rules. In terms of the chiral vertex operators, the duality properties lead to a so-called *exchange algebra* that describes how the product $\Phi_{s_2}(z_2)\Phi_{s_1}(z_1)$ with s_1, s_2 fixed can be expanded with respect to the products $\Phi_{t_1}(z_1)\Phi_{t_2}(z_2)$ with t_1, t_2 running over all possible values.

5.7 Fusing and braiding

The analytic continuation that is implicit in the duality transformations is a path dependent procedure. In other words, the duality transformation does not just amount to a permutation of the external legs, but rather to a "path dependent permutation", i.e. a braiding operation. As a consequence, the space of chiral blocks of an N point function of a conformal field theory carries a representation not just of the permutation group $\mathcal{S}_N$, but rather of the braid group $\mathcal{B}_N$. Recall from section 4.1 (p. 257) that the braid group $\mathcal{B}_N$ on N strands is generated by a unit e and $N-1$ elements σ_i, $i = 1, 2, \ldots, N-1$, subject to the relations $\sigma_i\sigma_j = \sigma_j\sigma_i$ for $|i-j| \geq 2$, and $\sigma_i\sigma_{i+1}\sigma_i = \sigma_{i+1}\sigma_i\sigma_{i+1}$.

As long as one is only interested in the properties of a conformal field theory at genus zero, one must actually consider the *sphere braid group* $\mathcal{B}_N(\mathcal{S}^2)$ rather than $\mathcal{B}_N$, which is characterized by the extra relation

$$\sigma_1\sigma_2\ldots\sigma_{N-2}(\sigma_{N-1})^2\sigma_{N-2}\ldots\sigma_2\sigma_1 = e. \tag{5.7.1}$$

As is clear from the crossing symmetry relations for the four-point functions, there exist two basic moves for interchanging the external legs of chiral blocks. The first, called *fusing*, is given by

$$\mathsf{F}: \qquad \mathcal{F}^{(\vec{m})}_{\ldots ijkl\ldots} \longrightarrow \mathcal{F}^{(\vec{m}')}_{\ldots ilkj\ldots}. \tag{5.7.2}$$

The second, called *braiding*, acts as

$$\mathsf{B}: \qquad \mathcal{F}^{(\vec{m})}_{\ldots ijkl\ldots} \longrightarrow \mathcal{F}^{(\vec{m}')}_{\ldots ikjl\ldots}. \tag{5.7.3}$$

The resulting linear transformations among the chiral blocks are expressed through the *fusing matrix* $\mathsf{F}[{}^{jk}_{il}]$ and the *braiding matrix* $\mathsf{B}[{}^{jk}_{il}]$, i.e. one writes

$$\mathcal{F}^{(\dots p\dots)}_{\dots ilkj\dots}(\dots,z_i,z_l,z_k,z_j,\dots)=\sum_m \mathsf{F}_{pm}[{}^{jk}_{il}]\mathcal{F}^{(\dots m\dots)}_{\dots ijkl\dots}(\dots,z_i,z_j,z_k,z_l,\dots),$$
$$\mathcal{F}^{(\dots n\dots)}_{\dots ikjl\dots}(\dots,z_i,z_k,z_j,z_l,\dots)=\sum_m \mathsf{B}_{nm}[{}^{jk}_{il}]\mathcal{F}^{(\dots m\dots)}_{\dots ijkl\dots}(\dots,z_i,z_j,z_k,z_l,\dots). \tag{5.7.4}$$

(In the literature, the fusing and braiding transformations are sometimes defined as going in the opposite direction; in that case here and below one has to replace $\mathsf{F}_{mn}[{}^{jk}_{il}]$ by $\mathsf{F}_{nm}[{}^{jk}_{il}]$, and analogously for B.) If the fusion rule coefficients are larger than one, then the matrix indices of F and B label not just the primary fields of the theory, but also the various possible couplings of the relevant triple of primaries, i.e. one would better write $\mathsf{F}_{mn;\alpha\beta}$ for F_{mn}; for the moment we will suppress this subtlety. In the above equation the numbering of internal lines is the same as in the picture (5.6.6); using Feynman like diagrams to denote the chiral blocks $\mathcal{F}^{(m)}$ (rather than, as was done there, the full contribution from the mth intermediate family of fields to the correlator F), one obtains a pictorial representation of the braiding and fusing procedures, as shown in figure (5.7.5). This graphical representation justifies the names fusing and braiding. Actually, for non-selfconjugate fields one should attach arrows to the lines whose direction identifies whether the field corresponds to an upper or to a lower index of $\mathcal{N}_{ij}{}^{k}$; for simplicity, these arrows are suppressed in the pictures.

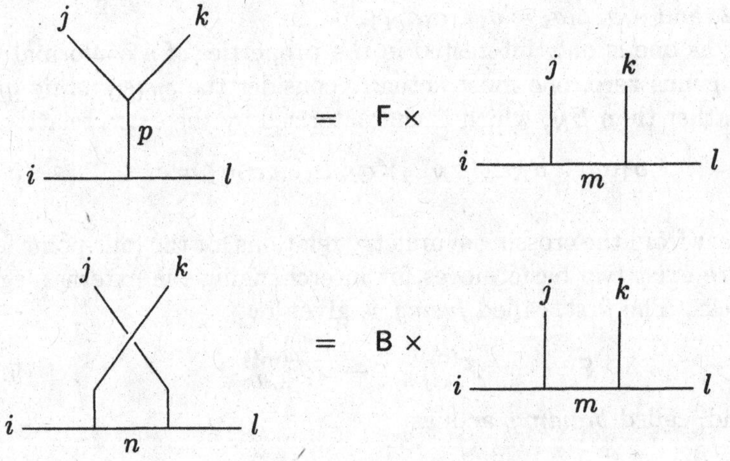

(5.7.5)

As an example, let us present the duality matrices of the $A_1^{(1)}$ WZW theory. One finds (in a normalization different from the canonical normalization (5.6.4))

$$\mathsf{F}_{\mu\nu}\left[\begin{smallmatrix}\lambda_2\lambda_3\\ \lambda_1\lambda_4\end{smallmatrix}\right] = \begin{Bmatrix} \frac{1}{2}\lambda_1 & \frac{1}{2}\lambda_2 & \frac{1}{2}\mu \\ \frac{1}{2}\lambda_3 & \frac{1}{2}\lambda_4 & \frac{1}{2}\nu \end{Bmatrix},$$

$$\begin{aligned}\mathsf{B}_{\mu\nu}\left[\begin{smallmatrix}\lambda_2\lambda_3\\ \lambda_1\lambda_4\end{smallmatrix}\right] &\\ = \exp&\left(\frac{\pi i}{4(k^\vee+2)}\left[\lambda_1(\lambda_1+2)+\lambda_4(\lambda_4+2)-\mu(\mu+2)-\nu(\nu+2)\right]\right)\\ &\cdot(-1)^{(\mu+\nu-\lambda_1-\lambda_4)/2}\begin{Bmatrix} \frac{1}{2}\lambda_2 & \frac{1}{2}\lambda_1 & \frac{1}{2}\mu \\ \frac{1}{2}\lambda_3 & \frac{1}{2}\lambda_4 & \frac{1}{2}\nu \end{Bmatrix},\end{aligned} \tag{5.7.6}$$

where the fields are denoted by their A_1 Dynkin label (written for brevity without an overbar), and where

$$\begin{aligned}\begin{Bmatrix} a & b & c \\ d & e & f \end{Bmatrix} &= \delta(a,b,c)\,\delta(d,e,c)\,\delta(a,e,f)\,\delta(d,b,f)\textstyle\sum_{n\geq 0}(-1)^n\lfloor n+1\rfloor!\\ &\cdot(\lfloor n-a-b-c\rfloor!\lfloor n-d-e-c\rfloor!\lfloor n-a-e-f\rfloor!\lfloor n-d-b-f\rfloor!\\ &\quad\lfloor a+b+d+e-n\rfloor!\lfloor a+c+d+f-n\rfloor!\lfloor b+c+e+f-n\rfloor!)^{-1},\\ \delta(a,b,c) &= (\lfloor -a+b+c\rfloor!\lfloor a-b+c\rfloor!\lfloor a+b-c\rfloor!/\lfloor a+b+c+1\rfloor!)^{1/2}\end{aligned} \tag{5.7.7}$$

(the q-numbers $\lfloor x\rfloor$ are defined as in (1.7.43) with $q=\exp(2\pi i/(k^\vee+2))$.

Despite its name, the braiding manipulation described by the matrix B does not simply correspond to a braid operation in the sense of representing an element of the braid group $\mathcal{B}$. The reason for this is that the points z_i on which the braid group representation acts do not exist as isolated points in their own right, but rather are the arguments of chiral blocks subjected to analytic continuation and hence must be thought of as being the origin of a local coordinate system. As a consequence, the manipulation B braids framed strands or "ribbons" instead of strands, and these ribbons can be left twisted even after a manipulation that as far as the braid group is concerned is the identity. Mathematically, the braiding of ribbons is described by so-called *Dehn twists*, which in addition to the braid group part are locally rotations around the origin. The latter operation is generated infinitesimally by $L_0-\bar{L}_0$, and hence a finite operation gives rise to phases of $\exp(\pi i\Delta)$ for the holomorphic blocks, and to $\exp(-\pi i\bar{\Delta})$ for the antiholomorphic ones. In more detail, the phases are obtained up to a sign from the sl_2-covariant prefactor that according to the general solution (3.4.16) of the projective Ward identities stands in front of the sl_2-invariant part of the correlation function. Thus exchanging two arguments z_j and z_k of a (holomorphic) block amounts to

$$(z_j - z_k) \to \mathrm{e}^{\pi i}\,(z_k - z_j), \tag{5.7.8}$$

while twisting them by an angle ϑ yields a phase $\exp(\pi i \vartheta)$. In addition to the phase obtained this way, there arises a sign factor $\varepsilon_{jk}^{\;\;m}$ equal to $+1$ or -1, depending on whether the primary fields corresponding to the lines j and k couple symmetrically or antisymmetrically to the field that belongs to the third line m entering the relevant vertex. (To be precise, what matters are not the primaries themselves, but rather the modules which they carry with respect to the zero mode subalgebra of the chiral symmetry algebra; if the representation theory of this zero mode subalgebra is not known, then the only possibility of determining $\varepsilon_{jk}^{\;\;m}$ is to look for a self-consistent solution of the polynomial equations that will be introduced below.)

Summarizing, the braiding operation B involves, in addition to the braid group part, an operation π that acts on the chiral blocks as multiplication by a phase. For the three-point function, the action is

$$\begin{aligned} \pi_{jk}: \qquad F_{jk}^{\;\;m} &= \mathsf{B}_{jk}\big[{}^{jk}_{m0}\big]\, F_{kj}^{\;\;m} \\ &= \varepsilon_{jk}^{\;\;m}\, \mathrm{e}^{\pi i(\Delta_m - \Delta_j - \Delta_k)}\, F_{kj}^{\;\;m}, \end{aligned} \tag{5.7.9}$$

and for the chiral blocks of any correlator the phase consists of a product of such factors, one for each vertex in the Feynman graph for the block (note that this way the phase is well defined: the conformal dimensions of all fields corresponding to a given line in the diagram differ only by integers; therefore the $\exp(\pi i \Delta)$ part of the phase is unique up to a sign, and a change in this sign is always compensated by a sign change of ε).

By considering a braiding operation which for the braid group part is the identity but is nevertheless non-trivial, one gets relations between various phases of the type just described, and hence constraints on the conformal dimensions, modulo integers, of the primary fields of the theory. These reproduce precisely the equation (5.6.8) for the four-point function and its generalizations for higher n-point functions, except for the identification of the integer M which in the present context is left arbitrary.

The operation π_{jk} also describes the exchange of external legs j and k that are connected to the same vertex of the Feynman graph, because in this case the braiding operation acts diagonally. As a consequence, one obtains a simple relation between the braiding and fusing operations, namely

$$\mathsf{F}[{}^{jk}_{il}] = \pi_{il} \circ \mathsf{B}[{}^{jl}_{ik}] \circ \pi_{kl}, \tag{5.7.10}$$

or, in terms of matrix elements,

$$\mathsf{F}_{mn}[{}^{jk}_{il}] = \varepsilon_{kl}{}^{m} \varepsilon_{nl}{}^{i}\, \mathrm{e}^{\pi i(\Delta_m+\Delta_n-\Delta_i-\Delta_k)}\, \mathsf{B}_{mn}[{}^{jl}_{ik}]. \tag{5.7.11}$$

The braiding and fusing matrices are thus not independent, but just related by phases which up to signs are already determined by the conformal dimensions of the relevant fields. Nevertheless it is often convenient to keep both matrices so that one need not write out the phases explicitly.

In terms of the fusing and braiding matrices, the crossing symmetry relations for the conformal blocks amount to a set of *polynomial equations.* With some effort (e.g. from a category theoretic treatment) it can be shown that, for a conformal field theory in the complex plane, a maximal set of independent equations of this type is given by only two identities. These stem from the consideration of five point functions, and they involve five respectively six of the duality matrices F and B and are consequently called the *pentagon identity* and the *hexagon identity.* The pentagon identity for F reads

$$\mathsf{F}_{ps}[{}^{ij}_{mn}]\mathsf{F}_{qn}[{}^{pk}_{ml}] = \sum_r \mathsf{F}_{rn}[{}^{jk}_{sl}]\mathsf{F}_{qs}[{}^{ir}_{ml}]\mathsf{F}_{pr}[{}^{ij}_{qk}], \tag{5.7.12}$$

while the hexagon identity for B is

$$\sum_p \mathsf{B}_{ps}[{}^{ij}_{mn}]\mathsf{B}_{qn}[{}^{ik}_{pl}]\mathsf{B}_{rp}[{}^{jk}_{mq}] = \sum_t \mathsf{B}_{tn}[{}^{jk}_{sl}]\mathsf{B}_{rs}[{}^{ik}_{mt}]\mathsf{B}_{qt}[{}^{ij}_{rl}]. \tag{5.7.13}$$

In Feynman diagrammatic notation, these polynomial equations look as shown in figures (5.7.14) and (5.7.15). Thus in particular the hexagon identity is nothing but the implementation of the braid group relation $\sigma_i\sigma_{i+1}\sigma_i = \sigma_{i+1}\sigma_i\sigma_{i+1}$.

For any rational conformal field theory, the pentagon and hexagon identities together with the relation (5.7.11) between the braiding and fusing matrices provide a finite set of equations. This offers a new possibility to find a classification of all rational conformal field theories: First one enumerates all possible solutions to the pentagon and hexagon identity. For each solution one then calculates the chiral blocks; this amounts to a solution of the so-called *Riemann monodromy problem* which is the problem of finding the multivalued functions that transform in a prescribed way under analytic continuation around given regular singular points. Unfortunately, just as for the previously mentioned classification attempt that involves the construction of the differential equations obeyed by the chiral

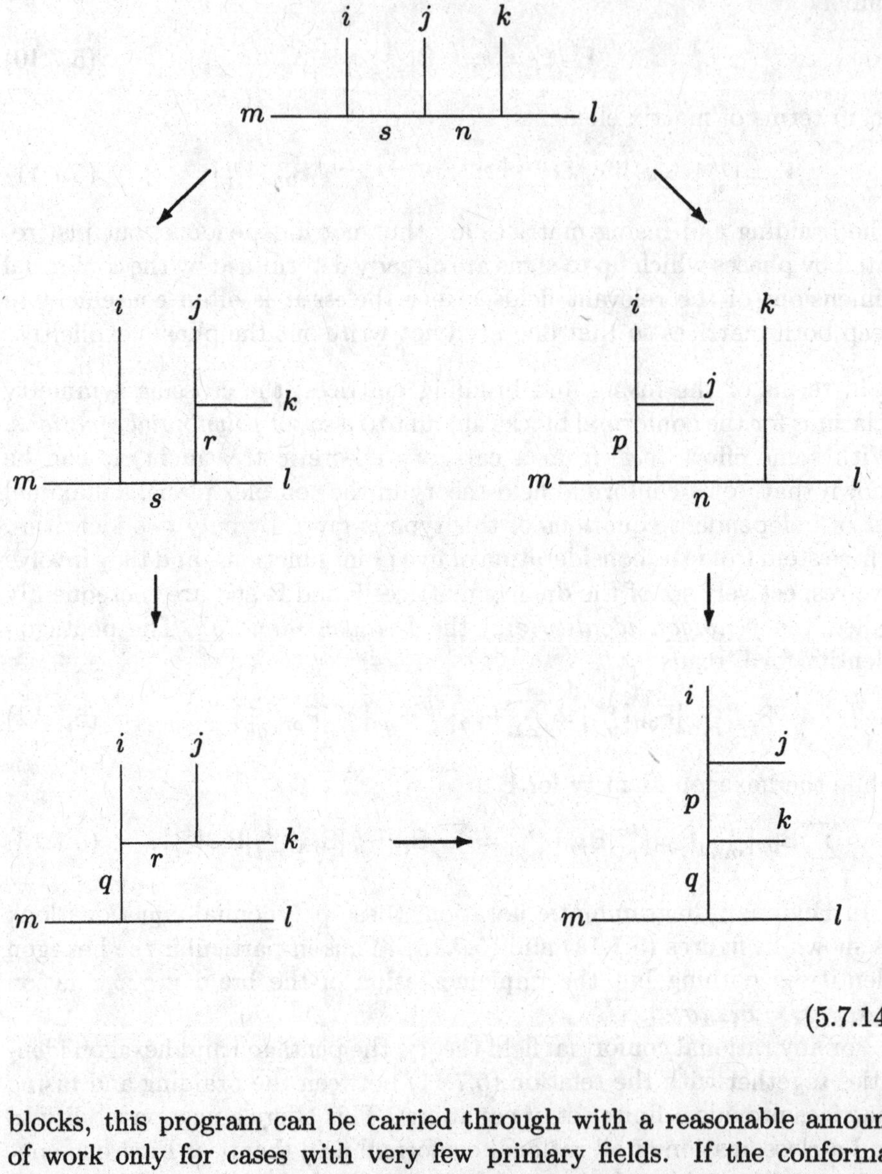

(5.7.14)

blocks, this program can be carried through with a reasonable amount of work only for cases with very few primary fields. If the conformal field theory is considered on higher genus Riemann surfaces, then new polynomial equations arise. It can be shown that there is just a single new independent equation, namely the one obtained from the genus one partition function which diagrammatically is given by a closed loop with a single two-point vertex. By considering the phases involved in the single possible braiding, one obtains again a constraint on the conformal dimensions of primary fields, which now however is coupled to the

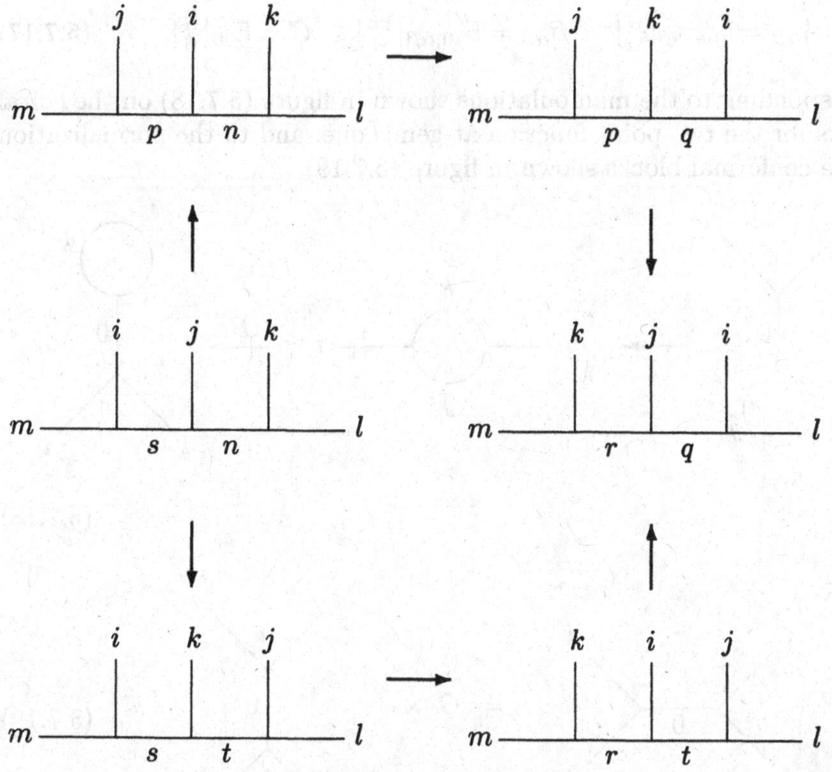

(5.7.15)

value of the conformal central charge c. Namely,

$$6\sum_{i=1}^{M}\Delta_i = \frac{M}{4}\cdot c \mod \mathbb{Z}, \tag{5.7.16}$$

where the sum extends over all primary fields of the theory. As has been shown in section 5.3 (p. 346), formulae of this type can also be obtained by considering the relation $(ST)^3 = C$ and implementing the relation between fusion rules and modular transformations. Moreover, just as was the case there, the result can again can be strengthened to an inequality by analysing the relevant differential equation, which is now a differential equation in the modular parameter τ for the zero-point blocks, i.e. for the characters of the theory.

As an important application of the pentagon identity, we are now in a position to complete the proof of the Verlinde formula (5.3.14). Let us introduce the notations

$$A_{\alpha\beta} = \mathsf{F}_{0k;\alpha\beta}\left[\begin{smallmatrix} ji \\ ji \end{smallmatrix}\right], \quad B_{\alpha\beta} = \mathsf{F}_{j0;\alpha\beta}\left[\begin{smallmatrix} kk \\ ii \end{smallmatrix}\right], \quad C = \mathsf{F}_{00}\left[\begin{smallmatrix} ii \\ ii \end{smallmatrix}\right], \tag{5.7.17}$$

corresponding to the manipulations shown in figure (5.7.18) on the chiral blocks for the two-point function at genus one, and to the normalization of the conformal blocks shown in figure (5.7.19).

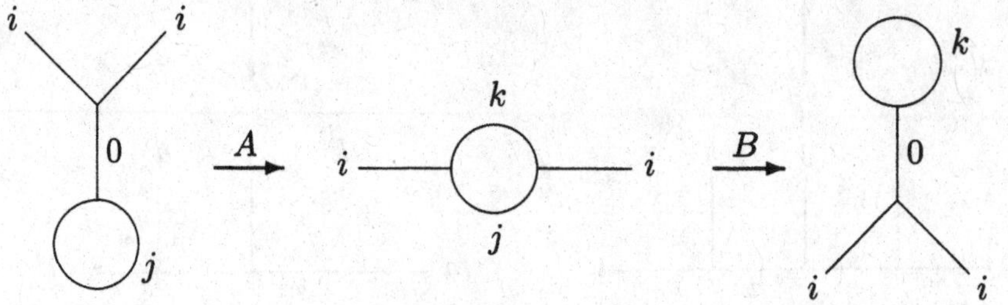

(5.7.18)

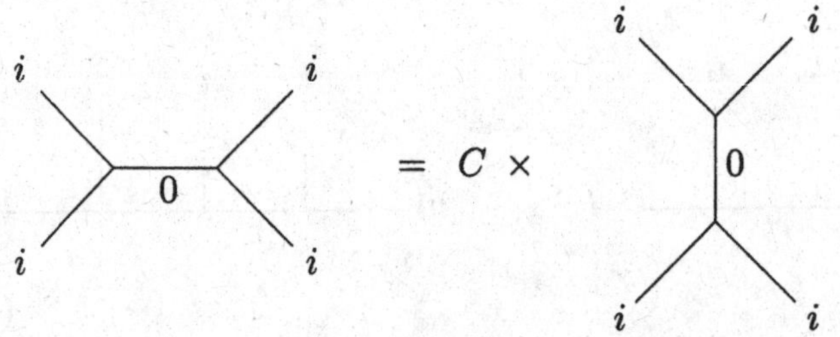

(5.7.19)

In (5.7.17) we have written out the multiplicity indices arising for fusion rule coefficients larger than one, and as usual the index 0 refers to the identity primary field (whose couplings are unique so that no multiplicity index is needed for the quantity C). It is easily checked that the assumption $\Psi_i(b)\tilde{\chi}_j = \sum_k \mathcal{N}_{ij}{}^k\, \tilde{\chi}_k$ that was the starting point of the formula (5.3.14) can be written in terms of these quantities as

$$\sum_{\alpha,\beta=1}^{\mathcal{N}_{ij}{}^k} C^{-1} A^{\alpha\beta} B^{\alpha\beta} = \mathcal{N}_{ij}{}^k. \tag{5.7.20}$$

If we now apply the pentagon identity to the chiral block (figure (5.7.21)) of the three-point function on the torus, two of the fusing transformations act trivially because they involve the identity field, while the other three fusings are precisely given by the operations A, B, C just defined. The pentagon identity then reduces to the relation (5.7.20), and hence provides the missing piece in the proof of the fusion rule formula (5.3.14).

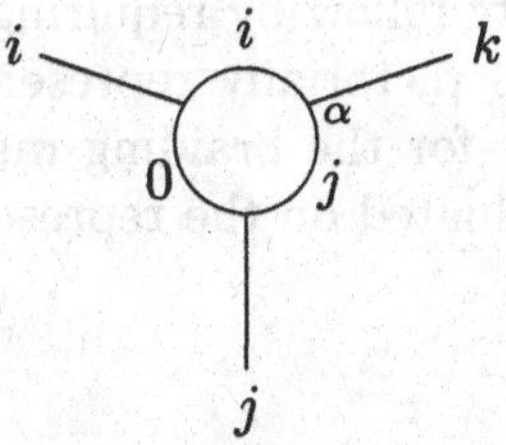

(5.7.21)

The braiding and fusing operations on chiral blocks bear a close resemblance to the manipulations that implement the associativity of Kronecker tensor products of modules of a Lie or Hopf algebra. It is therefore tempting to interpret the chiral blocks for a correlator $\langle \phi_{i_1}\phi_{i_2}\ldots\phi_{i_N}\rangle$ as describing a tensor product decomposition

$$R_{i_1}\times R_{i_2}\times\ldots\times R_{i_N}=\bigoplus_j \mathcal{F}_{i_1i_2\ldots i_Nj}\otimes R_j \tag{5.7.22}$$

with respect to an underlying symmetry object $\mathcal{G}$. Here R_j denotes the $\mathcal{G}$-module carried by the primary field ϕ_j, and in order to make sense the decomposition must be compatible with the fusion rules (and hence in particular be associative) and with the duality properties of the chiral blocks. By reading the decomposition (5.7.22) from right to left, one can define an action of $\mathcal{G}$ on tensor products in terms of the action on single modules. Technically this amounts, for $N=2$, to the definition of a co-multiplication $\mathbf{\Delta}$ for $\mathcal{G}$. The decomposition (5.7.22) then also implies that the trace of the co-product $\mathbf{\Delta}(x)$ in a tensor product $R_j\times R_k$ is expressible through the trace of x in the various representations appearing on the right hand side,

$$\mathrm{tr}\,[(R_j\times R_k)(\mathbf{\Delta}(x))]=\sum_m \mathcal{N}_{jk}{}^m\,\mathrm{tr}\,(R_m(x))\,. \tag{5.7.23}$$

According to (5.7.22) the multiplicity of the irreducible module R_j in the product is given by the dimensionality of the space $\mathcal{F}_{i_1i_2\ldots i_Nj}$, i.e. by the number of blocks $\mathcal{F}^{(\vec m)}_{i_1i_2\ldots i_Nj}$. As a consequence, the blocks $\mathcal{F}^{(\vec m)}_{i_1i_2\ldots i_Nj}$ provide a representation of the centralizer $\mathcal{C}$ of $\mathcal{G}$ in the Kronecker product $R_{i_1}\times R_{i_2}\times\ldots\times R_{i_N}$. Assuming that $\mathcal{C}$ and $\mathcal{G}$ are finite-dimensional semisimple algebras, it follows from the reciprocity (1.1.40) that, conversely, $\mathcal{G}$ is the centralizer of $\mathcal{C}$ in the Kronecker product. Thus one can identify the symmetry object $\mathcal{G}$ by looking for the centralizer of the braid group action on the chiral blocks. When this program is carried through, it turns out that $\mathcal{G}$ possesses all the properties (4.1.38) of a quasitriangular Hopf algebra. In particular, from the braid group representation furnished by the chiral blocks one can construct a matrix $\mathcal{R}^{(R_j,R_k)}$ that acts on contiguous spaces in the tensor product on the left hand side of

(5.7.22) and obeys $\mathcal{R}\boldsymbol{\Delta} = \boldsymbol{\Delta}'\mathcal{R}$, thus reproducing (4.1.38a). The other axioms of quasitriangularity follow by requiring the compatibility of fusing and braiding, which is pictorially represented as in figure (5.7.24). Also, the hexagon identity for the braiding matrices translates into the Yang–Baxter equation evaluated on the representation $R_i \times R_j \times R_k$.

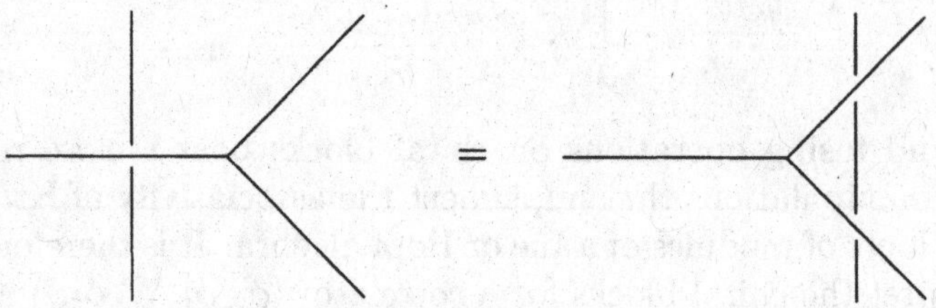

(5.7.24)

Summarizing, the duality properties of a rational conformal field theory can be translated into the defining properties of a quasitriangular Hopf algebra that can be constructed as the centralizer of the action of the braid group on the chiral blocks of the theory.

In the case of the WZW theory based on an untwisted affine Lie algebra $\mathfrak{g}$, $\mathcal{G}$ is the Hopf algebra $\mathsf{U}_q(\bar{\mathfrak{g}})$, with the product in (5.7.22) being the truncated Kronecker product defined in section 4. That one has to use the truncated Kronecker product rather than the ordinary Kronecker product arises from the requirement that the representation theory should be semisimple, which is needed in order to apply the reciprocity theorem for the centralizer algebras. The identity (5.7.23) is then the source of the quantum dimension sum rule that is valid for the truncated Kronecker products of $\mathsf{U}_q(\bar{\mathfrak{g}})$. The relevant value of the deformation parameter q is obtained as follows. By considering a fusion rule with only two primary fields on the right hand side, i.e. $\phi * \phi = \phi_1 + \phi_2$, one gets in addition to the braid group relations a further quadratic relation for the generators σ_i, and as a consequence the chiral blocks carry a representation of the Hecke algebra (4.3.37) rather than just of the braid group. The parameter q appearing in the Hecke algebra turns out to be

$$q = -\varepsilon_{\phi\phi}^{\phi_1}\, \varepsilon_{\phi\phi}^{\phi_2}\, \mathrm{e}^{\pi i(\Delta_1 - \Delta_2)} \tag{5.7.25}$$

(note that interchanging $\phi_1 \leftrightarrow \phi_2$ amounts to replacing q by q^{-1}, but this is irrelevant because the corresponding quantum groups are isomorphic). As an example, take ϕ to carry the defining module of $\bar{\mathfrak{g}} = sl_n$. Then ϕ_1 and ϕ_2 carry the symmetric and antisymmetric tensor module of sl_n, respectively, and therefore $\varepsilon_{\phi\phi}^{\phi_1} = -\varepsilon_{\phi\phi}^{\phi_2} = 1$, and $\Delta_1 = (n-1)(n+2)/n(k^\vee + n)$, $\Delta_2 = (n+1)(n-2)/n(k^\vee + n)$. Thus one obtains the correct value $q = \exp(2\pi i/(k^\vee + n))$.

From the interpretation (5.7.22) of the chiral blocks it follows that the duality operations F and B play the role of the $6j$-symbols of the quantum group $\mathcal{G}$ which according to section 1.2 describe the basic associativity property of Kronecker products. In particular, for the WZW theory based on some untwisted affine Lie algebra g, the duality matrices are given (up to a basis transformation) by the $6j$-symbols of the deformed universal enveloping algebra $\mathsf{U}_q(\bar{\mathsf{g}})$. For the case of $\bar{\mathsf{g}} = A_1$, for which both the duality matrices (compare (5.7.6)) and the quantum $6j$-symbols have been computed, the results do agree as they should. Another case where the coincidence has been checked is the defining module of $\mathsf{U}_q(A_r)$ for arbitrary rank r.

According to (5.3.8), the operator $\Psi_i(a)$ that was introduced in section 5.3 acts on the characters diagonally, namely as a multiplication by the fusion rule eigenvalues $\ell_i^{(j)}$. As a consequence, from the proof of the Verlinde formula that was presented above, it follows that the fusion rule eigenvalues can be expressed through the fusing and braiding matrices by the formula

$$\ell_i^{(j)} = \frac{1}{f_i} \sum_m \mathsf{B}_{0m}[{}^{ij^+}_{i\,j}]\, \mathsf{B}_{m0}[{}^{j^+i}_{i\,j}] \tag{5.7.26}$$

with

$$f_i = \varepsilon^{0}_{ii^+} F_{00}[{}^{ii^+}_{i\,i}]. \tag{5.7.27}$$

For the special value $j = 0$, this formula simplifies because $\mathsf{B}_{m0}[{}^{i0}_{i0}]$ is nonzero only for $m = i$, and is inverse to $\mathsf{B}_{0m}[{}^{0i}_{i0}]$; upon re-inserting this result in the original formula, one finds

$$S_{ij} = \frac{S_{0i}S_{0j}}{S_{00}} \sum_m \mathsf{B}_{0m}[{}^{ij^+}_{i\,j}]\, \mathsf{B}_{m0}[{}^{j^+i}_{i\,j}]. \tag{5.7.28}$$

Also, after expressing B through F, the pentagon equation can be employed to obtain

$$S_{ij} = \sum_k \mathrm{e}^{-2\pi i(\Delta_k - \Delta_i - \Delta_j)} S_{0k} \, {\mathcal{N}_{ij}}^k. \tag{5.7.29}$$

In particular, for the $A_1^{(1)}$ WZW theory for which the elements $S_{ij} \equiv S_{\Lambda,\Lambda'}$ of the modular matrix S are given by (2.7.41), one obtains this way the following identity among q-integers:

$$\lfloor(\bar\lambda+1)(\bar\mu+1)\rfloor = \sum_\nu q^{[\bar\nu(\bar\nu+2)-\bar\lambda(\bar\lambda+2)-\bar\mu(\bar\mu+2)]/4} {\mathcal{N}_{\lambda\mu}}^{\nu} \lfloor\bar\nu+1\rfloor. \tag{5.7.30}$$

(Note that this differs from the sum rule for the quantum dimensions, which reads $\lfloor\bar\lambda+1\rfloor\lfloor\bar\mu+1\rfloor = \sum_\nu \lfloor\bar\nu+1\rfloor$.) It turns out that the result for

S_{ij} obtained above can also be understood by considering the particular element

$$u := \mathcal{M} \circ (\gamma \times id)(\pi(\mathrm{R})) \tag{5.7.31}$$

of the quantum group $\mathcal{G}$. In the representation $R_{\bar\Lambda}$, the special element u of $\mathcal{G} = \mathsf{U}_q(A_1)$ reads

$$R_{\bar\Lambda}(u) = q^{-\bar\Lambda(\bar\Lambda+2)/4}\, q^{R_{\bar\Lambda}(H)/2}, \tag{5.7.32}$$

so that

$$\mathrm{tr}\,(R_{\bar\Lambda}(u)) = q^{-\bar\Lambda(\bar\Lambda+2)/4}\, \lfloor \bar\Lambda + 1 \rfloor . \tag{5.7.33}$$

As a consequence, one finds

$$\begin{aligned} S_{\Lambda,\Lambda'}/S_{k^\vee\Lambda_0,k^\vee\Lambda_0} &= q^{[\bar\Lambda(\bar\Lambda+2)+\bar\Lambda'(\bar\Lambda'+2)]/4} \\ &\quad\cdot \mathrm{tr}\left[(R_{\bar\Lambda} \times R_{\bar\Lambda})\,(\mathrm{R}^{-1}\circ\pi(\mathrm{R}^{-1})\circ(u \otimes u))\right] . \end{aligned} \tag{5.7.34}$$

There is also an interpretation of the underlying quantum group structure of a rational conformal field theory analogous to the idea (see section 4.7) of quantum groups as quantized groups. Consider first the case of WZW theories. When the level $k^\vee$ of the relevant affine algebra $\mathfrak{g}$ tends to infinity, the existence of null vectors in the affine modules becomes less and less important, and in the limit the theory just describes the classical mechanical system whose configuration space is a Lie group G with Lie algebra $\bar{\mathfrak{g}}$. The latter theory may then be called the *classical limit* of the WZW theory. The role of a kind of "Planck constant" $\hbar$ is then played by $(k^\vee + g^\vee)^{-1}$, which goes to zero for $k^\vee \to \infty$.

It is expected that also for most if not all other rational conformal field theories there exists some parameter $\hbar$ such that for $\hbar \to 0$ one recovers a classical theory, and that this classical limit is describable in terms of (finite or continuous) groups.

5.8 Screened vertex operators and quantum groups

In section 4.8 we have proposed some ideas on how to express the quantum group generators in terms of those of the chiral algebra of a conformal field theory. While we were able to obtain some suggestive results, we did not succeed in carrying this program to the end. One possible explanation for this failure is that the modules on which the quantum group of a conformal field theory acts are not directly related to the highest weight modules of the chiral algebra (i.e. the Verma modules, respectively their irreducible quotients), but rather to some other type of vector spaces that can be constructed from the fields of the theory. It has indeed been possible to identify such vector spaces; they are constructed in terms of so-called *screened vertex operators*. In order to understand these constructions, it is necessary to introduce some facts about free field realizations of conformal field theories.

Let us begin with the case of WZW theories. As mentioned at the beginning of section 3.6, a free field realization of WZW theories is provided by the Wakimoto construction. This construction expresses the currents and the primary fields of the WZW theory based on a simple Lie algebra $\mathfrak{g}$ in terms of $r = \operatorname{rank}\mathfrak{g}$ free bosons $\varphi^i(z)$, $i = 1, \ldots, r$, and of $(d-r)/2$ pairs $(\beta^{\bar{\alpha}}, \gamma^{\bar{\alpha}})$ of (bosonic) *ghosts*, where $d = \dim\mathfrak{g}$, i.e. there is one pair for each positive root $\bar{\alpha}$ of $\mathfrak{g}$. As already mentioned in section 4.8, free bosons are not proper conformal fields, as is demonstrated by their operator product; in the present case, the contraction of two free bosons has to be taken as

$$\underbrace{\varphi_i(z)\,\varphi_j(w)} = -G_{ij}\,\ln(z-w), \tag{5.8.1}$$

where G_{ij} is the quadratic form matrix of $\mathfrak{g}$, which together with its inverse G^{ij} is used to lower and raise the indices of the boson fields (here and below, we simplify the notation by fixing the normalization of inner products such that the length squared of the highest root is $(\bar{\theta}, \bar{\theta}) = 2$). The ghost pairs $(\beta^{\bar{\alpha}}, \gamma^{\bar{\alpha}})$ are by definition primary fields of the Virasoro algebra with conformal dimension 1 (for $\beta^{\bar{\alpha}}$) and 0 (for $\gamma^{\bar{\alpha}}$), which possess the operator products

$$\underbrace{\gamma^{\bar{\alpha}}(z)\,\beta^{\bar{\alpha}'}(w)} = \frac{\delta^{\bar{\alpha},\bar{\alpha}'}}{z-w} \tag{5.8.2}$$

and

$$\underbrace{\gamma^{\bar{\alpha}}(z)\,\gamma^{\bar{\alpha}'}(w)} = 0 = \underbrace{\beta^{\bar{\alpha}}(z)\,\beta^{\bar{\alpha}'}(w)}. \tag{5.8.3}$$

In addition, all the free fields φ_i, $\beta^{\bar{\alpha}}$, $\gamma^{\bar{\alpha}}$ are supposed to depend only on the holomorphic coordinate z, and not on its antiholomorphic counterpart $\bar{z}$, and hence they transform trivially under the antiholomorphic Virasoro algebra of the theory.

The WZW currents $J^a(z)$ are given by rather complicated expressions in the free bosons and the ghosts. They will not be needed for the considerations below, and hence we present them only for a particular case, namely for $\mathfrak{g} = A_r$. Denoting the $r(r+1)/2$ ghost pairs for the roots $\bar{\alpha}^{(i)} + \ldots + \bar{\alpha}^{(j)}$, $1 \le i \le j \le r$, by $(\beta^{\bar{\alpha}}, \gamma^{\bar{\alpha}}) \equiv (\beta^{i,j+1}, \gamma^{i,j+1})$, a Chevalley–Serre basis for $A_r^{(1)}$ is given by

$$\begin{aligned}
h^j(z) &= \kappa\,(\bar\alpha^{(j)}, i\partial\varphi(z)) + \sum_{\bar\alpha>0} \bar\alpha^j \;:\gamma^{\bar\alpha}(z)\beta^{\bar\alpha}(z): \\
&= i\kappa\,\partial\varphi^j(z) + 2\;:\gamma^{j,j+1}(z)\beta^{j,j+1}(z): \\
&\quad - \sum_{i<j}(:\gamma^{i,j}(z)\beta^{i,j}(z): - :\gamma^{i,j+1}(z)\beta^{i,j+1}(z):) \\
&\quad + \sum_{i>j+1}(:\gamma^{j,i}(z)\beta^{j,i}(z): - :\gamma^{j+1,i}(z)\beta^{j+1,i}(z):)\,, \\
e^j_+(z) &= \beta^{j,j+1}(z) - \sum_{i<j}\gamma^{i,j}(z)\beta^{i,j+1}(z)\,, \\
e^j_-(z) &= -i\kappa\,\gamma^{j,j+1}(z)\,\partial\varphi^j(z) - (k^\vee + j - 1)\,\partial\gamma^{j,j+1}(z) \\
&\quad - \sum_{i<j}\gamma^{i,j+1}(z)\beta^{i,j}(z) + \sum_{i>j+1}\gamma^{j,i}(z)\beta^{j+1,i}(z) \\
&\quad - :\gamma^{j,j+1}(z)\,[\sum_{i>j}\gamma^{j,i}(z)\beta^{j,i}(z) - \sum_{i>j+1}\gamma^{j+1,i}(z)\beta^{j+1,i}(z)]:\,.
\end{aligned} \tag{5.8.4}$$

(Note that in some of the terms no normal ordering is necessary, because the contractions of the relevant fields vanish.)

In contrast, the expression for the Sugawara type energy–momentum tensor of the theory in terms of the free fields looks more simple:

$$T(z) = T_{\varphi,1}(z) + T_{\varphi,2}(z) + T_{\beta,\gamma}(z), \tag{5.8.5}$$

where the various terms are as follows. $T_{\varphi,1}$ is the ordinary energy–momentum tensor for a system of free bosons, which involves only the first derivatives of the bosons,

$$T_{\varphi,1}(z) = -\tfrac{1}{2}\sum_{j=1}^{r} :\partial\varphi^j(z)\partial\varphi_j(z):; \tag{5.8.6}$$

$T_{\varphi,2}$ is an additional bosonic term which involves the second derivatives of the bosons

$$T_{\varphi,2}(z) = -i\kappa\sum_{j=1}^{r}\partial^2\varphi^j(z), \tag{5.8.7}$$

where

$$\kappa = (k^\vee + g^\vee)^{-1/2} \tag{5.8.8}$$

with $g^\vee$ the dual Coxeter number of $\bar{\mathfrak{g}}$ and $k^\vee$ the level of the untwisted affine Lie algebra $\mathfrak{g}$ (having $\bar{\mathfrak{g}}$ as its horizontal subalgebra) that underlies the WZW theory; finally, $T_{\beta,\gamma}$ is the ghost contribution, reading

$$T_{\beta,\gamma}(z) = -\sum_{\bar{\alpha}>0} :\beta^{\bar{\alpha}}(z)\partial\gamma^{\bar{\alpha}}(z): . \tag{5.8.9}$$

The operator products for the various contributions to $T(z)$ are easily calculated from the operator products of the free fields with the help of the standard Wick contractions of free field theory which are special cases of contraction formulae such as (3.1.38). For the contributions to the central term of the Virasoro algebra, one finds $c_{\varphi,1} = r$ (namely a unit contribution by each free boson φ^i), $c_{\beta,\gamma} = d - r$ (namely a contribution 2 by each (β, γ) ghost pair), as well as

$$c_{\varphi,2} = -12\,\kappa^2 \sum_{i,j=1}^{r} G_{ij} = -\frac{g^\vee d}{k^\vee + g^\vee}, \tag{5.8.10}$$

where in the second equality the strange formula (1.4.74) has been used. These contributions add up to

$$c = c_{\varphi,1} + c_{\varphi,2} + c_{\beta,\gamma} = d - \frac{g^\vee d}{k^\vee + g^\vee} = \frac{k^\vee d}{k^\vee + g^\vee}, \tag{5.8.11}$$

i.e. give the correct central charge value for the WZW theory at level $k^\vee$.

The primary WZW fields also look quite simple; they are just normal ordered exponentials in the free bosons, or in other words vertex operators. (For the secondary fields, the situation is more involved: from the vertex operators, one can construct *Fock spaces* by acting on the vertex operator with the modes of the fields $\partial\varphi$, which leads to so-called *Fock modules* of the chiral algebra, but these Fock modules do not coincide with the (Verma or irreducible) highest weight modules of the algebra. This important issue is however not relevant to the questions considered here.) Namely, writing $\varphi(z)$ for the one-form $(\varphi^1(z), \varphi^2(z), \ldots, \varphi^r(z))$, one has, for the primary field carrying a highest weight Λ of $\mathfrak{g}$,

$$\phi_\Lambda(z) = \,:\exp\left[i\kappa\,(\bar{\Lambda}, \varphi(z))\right]: \,. \tag{5.8.12}$$

Let us verify that this field has the correct conformal dimension. This is again accomplished by calculating operator products with the help of the conventional quantum field theoretic Wick theorem, and yields

$$\Delta_\Lambda = \tfrac{1}{2}\kappa^2\,(\bar{\Lambda}, \bar{\Lambda}) + \kappa^2 \sum_{j=1}^{r} \bar{\Lambda}_j, \tag{5.8.13}$$

where the first term comes from the $T_{\varphi,1}$-contribution to $T(z)$, and the second term from $T_{\varphi,2}$. Using the identity $\sum_j \bar{\Lambda}_j = (\bar{\Lambda}, \bar{\rho})$, this indeed reproduces the conformal dimension

$$\Delta_\Lambda = \tfrac{1}{2}\,\kappa^2(\bar{\Lambda}, \bar{\Lambda} + 2\bar{\rho}) \tag{5.8.14}$$

as given by (3.4.20). (In addition, one should of course multiply the vertex operator by an analogous vertex operator depending on antiholomorphic bosons in order to reproduce also the antiholomorphic conformal dimension $\bar\Delta_{\bar\Lambda}$, but here we are always interested in only one chiral half of the conformal field theory.)

Note, however, that in the latter forms Λ can in fact be any weight of $\mathfrak{g}$, not just the highest weight of an integrable highest weight module. In particular one may consider the field $\Phi^{(j)}(z) := \phi_{-\alpha^{(j)}}(z)$, i.e., restricting from now on for simplicity to simply laced $\mathfrak{g}$,

$$\Phi^{(j)}(z) = :\exp[-i\kappa\varphi^j(z)]. \tag{5.8.15}$$

Now from $(\bar\alpha^{(j)},\bar\alpha^{(j)}) = 2$ and $(\bar\alpha^{(j)},\bar\rho) = \sum_{i=1}^r \bar\alpha_i^{(j)} = 1$, it follows that the conformal dimension of these fields vanishes, $\Delta(\Phi^{(j)}) = 0$, and hence the fields

$$\mathcal{J}^i(z) = \mathcal{P}^i(\beta^{\bar\alpha},\gamma^{\bar\alpha})\,\Phi^{(i)}(z) \tag{5.8.16}$$

have unit conformal dimension if $\mathcal{P}^i$ is any polynomial in the ghosts (without derivatives) which is of degree one in the $\beta^{\bar\alpha}$,

$$\Delta(\mathcal{J}^i) = 1. \tag{5.8.17}$$

In addition, it turns out to be possible to choose the polynomials $\mathcal{P}^i$ such that the operator product between the fields $\mathcal{J}^i(z)$ and the currents $J^a(z)$ is of the following simple form: for the raising currents $e^{\bar\alpha}(z)$, $\bar\alpha > 0$, and for the Cartan subalgebra currents $h^j(z)$ it is regular (i.e. the contraction vanishes), and for the lowering currents $e^{\bar\alpha}(z)$, $\bar\alpha < 0$, it is a total derivative. This choice of polynomials $\mathcal{P}^i$ is essentially unique; the corresponding fields $\mathcal{J}^i(z)$ are called the *screening currents* of the WZW theory. The precise form of the polynomials $\mathcal{P}^i$ will not be needed in the following; one only has to keep in mind the operator product properties of the screening currents that were just mentioned.

The origin of the name screening current is as follows. Owing to the contribution $T_{\varphi,2}$ to the energy–momentum tensor, the conformal Ward identities for correlation functions cannot be satisfied by the primary fields alone, and this situation can be remedied by including also *screening charges*, i.e. contour integrals over the screening currents, into the correlators; this is often expressed by saying that the screening charges neutralize, or screen, a *background charge* that is described by $T_{\varphi,2}$. The property that the operator products of the screening currents with the affine current fields are at most total derivatives ensures that the insertion of the screening charges does not spoil the affine Ward identities. The relevant integration contours appearing in the definition of the screening charges are however not the contours of our present interset, which will be defined in (5.8.22) below, but other types of contours that will be briefly discussed at the end of this section.

By employing the free field Wick theorem, it is also straightforward to compute the operator products among the vertex operators ϕ_Λ and the screening currents $\mathcal{J}^i$, and hence in particular their braiding properties. One finds

$$\begin{aligned} \phi_\lambda(z)\,\phi_\mu(w) &= q^{(\bar\lambda,\bar\mu)/2}\,\phi_\mu(w)\,\phi_\lambda(z), \\ \mathcal{J}^i(z)\,\phi_\lambda(w) &= q^{-\bar\lambda^i/2}\,\phi_\lambda(w)\,\mathcal{J}^i(z), \\ \mathcal{J}^i(z)\,\mathcal{J}^j(w) &= q^{G^{ij}/2}\,\mathcal{J}^j(w)\,\mathcal{J}^i(z). \end{aligned} \tag{5.8.18}$$

Here q denotes the following number:

$$q = \exp{(2\pi i\,\kappa^2)} = \exp\left(\frac{2\pi i}{k^\vee + g^\vee}\right). \tag{5.8.19}$$

The appearance of this number in the braiding relations suggests that there might exist an action of the quantum group $\mathsf{U}_q(\mathfrak{g})$ that is present in the WZW theory on some vector space which can be constructed out of the vertex operators and the screening currents. Indeed this turns out to be true. What one has to do is to introduce to any vertex operator a family of so-called *screened vertex operators*; these are obtained as appropriate contour integrals of products of the vertex operator with the screening currents.

Let us first consider the case of $\mathfrak{g} = A_1$. There is only a single positive root, and hence a single screening current $\mathcal{J}(z)$, and the braiding relations read

$$\begin{aligned} \phi_\lambda(z)\,\phi_\mu(w) &= q^{\bar\lambda\bar\mu/4}\,\phi_\mu(w)\,\phi_\lambda(z), \\ \mathcal{J}(z)\,\phi_\lambda(w) &= q^{-\bar\lambda/2}\,\phi_\lambda(w)\,\mathcal{J}(z), \\ \mathcal{J}(z)\,\mathcal{J}(w) &= q\,\mathcal{J}(w)\,\mathcal{J}(z) \end{aligned} \tag{5.8.20}$$

with $q = \exp(2\pi i/(k^\vee + 2))$. The screened vertex operators are then introduced as

$$U^s_\lambda(z) := \left(\prod_{j=1}^{s} \int_{C_j} \mathrm{d}w_j\,\mathcal{J}(w_j)\right)\,\phi_\lambda(z). \tag{5.8.21}$$

In this expression, the fields in the integrand should be taken in the given order, i.e. must *not* be normal ordered. Also, the contours C_j have to be chosen as follows. They surround the point z counterclockwise, are open at infinity, and are ordered in such a way that C_j lies "inside" C_{j-1}. This is shown in figure (5.8.22).

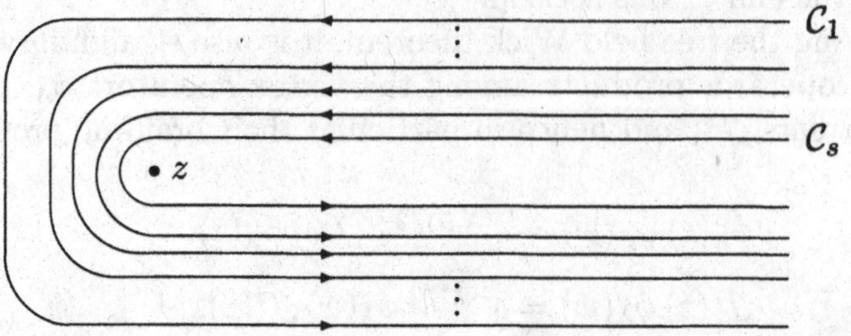

(5.8.22)

In the following we will refer to contours of the type (5.8.22) simply as $\mathcal{C}$-*contours*. This particular choice of contours turns out to be very important. Namely, hidden in the contours is the fact that for $\bar\Lambda$ a dominant integral weight, i.e. $\bar\Lambda \in \mathbb{Z}_{\geq 0}$, only screened vertex operators with some maximal number of screenings are nonvanishing, and hence the vector space generated by the screened vertex operators associated to ϕ_Λ is finite-dimensional. To see this, one has to rewrite the screened vertex operators in terms of the fields

$$\hat U^s_\lambda(z) := \left(\prod_{j=1}^{s} \int_\infty^z \mathrm{d}w_j\, \mathcal{J}(w_j)\right) \phi_\lambda(z), \tag{5.8.23}$$

where the integrand is again not normal ordered. Using standard manipulations of complex analysis, one can show that

$$U^s_\lambda(z) = \left(\prod_{j=0}^{s-1} (1 - q^{j-\bar\lambda})\right) \hat U^s_\lambda(z). \tag{5.8.24}$$

Thus for $\bar\Lambda \in \mathbb{Z}_{\geq 0}$ one has

$$U^s_\Lambda(z) = 0 \qquad \text{for} \qquad s \geq \bar\Lambda + 1; \tag{5.8.25}$$

this means that s effectively can take only the values $s = 0, 1, \ldots, \bar\Lambda$, i.e. the vector space generated by the screened vertex operators has dimension $\bar\Lambda + 1$. Note that this result holds for any value of q, or in other words for any complex value of the parameter $k^\vee$.

It must also be mentioned that the screened vertex operators $U^s_\lambda(z)$ depend not just on the coordinate z, but implicitly also on a second coordinate $z_\infty \simeq \infty$ that describes the precise manner in which the $\mathcal{C}$-contours behave at infinity. As a consequence, the conformal transformation properties of screened vertex operators involve also boundary terms depending on z_∞, so that in particular they may *not* be thought of as describing the secondary fields associated to the primary $\phi_\lambda(z)$.

Our next task is to find expressions for the $\mathsf{U}_q(A_1)$-generators that are supposed to act on the vector space just constructed. Since the individual generators of the vector space are labelled by the number of screenings, it is natural to expect that the number of screenings is essentially the A_1-weight of a state; hence one should try to express the ladder operators as operators that create respectively destroy screenings, whereas the Cartan subalgebra generator should not change the number of screenings. Thus let us require that the ladder operator E_- acts on the family of screened vertex operators as

$$E_- \cdot U^s_\lambda(z) = \int_{\mathcal{C}} \mathrm{d}w\, \mathcal{J}(w)\, U^s_\lambda(z) = U^{s+1}_\lambda(z). \tag{5.8.26}$$

Before discussing the other generators, one may already note that this definition implies a non-trivial co-multiplication property of E_-. Namely, when acting with E_- on a tensor product of states, one has to add a $\mathcal{C}$-contour around two screened vertex operators. This contour can then be deformed into the union of two $\mathcal{C}$-contours, one around each screened vertex, according to

$$\begin{aligned} \Delta(E_-)\cdot(U^s_\lambda(z)\otimes U^t_\mu(w)) &:= \int_{\mathcal{C}} \mathrm{d}y\, \mathcal{J}(y)\,(U^s_\lambda(z)\otimes U^t_\mu(w)) \\ = \int_{\mathcal{C}_1} \mathrm{d}y\, \mathcal{J}(y)\,(U^s_\lambda(z)\otimes U^t_\mu(w)) &+ \int_{\mathcal{C}_2} \mathrm{d}y\, \mathcal{J}(y)\,(U^s_\lambda(z)\otimes U^t_\mu(w)). \end{aligned} \tag{5.8.27}$$

Now in the first term, $\int \mathrm{d}y\, \mathcal{J}(y)$ just acts as the lowering operator E_- on U^s_λ; in contrast, in the second term, $\int \mathrm{d}y\, \mathcal{J}(y)$ acts as E_- on U^t_μ only after it has been braided through U^s_λ and its screenings. According to (5.8.20), the latter manipulation produces a $\bar\lambda$- and s-dependent phase; explicitly,

$$\begin{aligned} &\Delta(E_-)\cdot(U^s_\lambda(z)\otimes U^t_\mu(w)) \\ &\qquad = U^{s+1}_\lambda(z)\otimes U^t_\mu(w) + q^{s-\bar\lambda/2}\,U^s_\lambda(z)\otimes U^{t+1}_\mu(w). \end{aligned} \tag{5.8.28}$$

This result must be compared with the co-multiplication formula (4.2.17). Identifying the currently used generator E_- as the combination $q^{-H/4}E_-$ of the generators used in section 4.2, that equation becomes (using the homomorphism property of the co-multiplication)

$$\Delta(E_-) = E_-\otimes 1 + q^{-H/2}\otimes E_-. \tag{5.8.29}$$

This formula is correctly reproduced if one defines the action of the Cartan subalgebra generator H as

$$H\cdot U^s_\lambda(z) = (\bar\lambda - 2s)\, U^s_\lambda(z). \tag{5.8.30}$$

From this multiplicative action of H, it also follows immediately that on tensor products H acts according to

$$\Delta(H) = H \otimes 1 + 1 \otimes H. \tag{5.8.31}$$

Finally we have to define the action of the ladder operator E_+. Since $[E_+, E_-]$ should act diagonally on the modules, it is clear that this involves the destruction of a contour. However, in order to get the correct commutation relation, one must also include a multiplicative part. As we will verify in a moment, the following choice does the job:

$$E_+ \cdot U_\lambda^s = \frac{1-q^{s-\bar\lambda-1}}{1-q^{-1}} \frac{1-q^s}{1-q} U_\lambda^{s-1}(z). \tag{5.8.32}$$

Again it is not difficult to derive the action on tensor products: imagining that the removed $\mathcal{C}$-contour around the two screened vertex operators had been obtained by the deformation of two contours, one around each of the screened vertices, one finds

$$\Delta(E_+) = E_+ \otimes 1 + q^{-H/2} \otimes E_+, \tag{5.8.33}$$

which is the correct result if one identifies the generator E_+ used here with the product $q^{-H/4} E_+$ of the generators used in section 4.2. The co-algebra structure of the operators $E_\pm$ and H is thus indeed the one of $\mathsf{U}_q(A_1)$. That the same is true for the algebra structure is seen as follows. By the definition of the operators, one has

$$\begin{aligned}[][H, E_-] \cdot U_\lambda^s(z) &= H \cdot U_\lambda^{s+1}(z) - E_- \cdot ((\bar\lambda - 2s)\, U_\lambda^s(z)) \\ &= -2\, U_\lambda^{s+1}(z) = -2\, E_- \cdot U_\lambda^s(z),\end{aligned} \tag{5.8.34}$$

and similarly

$$[H, E_+] \cdot U_\lambda^s(z) = 2\, E_+ \cdot U_\lambda^s(z), \tag{5.8.35}$$

as well as

$$\begin{aligned}(E_+E_- - q\, E_-E_+) &\cdot U_\lambda^s(z) \\ &= E_+ \cdot U_\lambda^{s+1}(z) - q\, E_- \cdot \left(\frac{1-q^{s-\bar\lambda-1}}{1-q^{-1}} \frac{1-q^s}{1-q} U_\lambda^{s-1}(z) \right) \\ &= \left(\frac{1-q^{s-\bar\lambda}}{1-q^{-1}} \frac{1-q^{s+1}}{1-q} - q \frac{1-q^{s-\bar\lambda-1}}{1-q^{-1}} \frac{1-q^s}{1-q} \right) U_\lambda^s(z) \\ &= \frac{1-q^{2s-\bar\lambda}}{1-q^{-1}} U_\lambda^s(z).\end{aligned} \tag{5.8.36}$$

Thus one has

$$\begin{aligned}[][H, E_\pm] &= \pm 2\, E_\pm, \\ E_+E_- - q\, E_-E_+ &= \frac{1-q^{-H}}{1-q^{-1}}.\end{aligned} \tag{5.8.37}$$

These are indeed the correct relations for the basis chosen here (the second formula follows from the corresponding one for the generators used in section 4.2 with the help of the identity (4.2.21)).

In summary, the generators $E_\pm$ and H reproduce the correct algebra and co-algebra structure of the quantum group $\mathsf{U}_q(A_1)$. In addition, it is compatible with the definitions given above to define the antipode of these generators by the usual formula for $\mathsf{U}_q(A_1)$, and hence one has the full Hopf algebra structure of $\mathsf{U}_q(A_1)$. The final task is then to identify the quasitriangularity, or in other words the R-matrix of $\mathsf{U}_q(A_1)$. This is done as follows. One defines the finite-dimensional R-matrix in a given tensor product of representations as measuring the non-triviality of the braiding of the screened vertex operators under consideration:

$$U^s_\lambda(z) \otimes U^t_\mu(w) = \sum_{s',t'} \left(\mathcal{R}^{\lambda\mu}\right)^{st}_{s't'} U^{s'}_\mu(w) \otimes U^{t'}_\lambda(z). \qquad (5.8.38)$$

The matrix $\mathcal{R}^{\lambda\mu}$ can then be computed by performing the various contour deformations that effect the necessary braidings of the vertex operators and their screenings (in doing so, one must not forget to redistribute the contours obtained after the braiding back to the correct order required for the $\mathcal{C}$-contours of the screened vertices). The simplest way to perform the calculation is to work with the screened vertices $\hat{U}^s_\lambda$. One then has to perform the following three types of manipulations: braiding $\hat{U}^s_\lambda(z)$ around $\hat{U}^t_\mu(w)$, accompanied by the corresponding displacement of the contours that are attached to $\hat{U}^s_\lambda$; splitting these displaced contours into pieces ending at w and pieces going from w to z; and deforming the latter pieces to infinity, thereby splitting them into pieces ending at w and pieces ending at z. Each of these operations requires only standard tricks of complex analysis and hence is straightforward, although the procedure involves some combinatorics and gets quite lengthy. The result of the computation is in full agreement with the known finite-dimensional R-matrices of $\mathsf{U}_q(A_1)$.

So far the parameter $k^\vee$ has been completely arbitrary. It must therefore be asked what special features arise if $k^\vee$ is restricted to being a non-negative integer, which is the choice relevant to WZW theories. To enter this discussion, one may note that from (5.8.14) it is evident that

$$\Delta_{-\lambda-2\rho} = \Delta_\lambda. \qquad (5.8.39)$$

For $\mathfrak{g} = A_1$ this means that Δ_λ is invariant under the transformation $\bar{\lambda} \mapsto -\bar{\lambda} - 2$. In addition, this extends to a symmetry of the action of the quantum group, provided that it is supplemented by $s \mapsto s - \bar{\lambda} - 1$, which is needed to have the multiplicative prefactor involved in the action of E_+

invariant. Now we have already seen that the vector space generated by the fields $U^s_{\bar\lambda}$ is finite-dimensional if $\bar\lambda$ is dominant integral. In this case of course $-\bar\lambda - 2$ is still integral, but not dominant, and hence for generic q the associated vector space is not finite-dimensional; it becomes finite-dimensional, however, if q is a root of unity, i.e. if $k^\vee$ is rational. Moreover, for rational $k^\vee$ there is only a finite number of dominant integral weights $\bar\lambda$ that give rise to non-trivial screened vertex operators. Namely, writing $k^\vee + 2 = p/p'$ with p and p' coprime integers, one sees from the relation (5.8.24) between U and $\hat U$ that one needs $\bar\lambda < p$.

For integer values of $k^\vee$, this means that $\bar\lambda \le k^\vee + 1$. If in addition one formally continues the prefactor in (5.8.24) to nonpositive values of s, it follows that for the limiting value $\bar\lambda = k^\vee + 1$ the dimensionality is actually zero. Thus for $k^\vee \in \mathbb{Z}_{\ge 0}$ the vector space for $-\bar\lambda - 2$ has positive dimension iff $\bar\lambda$ obeys $\bar\lambda \le k^\vee$ in addition to $\bar\lambda \in \mathbb{Z}_{\ge 0}$. This provides certainly a motivation to restrict $k^\vee$ and $\bar\lambda$ to these particular values which are just the ones appearing in the $A_1^{(1)}$ WZW theory. Of course, these values should be *selected* by requiring that the quantum group modules to be constructed from the vector spaces must be unitarizable, but this analysis has not yet been carried out. Thus in the following we will *by assumption* restrict the allowed values of $k^\vee$ and $\bar\lambda$ to the values relevant for the WZW theory.

Now let us consider the case of arbitrary simple $\mathfrak{g}$. Then a construction completely analogous to the A_1-case is possible. One now has $r = \operatorname{rank}\mathfrak{g}$ screening currents $\mathcal{J}^i$, and the operators $E^i_\pm$ create or destroy the corresponding screenings $\int_C \mathrm{d}w\, \mathcal{J}^i(w)$, while the Cartan subalgebra generators do not change the number of screenings. The only new ingredient is that the various screenings no longer commute, so that not only does the index s become a multi-index, but also the ordering of the screenings is relevant, i.e.

$$U^s_\lambda(z) \equiv U_\lambda^{(i_1 i_2 \ldots i_s)} := \left(\prod_{j=1}^{s} \int_{C_j} \mathrm{d}w_j\, \mathcal{J}^{i_j}(w_j) \right) \phi_\lambda(z). \tag{5.8.40}$$

The non-commutativity of the screenings should in turn give rise to the new ingredient in the relations of $\mathsf{U}_q(\mathfrak{g})$, that is to the quantum Serre relations. This can indeed be shown; it requires again a quite lengthy calculation, and we only present some relevant forms. First one introduces radially ordered versions of the screened vertex operators,

$$\tilde U^s_\lambda := \mathcal{R}(\hat U^s_\lambda), \tag{5.8.41}$$

where the symbolic notation means that the radial ordering prescription must be applied to the integrand of the object on which $\mathcal{R}$ is acting. By contour manipulations, one can show that

$$E^i_- \cdot \phi_\lambda(z) = (1 - q^{-\bar\lambda^i})\, \tilde U^{(i)}(z) \tag{5.8.42}$$

and

$$\begin{aligned} E^i_- E^j_- \cdot \phi_\lambda(z) = {} & (1 - q^{-\bar\lambda^j})\,(1 - q^{-\bar\lambda^i + (\bar\alpha^{(i)}, \bar\alpha^{(j)})})\, \tilde U^{(ij)}(z) \\ & + (1 - q^{-\bar\lambda^i})\,(1 - q^{-\bar\lambda^j})\, q^{(\bar\alpha^{(i)}, \bar\alpha^{(j)})/2}\, \tilde U^{(ji)}(z), \end{aligned} \tag{5.8.43}$$

and with some more effort that in general

$$\begin{aligned} & E^{i_1}_- \dots E^{i_s}_- \cdot \phi_\lambda(z) \\ & \quad = \sum_{j=1}^{s} q^{\left[\sum_{k=j+1}^{s} (\bar\alpha^{(i_j)}, \bar\alpha^{(i_k)})\right]} \left(1 - q^{\left[-\bar\lambda^{i_j} + \sum_{k=1}^{j-1} (\bar\alpha^{(i_j)}, \bar\alpha^{(i_k)})\right]} \right) \\ & \qquad \cdot \mathcal{R} \left(\textstyle\int_\infty^z \mathrm{d}w\, \mathcal{J}^{i_j}(w)\, E^{i_1}_- \dots E^{i_{j-1}}_-\, E^{i_{j+1}}_- \dots E^{i_s}_- \cdot \phi_\lambda(z) \right). \end{aligned} \tag{5.8.44}$$

With these forms at hand, one can verify the Serre relations involving the lowering operators E^i_-. For the raising operators E^i_+ one proceeds similarly, but the calculation is even more complicated because one must now consider various different orderings of the screenings appearing in the operators $\tilde U^s_\lambda$ on which the raising operators act.

The construction just outlined for WZW theories can also be used for other types of conformal field theories, and in fact it was invented in the context of the minimal conformal field theories having conformal central charge $c < 1$. In the following we briefly describe the construction for this class of models as well. The screening procedure turns out to be more complicated than in the case of the $A_1^{(1)}$ WZW theory, as two types of screenings are involved, and as a consequence a more complicated quantum group arises. But the free field realization is even simpler than in the $A_1^{(1)}$ case: it involves only a single free boson $\varphi(z)$ with contraction

$$\underbrace{\varphi(z)\,\varphi(w)} = -\,2\,\ln(z - w). \tag{5.8.45}$$

The energy–momentum tensor for this free boson consists again of two contributions, the conventional term T_1 that contains only first derivatives of the boson, and a term T_2 proportional to the second derivative of φ:

$$T(z) = -\tfrac{1}{4}\, :\!\partial\varphi(z)\,\partial\varphi(z)\!:\, + i\alpha_0\, \partial^2\varphi(z). \tag{5.8.46}$$

Note that T_2 is a total derivative. In terms of the path integral formulation of quantum field theory, it corresponds to a boundary term in the action. While this term leaves the field equations unaltered, it does change the boundary conditions that are to be imposed on their solutions. The logarithmic behavior of the φ-φ-contraction is analogous to the potential for a two-dimensional gas of electric (or magnetic) charges interacting via Coulomb forces. The description of an interacting conformal field theory in terms of free bosons is therefore often called the *Coulomb gas* formulation of the theory; in this picture, the boundary term T_2 corresponds to a "background charge at infinity".

Just as in the WZW case, the operator product of $T(z)$ with itself is calculable with the help of the Wick theorem. One obtains the Virasoro algebra with the value of the central charge given by

$$c = 1 - 24\,\alpha_0^2, \tag{5.8.47}$$

where the unit contribution comes from $T_1(z)T_1(w)$ and the second term from $T_2(z)T_2(w)$.

The primary fields of the theory are given by vertex operators

$$V_\alpha(z) = :\exp\,(i\alpha\varphi(z)):, \tag{5.8.48}$$

where α is an arbitrary real number. One easily calculates the operator products

$$\begin{aligned} \underbrace{T_1(z)\,V_\alpha(w)} &= \alpha^2\,(z-w)^{-2}\,V_\alpha(w) + (z-w)^{-1}\,\partial V_\alpha(w), \\ \underbrace{\partial^2\varphi(z)\,V_\alpha(w)} &= 2i\alpha\,(z-w)^{-2}\,V_\alpha(w), \end{aligned} \tag{5.8.49}$$

from which one concludes that V_α is indeed primary, with conformal dimension

$$\Delta_\alpha = \alpha\,(\alpha - 2\alpha_0). \tag{5.8.50}$$

In particular, for the values

$$\alpha_\pm = \alpha_0 \pm \sqrt{\alpha_0^2 + 1} = \frac{1}{2\sqrt{6}}\left(\sqrt{1-c} \pm \sqrt{25-c}\right), \tag{5.8.51}$$

one has

$$\Delta_\pm \equiv \Delta_{\alpha_\pm} = -\alpha_+\alpha_- = 1. \tag{5.8.52}$$

The special vertex operators

$$\mathcal{J}^\pm(z) := V_{\alpha_\pm}(z) \tag{5.8.53}$$

then play again the role of screening currents. Analogously to the WZW case, the vector spaces that are to be identified as the modules of some quantum group are generated by screened vertex operators $U_\alpha^{s,t}$ defined as

$$U_\alpha^{s,t}(z) := \left(\prod_{j=1}^{s} \int_{\mathcal{C}_j} \mathrm{d}w_j\,\mathcal{J}^+(w_j)\right)\left(\prod_{j=1}^{t} \int_{\mathcal{C}_{s+j}} \mathrm{d}y_j\,\mathcal{J}^-(y_j)\right) V_\alpha(z), \tag{5.8.54}$$

where the contours are again the $\mathcal{C}$-contours of (5.8.22).

The operator product of two of the vertex operators is simply

$$\begin{aligned} \underbrace{V_\alpha(z)\, V_\beta(w)} &= \exp\left(-\, i\alpha\, i\beta\, \ln(z-w)\right)\, V_{\alpha+\beta}(w) \\ &= (z-w)^{\alpha\beta}\, V_{\alpha+\beta}(w). \end{aligned} \tag{5.8.55}$$

As a consequence, the braiding of two vertex operators is given by

$$V_\alpha(z)\, V_\beta(w) = \mathrm{e}^{2\pi i\alpha\beta}\, V_\beta(w)\, V_\alpha(z). \tag{5.8.56}$$

Owing to $\alpha_+\alpha_- = -1$, it follows that $\mathcal{J}^+$ and $\mathcal{J}^-$ have no relative braiding.

For special values of the number α, the vector space constructed from the screenings of V_α is again finite-dimensional, which is best seen by rewriting the screened vertices in terms of the integrals

$$\hat U_\alpha^{s,t}(z) := \left(\prod_{j=1}^{s} \int_\infty^z \mathrm{d}w_j\, \mathcal{J}^+(w_j)\right) \left(\prod_{j=1}^{t} \int_\infty^z \mathrm{d}y_j\, \mathcal{J}^-(y_j)\right) V_\alpha(z). \tag{5.8.57}$$

After the usual lengthy contour manipulations one finds

$$\begin{aligned} U_\alpha^{s,t}(z) = {} & \prod_{j=0}^{s-1} \left[1-\exp\left(2\pi i\, \alpha_+(2\alpha+j\alpha_+)\right)\right] \\ & \cdot \prod_{j=0}^{t-1} \left[1-\exp\left(2\pi i\, \alpha_-(2\alpha+j\alpha_-)\right)\right] \hat U_\alpha^{s,t}(z). \end{aligned} \tag{5.8.58}$$

Together with $\alpha_+\alpha_- = -1$, this result implies that $U_\alpha^{s,t} \equiv 0$ for $s \geq m$ or $t \geq n$ if the number α takes one of the special values

$$\alpha = \alpha_{m.n} := \tfrac{1}{2}\,(1-m)\,\alpha_+ + \tfrac{1}{2}\,(1-n)\,\alpha_-, \tag{5.8.59}$$

i.e. for these values the vector space spanned by the fields $U_\alpha^{s,t}$ is of finite dimension mn. Considering now the formula (5.8.55) for the braiding of vertex operators and comparing it with the analogous result (5.8.18) of the WZW case, one is led to expect that the numbers $q_\pm$ defined as

$$q_\pm := \exp\left(-2\pi i\, \alpha_\pm^2\right) \tag{5.8.60}$$

play the role of deformation parameters of the quantum group we are looking for. Note that, analogously to the WZW case, the finite-dimensionality of the vector space generated by the screened vertices does not require that $q_\pm$ takes any special values, but restricts the possible values of α for a given choice of $q_\pm$.

However, again rational values of $\alpha_\pm^2$, i.e. values for which the numbers $q_\pm$ are roots of unity, play a special role. For α_+^2 rational, let us set $\alpha_+^2 = p'/p$ with p and p' coprime integers. Then from the relation (5.8.58) between U and $\hat U$ it can be read off that the finite-dimensional vector spaces based on $V_{m,n}$ are nonzero only for $1 \leq m \leq p$ and $1 \leq m \leq p'$. Thus for rational α_+^2 there is a finite number of finite-dimensional vector spaces, corresponding to a rational conformal field theory. Moreover from the result (5.8.50) for the conformal dimension of V_α it directly follows that

$$\Delta_{2\alpha_0-\alpha} = \Delta_\alpha. \tag{5.8.61}$$

(compare the analogous identity (5.8.39) of the WZW case). One can consistently identify the vector spaces based on V_α and $V_{2\alpha_0-\alpha}$. If one formally continues the prefactor in (5.8.58) to $s, t = 0$, this means that for $m = p$ or $n = p'$ the vector space based on $V_{m,n}$ gets identified with a zero-dimensional space. As a consequence the allowed values of m and n get restricted to

$$1 \le m \le p-1, \quad 1 \le n \le p'-1. \tag{5.8.62}$$

Of particular interest is the case $p = \ell{+}3$, $p' = \ell{+}2$: this corresponds to the unitary minimal series (3.2.30), with ℓ coinciding with the parameter chosen there. Consider e.g. the conformal dimension of $V_{m,n} \equiv V_{\alpha_{m,n}}$: inserting (5.8.59) into the formula (5.8.50) for conformal dimensions, and using the relation between $\alpha_\pm$ and c, one obtains

$$\begin{aligned}\Delta_{m,n} = {} & \tfrac{1}{24}\,(c-1) - \tfrac{1}{2}\,mn + \tfrac{1}{48}\,(m^2+n^2)(13-c) \\ & + \tfrac{1}{48}\,(m^2-n^2)\sqrt{(1-c)(25-c)},\end{aligned} \tag{5.8.63}$$

and this reduces to the formula given in (3.2.30) upon use of the relation between c and the integer ℓ which reads

$$\ell = \frac{1}{2}\left(\sqrt{\frac{25-c}{1-c}} - 5\right). \tag{5.8.64}$$

To enter the construction of the quantum group generators, let us define operators that create screenings by

$$E_-^\pm \cdot U_\alpha^{s,t}(z) = \int_C \mathrm{d}w\, \mathcal{J}^\pm(w)\, U_\alpha^{s,t}(z), \tag{5.8.65}$$

where C is a $\mathcal{C}$-contour that surrounds the whole twisted vertex operator, so that

$$\begin{aligned} E_-^+ \cdot U_\alpha^{s,t} &= E_\alpha^{s+1,t}, \\ E_-^- \cdot U_\alpha^{s,t} &= E_\alpha^{s,t+1} \end{aligned} \tag{5.8.66}$$

(here for the second line it is important that the relative braiding of $\mathcal{J}^+$ and $\mathcal{J}^-$ is trivial). Similarly, for the product of two screened vertices, one defines the action of these operators as providing an additional screening around the two screened vertex operators; deforming this contour into a sum of two, one around each of the screened vertices, one obtains the co-multiplication

$$\begin{aligned}\mathbf{\Delta}(E_-^+)\cdot\left(U_\alpha^{s,t}(z)\otimes U_\beta^{u,v}(w)\right) &= U_\alpha^{s+1,t}(z)\otimes U_\beta^{u,v}(w)\\ &+\exp\left[\pi i\,\alpha_\pm\,(\alpha+s\alpha_+ + t\alpha_-)\right]U_\alpha^{s,t}(z)\otimes U_\beta^{u+1,v}(w),\end{aligned} \tag{5.8.67}$$

and analogously for $\mathbf{\Delta}(E_-^-)$. This can be turned into an operator equation by defining

$$K^\pm\cdot U_\alpha^{s,t}(z) := \exp\left(\pi i\,\alpha_\pm\oint_{C_z}\mathrm{d}w\,\partial\varphi(w)\right)U_\alpha^{s,t}(z), \tag{5.8.68}$$

where C_z is a contour encircling z infinitesimally close to z, and $(K^\pm)^{-1}$ analogously with an additional minus sign in the exponent. With this definition, the above result reads

$$\mathbf{\Delta}(E_-^\pm) = E_-^\pm\otimes 1 + (K^\pm)^2\otimes E_-^\pm. \tag{5.8.69}$$

It is also not difficult to derive that

$$\mathbf{\Delta}(K^\pm) = K^\pm\otimes K^\pm, \tag{5.8.70}$$

and

$$\begin{aligned}&[E_-^+,E_-^-]=0,\\ &[K^+,K^-]=0,\\ &K^\pm E_-^\pm = q_\pm^{-1/2}\,E_-^\pm K^\pm,\\ &K^\pm E_-^\mp = -E_-^\mp K^\pm,\end{aligned} \tag{5.8.71}$$

and also that the latter relations are preserved by $\mathbf{\Delta}$. This means that the vector space generated by polynomials in $E_-^\pm$ and $K^\pm$ (and also in $(K^\pm)^{-1}$) carries the structure of a bi-algebra. Moreover, this can be extended to a Hopf algebra structure by defining a co-unit and an antipode as

$$\begin{aligned}&\varepsilon(K^\pm)=1, &&\varepsilon(E_-^\pm)=0,\\ &\gamma(K^\pm)=(K^\pm)^{-1}, &&\gamma(E_-^\pm)=-(K^\pm)^{-2}\,E_-^\pm.\end{aligned} \tag{5.8.72}$$

When acting on screened vertex operators $U_{m,n}^{s,t}\equiv U_{\alpha_{m,n}}^{s,t}$, the phase factor obtained by the action of $K^\pm$ splits naturally into two parts corresponding to the terms $\alpha_\pm^2$ and $\alpha_+\alpha_-=-1$ in the exponent, so that one may write

$$K^\pm = (q_\pm)^{H^\pm/4}\mathrm{e}^{\pi i\,H^\mp/2}, \tag{5.8.73}$$

where $H^\pm$ act as

$$\begin{aligned}H^+\cdot U_{m,n}^{s,t} &= (m-2s-1)\,U_{m,n}^{s,t},\\ H^-\cdot U_{m,n}^{s,t} &= (n-2t-1)\,U_{m,n}^{s,t}.\end{aligned} \tag{5.8.74}$$

For the special case where in addition $n = 1$, this means that not only E^-_-, but also H^- acts as zero. The relations among the remaining operators E^+_- and H^+ are then precisely those of a Borel subalgebra of $\mathsf{U}_{q_+}(A_1)$. Analogously, for $m = 1$ the operators E^+_- and H^+ act as zero, while the polynomials E^-_- and H^- generate a Borel subalgebra of $\mathsf{U}_{q_-}(A_1)$.

It is natural to expect that for $m = 1$ or $n = 1$ one gets the full quantum group by including the raising operator of $\mathsf{U}_{q_\pm}(A_1)$. Thus one defines further operators $E^\pm_+$ as annihilation of screenings, combined with a multiplicative factor similar to the one in (5.8.32):

$$\begin{aligned} E^+_+ \cdot U^{s,t}_\alpha &= \frac{1-q_+^{1-s} \mathrm{e}^{4\pi i\,\alpha\alpha_+}}{1-q_+} \, \frac{1-q_+^{-s}}{1-q_+^{-1}} \, U^{s-1,t}_\alpha(z), \\ E^-_+ \cdot U^{s,t}_\alpha &= \frac{1-q_-^{1-t} \mathrm{e}^{4\pi i\,\alpha\alpha_-}}{1-q_-} \, \frac{1-q_-^{-t}}{1-q_-^{-1}} \, U^{s,t-1}_\alpha(z). \end{aligned} \tag{5.8.75}$$

With these definitions, it is straightforward to verify the relations

$$\begin{aligned} &[E^+_+, E^-_+] = 0, \\ &K^\pm E^\pm_+ = q_\pm^{1/2}\, E^\pm_+ K^\pm, \\ &K^\pm E^\mp_+ = -E^\mp_+ K^\pm, \end{aligned} \tag{5.8.76}$$

and

$$\begin{aligned} &[E^\pm_+, E^\mp_-] = 0, \\ &E^\pm_+ E^\pm_- - q_\pm^{-1}\, E^\pm_- E^\pm_+ = \frac{1-(K^\pm)^4}{1-q_\pm}. \end{aligned} \tag{5.8.77}$$

Also, by complicated contour manipulations one can derive

$$\Delta(E^\pm_+) = E^\pm_+ \otimes 1 + (K^\pm)^2 \otimes E^\pm_+, \tag{5.8.78}$$

and this co-multiplication preserves the above relations. Finally, one can also extend the definition of the co-unit and the antipode to $E^\pm_+$, and one can construct finite-dimensional R-matrices describing the braiding properties of the screened vertices in analogy with (5.8.38).

In summary, what we have identified as the quantum group of the $c < 1$ conformal field theories is a non-direct sum of the deformed enveloping algebras $\mathsf{U}_{q_+}(A_1)$ and $\mathsf{U}_{q_-}(A_1)$, where the result that the sum is non-direct manifests itself only in the fact that $K^\pm$ anticommutes rather than commutes with $E^\mp_+$ and with $E^\mp_-$. In addition, for the unitary minimal series of $c < 1$ conformal field theories, the quantum group reduces to $\mathsf{U}_{q_+}(A_1)$, respectively to $\mathsf{U}_{q_-}(A_1)$, for the subsets $U^{s,0}_{m,1}$ and $U^{0,t}_{1,n}$ of screened vertices.

Owing to the fact that the conformal transformation law for screened vertex operators contains boundary terms, the action of the quantum group (more precisely, of the Borel subalgebra generated by $K^{\pm}$ and $E_{-}^{\pm}$) does not commute with the action of the Virasoro algebra. However, one can argue that the conformal blocks of the theory can be interpreted as the invariant tensors of the quantum group, and that on the space of invariant tensors the actions of the quantum group and the Virasoro algebra commute.

To conclude this section, let us point out the way in which the previous results are related to the identification of the truncated Kronecker products of the quantum group with the fusion rules of the conformal field theory that was made in section 5.5. The key observation is that the chiral block functions of the (Coulomb gas representation of the) conformal field theory can be written as correlators of a different type of screened vertex operators, namely of combinations of vertex operators and the screening charges that were already mentioned previously. This is known as the *Feigin–Fuks* or *Dotsenko–Fateev* integral representation of correlation functions. The screening charges are defined as contour integrals over the screening currents and are hence similar to the screenings used in the definition (5.8.54) of screened vertex operators, but the contours must be chosen differently. Namely, they are those used in the definition of hypergeometric functions and generalizations thereof; e.g. they may be chosen as encircling several points z_i at which the vertex operators are sitting, topologically being circles or the "double-eight formed" so-called Pochhammer contours, or also as open contours that connect the points z_i (the latter choice leads to contour integrals such as the ones appearing in the formula (3.4.57)).

The chiral blocks of a conformal field theory have built into them the null vector structure of the modules of the symmetry algebra. They can therefore be used to determine the fusion rule algebra of a conformal field theory: the fusion rules are obtained through the construction of the full correlation functions of the theory by combining the holomorphic and antiholomorphic blocks in a crossing symmetric way. (In fact, as discussed in section 3.4 (p. 183), this provides not only the fusion rules, but also the full operator algebra of the theory; with the help of the Feigin–Fuks construction this calculation of the operator product algebra has been completed for the $c < 1$ minimal series and also for several other series of conformal field theories.) Now different possible configurations of contours for the screening integrals produce different analytic functions, but these different functions are related to each other by a change of basis. Thus in particular the conformal blocks should be obtainable as linear combinations of tensor products of screened vertex operators; in fact, the coefficients of this linear transformation can be shown to be Clebsch–Gordan coefficients of the quantum group. As a consequence, the conformal blocks, although being invariant under the quantum group, have in a sense built in the full quantum group structure. The relation between screened vertices and chiral blocks therefore seems to be able to explain in an intuitive manner the coincidence of the fusion rules of the conformal field theory and the truncated Kronecker products of the quantum group.

5.9 Literature

The concept of fusion rules is already implicit in [47], but in the concise form used here it is due to Verlinde [556]. The relation between fusion rules and modular transformations was proposed by Verlinde [557], and proved by Moore and Seiberg [434, 435]; see also [92, 155, R1]. In [149, 239] it is discussed how one can employ the Verlinde formula to compute the modular matrix S. As abstract algebras, fusion rule algebras have been discussed in [101, 102].

The constraints arising for conformal dimensions and c-values of rational conformal field theories were discussed in [557, 550, 19]. The classification of rational conformal field theories is so far only complete for $c < 1$ [47, 218, 219] and for $c = 1$ [156, 257, 346]. Various steps in the program of classifying *all* rational conformal field theories have been made in [435, 436, 14, 437, 384, 412, 15, 413, 433].

The importance of modular invariance in string theory has been known for a long time; the exposition in section 5.2 follows [R24] and [556]. In the context of statistical mechanics, the relevance of modular invariance was pointed out by Cardy [103]; see also [293]. For the particular case of WZW theories, the modular invariance constraints were first considered in [251]. This led in particular to conjectures about the modular invariants for $A_1^{(1)}$ [248, 99], namely the classification (5.2.2). The proof of this classification was obtained in [100]; for an explanation of the relation between these modular invariants and the exponents of the simply laced simple Lie algebras, see [447]. The $A_1^{(1)}$ spectra of (5.2.3) which are not modular invariant but still lead to a closed operator product algebra, were classified in [122].

The observation that in terms of the maximally extended symmetry algebra, non-diagonal modular invariants are due to fusion rule automorphisms was made in [435, 556]. The relevance of conformal embeddings for the construction of modular invariants was pointed out by various authors; see e.g. [76, 21, 320, 322, 566, 126]; modular invariants for various WZW theories can be found in these papers as well as in [53, 12, 202, 509, 3, 4, 502, 558, 559]. The fundamental role played by simple currents was realized in full generality by Schellekens and Yankielowicz [509, 510, 511, 512]; however, many modular invariants associated to simple currents of WZW theories had been known before [53, 202, 12]. The interpretation of the non-diagonal $A_1^{(1)}$ modular invariants in terms of simple currents and/or extended symmetry algebras was found in [1, 92, 435] for the D-invariants, in [81, 123] for the E_6- and E_8-invariants, and in [155] for the E_7-invariant. For a review of what is currently known about modular invariants of affine Lie algebras, see [R24] and [558]. The remarks in section 5.4 about the modular invariants for tensor product and coset theories are also taken from [R24].

The classification of the cominimal simple currents of WZW theories was obtained in [235]; the single exceptional simple WZW current (for $E_8^{(1)}$ at level two) was found in [208]; the proof that together these are all simple WZW currents is given in [233]. The transformation behavior of WZW fusion rules under automorphisms of the affine Dynkin diagram were first observed in [228]; see also [53, 238, 567, 241].

The extended Racah–Speiser algorithm leading to formula (5.5.24) is due to Kac [314, B20], and was independently conjectured by Walton [567]. Detailed proofs of the algorithm were independently written down in [568, 241, 239]. The examples presented in section 5.5 are taken from [239] and from unpublished notes by those authors. Arguments for the isomorphism between fusion rules and truncated Kronecker products are contained in [14, 241]. The proof of the isomorphism for the case of $\mathfrak{g} = A_r$ is due to Goodman and Nakanishi [268, 271].

The connection between the fusion rules and differential equations, leading e.g. to the sum rule (5.6.8) for conformal dimensions, is made in [125, 384, 345, 232]; differential equations for correlation functions and for characters of rational conformal field theories are also discussed in [180, 283, 411, 412]. The Riemann monodromy problem is discussed in the context of differential equations for chiral blocks in [69, 347, 345].

For the mathematical treatment of the braid group, the standard reference is [B5]. For introductions to category theory, see [R9, 436, R21].

Chiral vertex operators were introduced in [544, 488]; see also [545, 487, 515, 484, R12, 436, 202]. The polynomial equations (5.7.12) and (5.7.13) have been found by Moore and Seiberg [434] in the course of proving the Verlinde formula for the fusion rules. They can also be understood in the context of the topological ("Chern-Simons") 2+1-dimensional field theories that were introduced by Witten [583, 584]. The relation between topological field theory and rational conformal field theories was analyzed in [583, 71, 176] from the point of view of canonical quantization; various other aspects of this relation are treated in [437, 183, 134, 16]; for a review, see [R29]. The relation between topological field theory and quantum groups is discussed in [585, 15, 272, 273]. For the connection between knot theory [B30, B5, 304, 305, 306, 332, 215, 439] and rational conformal field theory and/or topological field theory see also [546, 489, 490, 5, 385, R12, 147, 223].

Quantum groups have been identified as the natural structure underlying rational conformal field theories by Alvarez-Gaumé, Gomez, and Sierra in [13, 14, 15], and by others from different points of view in [254, 433, 402, 279, 28, 6]. In these papers mainly the case of minimal models and WZW theories is treated; quantum groups underlying orbifold models are discussed in [154, 36]. The action of a quantum group on the space of chiral blocks of the Coulomb gas picture of conformal field theories was considered in [14, 433, 201, 78, 79, 8, 505]. The identification of the quantum group structure acting on the space of screened vertex operators is due to Gomez and Sierra who introduced the contours (5.8.22) and worked out the case of the $c < 1$ conformal field theories [266, 267]; the corresponding analysis for WZW theories was performed in [478]. From the point of view of R-matrices, the relation between conformal field theory and quantum groups was described by Gervais and collaborators (for the case of the Virasoro algebra in [254], and for extended algebras in [65, 138]). In the context of the algebraic approach to field theory structures related to quantum groups were discovered by Fredenhagen, Rehren, and Schroer [209, 210, 485], and the quantum group structure of the Ising model was displayed by Mack and Schomerus [402].

The Clebsch–Gordan coefficients for $\mathsf{U}_q(A_1)$ were obtained in [342, 455]; see also [460, 275, 347]. The quantum $6j$-symbols for $\mathsf{U}_q(A_1)$ are discussed in [342, 243, 82, 455, 243, 500, 85]. (These results can be understood in terms of q-deformations of hypergeometric functions; these and q-deformations of other special functions are discussed in [24, 342, 414, 355, 551, 457, 504].) The corresponding results for q a root of unity were obtained in in [13, 552]. The duality matrices of the $A_1^{(1)}$ WZW theory which coincide (up to a similarity transformation) with the $6j$-symbols were computed in [597, 124], see also [350, 544]. A few $6j$-symbols for $\mathsf{U}_q(E_6)$ are calculated in [351].

References

A Books

[B1] S. Adler and R. Dashen, *Current Algebras and Applications to Particle Physics* (Benjamin/Cummings, New York 1968)

[B2] E. Abe, *Hopf Algebras* (Cambridge University Press, Cambridge 1977)

[B3] O. Babelon and C.-M. Viallet, *Integrable Models and Yang–Baxter equations* (Cambridge University Press, to appear), Trieste preprint SISSA 54 EP (May 1989)

[B4] L.C. Biedenharn and H. Van Dam, eds., *Quantum Theory of Angular Momentum* (Academic Press, New York 1965)

[B5] J. Birman, *Links, Braids, and Mapping Class Groups* (Princeton University Press, Princeton 1974)

[B6] N. Bourbaki, *Groupes et Algèbres de Lie* (Masson, Paris 1982)

[B7] M.R. Bremner, R.V. Moody, and J. Patera, *Tables of Dominant Weight Multiplicities for Representations of Simple Lie Algebras* (Marcel Dekker, New York 1985)

[B8] R.N. Cahn, *Semi-simple Lie Algebras and Their Representations* (Benjamin/Cummings, Menlo Park 1984)

[B9] J. Dixmier, *Enveloping Algebras* (North Holland, New York 1977)

[B10] U. Fano and G. Racah, *Irreducible Tensorial Sets* (Academic Press, New York 1959)

[B11] H. Freudenthal and H. de Vries, *Linear Lie Groups* (Academic Press, New York 1969)

[B12] R. Gilmore, *Lie Groups, Lie Algebras and Some of Their Representations* (John Wiley, New York 1974)

[B13] M. Gourdin, *Basics of Lie Groups* (Editions Frontières, Gif sur Yvette 1982)

[B14] M. Green, J. Schwarz, and E. Witten, *Superstring Theory* (Cambridge University Press, Cambridge 1987)

[B15] M. Hamermesh, *Group Theory and its Application to Physical Problems* (Addison-Wesley, Reading 1962)

[B16] S.S. Horuzhy, *Introduction to Algebraic Quantum Field Theory* (Kluwer Academic Publishers, Dordrecht 1990)

[B17] J.E. Humphreys, *Introduction to Lie Algebras and Representation Theory* (Springer Verlag, Berlin 1972)

[B18] N. Jacobson, *Lie Algebras* (Wiley Interscience, New York 1962)

[B19] V.G. Kac, *Infinite-dimensional Lie algebras* (Birkhäuser, Boston 1983)

[B20] V.G. Kac, *Infinite-dimensional Lie algebras* (third edition, Cambridge University Press, Cambridge 1990)

[B21] V.G. Kac and A.K. Raina, Bombay Lectures on *Highest Weight Representations of Infinite Dimensional Lie Algebras* (World Scientific, Singapore 1987)

[B22] D. Kastler, ed., *The Algebraic Theory of Superselection Sectors. Introduction and Recent Results* (World Scientific, Singapore 1990)

[B23] D.E. Knuth, *The TEXbook* (Addison-Wesley, Reading 1984)

[B24] L. Lamport, LATEX, *A Document Preparation System, User's Guide and Reference Manual* (Addison-Wesley, Reading 1985)

[B25] D.E. Littlewood, *The Theory of Group Characters and Matrix Representations of Groups* (Oxford University Press, London 1950)

[B26] D. Lüst and S. Theisen, *Lectures on String Theory*, Springer Lecture Notes in Physics 346 (Springer Verlag, Berlin 1989)

[B27] I. MacDonald, *Symmetric Functions and Hall Polynomials* (Oxford University Press, London 1979)

[B28] W.G. McKay and J. Patera, *Tables of Dimensions, Indices and Branching Rules for Representations of Simple Lie Algebras* (Marcel Dekker, New York 1981)

[B29] A. Pressley and G. Segal, *Loop Groups* (Clarendon, Oxford 1986)

[B30] D. Rolfsen, *Knots and Links* (Publish or Perish, Berkeley 1976)

[B31] D.E. Rutherford, *Substitutional Analysis* (Edinburgh University Press, Edinburgh 1948)

[B32] M.E. Sweedler, *Hopf Algebras* (Benjamin/Cummings, Menlo Park 1969)

[B33] H. Weyl, *The Theory of Groups and Quantum Mechanics* (Dover Publications, New York 1931)

[B34] H. Weyl, *The Classical Groups* (Princeton University Press, Princeton 1939)

[B35] B. Wybourne, *Classical Groups for Physicists* (John Wiley, New York 1974)

[B36] D. Zhelobenko, *Compact Lie Groups and Their Representations* (American Mathematical Society, Providence 1973)

B Reviews

[R1] L. Alvarez-Gaumé, C. Gómez and G. Sierra, in: *The Physics and Mathematics of Strings, Memorial Volume for V.G. Knizhnik*, L. Brink et al., eds. (World Scientific, Singapore 1990), p. 16

[R2] J. Bagger, in: *Particles and Fields* 3, Proceedings of the Banff Summer Institute 1988, A.N. Kamal and F.C. Khanna, eds. (World Scientific, Singapore 1989), p. 556

[R3] T. Banks, Lectures at the Theoretical Advanced Studies Institute, Santa Fe 1987

[R4] R. Behrends, J. Dreitlein, C. Fronsdal, and W. Lee, Rev.Mod.Phys. 34 (1962) 1

[R5] D. Bernard and J. Thierry-Mieg, Commun.Math.Phys. 111 (1987) 181

[R6] J. Cardy, in: Les Houches Summer Session 1988 on *Fields, Strings, and Critical Phenomena*, E. Brézin and J. Zinn-Justin, eds. (North Holland, Amsterdam 1989), p. 169

[R7] A. Connes, in: 1987 Cargèse Lectures on *Nonperturbative Quantum Field Theory*, G. t'Hooft et al., eds. (Plenum, New York 1988), p. 33

[R8] R. Coquereaux, J.Geom. and Phys. 6 (1989) 425

[R9] P. Deligne and J.S. Milne, Springer Lecture Notes in Mathematics 900 (1982) 101

[R10] J.J. de Swart, Rev.Mod.Phys. 35 (1963) 916

[R11] H.J. de Vega, Int.J.Mod.Phys. A 4 (1989) 2371

[R12] J. Fröhlich, in: Cargèse Lectures 1987 on *Nonperturbative Quantum Field Theory*, G. t'Hooft et al., eds. (Plenum, New York 1988), p. 71

[R13] P. Furlan, G. Sotkov, and I. Todorov, Riv.Nuovo Cim. 12 No. 4 (1989) 1

[R14] P. Ginsparg, in: Les Houches Summer Session 1988 on *Fields, Strings, and Critical Phenomena*, E. Brézin and J. Zinn-Justin, eds. (North Holland, Amsterdam 1989), p. 1

[R15] P. Goddard and D. Olive, Int.J.Mod.Phys. A 1 (1986) 303

[R16] C. Itzykson, H. Saleur, and J.-B. Zuber, eds., *Conformal Invariance and Applications to Statistical Mechanics* (World Scientific, Singapore 1988)

[R17] D. Kastler, M. Mebkhout, and K.-H. Rehren, in: [B22], p. 356

[R18] J. Lepowski, Springer Lecture Notes in Mathematics 933 (1982) 130

[R19] I.G. MacDonald, Springer Lecture Notes in Mathematics 901 (1981) 258

[R20] G. Mack, in: Cargèse Lectures 1987 on *Nonperturbative Quantum Field Theory*, G. t'Hooft et al., eds. (Plenum, New York 1988), p. 353

[R21] S. Majid, Int.J.Mod.Phys. A 5 (1990) 1

[R22] G. Racah, *Group Theory and Spectroscopy* (IAS Lecture Notes, Princeton 1951, reprinted in Ergebnisse der exakten Naturwissenschaften 37 (1965) 28)

[R23] W. Rühl, unpublished lecture notes (Kaiserslautern University, 1989)

[R24] A.N. Schellekens, Nucl.Phys. B (Proc.Suppl.) 15 (1990) 3

[R25] B. Schroer, in: Proceedings of the 16th International Conference on *Differential Geometric Methods in Theoretical Physics*, K. Bleuler and M. Werner, eds. (Kluwer Academic Publishers, Dordrecht 1988), p. 289

[R26] J. Schwarz, ed., *Superstrings, the First Fifteen Years* (World Scientific, New York 1985)

[R27] R. Slansky, Phys.Rep. 79 (1981) 1

[R28] R. Slansky, Comm.Nucl.Part.Phys. XVIII (1988) 175

[R29] P. van Baal, Acta Phys.Pol. B 28 (1990) 73

[R30] J.-B. Zuber, in: Les Houches Summer Session 1988 on *Fields, Strings, and Critical Phenomena*, E. Brézin and J. Zinn-Justin, eds. (North Holland, Amsterdam 1989), p. 247

C Articles

[1] A. Abouelsaood and D. Gepner, Phys.Lett. B 176 (1986) 380

[2] V. Agrawala and J. Belinfante, Ann.Phys. 49 (1968) 130

[3] C. Ahn and M. Walton, Phys.Lett. B 223 (1989) 343

[4] C. Ahn and M. Walton, Phys.Rev. D 41 (1990) 2558

[5] Y. Akutsu and M. Wadati, Commun.Math.Phys. 117 (1988) 243

[6] A. Alekseev, L.D. Faddeev, M. Semenov-Tian-Shanskii, and A. Volkov, CERN preprint CERN-TH.5981/91 (January, 1991)

[7] A. Alekseev and S. Shatashvili, Nucl.Phys. B 323 (1989) 719

[8] A. Alekseev and S. Shatashvili, Commun.Math.Phys. 133 (1990) 353

[9] D. Altschüler, Phys.Lett. B 163 (1985) 193

[10] D. Altschüler, K. Bardakci, and E. Rabinovici, Commun.Math. Phys. 118 (1988) 241

[11] D. Altschüler, M. Bauer, and C. Itzykson, Commun.Math.Phys. 132 (1990) 349

[12] D. Altschüler, J. Lacki, and P. Zaugg, Phys.Lett. B 205 (1988) 281

[13] L. Alvarez-Gaumé, C. Gómez, and G. Sierra, Nucl.Phys. B 319 (1989) 155

[14] L. Alvarez-Gaumé, C. Gómez, and G. Sierra, Phys.Lett. B 220 (1989) 142

[15] L. Alvarez-Gaumé, C. Gómez, and G. Sierra, Nucl.Phys. B 330 (1990) 347

[16] L. Alvarez-Gaumé, J. Labatisda, and A. Ramallo, Nucl.Phys. B 334 (1990) 103

[17] O. Alvarez, P. Windey, and M. Mangano, Nucl.Phys. B 277 (1986) 317

[18] H. Andersen, P. Polo, and W. Kexin, Invent.math. 104 (1991) 1

[19] G. Anderson and G. Moore, Commun.Math.Phys. 117 (1988) 441

[20] I. Antoniadis and C. Bachas, Nucl.Phys. B 278 (1986) 343

[21] R. Arcuri, J. Gomez, and D. Olive, Nucl.Phys. B 285 (1987) 327

[22] D. Arnaudon and A. Chakrabarti, Commun.Math.Phys. 139 (1991) 461

[23] R. Ashworth, Int.J.Mod.Phys. A 4 (1989) 1427

[24] R. Askey and J. Wilson, SIAM J.Math.Anal. 10 (1979) 1008

[25] H. Au-Yang, B. McCoy, J. Perk, S. Tang, and M.-L. Yan, Phys.Lett. A 123 (1987) 289

[26] J. Avan and M. Talon, Phys.Lett. B 241 (1990) 77

[27] O. Babelon, Lett.Math.Phys. 15 (1988) 111

[28] O. Babelon, Commun.Math.Phys. 139 (1991) 619

[29] O. Babelon and C. Viallet, Phys.Lett. B 237 (1990) 411

[30] J. Bagger, D. Nemeschansky, and S. Yankielowicz, Phys.Rev.Lett. 60 (1988) 389

[31] J. Bagger, D. Nemeschansky, and J.-B. Zuber, Phys.Lett. B 286 (1989) 320

[32] F.A. Bais and P. Bouwknegt, Nucl.Phys. B 279 (1987) 561

[33] F.A. Bais, P. Bouwknegt, K. Schoutens, and M. Surridge, Nucl.Phys. B 304 (1988) 348

[34] F.A. Bais, P. Bouwknegt, K. Schoutens, and M. Surridge, Nucl.Phys. B 304 (1988) 371

[35] T. Banks, D. Horn, and H. Neuberger, Nucl.Phys. B 108 (1976) 119

[36] P. Bantay, Phys.Lett. B 245 (1990) 477

[37] K. Bardakci and M. Halpern, Phys.Rev. D 3 (1971) 2493

[38] A. Barut and R. Racka, Proc.Roy.Soc. Ser. A 287 (1965) 519

[39] M. Bauer, P. di Francesco, C. Itzykson, and J.-B. Zuber, Nucl.Phys. B 362 (1991) 515

[40] R.J. Baxter, Ann.Phys. 70 (1972) 193

[41] R.J. Baxter, Ann.Phys. 76 (1973) 1

[42] R.J. Baxter, J. Perk, and H. Au-Yang, Phys.Lett. A 128 (1988) 138

[43] V. Bazhanov, Commun.Math.Phys. 113 (1987) 471

[44] A.A. Belavin and V.G. Drinfeld, Funct.Anal.Appl. 16 (1982) 159

[45] A.A. Belavin and V.G. Drinfeld, Funct.Anal.Appl. 17 (1983) 220

[46] A.A. Belavin and V.G. Drinfeld, Sov.Sci.Rev. C 4 (1984) 93

[47] A.A. Belavin, A.M. Polyakov, and A.B. Zamolodchikov, Nucl.Phys. B 241 (1984) 333

[48] E. Beltrametti and A. Blasi, Phys.Lett. 20 (1966) 62

[49] G. Benkart, Can.Math.Soc.Proc. 5 (1986) 111

[50] L. Benoit and Y. Saint-Aubin, Phys.Lett. B 215 (1988) 517

[51] B. Berg, M. Karowkski, P. Weisz, and V. Kurak, Nucl.Phys. B 288 (1987) 628

[52] S. Berman and R.V. Moody, Proc.Amer.Math.Soc. 76 (1979) 223

[53] D. Bernard, Nucl.Phys. B 288 (1987) 628

[54] D. Bernard and G. Felder, Commun.Math.Phys. 127 (1990) 145

[55] I. Bernstein, I. Gelfand, and S. Gelfand, Funct.Anal.Appl. 5 (1971) 8

[56] M. Bershadsky, Phys.Lett. B 174 (1986) 285

[57] M. Bershadsky, V. Knizhnik, and M. Teitelman, Phys.Lett. B 151 (1985) 31

[58] L.C. Biedenharn, Phys.Lett. 3 (1962) 69

[59] L.C. Biedenharn, J.Math.Phys. 4 (1963) 436

[60] L.C. Biedenharn, J.Phys. A 22 (1989) L873

[61] L.C. Biedenharn, J. Blatt, and M. Rose, Rev.Mod.Phys. 24 (1952) 249

[62] L.C. Biedenharn and M. Tarlini, Lett.Math.Phys. 20 (1990) 271

[63] A. Bilal, Int.J.Mod.Phys. A 5 (1990) 1881

[64] A. Bilal and J.-L. Gervais, Nucl.Phys. B 314 (1989) 646

[65] A. Bilal and J.-L. Gervais, Nucl.Phys. B 318 (1989) 579

[66] G. Birkhoff, Ann.Math. 38 (1937) 526

[67] J.S. Birman and H. Wenzl, Trans.Amer.Math.Soc. 313 (1989) 249

[68] B. Blok and S. Yankielowicz, Nucl.Phys. B 315 (1989) 25

[69] B. Blok and S. Yankielowicz, Nucl.Phys. B 321 (1989) 717

[70] R. Borcherds, Adv.Math. 83 (1990) 30

[71] M. Bos and V. Nair, Phys.Lett. B 223 (1989) 61

[72] A. Bose and J. Patera, J.Math.Phys. 11 (1970) 2231

[73] H. Borchers, Commun.Math.Phys. 4 (1967) 315

[74] R. Bott, Enseign.Math. 23 (1977) 209

[75] W. Boucher, D. Friedan, and A. Kent, Phys.Lett. B 172 (1986) 316

[76] P. Bouwknegt, Nucl.Phys. B 290 (1987) 507

[77] P. Bouwknegt, in: Proceedings of the 1988 Luminy Conference on *Infinite-dimensional Lie Algebras and Groups*, V.G. Kac, ed. (World Scientific, Singapore 1989), p. 527

[78] P. Bouwknegt, J. McCarthy, and K. Pilch, Phys.Lett. B 234 (1990) 297

[79] P. Bouwknegt, J. McCarthy, and K. Pilch, Commun.Math.Phys. 131 (1990) 125

[80] P. Bouwknegt, J. McCarthy, and K. Pilch, Progr.Theor.Phys.Suppl. 102 (1990) 67

[81] P. Bouwknegt and W. Nahm, Phys.Lett. B 184 (1987) 359

[82] Bo-Yu Hou, Bo-Yuan Hou, and Zhong-Qi Ma, Commun.Theor.Phys. 13 (1990) 341

[83] P. Bowcock, Nucl.Phys. B 316 (1989) 80

[84] P. Bowcock and P. Goddard, Nucl.Phys. B 285 (1987) 651

[85] Bo-Yu Hou, Kang-Jie Shi, Pei Wang, and Rui-Hong Yue, Nucl.Phys. B 345 (1990) 659

[86] R. Brauer, Math.Z. 41 (1936) 330

[87] R. Brauer, Ann.Math. 38 (1937) 857

[88] R. Brauer and H. Weyl, Amer.J.Math. 57 (1935) 425

[89] J. Briggs, Rev.Mod.Phys. 43 (1971) 189

[90] L. Brink, D. Olive, and J. Scherk, Nucl.Phys. B 61 (1973) 173

[91] R. Brower and C. Thorn, Nucl.Phys. B 31 (1971) 163

[92] R. Brustein, S. Yankielowicz, and J.-B. Zuber, Nucl.Phys. B 313 (1989) 328

[93] D. Buchholz and K. Fredenhagen, Commun.Math.Phys. 84 (1982) 1

[94] D. Buchholz, G. Mack, and I. Todorov, Nucl.Phys. B (Proc.Suppl.) 5B (1988) 20

[95] N. Burroughs, Commun.Math.Phys. 127 (1990) 109

[96] N. Burroughs, Commun.Math.Phys. 133 (1990) 91

[97] P. Butler and B. Wybourne, J.Math.Phys. 11 (1970) 2512

[98] G. Canning, Phys.Rev. D 8 (1973) 1151

[99] A. Cappelli, C. Itzykson, and J.-B. Zuber, Nucl.Phys. B 280 (1987) 445

[100] A. Cappelli, C. Itzykson, and J.-B. Zuber, Commun.Math.Phys. 113 (1987) 1

[101] M. Caselle and G. Ponzano, Phys.Lett. B 224 (1989) 303

[102] M. Caselle, G. Ponzano, and F. Ravanini, Phys.Lett. B 251 (1990) 260

[103] L. Cardy, Nucl.Phys. B 270 [FS16] (1986) 186

[104] U. Carow-Watamura, M. Schlieker, M. Scholl, and S. Watamura, Z.Physik C 48 (1990) 159

[105] E. Cartan, thesis (Paris 1894), reprinted in: *E. Cartan, Oeuvres complètes* (Gauthiers-Villars, Paris 1984), vol. I, p. 137

[106] E. Cartan, Bull.Soc.Math.de France 41 (1913) 53

[107] E. Cartan, Ann.Sci.Ec.Norm.Supér. 31 (1914) 263

[108] E. Cartan, C.R.Acad.Sci. Paris 186 (1928) 1595

[109] E. Cartan, C.R.Acad.Sci. Paris 190 (1930) 610

[110] H. Casimir, Proc.Kon.Ned.Akad.Wetensch. XXXIV (1931) 844

[111] H. Casimir and B.L. van der Waerden, Math.Ann. 111 (1935) 1

[112] A. Ceresole, A. Lerda, P. Pizzochero, and P. van Nieuwenhuizen, Phys. Lett. B 189 (1987) 34

[113] M. Chaichian and D. Ellinas, J.Phys. A 23 (1990) L291

[114] M. Chaichian and P.P. Kulish, Phys.Lett. B 78 (1978) 413

[115] M. Chaichian and P.P. Kulish, Phys.Lett. B 234 (1990) 72

[116] M. Chaichian, P.P. Kulish, and J. Lukierski, Phys.Lett. B 237 (1990) 401

[117] Ho-Jui Chang, Hamburger Abhandlungen 14 (1941) 151

[118] I.V. Cherednik, Theor.Math.Phys. 43 (1980) 356

[119] I.V. Cherednik, Funct.Anal.Appl. 19 (1985) 77

[120] I.V. Cherednik, Sov.Math.Dokl. 33 (1986) 507

[121] C. Chevalley, Tohoku Math.J. (Ser. 2) I (1955) 14

[122] P. Christe, Ph.D. thesis (preprint BONN-IR-86-32, Bonn 1986)

[123] P. Christe, Phys.Lett. B 198 (1987) 285

[124] P. Christe and R. Flume, Nucl.Phys. B 282 (1987) 466

[125] P. Christe and F. Ravanini, Phys.Lett. B 287 (1989) 252

[126] P. Christe and F. Ravanini, Int.J.Mod.Phys. A 4 (1989) 897

[127] A. Coleman, Invent.math. 95 (1989) 447

[128] S. Coleman, Phys.Rev. D 11 (1975) 2088

[129] A. Coleman and M. Howard, C.R.Math.Rep.Acad.Sci.Canada XI (1989) 15

[130] A. Connes, Ann.Math. 104 (1976) 73

[131] A. Connes, Publ.Math.I.H.E.S. 62 (1986) 257

[132] E. Corrigan, D.B. Fairlie, P. Fletcher, and R. Sasaki, J.Math.Phys. 31 (1990) 776

[133] S. Cordes and Y. Kikuchi, Texas A&M preprint CTP-TAMU-92/88 (December 1988, unpublished)

[134] P. Cotta-Ramusino, E. Guadagnini, M. Martellini, and M. Mintchev, Nucl.Phys. B 330 (1990) 557

[135] H. Coxeter, Ann.Math. 35 (1934) 588

[136] H. Coxeter, in an appendix to notes on lectures of H. Weyl on *The structure and representations of continuous groups* (The Institute for Advanced Studies, Princeton 1935)

[137] H. Coxeter, Duke Math.J. 18 (1951) 765

[138] E. Cremmer and J.-L. Gervais, Commun.Math.Phys. 134 (1990) 619

[139] C. Cummins, J.Phys. A 24 (1991) 391

[140] T. Curtright, G. Ghandour, and C. Zachos, J.Math.Phys. 32 (1991) 676

[141] T. Curtright and C. Zachos, Phys.Lett. B 243 (1990) 237

[142] P. Cvitanovic, Phys.Rev. D 14 (1976) 1536

[143] R. Dashen and Y. Frishman, Phys.Rev. D 11 (1975) 278

[144] E. Date, M. Jimbo, and T. Miwa, in: *The Physics and Mathematics of Strings, Memorial Volume for V.G. Knizhnik*, L. Brink et al., eds. (World Scientific, Singapore 1990), p. 185

[145] E. Date, M. Jimbo, K. Miki, and T. Miwa, Publ. RIMS 27 (1991) 347

[146] J. de Azcárraga, J. Izquierdo, and A. Macfarlane, Ann.Phys. 202 (1990) 1

[147] J. de Boer and J. Goeree, Commun.Math.Phys. 139 (1991) 267

[148] C. De Concini and V.G. Kac, Colloque Dixmier (1990) 471 (Progress in Mathematics 92, Birkhäuser, Basel 1990)

[149] P. Degiovanni, Commun.Math.Phys. 127 (1990) 71

[150] J.-R. Derome and W. Sharp, J.Math.Phys. 6 (1965) 1584

[151] H.J. de Vega and H. Nicolai, Phys.Lett. B 244 (1990) 295

[152] P. di Francesco, Phys.Lett. B 285 (1988) 124

[153] P. di Francesco, H. Saleur, and J.-B. Zuber, J.Stat.Phys. 49 (1987) 57

[154] R. Dijkgraaf, V. Pasquier, and P. Roche, in: Proceedings of the *International Colloquium on Modern Quantum Field Theory*, Bombay, January 1990 (World Scientific, Singapore 1991)

[155] R. Dijkgraaf and E. Verlinde, Nucl.Phys. B (Proc.Suppl.) 5 (1988) 87

[156] R. Dijkgraaf, E. Verlinde, and H. Verlinde, Commun.Math.Phys. 115 (1988) 649

[157] J. Distler and Z. Qiu, Nucl.Phys. B 336 (1990) 533

[158] P. DiVecchia, J. Petersen, and H. Zheng, Phys.Lett. B 162 (1985) 327

[159] P. DiVecchia, V. Knizhnik, J. Petersen, and P. Rossi, Nucl.Phys. B 253 (1985) 701

[160] V. Dobrev, in: Proceedings of the 13th Johns Hopkins Workshop on *Knots, Topology, and Field Theory*, L. Lusanna, ed. (World Scientific, Singapore 1990), p. 539

[161] V. Dobrev, in: Proceedings of the International Symposium on *Symmetries in Science V*, B. Gruber, ed. (Plenum, New York 1991)

[162] R. Dodd, J.Math.Phys. 31 (1990) 533

[163] H. Doebner, J. Hennig, and W. Lücke, Springer Lecture Notes in Physics 370 (1990) 29

[164] S. Doplicher, R. Haag, and J. Roberts, Commun.Math.Phys. 13 (1969) 1

[165] S. Doplicher, R. Haag, and J. Roberts, Commun.Math.Phys. 23 (1971) 199

[166] Vl.S. Dotsenko and V.A. Fateev, Nucl.Phys. B 240 [FS12] (1984) 312
[167] Vl.S. Dotsenko and V.A. Fateev, Nucl.Phys. B 251 [FS13] (1985) 691
[168] Vl.S. Dotsenko and V.A. Fateev, Phys.Lett. B 154 (1985) 291
[169] M. Douglas, Ph.D. thesis, Caltech preprint CALT-68-1453 (September 1987, unpublished)
[170] M. Douglas and S. Trivedi, Nucl.Phys. B 320 (1989) 461
[171] V.G. Drinfeld, Sov.Math.Dokl. 27 (1983) 68
[172] V.G. Drinfeld, Sov.Math.Dokl. 32 (1985) 254
[173] V.G. Drinfeld, in: Proceedings of the *International Congress of Mathematicians* 1986, A.M. Gleason, ed. (American Mathematical Society, Providence 1987), p. 798
[174] V.G. Drinfeld, Leningrad Math.J. 1 (1990) 1419
[175] M. Dubois-Violette, Lett.Math.Phys. 19 (1990) 128
[176] G. Dunne, R. Jackiw, and C. Trugenberger, Ann.Phys. 194 (1989) 197
[177] E.B. Dynkin, Amer.Math.Soc.Transl. (2) 6 (1957) 111
[178] E.B. Dynkin, Amer.Math.Soc.Transl. (2) 6 (1957) 245
[179] E.B. Dynkin, Amer.Math.Soc.Transl. (1) 9 (1962) 308
[180] T. Eguchi and H. Ooguri, Phys.Lett. B 203 (1988) 44
[181] H. Eichenherr, Phys.Lett. B 151 (1985) 26
[182] E. El-Baz and B. Castel, Amer.J.Phys. 39 (1971) 868
[183] S. Elitzur, G. Moore, A. Schwimmer, and N. Seiberg, Nucl.Phys. B 326 (1989) 108
[184] L.D. Faddeev, Bol.Soc.Bras.Mat. 20 (1989) 47
[185] L.D. Faddeev, N.Yu. Reshetikhin, and L. Takhtajan, Leningrad Math. J. 1 (1990) 193
[186] L.D. Faddeev and L. Takhtajan, Russ.Math.Surveys 34 (1979) 11
[187] D. Fairlie, J.Phys. A 23 (1990) L183
[188] D. Fairlie and C. Zachos, Phys.Lett. B 256 (1991) 43
[189] V.A. Fateev and S. Lykyanov, Int.J.Mod.Phys. A 3 (1988) 507
[190] V.A. Fateev and S. Lykyanov, Sov.Phys. JETP 67 (1988) 447
[191] V.A. Fateev and A.B. Zamolodchikov, Sov.Phys. JETP 62 (1986) 285
[192] V.A. Fateev and A.B. Zamolodchikov, Nucl.Phys. B 280 [FS18] (1987) 644
[193] B.L. Feigin and E. Frenkel, Russ.Math.Surveys 43 (1989) 228
[194] B.L. Feigin and E. Frenkel, Commun.Math.Phys. 128 (1990) 161
[195] B.L. Feigin and D.B. Fuks, Funct.Anal.Appl. 16 (1982) 114
[196] B.L. Feigin and D.B. Fuks, Funct.Anal.Appl. 17 (1983) 241
[197] B.L. Feigin and D.B. Fuks, Springer Lecture Notes in Mathematics 1060 (1984) 230

[198] A. Feingold and I. Frenkel, Adv.Math. 56 (1985) 117

[199] A. Feingold and J. Lepowsky, Adv.Math. 29 (1978) 271

[200] G. Felder, Nucl.Phys. B 317 (1989) 285; *ibid.* 324 (1989) 548 (E)

[201] G. Felder, J. Fröhlich and G. Keller, Commun.Math.Phys. 130 (1990) 1

[202] G. Felder, K. Gawedzki, and A. Kupiainen, Commun.Math.Phys. 117 (1988) 127

[203] G. Felder, K. Gawedzki, and A. Kupiainen, Nucl.Phys. B 299 (1988) 355

[204] G. Felder and R. Silvotti, Commun.Math.Phys. 123 (1989) 1

[205] J. Figueroa-O'Farrill and S. Schrans, Phys.Lett. B 245 (1990) 471

[206] M. Flato and D. Sternheimer, Lett.Math.Phys. 22 (1991) 155

[207] E.G. Floratos, Phys.Lett. B 233 (1989) 395

[208] P. Forgács, Z. Horváth, L. Palla, and P. Vecsernyés, Nucl.Phys. B 308 (1988) 477

[209] K. Fredenhagen, K.-H. Rehren, and B. Schroer, Commun.Math. Phys. 125 (1989) 201

[210] K. Fredenhagen, K.-H. Rehren and B. Schroer, preprint in preparation

[211] I. Frenkel, Proc.Natl.Acad.Sci.USA 77 (1980) 6303

[212] I. Frenkel, J.Funct.Anal. 44 (1981) 259

[213] I. Frenkel, talk at the 1989 Luminy Conference on *Infinite-dimensional Lie Algebras and Groups*

[214] I. Frenkel and V.G. Kac, Invent.math. 62 (1980) 23

[215] P. Freyd, D. Yetter, W. Lickorish, K. Millet, A. Ocneanu, and J. Hoste, Bull.Amer.Math.Soc. 12 (1985) 239

[216] D. Friedan, E. Martinec, and S. Shenker, Nucl.Phys. B 271 (1986) 93

[217] D. Friedan, Z. Qiu, and S. Shenker, in: *Vertex Operators in Mathematics and Physics*, M.S.R.I. publication No. 3 (Springer Verlag, New York 1984), p. 491

[218] D. Friedan, Z. Qiu, and S. Shenker, Phys.Rev.Lett. 52 (1984) 1575

[219] D. Friedan, Z. Qiu, and S. Shenker, Phys.Lett. B 151 (1985) 37

[220] D. Friedan, Z. Qiu, and S. Shenker, Commun.Math.Phys. 107 (1986) 179

[221] D. Friedan and S. Shenker, Nucl.Phys. B 281 (1987) 509

[222] G. Frobenius, Sitz.Preuss.Akad.Wiss. (1903) 328

[223] J. Fröhlich and C. King, Int.J.Mod.Phys. A 4 (1989) 5328

[224] J. Fuchs, Nucl.Phys. B 286 (1987) 455

[225] J. Fuchs, Z.Physik C 35 (1987) 89

[226] J. Fuchs, Nucl.Phys. B 290 (1987) 392

[227] J. Fuchs, talk at the 1987 Poiana Brasov summer school on *Conformal Field Theory*, unpublished manuscript

[228] J. Fuchs, Nucl.Phys. B (Proc.Suppl.) 6 (1989) 157

[229] J. Fuchs, Nucl.Phys. B 318 (1989) 631

[230] J. Fuchs, Phys.Rev.Lett. 62 (1989) 1705

[231] J. Fuchs, Phys.Lett. B 222 (1989) 411

[232] J. Fuchs, Nucl.Phys. B 328 (1989) 585

[233] J. Fuchs, Commun.Math.Phys. 136 (1991) 345

[234] J. Fuchs, Commun.Theor.Phys. (Allahabad) 1 (1991) 59

[235] J. Fuchs and D. Gepner, Nucl.Phys. B 294 (1987) 30

[236] J. Fuchs and A. Klemm, Ann.Phys. 194 (1989) 303

[237] J. Fuchs, A. Klemm, and C. Scheich, Z.Physik C 46 (1990) 71

[238] J. Fuchs and P. van Driel, J.Math.Phys. 31 (1990) 1770

[239] J. Fuchs and P. van Driel, Nucl.Phys. B 346 (1990) 632

[240] D.B. Fuks, Funct.Anal.Appl. 23 (1989) 154

[241] P. Furlan, A.Ch. Ganchev, and V.B. Petkova, Nucl.Phys. B 343 (1990) 205

[242] O. Gabber and V.G. Kac, Bull.Amer.Math.Soc. 5 (1981) 185

[243] A.Ch. Ganchev and V.B. Petkova, Phys.Lett. B 233 (1989) 374

[244] C. Gardner, J. Green, M. Kruskal, and R. Muira, Phys.Rev.Lett. 19 (1967) 1095

[245] H. Garland and J. Lepowsky, Invent.math. 34 (1976) 37

[246] K. Gawedzki and A. Kupiainen, Nucl.Phys. B 320 (1989) 625

[247] I.M. Gelfand and D.B. Fuks, Funct.Anal.Appl. 2 (1968) 342

[248] D. Gepner, Nucl.Phys. B 287 (1987) 111

[249] D. Gepner, Nucl.Phys. B 290 (1987) 10

[250] D. Gepner, Nucl.Phys. B 296 (1988) 757

[251] D. Gepner and E. Witten, Nucl.Phys. B 278 (1986) 493

[252] A. Gerasimov, A. Marshakov and A. Morozov, Nucl.Phys. B 328 (1989) 664

[253] A. Gerasimov, A. Marshakov, A. Morozov, M. Olshanetsky, and S. Shatashvili, Int.J.Mod.Phys. A 5 (1990) 2495

[254] J.-L. Gervais, Commun.Math.Phys. 130 (1990) 257

[255] R. Gilmore, J.Math.Phys. 11 (1970) 513

[256] R. Gilmore, J.Math.Phys. 11 (1970) 1855

[257] P. Ginsparg, Nucl.Phys. B 295 [FS28] (1988) 153

[258] P. Goddard, A. Kent, and D. Olive, Phys.Lett. B 152 (1985) 88

[259] P. Goddard, A. Kent, and D. Olive, Commun.Math.Phys. 103 (1986) 105

[260] P. Goddard, W. Nahm, and D. Olive, Phys.Lett. B 160 (1985) 111

[261] P. Goddard, W. Nahm, D. Olive, and A. Schwimmer, Commun.Math. Phys. 107 (1986) 179

[262] P. Goddard and D. Olive, Nucl.Phys. B 257 [FS14] (1985) 226

[263] P. Goddard and A. Schwimmer, Phys.Lett. B 206 (1988) 62

[264] M. Golubitsky and B. Rothschild, Bull.Amer.Math.Soc. 77 (1971) 983

[265] F. Gomes, Phys.Lett. B 171 (1986) 75

[266] C. Gómez and G. Sierra, Phys.Lett. B 240 (1990) 149

[267] C. Gómez and G. Sierra, Nucl.Phys. B 352 (1991) 791

[268] F.M. Goodman and T. Nakanishi, Phys.Lett. B 262 (1991) 259

[269] R. Goodman and N. Wallach, J.reine angew.Math. 347 (1984) 69

[270] R. Goodman and N. Wallach, Trans.Amer.Math.Soc. 315 (1989) 1

[271] F.M. Goodman and H. Wenzl, Adv.Math. 82 (1990) 244

[272] E. Guadagnini, M. Martellini, and M. Mintchev, Phys.Lett. B 228 (1989) 489

[273] E. Guadagnini, M. Martellini, and M. Mintchev, Nucl.Phys. B 330 (1990) 575

[274] V. Gribov and A. Migdal, Sov.Phys. JETP 55 (1968) 1498

[275] V. Groza, I. Kachurik, and A. Klimyk, J.Math.Phys. 31 (1990) 2769

[276] B. Gruber and L. O'Raifeartaigh, J.Math.Phys. 5 (1964) 1796

[277] R. Haag and D. Kastler, J.Math.Phys. 5 (1964) 848

[278] R. Haag and B. Schroer, J.Math.Phys. 3 (1962) 248

[279] L. Hadjiivanov, R. Paunov, and I. Todorov, Nucl.Phys. B 356 (1991) 387

[280] M. Halpern, Phys.Rev. D 12 (1975) 1684

[281] M. Halpern and C. Thorn, Phys.Rev. D 4 (1971) 3084

[282] Harish-Chandra, Ann.Math. 50 (1949) 900

[283] J. Harvey and S. Naculich, Nucl.Phys. B 305 (1988) 417

[284] K. Hasegawa, Publ. RIMS 25 (1989) 741

[285] T. Hayashi, Invent.math. 94 (1988) 13

[286] T. Hayashi, Commun.Math.Phys. 127 (1990) 129

[287] N. Iem and I.S. t'Vollkommen, Phys.Bull.Proc.Ann. 5 (1957) 19

[288] K. Intriligator, Nucl.Phys. B 332 (1990) 541

[289] K. Ito, Phys.Lett. B 230 (1989) 71

[290] H. Itoyama, Phys.Lett. A 140 (1989) 391

[291] H. Itoyama and P. Moxhay, Nucl.Phys. B 338 (1990) 759

[292] H. Itoyama and A. Sevrin, Int.J.Mod.Phys. A 5 (1990) 281

[293] C. Itzykson and J.-B. Zuber, Nucl.Phys. B 275 [FS17] (1986) 580

[294] K. Iwahori, J.Math.Soc.Japan 10 (1958) 145

[295] K. Iwasawa, Jap.J.Math. 19 (1948) 405

[296] A. Izergin and V. Korepin, Commun.Math.Phys. 79 (1981) 303

[297] M. Jimbo, Lett.Math.Phys. 10 (1985) 63

[298] M. Jimbo, Lett.Math.Phys. 11 (1986) 247

[299] M. Jimbo, Commun.Math.Phys. 102 (1986) 537

[300] M. Jimbo, Springer Lecture Notes in Physics 246 (1986) 335

[301] M. Jimbo, A. Kuniba, T. Miwa, and M. Okado, Commun.Math. Phys. 119 (1988) 543

[302] M. Jimbo, T. Miwa, and M. Okado, Commun.Math.Phys. 116 (1988) 507

[303] V.F.R. Jones, Invent.math. 72 (1983) 1

[304] V.F.R. Jones, Bull.Amer.Math.Soc. 12 (1985) 103

[305] V.F.R. Jones, Ann.Math. 126 (1987) 335

[306] V.F.R. Jones, Commun.Math.Phys. 125 (1989) 459

[307] V.G. Kac, Funct.Anal.Appl. 1 (1967) 328

[308] V.G. Kac, Funct.Anal.Appl. 2 (1968) 182

[309] V.G. Kac, Funct.Anal.Appl. 3 (1969) 252

[310] V.G. Kac, Math.USSR Izv. 2 (1968) 1271

[311] V.G. Kac, Funct.Anal.Appl. 8 (1974) 68

[312] V.G. Kac, Adv.Math. 30 (1978) 85

[313] V.G. Kac, Springer Lecture Notes in Physics 94 (1979) 441

[314] V.G. Kac, talk at the Canadian Mathematical Society meeting on *Lie Algebras and Lie Groups* (Montréal, August 1989)

[315] V.G. Kac, D. Kazhdan, J. Lepowsky, and R. Wilson, Adv.Math. 42 (1981) 83

[316] V.G. Kac and D.H. Peterson, Proc.Natl.Acad.Sci.USA 78 (1981) 3308

[317] V.G. Kac and D.H. Peterson, Proc.Natl.Acad.Sci.USA 80 (1983) 1778

[318] V.G. Kac and D.H. Peterson, Adv.Math. 53 (1984) 125

[319] V.G. Kac and D.H. Peterson, in: Proceedings of the *Symposium on Anomalies, Geometry and Topology*, W. Bardeen and A. White, eds. (World Scientific, Singapore 1985), p. 276

[320] V.G. Kac and M.N. Sanielevici, Phys.Rev. D 37 (1988) 2231

[321] V.G. Kac and I. Todorov, Commun.Math.Phys. 102 (1985) 337; *ibid.* 104 (1986) 175 (E)

[322] V.G. Kac and M. Wakimoto, Adv.Math. 70 (1988) 156

[323] L. Kadanoff, Phys.Rev.Lett. 23 (1969) 1430

[324] I. Kantor, Sov.Math.Dokl. 9 (1968) 409

[325] L. Kaplan and M. Resnikoff, J.Math.Phys. 11 (1967) 2894

[326] D. Karabali and H.J. Schnitzer, Nucl.Phys. B 299 (1988) 548

[327] D. Karabali and H.J. Schnitzer, Nucl.Phys. B 329 (1990) 649

[328] M. Karowski, H. Thun, T. Truong, and P. Weisz, Phys.Lett. B 67 (1977) 328

[329] M. Kashiwara, Commun.Math.Phys. 133 (1990) 249

[330] D. Kastor, E. Martinec, and Z. Qiu, Phys.Lett. B 200 (1988) 434

[331] A. Kato and Y. Kitazawa, Nucl.Phys. B 319 (1989) 474

[332] L. Kauffman, Topology 26 (1987) 395

[333] G. Keller, Lett.Math.Phys. 21 (1991) 273

[334] S. Kerov, J.Sov.Math. 46 (1989) 2148

[335] S. Kerov, J.Sov.Math. 47 (1989) 2503

[336] W. Killing, Math.Ann. 31 (1888) 252

[337] W. Killing, Math.Ann. 33 (1889) 1

[338] W. Killing, Math.Ann. 36 (1890) 161

[339] R. King, J.Math.Phys. 12 (1971) 1588

[340] R. King, J.Phys. A 8 (1975) 429

[341] A.N. Kirillov, J.Sov.Math. 53 (1991) 264

[342] A.N. Kirillov and N.Yu. Reshetikhin, in: Proceedings of the 1988 Luminy Conference on *Infinite-dimensional Lie Algebras and Groups*, V.G. Kac, ed. (World Scientific, Singapore 1989), p. 285

[343] A.N. Kirillov and N.Yu. Reshetikhin, Commun.Math.Phys. 134 (1990) 421

[344] A.N. Kirillov and N.Yu. Reshetikhin, J.Sov.Math. 52 (1991) 3156

[345] E. Kiritsis, Nucl.Phys. B 324 (1989) 475

[346] E. Kiritsis, Phys.Lett. B 287 (1989) 427

[347] E. Kiritsis, Nucl.Phys. B 329 (1990) 591

[348] E. Kiritsis and G. Siopsis, Phys.Lett. B 184 (1987) 353; *ibid.* 189 (1987) 489 (E)

[349] V. Knizhnik and A.B. Zamolodchikov, Nucl.Phys. B 247 (1984) 83

[350] T. Kohno, Ann.Inst.Fourier 37 (1987) 139

[351] I.G. Koh and Z.-Q. Ma, Phys.Lett. B 234 (1990) 480

[352] K. Koike, Adv.Math. 74 (1989) 57

[353] K. Koike and I. Terada, J.Algebra 107 (1987) 466

[354] K. Koike and I. Terada, Adv.Studies in Pure Math. 11 (1987) 147

[355] T. Koornwinder, Proc.Kon.Ned.Akad.Wetensch. A 92 (1989) 97

[356] H. Kosaki, J.Funct.Anal. 66 (1986) 123

[357] Y. Kosmann-Schwarzbach, Mod.Phys.Lett. A 5 (1990) 981; *ibid.* 6 (1991) 3373 (E)

[358] B. Kostant, Trans.Amer.Math.Soc. 93 (1959) 53

[359] V.A. Kostelecký, O. Lechtenfeld, W. Lerche, S. Samuel, and S. Watamura, Nucl.Phys. B 288 (1987) 173

[360] P.P. Kulish and E. Damashinsky, J.Phys. A 23 (1990) L415

[361] P.P. Kulish and N.Yu. Reshetikhin, J.Sov.Math. 23 (1983) 2435

[362] P.P. Kulish, N.Yu. Reshetikhin, and E.K. Sklyanin, Lett.Math.Phys. 5 (1981) 393

[363] P.P. Kulish and E.K. Sklyanin, Phys.Lett. A 70 (1979) 461

[364] P.P. Kulish and E.K. Sklyanin, J.Sov.Math. 95 (1982) 1596

[365] S. Kumar, Invent.math. 89 (1987) 395

[366] A. Kuniba and T. Nakanishi, in: Proceedings of the *International Colloquium on Modern Quantum Field Theory*, Bombay, January 1990 (World Scientific, Singapore 1991)

[367] J. Kupsch, W. Rühl, and B. Yunn, Ann.Phys. 89 (1975) 115

[368] M. Kuwahara, N. Ohta, and H. Suzuki, Phys.Lett. B 235 (1990) 57

[369] P. Lardy, Comment.Math.Helv. 8 (1935) 189

[370] R. Le Blanc, K. Hecht, and L.C. Biedenharn, J.Phys. A 24 (1991) 1393

[371] F. Lemire and J. Patera, J.Math.Phys. 12 (1980) 2026

[372] J. Lepowsky, Adv.Math. 27 (1978) 230

[373] J. Lepowsky, in: *Vertex Operators in Mathematics and Physics*, M.S.R.I. publication No. 3 (Springer Verlag, New York 1984), p. 1

[374] J. Lepowsky, Proc.Natl.Acad.Sci.USA 82 (1985) 8295

[375] J. Lepowsky and S. Milne, Adv.Math. 29 (1978) 15

[376] J. Lepowsky and R. Wilson, Commun.Math.Phys. 62 (1978) 43

[377] J. Lepowsky and R. Wilson, Proc.Natl.Acad.Sci.USA 78 (1981) 7254

[378] J. Lepowsky and R. Wilson, Adv.Math. 45 (1982) 28

[379] J. Lepowsky and R. Wilson, Invent.math. 77 (1984) 199

[380] S. Levendorskii and Ya.S. Soibelman, Commun.Math.Phys. 139 (1991) 141

[381] S. Levendorskii and Ya.S. Soibelman, J.Geom. and Phys. 7 (1991) 1

[382] F. Levol and I.S. Mooi, Phys.Bull.Proc.Ann. 10 (1988) 1

[383] D. Levy, Phys.Rev.Lett. 64 (1990) 499

[384] D. Lewellen, Nucl.Phys. B 320 (1989) 345

[385] M. Li and M. Yu, Commun.Math.Phys. 127 (1990) 195

[386] S. Lie, Gött.Nachr. (1874) 529; reprinted in: *Gesammelte Abhandlungen* (B.G. Teubner, Leipzig 1924) Vol. V, p. 1

[387] S. Lie, Arch.Math.Naturvid. (1876) 19, etc.; reprinted in: *Gesammelte Abhandlungen* (B.G. Teubner, Leipzig 1924) Vol. V, pp. 9, 42, 78, 136, 199

[388] D.E. Littlewood and A. Richardson, Phil.Trans.Roy.Soc.London Ser.A 233 (1934) 99

[389] R. Longo, Commun.Math.Phys. 126 (1989) 217

[390] R. Longo, Commun.Math.Phys. 130 (1990) 285

[391] M. Lüscher, Commun.Math.Phys. 50 (1976) 23

[392] M. Lüscher and G. Mack, Commun.Math.Phys. 41 (1975) 203

[393] M. Lüscher and G. Mack, unpublished manuscript (1976)

[394] G. Lusztig, Adv.Math. 70 (1988) 237

[395] G. Lusztig, Contemp.Math. 82 (1989) 59

[396] G. Lusztig, J.Algebra 131 (1990) 466

[397] G. Lusztig, J.Amer.Math.Soc 3 (1990) 447

[398] I.G. MacDonald, Invent.math. 15 (1972) 91

[399] I.G. MacDonald, Can.Math.Soc.Proc. 5 (1986) 69

[400] A. MacFarlane, J.Phys. A 22 (1989) 4581

[401] G. Mack, Commun.Math.Phys. 53 (1977) 155

[402] G. Mack and V. Schomerus, Commun.Math.Phys. 134 (1990) 139

[403] G. Mack and K. Symanzik, Commun.Math.Phys. 27 (1972) 247

[404] S. Majid, J.Algebra 130 (1990) 17

[405] S. Majid, in: Proceedings of the XVIIIth International Conference on *Differential Geometric Methods in Theoretical Physics* (Plenum, New York 1989)

[406] A. Malcev, Amer.Math.Soc.Transl. (1) 9 (1962) 172

[407] S. Mandelstam, Phys.Rev. D 7 (1973) 3777

[408] S. Mandelstam, Phys.Rev. D 11 (1975) 3026

[409] Yu.I. Manin, Ann.Inst.Fourier 37 (1987) 191

[410] Yu.I. Manin, Montreal preprint CRM-1561 (1988)

[411] S. Mathur, S. Mukhi, and A. Sen, Nucl.Phys. B 312 (1989) 15

[412] S. Mathur, S. Mukhi, and A. Sen, Nucl.Phys. B 318 (1989) 483

[413] S. Mathur and A. Sen, Nucl.Phys. B 327 (1989) 725

[414] T. Masuda, K. Mimachi, Y. Nakagami, M. Noumi, and K. Ueno, C.R. Acad.Sci. Paris 307 (1988) 559

[415] T. Masuda, K. Mimachi, Y. Nakagami, M. Noumi, Y. Saburi, and K. Ueno, Lett.Math.Phys. 19 (1990) 187

[416] I.N. McArthur, Nucl.Phys. B 340 (1990) 148

[417] J.B. McGuire, J.Math.Phys. 5 (1964) 622

[418] W. McKay, J. Patera, and R. Sharp, J.Math.Phys. 17 (1976) 1371

[419] M. Mehta and P. Srivastava, J.Math.Phys. 7 (1966) 1833

[420] L. Mezincescu and R. Nepomechie, Phys.Lett. B 246 (1990) 412

[421] A. Migdal, Phys.Lett. 37 (1971) 356

[422] K. Misra, J.Algebra 88 (1984) 196

[423] K. Misra and T. Miwa, Commun.Math.Phys. 134 (1990) 79

[424] R. Mkrtchyan and L. Zurabyan, Yerevan preprint YERPHI-1149(26)-89 (1989, unpublished)

[425] E. Mlawer, S. Naculich, H. Riggs, and H.J. Schnitzer, Nucl.Phys. B 352 (1991) 863

[426] R.V. Moody, Bull.Amer.Math.Soc. 73 (1967) 287

[427] R.V. Moody, J.Algebra 10 (1968) 281

[428] R.V. Moody, Canad.J.Math. 28 (1969) 1432

[429] R.V. Moody, Proc.Amer.Math.Soc. 48 (1975) 43

[430] R.V. Moody and J. Patera, Bull.Amer.Math.Soc. 7 (1982) 237

[431] R.V. Moody and J. Patera, SIAM J.Alg.Disc.Meth. 5 (1984) 359

[432] R.V. Moody, J. Patera, and R. Sharp, J.Math.Phys. 24 (1983) 2387

[433] G. Moore and N.Yu. Reshetikhin, Nucl.Phys. B 328 (1990) 557

[434] G. Moore and N. Seiberg, Phys.Lett. B 212 (1988) 451

[435] G. Moore and N. Seiberg, Nucl.Phys. B 313 (1989) 16

[436] G. Moore and N. Seiberg, Commun.Math.Phys. 123 (1989) 177

[437] G. Moore and N. Seiberg, Phys.Lett. B 220 (1989) 422

[438] J. Murakami, Adv.Studies in Pure Math. 19 (1989) 1

[439] J. Murakami, Osaka J.Math. 24 (1987) 745

[440] J. Murakami, Publ. RIMS 26 (1990) 935

[441] F. Murray and J. von Neumann, Ann.Math. 44 (1943) 716

[442] G. Mussardo, G. Sotkov, and M. Stanishkov, Nucl.Phys. B 305 (1988) 69

[443] G. Mussardo, G. Sotkov, and M. Stanishkov, Int.J.Mod.Phys. A 4 (1989) 1135

[444] S. Naculich and H.J. Schnitzer, Phys.Lett. B 244 (1990) 235

[445] S. Naculich and H.J. Schnitzer, Nucl.Phys. B 347 (1990) 687

[446] W. Nahm, Davis preprint UCD-88-02 (1988, unpublished)

[447] W. Nahm, Commun.Math.Phys. 118 (1988) 171

[448] T. Nakashima, Publ. RIMS 26 (1990) 723

[449] S. Nam, Phys.Lett. B 172 (1986) 323

[450] S. Nam, Phys.Lett. B 187 (1987) 340

[451] A. Navon and J. Patera, J.Math.Phys. 8 (1967) 489

[452] D. Nemeschansky, Phys.Lett. B 224 (1989) 128
[453] A. Neveu and J. Schwarz, Nucl.Phys. B 31 (1971) 86
[454] B. Nienhuis, J.Stat.Phys. 34 (1984) 731
[455] M. Nomura, J.Math.Phys. 10 (1989) 2397
[456] S.P. Novikov, Sov.Math.Dokl. 24 (1981) 222
[457] M. Nuomi and K. Mimachi, Commun.Math.Phys. 128 (1990) 528
[458] D. Olive and N. Turok, Nucl.Phys. B 285 [FS7] (1983) 470
[459] L. Onsager, Phys.Rev. 65 (1944) 117
[460] V. Pasquier, Commun.Math.Phys. 118 (1988) 355
[461] V. Pasquier and H. Saleur, Nucl.Phys. B 330 (1990) 523
[462] J. Patera, R. Sharp, and P. Winternitz, J.Math.Phys. 17 (1976) 1972
[463] J. Patera, R. Sharp, and P. Winternitz, J.Math.Phys. 18 (1977) 1519
[464] A. Perelomov and V. Popov, Sov.Phys. JETP Lett. 1 (1966) 160
[465] A. Perelomov and V. Popov, Sov.Math.Dokl. 8 (1967) 631
[466] D.H. Peterson, Springer Lecture Notes in Mathematics 933 (1982) 168
[467] V.B. Petkova, Int.J.Mod.Phys. A 3 (1988) 2945
[468] V.B. Petkova, Phys.Lett. B 225 (1989) 357
[469] P. Podlès, Lett.Math.Phys. 14 (1987) 193
[470] P. Podlès and S. Woronowicz, Commun.Math.Phys. 130 (1990) 381
[471] A.M. Polyakov, Sov.Phys. JETP 59 (1970) 542
[472] A.M. Polyakov, Sov.Phys. JETP Lett. 12 (1970) 538
[473] A.M. Polyakov, Sov.Phys. JETP 66 (1974) 23
[474] A. M. Polyakov and P.B. Wiegmann, Phys.Lett. B 141 (1984) 223
[475] A. Polychronakos, Mod.Phys.Lett. A 5 (1990) 2325
[476] G. Racah, Lincei Rend.Sci.Fis.Mat.Nat. 8 (1950) 108
[477] G. Racah, in: Istanbul Summer School on *Group Theoretical Concepts and Methods in Elementary Particle Physics*, F. Gursey, ed. (Gordon and Breach, New York 1964), p. 1
[478] C. Ramírez, H. Ruegg, and M. Ruiz-Altaba, Phys.Lett. B 247 (1990) 499
[479] P. Ramond, Phys.Rev. D 3 (1971) 2415
[480] F. Ravanini, Mod.Phys.Lett. A 3 (1988) 397
[481] A. Redlich and H.J. Schnitzer, Phys.Lett. B 167 (1986) 315; *ibid.* B 193 (1987) 536 (E)
[482] A. Redlich and H.J. Schnitzer, Phys.Lett. B 193 (1987) 471
[483] R. Ree, Trans.Amer.Math.Soc. 83 (1956) 510
[484] K.-H. Rehren, Commun.Math.Phys. 116 (1988) 675
[485] K.-H. Rehren, Springer Lecture Notes in Physics 370 (1990) 318

[486] K.-H. Rehren, Commun.Math.Phys. 132 (1990) 461
[487] K.-H. Rehren and B. Schroer, Phys.Lett. B 198 (1987) 84
[488] K.-H. Rehren and B. Schroer, Nucl.Phys. B 312 (1989) 715
[489] N.Yu. Reshetikhin, Leningrad preprint LOMI E-4-87 (1987)
[490] N.Yu. Reshetikhin, Leningrad preprint LOMI E-17-87 (1987)
[491] N.Yu. Reshetikhin and V. Turaev, Commun.Math.Phys. 127 (1990) 1
[492] A. Rocha-Caridi, in: *Vertex Operators in Mathematics and Physics*, M.S.R.I. publication No. 3 (Springer Verlag, New York 1984), p. 451
[493] P. Roche and D. Arnaudon, Lett.Math.Phys. 17 (1989) 295
[494] M. Rodríguez-Plaza, J.Math.Phys. 32 (1991) 2020
[495] M. Rosso, C.R.Acad.Sci. Paris 304 (1987) 323
[496] M. Rosso, Commun.Math.Phys. 117 (1988) 581
[497] M. Rosso, Commun.Math.Phys. 124 (1989) 307
[498] M. Rosso, Duke Math.J. 61 (1990) 11
[499] M. Rosso, Séminaire Bourbaki 43 (1991) n° 744
[500] H. Ruegg, J.Math.Phys. 31 (1990) 1085
[501] W. Rühl, Ann.Phys. 206 (1991) 368
[502] Ph. Ruelle, E. Thiran, and J. Weyers, Commun.Math.Phys. 133 (1990) 305
[503] C. Saclioglu, J.Phys. A 22 (1989) 3753
[504] M. Salam and B. Wybourne, J.Phys. A 24 (1991) L317
[505] H. Saleur, Phys.Rep. 184 (1989) 177
[506] H. Saleur and D. Altschüler, Nucl.Phys. B 354 (1991) 579
[507] A.N. Schellekens, Phys.Lett. B 244 (1990) 255
[508] A.N. Schellekens and N.P. Warner, Phys.Rev. D 34 (1986) 3092
[509] A.N. Schellekens and S. Yankielowicz, Nucl.Phys. B 327 (1989) 673
[510] A.N. Schellekens and S. Yankielowicz, Phys.Lett. B 227 (1989) 387
[511] A.N. Schellekens and S. Yankielowicz, Nucl.Phys. B 334 (1990) 67
[512] A.N. Schellekens and S. Yankielowicz, Int.J.Mod.Phys. A 5 (1990) 2903
[513] M. Schlieker and M. Scholl, Z.Physik C 47 (1990) 625
[514] H.J. Schnitzer, Nucl.Phys. B 324 (1989) 412
[515] B. Schroer, Nucl.Phys. B 295 (1988) 4
[516] B. Schroer and J. Swieca, Phys.Rev. D 10 (1974) 480
[517] B. Schroer, J. Swieca, and A. Völkel, Phys.Rev. D 11 (1975) 11
[518] J. Schwinger, Phys.Rev.Lett. 3 (1959) 296
[519] G. Segal, Commun.Math.Phys. 80 (1981) 301
[520] M. Semenov-Tian-Shanskii, Funct.Anal.Appl. 17 (1983) 259

[521] A. Sirota and A. Solodovnikov, Russ.Math.Surveys 18 (1963) 85
[522] E.K. Sklyanin, Sov.Phys.Dokl. 23 (1979) 107
[523] E.K. Sklyanin, Funct.Anal.Appl. 16 (1983) 263
[524] E.K. Sklyanin, Funct.Anal.Appl. 17 (1983) 273
[525] E.K. Sklyanin, L. Takhtajan, and L.D. Faddeev, Theor.Math.Phys. 40 (1979) 194
[526] T. Skyrme, Proc.Roy.Soc. Ser. A 262 (1961) 237
[527] D.-J. Smit, Commun.Math.Phys. 128 (1990) 1
[528] Ya.S. Soibelman, Funct.Anal.Appl. 24 (1991) 253
[529] C.M. Sommerfield, Phys.Rev. 176 (1968) 2019
[530] D. Speiser, Helv.Phys.Acta 38 (1965) 73
[531] D. Speiser, in: Istanbul Summer School on *Group Theoretical Concepts and Methods in Elementary Particle Physics*, F. Gursey, ed. (Gordon and Breach, New York 1964), p. 237
[532] M. Spiegelglas, Phys.Lett. B 247 (1990) 36
[533] A. Stone, Proc.Camb.Philos.Soc. 57 (1961) 460
[534] R. Streater and I. Wilde, Nucl.Phys. B 24 (1970) 561
[535] H. Sugawara, Phys.Rev. 170 (1968) 1659
[536] Chang-Pu Sun and Hong-Chen Fu, J.Phys. A 22 (1989) L983
[537] Chang-Pu Sun, Jing-Fa Lu, and Mo-Lin Ge, J.Phys. A 24 (1991) 3731
[538] L. Takhtajan, Springer Lecture Notes in Physics 370 (1990) 3
[539] T. Tanisaki, Commun.Math.Phys. 127 (1990) 555
[540] H. Temperley and E.H. Lieb, Proc.Roy.Soc. Ser. A 322 (1971) 251
[541] F. ten Kroode and J. van der Leur, Commun.Math.Phys. 137 (1991) 67
[542] C. Thorn, Nucl.Phys. B 248 (1984) 551
[543] T. Tjin, University of Amsterdam preprint (1991)
[544] A. Tsuchiya and Y. Kanie, Lett.Math.Phys. 13 (1987) 303
[545] A. Tsuchiya and Y. Kanie, Adv.Studies in Pure Math. 16 (1988) 297
[546] V. Turaev, Invent.math. 92 (1988) 527
[547] K. Ueno, Lett.Math.Phys. 18 (1989) 215
[548] M. Umezawa, Nucl.Phys. B 57 (1964) 65
[549] M. Umezawa, Proc.Kon.Ned.Akad.Wetensch. 69 (1966) 620
[550] C. Vafa, Phys.Lett. B 206 (1988) 428
[551] L.L. Vaksman and Ya.S. Soibelman, Funct.Anal.Appl. 22 (1989) 170
[552] P. Valtancoli, Int.J.Mod.Phys. A 5 (1990) 3887
[553] L. van der Waerden, Math.Z. 37 (1933) 446
[554] D.-N. Verma, Bull.Amer.Math.Soc. 74 (1968) 160
[555] D.-N. Verma, Math.Reviews 50 (1975) 1371

[556] E. Verlinde, Ph.D. thesis (Utrecht 1988)

[557] E. Verlinde, Nucl.Phys. B 300 (1988) 360

[558] D. Verstegen, Nucl.Phys. B 346 (1990) 349

[559] D. Verstegen, Commun.Math.Phys. 137 (1991) 567

[560] M. Virasoro, Phys.Rev. D 1 (1970) 2933

[561] M. Virasoro, in: *Duality and Symmetry in Hadron Physics*, E. Gotsman, ed. (Weizmann Science Press, Jerusalem 1971)

[562] S. Vokos, J. Wess, and B. Zumino, in: *Symmetry in Nature. A volume in Honour of Luigi A. Radicati di Brozolo*, Vol. II (Scuola Normale Superiore, Pisa 1989)

[563] S. Vokos, J. Wess, and B. Zumino, Z.Physik C 48 (1990) 65

[564] L. Vretare, J.Math.Phys. 30 (1989) 18

[565] M. Wakimoto, Commun.Math.Phys. 104 (1986) 605

[566] M. Walton, Nucl.Phys. B 322 (1989) 775

[567] M. Walton, Nucl.Phys. B 340 (1990) 777

[568] M. Walton, Phys.Lett. B 241 (1990) 365; *ibid.* B 244 (1990) 580 (E)

[569] A. Wassermann, Oxford preprint (1990)

[570] H. Wenzl, Invent.math. 92 (1988) 349

[571] H. Wenzl, Ann.Math. 128 (1988) 173

[572] H. Wenzl, Commun.Math.Phys. 133 (1990) 383

[573] J. Wess and B. Zumino, Phys.Lett. B 37 (1971) 95

[574] J. Wess and B. Zumino, Nucl.Phys. B (Proc.Suppl.) 18 (1990) 302

[575] H. Weyl, Math.Z. 23 (1925) 271

[576] H. Weyl, Math.Z. 24 (1926) 377

[577] H. Weyl, Math.Z. 35 (1932) 300

[578] E. Wigner, Amer.J.Math. 63 (1941) 57

[579] K.G. Wilson, Phys.Rev. 179 (1969) 1499

[580] E. Witten, Nucl.Phys. B 223 (1983) 422

[581] E. Witten, Commun.Math.Phys. 92 (1984) 455

[582] E. Witten, Commun.Math.Phys. 114 (1988) 1

[583] E. Witten, Commun.Math.Phys. 117 (1988) 353

[584] E. Witten, Commun.Math.Phys. 128 (1989) 351

[585] E. Witten, Nucl.Phys. B 330 (1990) 285

[586] S. Woronowicz, Publ. RIMS 23 (1987) 117

[587] S. Woronowicz, Commun.Math.Phys. 111 (1987) 613

[588] S. Woronowicz, Commun.Math.Phys. 122 (1990) 125

[589] C.N. Yang, Phys.Rev.Lett. 19 (1967) 1312

[590] C.N. Yang and C.P. Yang, J.Math.Phys. 10 (1969) 1115

[591] A. Young, Proc.London Math.Soc. 33 (1900) 97

[592] A. Young, Proc.London Math.Soc. 34 (1902) 361

[593] M. Yu and H.B. Zheng, Nucl.Phys. B 288 (1987) 275

[594] C. Zachos, in: Proceedings of the International Symposium on *Symmetries in Science V*, B. Gruber, ed. (Plenum, New York 1991)

[595] A.B. Zamolodchikov, Theor.Math.Phys. 65 (1986) 1205

[596] Al.B. Zamolodchikov, Theor.Math.Phys. 73 (1988) 1088

[597] A.B. Zamolodchikov and V.A. Fateev, Sov.J.Nucl.Phys. 43 (1986) 657

[598] A.B. Zamolodchikov and Al.B. Zamolodchikov, Nucl.Phys. B 133 (1978) 525

[599] A.B. Zamolodchikov and Al.B. Zamolodchikov, Ann.Phys. 120 (1979) 253

[600] H. Zassenhaus, Hamburger Abhandlungen 13 (1939) 1

Index